Pre-publication REVIEWS, COMMENTARIES, EVALUATIONS . . .

"*Introduction to Fruit Crops* is well organized, well illustrated, scholarly, and reader-friendly. This book is a monumental contribution representing years of work by the author. It is well suited and highly recommended for classroom instruction, for the hobbyist, the home gardener, and the commercial fruit grower. Because the book includes tropical and temperate fruits, it is international in scope and thus differs from most traditional pomology texts. Each chapter includes an excellent bibliography. The book is encyclopedic, and any library, including those of hobbyists and commercial growers, will be incomplete without it."

Darrell Sparks, PhD
Professor of Horticulture,
University of Georgia

"*Introduction to Fruit Crops by* Mark Rieger provides a concisely written, extensive review of the world's major fruit crops. This is one of the few books that covers temperate, subtropical, and tropical fruit species. The text is technically strong but also provides interesting insights into unusual and nontraditional uses for this diverse group of fruit crops. The introductory chapter on general fruit crops is very well written and would be particularly useful in the classroom. This book will serve as an invaluable source of information on fruit crops for a wide horticultural audience. Dr. Rieger has done an excellent job of bringing together information on such a wide range of fruit crops in an interesting and entertaining manner."

Frederick S. Davies, PhD
Professor of Horticulture,
University of Florida, Gainesville

"Mark Rieger's *Introduction to Fruit Crops* offers a comprehensive look at the world's major fruit crops. The text, drawings, and photos provide a solid foundation for students and practitioners in fruit crops and an entry into the world arena. Figures provided on worldwide production by country are detailed and useful. Sections on the historical and current uses and nutritional status of each fruit address these increasingly important areas of fruit crops."

Richard J. Campbell, PhD
Senior Curator of Tropical Fruit,
Fairchild Tropical Botanic Garden
Coral Gables, Florida

"*Introduction to Fruit Crops* is an excellent reference for horticulture professionals as well as committed amateurs. Virtually every aspect of an individual fruit's characteristics, culture, history, and use is covered. The illustrations and color plates are first-rate. Mark Rieger's obvious love for the subject comes through in his many asides describing personal experiences with particular fruits. Don't let the title fool you! Mark includes quite a few nut species, even coffee and cacao, to thoroughly cover the subject. I found *Introduction to Fruit Crops* to be easy to understand, well organized, and comprehensive. It will definitely be useful in my work as a horticulture educator."

Walter Reeves
Radio and Television Host;
Co-author, *Georgia Gardener's Guide*

NOTES FOR PROFESSIONAL LIBRARIANS AND LIBRARY USERS

This is an original book title published by Haworth Food & Agricultural Products Press®, an imprint of The Haworth Press, Inc. Unless otherwise noted in specific chapters with attribution, materials in this book have not been previously published elsewhere in any format or language.

CONSERVATION AND PRESERVATION NOTES

All books published by The Haworth Press, Inc. and its imprints are printed on certified pH neutral, acid-free book grade paper. This paper meets the minimum requirements of American National Standard for Information Sciences-Permanence of Paper for Printed Material, ANSI Z39.48-1984.

DIGITAL OBJECT IDENTIFIER (DOI) LINKING

The Haworth Press is participating in reference linking for elements of our original books. (For more information on reference linking initiatives, please consult the CrossRef Web site at www.crossref.org.) When citing an element of this book such as a chapter, include the element's Digital Object Identifier (DOI) as the last item of the reference. A Digital Object Identifier is a persistent, authoritative, and unique identifier that a publisher assigns to each element of a book. Because of its persistence, DOIs will enable The Haworth Press and other publishers to link to the element referenced, and the link will not break over time. This will be a great resource in scholarly research.

Introduction to Fruit Crops

Text Edition

HAWORTH FOOD & AGRICULTURAL PRODUCTS PRESS®
Titles of Related Interest

The Lowland Maya Area: Three Millennia at the Human-Wildland Interface edited by A. Gómez-Pompa, M. F. Allen, S. Fedick, and J. J. Jiménez-Osornio

Biodiversity and Pest Management in Agroecosystems, Second Edition by Miguel A. Altieri and Clara I. Nicholls

Plant-Derived Antimycotics: Current Trends and Future Prospects edited by Mahendra Rai and Donatella Mares

Concise Encyclopedia of Temperate Tree Fruit edited by Tara Auxt Baugher and Suman Singha

Landscape Agroecology by Paul A. Wojtkowski

Concise Encyclopedia of Plant Pathology by P. Vidhyasekaran

Molecular Genetics and Breeding of Forest Trees edited by Sandeep Kumar and Matthias Fladung

Testing of Genetically Modified Organisms in Foods edited by Farid E. Ahmed

Fungal Disease Resistance in Plants: Biochemistry, Molecular Biology, and Genetic Engineering edited by Zamir K. Punja

Plant Functional Genomics edited by Dario Leister

Immunology in Plant Health and Its Impact on Food Safety by P. Narayanasamy

Abiotic Stresses: Plant Resistance Through Breeding and Molecular Approaches edited by M. Ashraf and P. J. C. Harris

Teaching in the Sciences: Learner-Centered Approaches edited by Catherine McLoughlin and Acram Taji

Handbook of Industrial Crops edited by V. L. Chopra and K. V. Peter

Durum Wheat Breeding: Current Approaches and Future Strategies edited by Conxita Royo, Miloudi M. Nachit, Natale Di Fonzo, José Luis Araus, Wolfgang H. Pfeiffer, and Gustavo A. Slafer

Handbook of Statistics for Teaching and Research in Plant and Crop Science by Usha Rani Palaniswamy and Kodiveri Muniyappa Palaniswamy

Handbook of Microbial Fertilizers edited by M. K. Rai

Eating and Healing: Traditional Food As Medicine edited by Andrea Pieroni and Lisa Leimar Price

Physiology of Crop Production by N. K. Fageria, V. C. Baligar, and R. B. Clark

Plant Conservation Genetics edited by Robert J. Henry

Introduction to Fruit Crops by Mark Rieger

Generations Gardening Together: Sourcebook for Intergenerational Therapeutic Horticulture by Jean M. Larson and Mary Hockenberry Meyer

Agricultural Sustainability: Principles, Processes, and Prospects by Saroja Raman

Introduction to Agroecology: Principles and Practice by Paul A. Wojtkowski

Handbook of Molecular Technologies in Crop Disease Management by P. Vidhyasekaran

Handbook of Precision Agriculture: Principles and Applications edited by Ancha Srinivasan

Dictionary of Plant Tissue Culture by Alan C. Cassells and Peter B. Gahan

Handbook of Potato Production, Improvement, and Postharvest Management edited by Jai Gopal and S. M. Paul Khurana

Carbon Sequestration in Soils of Latin America edited by Rattan Lal, Carlos C. Cerri, Martial Bernoux, Jorge Etchevers, and Eduardo Cerri

Introduction to Fruit Crops

Text Edition

Mark Rieger, PhD

Haworth Food & Agricultural Products Press®
An Imprint of The Haworth Press, Inc.
New York • London • Oxford

For more information on this book or to order, visit
http://www.haworthpress.com/store/product.asp?sku=5547

or call 1-800-HAWORTH (800-429-6784) in the United States and Canada
or (607) 722-5857 outside the United States and Canada

or contact orders@HaworthPress.com

Published by

Haworth Food & Agricultural Products Press®, an imprint of The Haworth Press, Inc., 10 Alice Street, Binghamton, NY 13904-1580.

An editorial quality review group at The Haworth Food & Agricultural Products Press (formerly Food Products Press) determined that certain photographs in the first printing of this book did not meet the high standards expected by the professional horticultural community. The first printing of this book has been replaced by a new softcover printing that has been called a "Text Edition," and a hardcover edition is also now available.

Cover design by Marylouise Doyle.

Cover photos (boxed inset) by Mark Rieger.

Library of Congress Cataloging-in-Publication Data

Rieger, Mark.
Introduction to fruit crops / Mark Rieger.
p. cm.
Includes bibliographical references and index.
ISBN-13: 978-1-56022-172-2 (hard : alk. paper)
ISBN-10: 1-56022-172-0 (hard : alk. paper)
ISBN-13: 978-1-56022-259-0 (pbk. : alk. paper)
ISBN-10: 1-56022-259-X (pbk. : alk. paper)
1. Fruit. I. Title.

SB354.8.R54 2005
634—dc22

2005019440

CONTENTS

ABOUT THE AUTHOR

Mark Rieger, PhD, is a Professor of Horticulture specializing in fruit crops and has been involved in teaching and research for over 20 years. Mark has conducted research on citrus, peach, blueberry, and other fruit crops and has authored over 70 scientific and popular articles and presented over 50 lectures on the topic of fruit crops. The course that inspired this book of the same name has been taught at the University of Georgia since 1990 and is also available on the Internet to students worldwide. This book summarizes over two decades of study, teaching, and worldwide travel in pursuit of the art and science of fruit culture. Mark's fruit crops Web site, which served as an outline for the book, was given the following review by *Netsurfer Digest* (Vol. 7, Issue 21—12 July 2001): "We could scarcely believe how much there is to be found over here. It's all on the subject of fruit crops, but the information just pours over you, rather like a tidal wave."

After 19 years at the University of Georgia, Mark became the Associate Dean for Academic Programs at the University of Florida's College of Agricultural and Life Sciences, where he continues to interact with faculty and students in agriculture.

Chapter 1

Introduction to Fruit Crops and Overview of the Text

Humankind's relationship with fruiting plants began long before the origins of agriculture in 8,000 to 10,000 BC, when all human beings practiced the hunter-gatherer lifestyle. Fruits were mainstays of our diet, being excellent sources of fiber, vitamins, and other healthful or medicinal compounds unbeknownst to us then. While cereal grains, such as wheat and barley, were probably the first crop plants domesticated by humans, several of today's fruit crops were not far behind, since they were native to the very same area—the Fertile Crescent of Asia Minor. Domestication of wild fruiting plants may have been inadvertent; the first groves of fruit trees probably sprang from seeds thrown in waste heaps at the edge of villages. Careful observation and selection for useful traits, such as larger size, better taste, and higher yield, started the transformation of those wild plants into the crops we cultivate and enjoy today. During the age of discovery, fruits, seeds, or live plants were often taken on transoceanic voyages, and exchanges in both directions helped spread many crops throughout the world. Christopher Columbus and his contemporaries may not have realized the impact they would have on agriculture and society when they brought crops such as coffee and citrus to the New World, returning to Europe with previously unknown, but now common, foods such as cocoa and pineapple.

Today, we have well-established world trade networks and sophisticated cultural and postharvest technologies that allow fruits to be enjoyed throughout much of the year, instead of mere weeks per year, as our ancestors experienced. Global trade has made formerly rare and exotic treats derived from fruit crops commonplace in countries with no hope of cultivating the plants. Fruit crops are important agricultural commodities; they add tens of billions of dollars per year to the global economy and are major sources of income for developing countries. Worldwide, over 100 million acres of land has been devoted to their production, and the livelihood of literally millions of farming families depends on continued global trade.

This text is designed to acquaint you with the basics of the botany, production, economic value, general culture, and food uses of the world's major fruit crops. It is formatted as a reference text to allow quick retrieval of essential facts and figures. The following outline is used throughout the text to facilitate information retrieval and to keep the discussion as uniform as possible from crop to crop:

- Taxonomy
- Origin, History of Cultivation
- Folklore, Medicinal Properties, Nonfood Usage
- Production
- Botanical Description
  - Plant
  - Flowers
  - Pollination
  - Fruit

Introduction to Fruit Crops

doi:10.1300/5547_01

General Culture
- Soils and Climate
- Propagation
- Rootstocks
- Planting Design, Training, Pruning
- Pest Problems

Harvest, Postharvest Handling
Contribution to Diet
Bibliography

This chapter describes the general concepts and terminology related to each section of the outline. The book assumes only a basic understanding of plant biology and horticulture, and the glossary of terms at the end of the book includes definitions for **words boldfaced** in this chapter.

"FRUIT CROP" DEFINED

One would think that the term **fruit crop** would be clearly defined—not so. In fact, one of the most frequently asked questions from my Web site is, "Is the tomato a fruit or a vegetable?" My usual reply is "both," as it is clearly a **fruit** in the botanical sense, but a vegetable from a culinary perspective. There are also legal definitions on the books, since in some instances vegetables are taxed and fruits are not (or vice versa). I think it wise to avoid culinary and legal definitions since these change over time and across regions, and the botanical definition of a fruit, being invariable, is a safer bet. I have chosen to define fruit crop as "a **perennial,** edible crop where the economic product is the true botanical fruit or is derived therefrom." The word *perennial* eliminates crops grown as annuals, such as tomato, pepper, melons, and corn, even though the harvested product is the true botanical fruit. Annual cultivation practices differ markedly from those of perennial crops, and to call these fruit crops would only increase the existing confusion. Note that strawberry, an herbaceous perennial, is included in the text even though, in recent decades, much of the acreage is replanted annually. The word *edible* eliminates perennial crops whose fruits are used for fiber, such as kapok *(Ceiba pentandra),* or strictly industrial oils, such as tung nuts *(Aluerites fordii).* The *true botanical fruit* includes the ripened ovary, plus any associated parts, and contains the seeds of the plant. While most would not consider coffee and cacao fruit crops, they fit my definition because they are perennials, and coffee and cocoa are just roasted, ground-up seeds of the fruit. African oil palm, coconut, and olive might not strike you as fruit crops either, but, again, they fit the definition because coconut, olive, and palm oils are edible and derived from a true botanical fruit (all **drupes**). A **nut** is a dry, indehiscent fruit with a hard shell, and, accordingly, several nut crops are included in the text.

TAXONOMY

Plant Names

The scientific or Latin name of a plant is extremely important, as common names vary with location and language spoken. Taxonomic classification places a plant in a large group of related plants (a family) and assigns to it an official name, based largely on Latin and Greek root words. Scientific names generally consist of two italicized words, the first denoting the genus, the second a species within that genus. For example, *Malus domestica* is the name for the cultivated apple, where *Malus* is the genus and *domestica* the species name. In Latin, *Malus* is a noun meaning "apple" or, alternatively, "evil," "bad," or "wrong." (The dual meaning probably stems from the biblical story of Eve and the forbidden fruit in the Garden of Eden.) The species name *domestica* is an adjective meaning "around the house"; thus, the entire name translates roughly to the domesticated apple. Last, but not least, someone always has to take credit for things, so the authority is tacked on to the scientific name, denoting the person who named the plant. In the case of the apple, it was a botanist named Borkhausen, so the precise, full name for the apple is *Malus domestica* Borkh.

A Brief History of Apple Naming

The name of the apple has been, and still is, the subject of much debate. Originally named *Pyrus malus* by Linnaeus in 1753, it was placed in the same genus as pear and quince. Later, Philip Miller moved apples into their own genus, *Malus*, and left it open as to whether the cultivated apple belonged to *M. sylvestris* Mill. or *M. pumila* Mill. (names still used by people today). In 1803, Borkhausen named the apple *Malus domestica* Borkh., only to be overlooked by two other botanists, who later named it *M. communis* Poiret and *M. malus* Britt. These names are incorrect because they came after Borkhausen's treatment of the species, and the first to name a plant correctly gets the credit. Thus, *M. domestica* Borkh. has prevailed. In 1984, a good argument was put forth to suggest that the apple is the product of interspecific hybridization, probably starting thousands of years ago, so the most recent idea is to use *Malus* ×*domestica* Borkh, where the "×" indicates the plant is an interspecific hybrid.

Linneaus, the father of botany, named many plants, which explains the "L." found at the end of many plant names, such as *Pyrus communis* L. (pear). The authority is often omitted in popular literature, on nursery tags, or in plant catalogs, so outside of scientific literature it is not often seen. The authority is not italicized but is typically abbreviated, and literally thousands of these abbreviations are in use today.

To make matters more complex, some plants have been named or renamed several times, so two or more scientific names may be found for the same plant. For example, the Asian pear, *Pyrus pyrifolia* (Burm. f.) Nak., also is named *Pyrus serotina* L., so you may see either name used. Yes, this somewhat defeats the purpose of using a unique, scientific name for a given plant, but opinions vary as to which is "right" (see also A Brief History of Apple Naming).

We are all familiar with different types of fruits within a species, such as 'Red Delicious', 'Golden Delicious', and 'Granny Smith' apples. Horticulturists refer to these subspecies as **cultivars,** which means "**culti**vated **var**ieties." The term **variety** often is used interchangeably with cultivar, although purists prefer the latter. Note the convention of enclosing the cultivar name in single quotes. The precise name of that green, crisp apple we see in grocery stores is therefore *Malus domestica* Borkh. 'Granny Smith'; this can also be written as *Malus domestica* Borkh. cv. Granny Smith.

An even finer level of detail is found in some plants, the **strain** or **sport.** This is equivalent to a sub-subspecies, or a form, where within a cultivar we have several slight variations with horticultural importance. An example would be 'UltraEarli Fuji' apple, which is a strain of 'Fuji' apple that ripens a bit earlier than the original 'Fuji'. As in this example, strains are often given a new cultivar name if they become commercially important. Another example is 'Red Max', a strain of 'McIntosh' apple with deeper red color. This may seem like splitting hairs, but it is useful to note that, in this example, 'Red Max' and 'McIntosh' are genetically more similar than are 'Fuji' and 'McIntosh'. A few more points will help clarify the use of plant names:

- After the first reference to a genus in the text, the genus name is often abbreviated using only the first letter. For example, in the chapter on apple, I discuss species related to *Malus domestica* and refer to these as *M. floribunda, M. sargentii, M. micromalus,* etc., hoping that you understand that the "*M.*" stands for "*Malus*" in each case; this saves valuable typing time for the author and a few drops of ink for publishers.

- A multiplication sign (×) preceding the species denotes an **interspecific hybrid** within that genus. An example is the name for the cultivated strawberry, *Fragaria* ×*ananassa*, which was originally developed by crossing *Fragaria chiloensis* with *Fragaria virginiana*.
- Some genera of plants are so poorly characterized or so diverse or contain hybrids with complex parentage derived from so many species that it is convenient to refer to them as a group using the abbreviation "spp." (species, plural). Among fruit crops, blackberries are a good example of this, often referred to as *Rubus* spp.

Figure 1.1 shows an abbreviated family tree for the Rosaceae (or rose) family (one of the most important families of horticultural plants). A family is the major botanical classification group above the genus level, often containing hundreds or thousands of species. Note also the subfamily and subgenus levels, which are useful in pointing out the finer details of genetic relationships among various crops. For example, apple, peach, pear, and plum are all members of the Rosaceae family, but apple is more closely related to pear (both in subfamily Pomoideae) than to peach or plum (Prunoideae). The text lists the family and sometimes the subfamily or subgenus to give you a sense of the broader genetic background and interrelationships for each crop.

Cultivars

For each crop, I discuss only the major cultivars or groups of cultivars. Even brief descriptions of the many cultivars grown would require a doubling of the length of the text. Cultivars change over time and vary across different fruit-growing regions, making it impossible to cover them adequately and keep the text as concise as possible. Several other texts or sources are devoted to cultivar descriptions, including some listed at the end of this chapter, and I have included one or two references for each crop that give more detail on cultivars.

ORIGIN, HISTORY OF CULTIVATION

The center of diversity of a plant species denotes the area of the world where the species evolved and is found growing in the wild. It is interesting to note that many of the fruit crops (and crop plants in general) grown in the United States are not native to North America. Perhaps more interesting is that several cultivated fruits are not found in the wild, indicating that humankind has hybridized or selected the species over time to make it very different from its wild progenitors. It is not surprising, therefore, that many of the world's most common fruits are native to areas where agriculture had its roots—Asia Minor, China, and Mesoamerica.

Knowledge of soil and climatic conditions occurring in the native range provides clues about site selection and cultural methods that should be employed when growing these crops in foreign areas. Also, plant breeders often return to these centers of diversity to collect **germplasm** that is useful in breeding for disease and pest resistance or other traits. A brief history of cultivation is provided for each crop in the text. Lessons from history have been valuable in shaping the way we grow crops today.

FOLKLORE, MEDICINAL PROPERTIES, NONFOOD USAGE

Fruits have more than just nutritional value to us; some may contain anticancer compounds while others may cause health problems or even death. In this section, I have compiled informa-

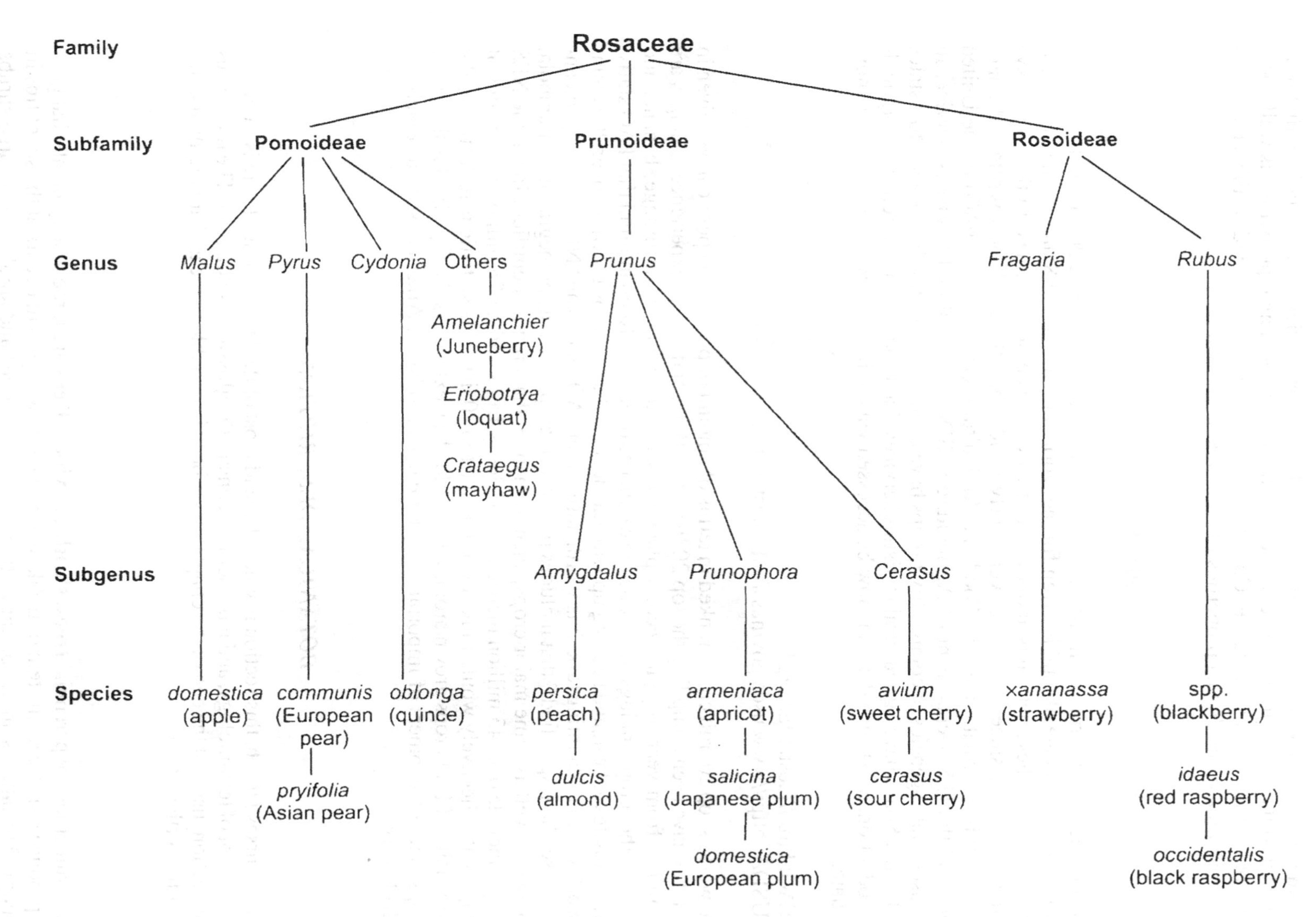

FIGURE 1.1. Family tree of the Rosaceae, showing the relationships among important fruit crops. The other three subfamilies—Spiraeoideae, Chrysobalanoideae, and Neuradoideae—are not shown as they do not contain major crops.

tion from various sources on the folklore, myths, healing properties, and symbolism that have surrounded fruits for centuries. Some of the information on medicinal properties is well documented, but some of it is speculative. Consult the sources listed at the end of this chapter for more information on medicinal properties of fruits and other plants.

PRODUCTION

The production in terms of metric tons (MT) per year (1 MT = 2,200 pounds or 1,000 kilograms) or percentage of total is given for the world and the United States. Generally, accurate publication of these data runs about one to two years behind the times. In the text, I have used the year 2004 since these data were fairly solid when the book was written. Primary sources for this information are the Food and Agricultural Organization (FAO) of the United Nations and the U.S. Department of Agriculture (USDA), Agricultural Statistics Service. In some cases, state agencies or commodity groups have been contacted for these data. Most states in the United States have agricultural statistics services that produce annual reports of agricultural production. Most such data can now be accessed via the Internet, which is generally more up-to-date:

FAO: http://faostat.fao.org
USDA: http://www.usda.gov/nass/pubs/agstats.htm

The world's top 20 fruit crops, ranked in terms of amount of production per year, are given in Table 1.1. I have been compiling the top 20 list since 1988, and, in my experience, changes are rather minor from year to year; second place has rotated among sweet orange, banana, and grape, and other crop rankings move one or two places over time. Keep in mind that FAO statistics are estimates at best, and keeping up with the dozens of fruit crops grown in over 200 countries worldwide is a daunting task. It is worth noting that FAO reports an "NES" category, meaning "not elsewhere specified," that includes minor fruit crops, such as pomegranate, carambola, and guava, as well as some major crops, such as mango, that are misidentified. In all, the NES categories include over 43 million metric tons of fleshy fruits and half a million metric tons of nut crops that collectively would rank seventh in the top 20 list. This statistic is reflective of the great diversity of fruit crops grown around the world. The text covers the top 20, plus the major nut crops, and a few others of importance that are native to North America, such as blueberry and blackberry.

BOTANICAL DESCRIPTION

In the courses I teach, this section is where I spend most of the lecture time. There is a considerable amount of terminology, and it is useful to consult the glossary as you read. The references for this section, listed at the end of the chapter, have been invaluable to me and are highly recommended for further study.

Plant

Most fruit crops are perennial **trees, shrubs,** or **vines.** Trees are large woody plants that generally produce a single main stem or trunk, where the renewal growth occurs at the shoot tips in the canopy. The latter is an important distinction between trees and shrubs, since large shrubs can be trained to a single stem but tend to produce new growth from the base or crown.

TABLE 1.1. The world's top 20 fruit crops ranked in terms of weight of production per year.

Crop	World production (MT)	Leading country	Land area (million acres)
1. African oil palm	153,578,600	Malaysia	30
2. Banana	70,629,047	India	11
3. Grape	65,486,235	Italy	19
4. Orange	63,039,736	Brazil	9
5. Apple	59,059,142	China	13
6. Coconut	53,473,584	Indonesia	26
7. Plantain	32,668,323	Uganda	13
8. Mango	27,043,155	India	9.4
9. Tangerine	22,198,791	China	4.3
10. Pear	17,909,496	China	4.3
11. Peach/nectarine	15,561,200	China	3.5
12. Olive	15,340,488	Spain	21
13. Pineapple	15,287,413	Thailand	2.1
14. Lemon/lime	12,126,233	Mexico	2
15. Plum	9,836,859	China	6.4
16. Coffee	7,719,600	Brazil	25
17. Date	6,772,068	Egypt	2.9
18. Papaya	6,504,369	Brazil	0.9
19. Grapefruit/pummelo	4,874,910	United States	0.6
20. Cacao	3,302,441	Cote d'Ivoire	17

Source: FAO statistics, 2004.

Vines or **lianas** are woody plants that are trained to have a single trunk at the base but use twining stems or **tendrils** to support the canopy. Vines rarely have large trunks, as do trees, since they support themselves by climbing on taller plants in nature or on trellises in cultivation. As a result, vines spend little of their energy on supportive wood, while growing as tall as trees and maximizing leaf exposure to sunlight.

Leaves take many forms, being **compound,** if composed of two or more leaflets, or **simple,** if just a single leaf blade (Figure 1.2). Characterizing the foliage is a great way to start the process of keying out a plant. Several terms are used to describe the overall shape, tip, and margins of leaves or leaflets (Figure 1.3).

Flowers

The floral morphology of a fruit crop is important in determining the mode of pollination (i.e., wind or insect) and type of **fruit** that will arise when the **ovary** matures. A **complete flower** is one possessing all four fundamental appendages: **sepals, petals, stamens,** and **pistils.**

An **incomplete flower** lacks one or more of these features. The position of the base of the pistil, or ovary, with respect to the other three appendages is important in identification and also partially determines what the fruit type will be. The two most common positions are **inferior (epigynous) ovary** and **superior (hypogynous) ovary.** A third possibility is **half-inferior (perigynous) ovary,** as found in the **stone fruits** (*Prunus* spp.). Figure 1.4 shows that a superior ovary sits *above* the point of attachment of the sepals, petals, and stamens, whereas an inferior ovary is embedded within the **receptacle,** *below* the point of attachment of the other floral organs (see also Plate 1.1). A perigynous ovary sits within a **hypanthium,** or floral cup (see Plate 1.1). Different fruit types arise from the flowers shown, partly as a result of the difference in ovary position.

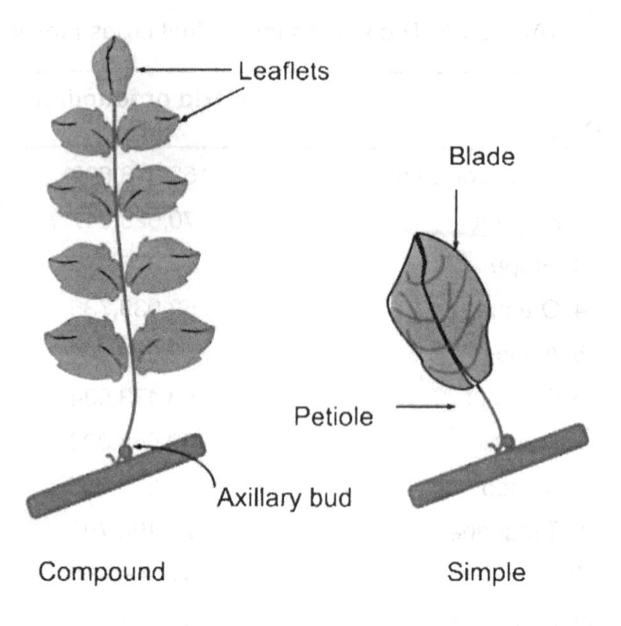

FIGURE 1.2. Compound and simple leaves and their associated parts. The compound leaf shown has a single terminal leaflet and therefore an uneven number of leaflets. This is termed **odd pinnate,** whereas leaves lacking the single terminal leaflet are **even pinnate.**

A **perfect flower** possesses both male (stamens) and female (pistils) parts, whereas an **imperfect flower** may be either **staminate** (functionally male, having only stamens) or **pistillate** (functionally female, having only a pistil or pistils). The **style** is very short or lacking in some pistillate flowers, as in pecan, but the **stigma** and ovary are always present if the pistil is functional. If staminate and pistillate flowers are borne on the same plant but in different locations, the species is termed **monoecious.** If staminate and pistillate flowers occur only on different plants, the species is termed **dioecious.** One can see the ramifications for **pollination** and orchard design: a dioecious species, such as pistachio or kiwifruit, must be planted in an orchard with male plants near females for pollination. Aside from pollination, the male plants are useless because they do not possess ovaries that will ripen into fruit.

An **inflorescence** is a cluster of flowers, and there are several terms for specific inflorescences (Figure 1.5). Generally, inflorescences fall into two categories: **determinate** and **indeterminate.** In a determinate inflorescence, the topmost flower is the most mature and generally opens first, whereas the topmost flower in an indeterminate inflorescence is the least mature and last to appear. The most common inflorescence types in fruit crops are indeterminate **(spikes, racemes, panicles, umbels, corymbs),** with the **cyme** being the most common determinate inflorescence.

Pollination

Pollination is the transfer of **pollen** from the **anther** to the stigma. This is usually mediated by the wind or an insect. In some cases, hummingbirds (wild pineapple), bats (wild bananas), or other animals may play the role of **pollinator,** but most fruits are pollinated by the familiar honeybee *(Apis mellifera).*

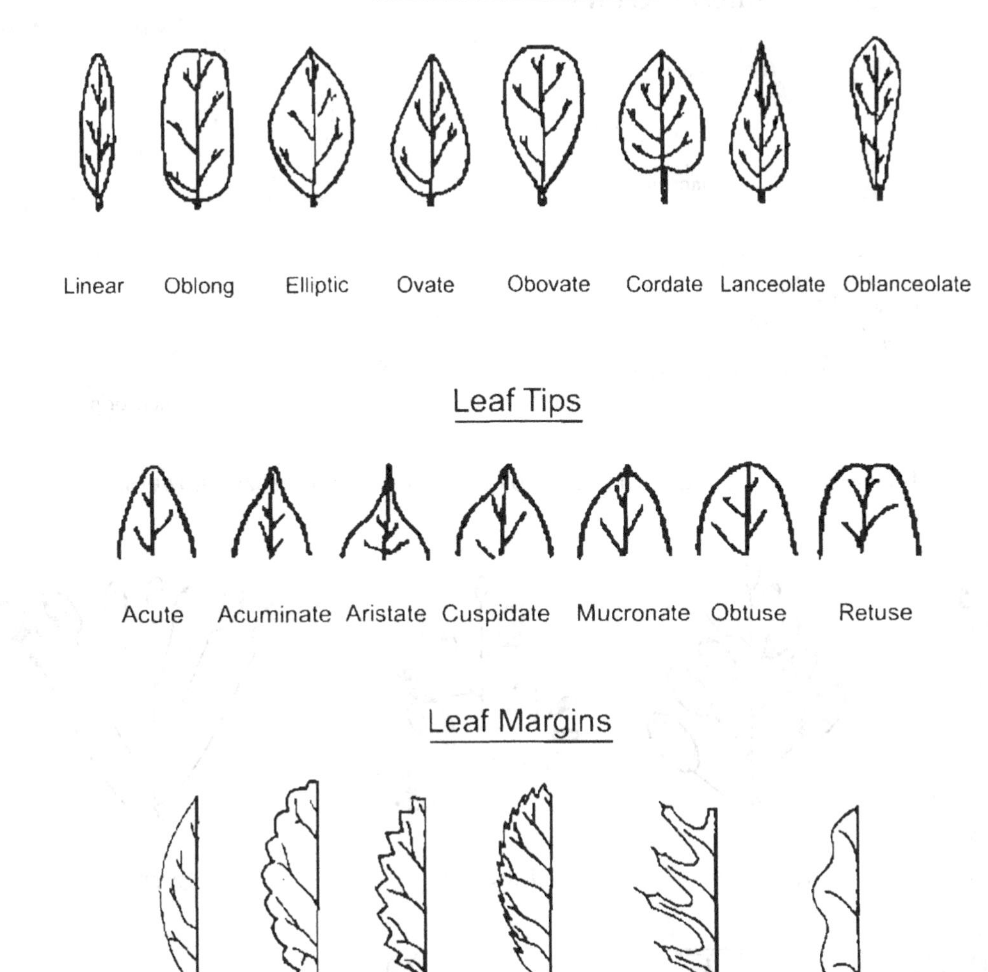

FIGURE 1.3. Terminology used to describe the overall shapes, tips, and margins of leaves.

Once the pollen is transferred, the pollen grain will germinate on the stigma, and the **pollen tube** will grow downward until it reaches an **ovule.** This may take a few days. The pollen tube, under control of the tube nucleus, allows for the movement of the two **generative nuclei** through the style and into the ovule. The processes of pollination and fertilization are depicted in Figure 1.6.

Upon release of the generative nuclei within the **embryo sac,** the process of double **fertilization** occurs, involving two fusion events (hence the name **double fertilization**): one generative nucleus unites with the **egg nucleus,** and the other generative nucleus unites with two **polar nuclei,** yielding the **zygote** and **endosperm,** respectively. You may want to review the details of double fertilization in one of the chapter-ending references listed for this section (which will show all of the other details left out here for the sake of clarity). Suffice it to say that the end

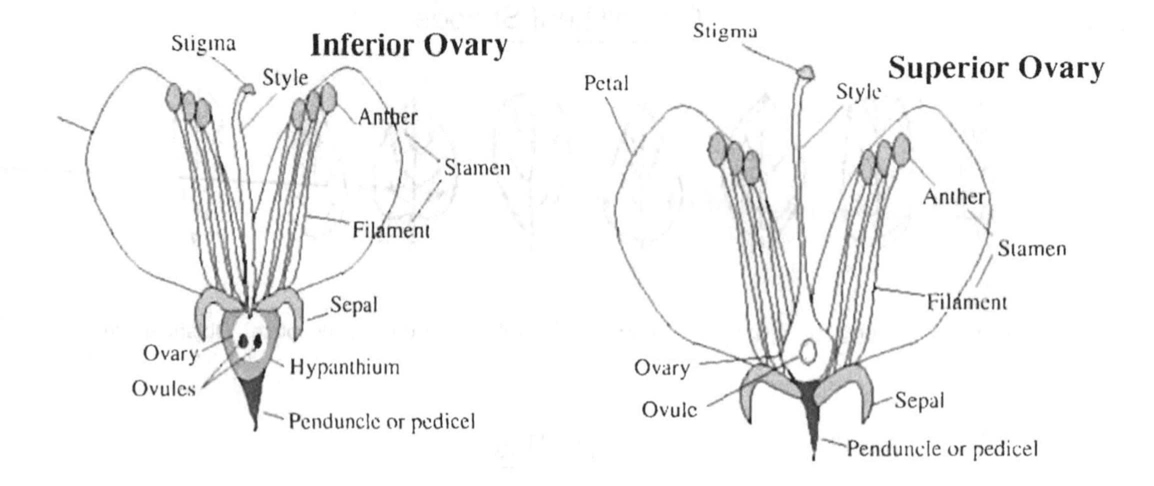

FIGURE 1.4. Superior and inferior ovary positions in idealized flowers.

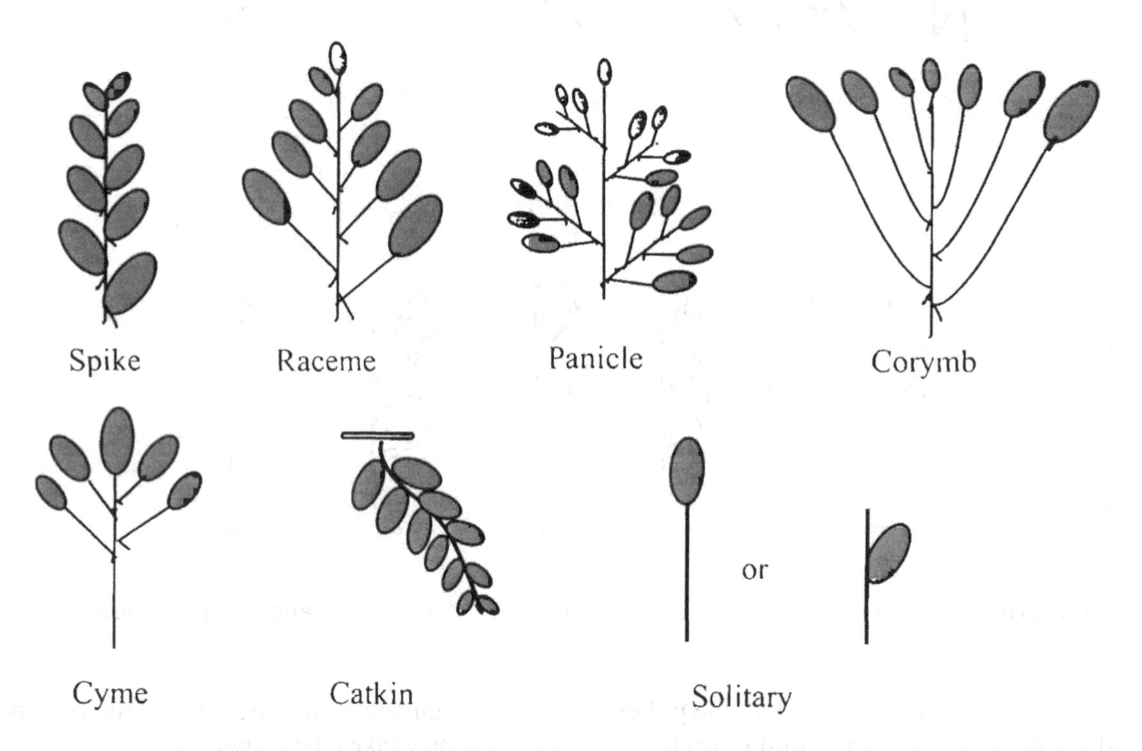

FIGURE 1.5. Stick figures of different inflorescence types commonly seen in fruit crops (see the glossary for complete definitions).

product of double fertilization is the **seed.** When a fruit containing seeds is deposited in a suitable place, the seeds germinate, and the life cycle of the plant is completed.

At this point, you might ask, "Why is pollination necessary for fruit culture, given that seeds are undesirable in fruits from a marketing standpoint?" The short answer is that seed development is the prerequisite for **fruit set.** From the plant's perspective, a fruit functions in seed dissemination. Therefore, investing a great quantity of photosynthetic energy into a fruit that does not contain viable seed is wasteful. Unpollinated and unfertilized ovaries are therefore dropped

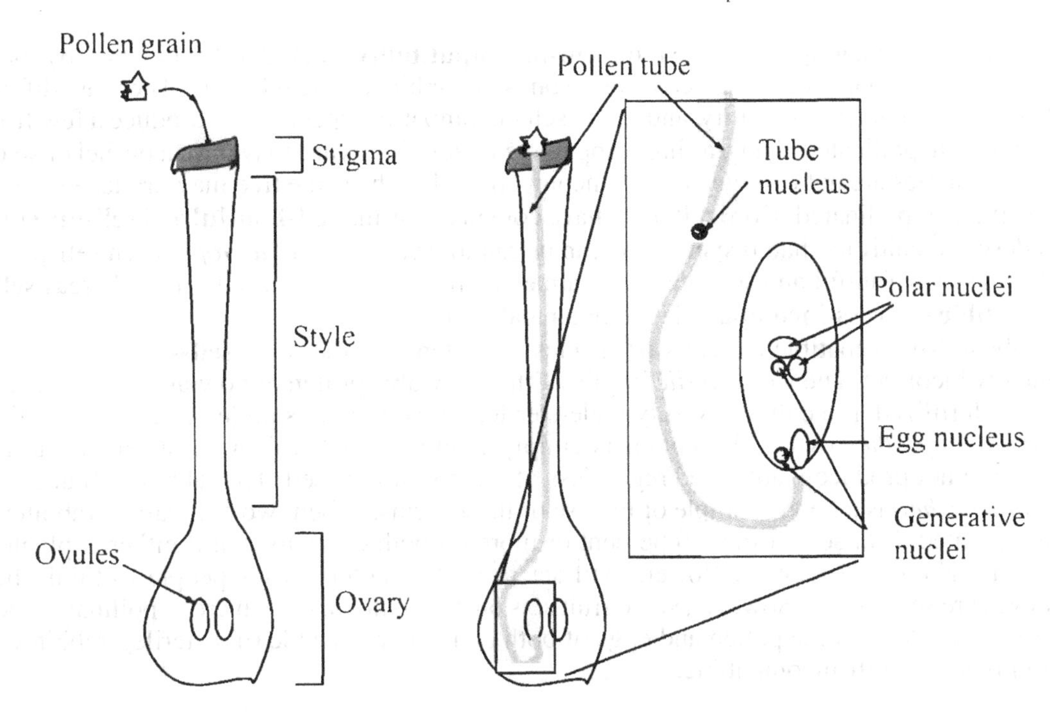

FIGURE 1.6. Simplified diagram of pollination and fertilization within a typical flower pistil. Some of the details within the embryo sac *(exploded view at right)* have been omitted for clarity (see text for explanation).

from the plant shortly after bloom, as there is no need to invest resources into fruits that cannot aid in the reproductive success of the species. Physiologically, the developing embryos within seeds produce growth regulators that prevent **abscission** of fruitlets and cause the fruit tissue in the vicinity of the seed to grow. Thus, dropping of unfertilized ovaries after bloom is "preprogrammed" and prevented only when one or more developing seeds are present. In fruits with multiple seeds, poor pollination and low seed set results in misshapen or asymmetrical fruits that are often unmarketable (Plate 1.2).

Having said that, it must have dawned on you by now that several types of seedless fruit do exist, and so the previous argument does not always hold true. Fruits that set and mature without seed are termed **parthenocarpic,** and these species, although rare in nature, have been exploited by horticulturists to a great extent. Examples include Persian lime, some seedless oranges and tangerines, banana, and pineapple. Seedless grapes are perhaps the best-known seedless fruits but are not truly parthenocarpic. They undergo pollination and fertilization, but seed development is aborted shortly afterward; this situation is termed **stenospermocarpy.**

Cross-pollination and **self-pollination** are important terms. Some fruit crops require genetically distinct pollen for fertilization and fruit set to take place and set fruit poorly if their own pollen is used. These species are planted in orchards with two or more **cross-compatible** cultivars to favor cross-pollination. On the other hand, some species set more than enough fruit when pollinated by themselves. Such self-pollinating species can be grown in large orchards composed of a single cultivar, which are easier to manage. The cultivar used as a source of compatible pollen is referred to as a **pollinizer** for cross-pollinating species. Note that transfer of pollen between one flower and another on the same tree is still considered self-pollination, as is the case when pollen is transferred between flowers on separate trees of the same genotype. The key to cross-pollination is having a genetic distinction between the pollen source and recipient.

Many cross-pollinating species exhibit **self-incompatibility,** such that fertilization by their own pollen is disfavored or prevented through physical or biochemical factors. There are different degrees of self-incompatibility, and many self-incompatible species will produce a few fruit even when self-pollinated. Thus, a single apple tree in your backyard may have a bushel or so of fruit, since apples are not *completely* self-incompatible, but the same tree may produce several bushels if cross-pollinated. Horticulturists have coined the terms **self-fruitful** and **self-unfruitful** to describe cultivars that, respectively, can or cannot set *commercial crops* when self-pollinated. Thus, self-fruitful and self-unfruitful are economic or horticultural terms, whereas **self-incompatible** or **cross-incompatible** are botanical terms.

Highly self-incompatible cultivars or species are often referred to as **self-sterile,** which is technically incorrect. The word *sterile* implies either no viable pollen or no viable ovules available to be fertilized. If a cultivar is truly male-sterile, for example, its pollen could not fertilize its own *or any other* ovule; the pollen is simply nonfunctional. Likewise, a female-sterile cultivar will not produce viable seed regardless of the pollinizer used. The rabbiteye blueberry (*Vaccinium ashei)* is a good example of misuse of these terms. When two cultivars of rabbiteye are interplanted, fruit set is often 50 percent or more on both cultivars, but if either is planted alone, only about 2 percent of flowers will set fruit. The latter causes people to think that rabbiteyes are self-sterile. However, since fruit sets on both cultivars when cross-pollination occurs, this indicates that the pollen and eggs of both cultivars are viable (not sterile); rabbiteyes are simply highly self-incompatible.

Fruit

Fruits are matured ovaries, plus any associated flower parts, that contain the seeds of the plant. The ovary may be subdivided into two or more **carpels** (then termed *compound ovary*), each bearing one to many ovules. Individual carpels develop into sections of a whole fruit, as with citrus, where each familiar segment of the fruit represents one matured carpel. If the ovary is not subdivided, then it is termed *simple.* The ovules will mature into seeds if fertilized.

The ovary has three layers of tissue: the **exocarp** (outermost), **mesocarp** (middle), and **endocarp** (innermost) (Figure 1.7). These layers may develop into distinct parts of the fruit. Generally, the exocarp becomes the fruit peel or skin, the mesocarp becomes the fruit flesh, and the endocarp becomes the innermost part of the flesh or a specialized tissue surrounding the seed(s), as with a **pit**. In many cases, however, the three layers are indistinguishable, and the term **pericarp** is applied to denote all ovarian tissues surrounding the seed(s).

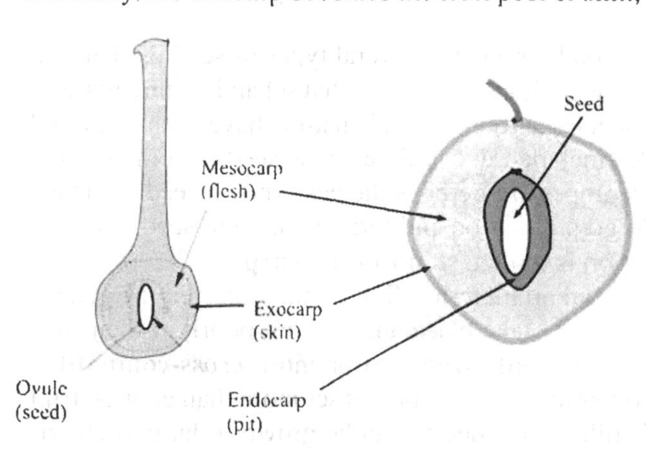

FIGURE 1.7. Fruits are matured ovaries that contain the seeds of the plant. Ovaries sometimes have distinct layers of tissues, as labeled in the figure, that develop into distinct structures of the fruit, as shown for a drupe. Not all fruits have such clearly demarcated tissues.

Fruit types are good identification criteria for plants and are often determined by floral morphology, particularly ovary position (i.e., superior or inferior). The major fruit types in commercial fruit crops are **pome, drupe, berry, hesperidium, aggregate, accessory, multiple fruit** or **syncarp,** and **nut.** Consult the glossary and/or the fruit key in Exhibit 1.1 for descriptions of fruit types.

EXHIBIT 1.1. Key to Common Fruit Types

1. Fruit developed from two or more separate flowers, or derived from an entire inflorescence
    2. Fruit consists mostly of receptacle tissue, with tiny, ripened ovaries borne along the inner wall of the hollow receptacle . . . . **Syconium** (Fig)
    2. Fruit consists of tightly clustered ripened ovaries plus the inflorescence axis . . . . **Multiple Fruit** or **Syncarp** (Pineapple, Mulberry)
1. Fruit developed from a single flower
    3. Fruit developed from two or more separate ovaries
        4. Fruit primarily ovarian tissue; an aggregation of fruitlets on a receptacle . . . . **Aggregate** (Brambles)
        4. Fruit primarily nonovarian tissue . . . . **Accessory** (Strawberry)
    3. Fruit developed from one ovary
        5. Fruit (mostly) fleshy at maturity
            6. Fruit with a thin skin, homogenous texture throughout (except seeds) . . **Berry** (Grape)
            6. Fruit with heterogenous texture
                7. Outer part of fruit tough and hard or leathery
                    8. Septa (partitions) present, several to many; outer rind leathery . . . . **Hesperidium** (Citrus)
                    8. No septa; outer rind tough, thick . . . . **Pepo** (Watermelon)
                7. Outer part of fruit soft; thin-skinned
                    9. Fruit with a hard, bony endocarp surrounding the seed . . . . **Drupe** (Peach, Mango)
                    9. Fruit with two or more seeds; center with papery or cartilaginous structure surrounding seeds . . . . **Pome** (Apple, Pear)
        5. Fruit dry at maturity
            10. Fruit indehiscent (not splitting open) at maturity
                11. Fruit winged . . . . **Samara** (Maple, Ash)
                11. Fruit not winged
                    12. Seed fused to fruit wall . . . . **Caryopsis, Grain** (Corn, Wheat)
                    12. Seed not fused to fruit wall
                        13. Fruit wall bladderlike, loose and free from seed . . . . . **Utricle** (Spinach)
                        13. Fruit wall not bladderlike, close fitting to seed
                            14. Fruit large, with hard, bony wall . . . . **Nut** (Walnut, Pecan)
                            14. Fruit small, with thin wall . . . . **Achene** (Sunflower)
            10. Fruit dehiscent (splitting open) at maturity
                15. Ovary compound; fruit developed from more than one carpel
                    16. Fruit splitting into one-seeded segments at maturity, but carpels not dehiscing to release seeds . . . . **Schizocarp** (Carrot)
                    16. Fruit splitting open to release seeds at maturity
                        17. Two carpels separated by a thin, translucent septum
                            18. Fruit less than twice as long as it is wide . . **Silicle** (Shepherd's Purse)
                            18. Fruit more than twice as long as it is wide . . . . . . **Silique** (Mustard)
                        17. More than two carpels, not separated by a thin, translucent septum . . . . **[Capsule]**
                            19. Capsule opening along a transverse circular line; top separating like a lid . . . . **Circumscissile Capsule** or **Pyxis** (Brazil Nut)
                            19. Capsule opening lengthwise or by pores, top not separating like a lid
                                20. Capsule opening by pores or flaps . . **Poricidal Capsule** (Poppy)
                                20. Capsule opening longitudinally, often lengthwise
                                    21. Capsule dehiscing through locules . . . . **Loculicidal Capsule** (Iris)
                                    21. Capsule dehiscing through septa . . . . **Septicidal Capsule** (Yucca)

(continued)

(continued)

15. Ovary simple; fruit developed from one carpel
    22. Fruit opening along a single suture . . . . **Follicle** (Milkweed)
    22. Fruit opening along two sutures
        23. Fruit not constricted between seeds . . . . **Legume** (Pea, Bean)
        23. Fruit constricted between seeds, breaking into one-seed segments at maturity . . . . **Loment** (Desmodium)

Note: This dichotomous key can be used to determine fruit types given some information on floral morphology and readily observable features of the mature fruit. Entries with the same number are mutually exclusive. Starting with number 1, choose the entry that best describes the fruit in question and proceed to the next higher pair of numbers immediately below each choice until you reach a specific fruit type.

The fruit **bearing habit** of a plant refers to the position and type of wood on which flower buds, and subsequently fruits, occur. This is important in pruning and training because we want to encourage the type of wood that bears the fruit and minimize unnecessary vegetative growth. Some species are **spur** bearing, where fruits are borne on very short, slow-growing, lateral branches (Figure 1.8). Spurs develop on 2-year-old and older wood and may grow only ¼ inch per year; thus, the fruits are borne at nearly the same points in the canopy from year to year. For spur-bearing species, it is important to keep good light exposure throughout the canopy because shaded spurs fail to form flower buds for next year's crop. Lateral-bearing species produce fruits from lateral buds on normal length branches. For these species, it is important to stimulate ample growth each year (by dormant pruning, fertilizing, etc.), so that enough fruiting shoots are available for the next year. A number of crops bear fruits on current season's growth, either laterally, as in grape, or terminally, as in walnut or mango. Some of the tropical crops exhibit **cauliflory,** where flowers and fruits are borne on large branches or trunks of trees (e.g., cacao).

Thinning refers to the partial removal of flowers or fruitlets to improve the size of the remaining fruits. Thinning is often practiced for large-fruited species that normally set too many fruits. By thinning, one directs the available **photosynthate** produced by the leaves into fewer, but ultimately larger, fruits, rather than many small, unmarketable fruits.

Thinning is accomplished by hand usually and is obviously labor intensive and expensive. Flowers or fruits are removed such that a certain number of fruits per tree, or a certain spacing between fruits on a limb is achieved. For example, apples are thinned to one fruit per spur, with spurs spaced about 4 to 6 inches apart, resulting in the removal of about 80 percent of the original number of fruitlets (Plate 1.3). Thinning should be uniform throughout the canopy, as fruits in clusters will remain small even if the correct total number of fruits is left. In some species, chemicals can be sprayed on trees to kill flowers or induce drop of fruitlets. Chemical thinning is less expensive, but riskier, since the degree of thinning depends not only on chemical and concentration but also on weather, cultivar, stage of fruit development, and skill of the orchardist.

In terms of timing, the earlier the tree is thinned, the better the result. When possible, thinning at bloom provides the greatest improvement in fruit size. However, many growers wait until the threat of frost has passed to thin to ensure there will be enough fruits for a full crop. Much of the benefit of thinning is lost if delayed more than about 45 days postbloom in most fruit crops.

GENERAL CULTURE

For each crop, the main aspects of cultivation can be found in this section. The subsections include soils and climate; propagation; rootstocks; planting design, training, and pruning; and pest problems.

Soils and Climate

Most fruit crops grow best on deep, well-drained, loamy soils, with pH of 6 to 7. The rare exceptions to this are noted. Climate is probably the strongest determinant of the success of fruit cultivation. Several aspects of climate are critical.

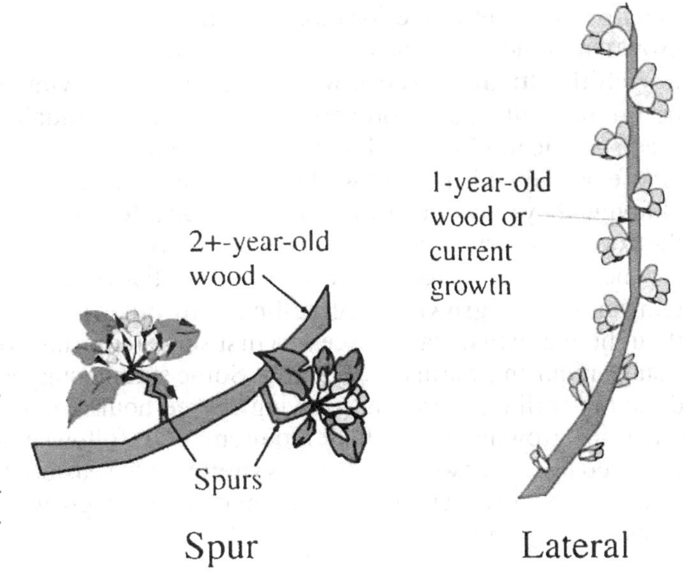

FIGURE 1.8. The two most common bearing habits of fruit crops: spur and lateral. Spurs are simply short, lateral branches that occur on 2-year-old and older wood. Lateral-bearing species produce fruits from lateral buds on elongated 1-year-old or current season's growth.

Coldhardiness

This is the minimum temperature tolerance for the plant, often quoted in degrees Fahrenheit or Celsius, causing 50 percent or greater mortality. The flower buds, vegetative buds, and wood often have different killing temperatures, with flower buds being the least hardy. Thus, a fruit tree may survive in a northern winter but produce little or no fruit. Maximum **coldhardiness** values are given for each species in the text. It is important to note that coldhardiness is not constant; in the summer, most fruit crops would be killed by relatively high temperatures (i.e., 10 to 30°F or –12 to –2°C). As they acclimate in fall and early winter, they obtain the ability to withstand temperatures well below 0°F (–18°C), in most cases. Conversely, as the buds begin to swell in late winter or spring, several degrees of hardiness are lost per week (Plate 1.4). In almost all fruit crops, open flowers or small fruitlets can withstand only 28 to 30°F (–1 to –2°C) without injury, making most crops vulnerable to relatively mild spring frosts. Fruit growers choose sites less prone to frost and/or use heaters, sprinkling, or wind machines to prevent crop losses when frost occurs (Plate 1.5). Tropical crops are exceptions; most have no capacity for acclimation and are killed by brief exposure to subfreezing temperatures. Subtropical crops, such as citrus and date, display a modest ability to acclimate and withstand temperatures 5 to 10°F below the freezing mark.

Chilling Requirement

This is the number of hours of exposure to about 45°F (7°C) required each winter to satisfy dormancy and allow normal growth the following spring. The basic components of dormancy are depicted in Figure 1.9. Most temperate woody plants require between 500 and 1,500 chill hours each winter, measured from leaf drop in autumn until February or March. If the winter is warm, and the chilling requirement is not met, then budbreak is sporadic and light and cropping is poor. If the plant receives much more chilling than needed, budbreak is accelerated and often premature, resulting in frost damage. This is why it is important to match the chilling requirement of the plant to the location, and also why temperate species, such as apples, cannot be grown in tropical lowlands, where temperatures rarely drop below 65°F. For most species, a few **low-chill cultivars** can be grown in areas with mild winters that would not support the growth of traditional cultivars. 'Flordaprince' peach and 'Flordahome' pear are examples of fruit trees bred specifically for low chilling requirement that can be grown successfully in Florida, southern Texas, and other warm winter locations. Tropical crops have no chilling requirements, and although they may often flower or break bud after winter in subtropical regions, they do not need the cool exposure to flower or grow normally.

Once the chilling requirement has been satisfied, temperate woody plants must receive a certain number of **growing degree-hours** in order to resume growth. Thus, dormancy can be thought of as a two-stage process: a first stage requiring cool-temperature exposure followed by a stage requiring warm temperatures. Some studies suggest that the two stages interact; that is, a deficit in chilling causes the growing degree-hour requirement to increase, and overchilling reduces the growing degree-hour requirement. It follows that bloom date in spring is strongly influenced by winter weather. In areas similar to the eastern United States, which experiences wild fluctuations in winter weather, bloom dates for a given fruit cultivar can vary by three to four weeks from year to year.

Growing Season Length

Some fruit crops require as few as 30 days for fruit maturation, whereas others require several months or over a year. Species such as pecan and kiwifruit require over 200 days between bloom and harvest for proper maturation; hence, they can be grown only where the growing season is long. The growing season is defined as the time interval between the last frost in spring and the first frost in autumn.

Sunlight

Sunlight drives not only photosynthesis but also pigment synthesis in fruit skins and flower bud formation for next year's crop. Red color development in apples is much greater in the sunny, desertlike climate of eastern Washington than in the cloudy, humid climate of New York,

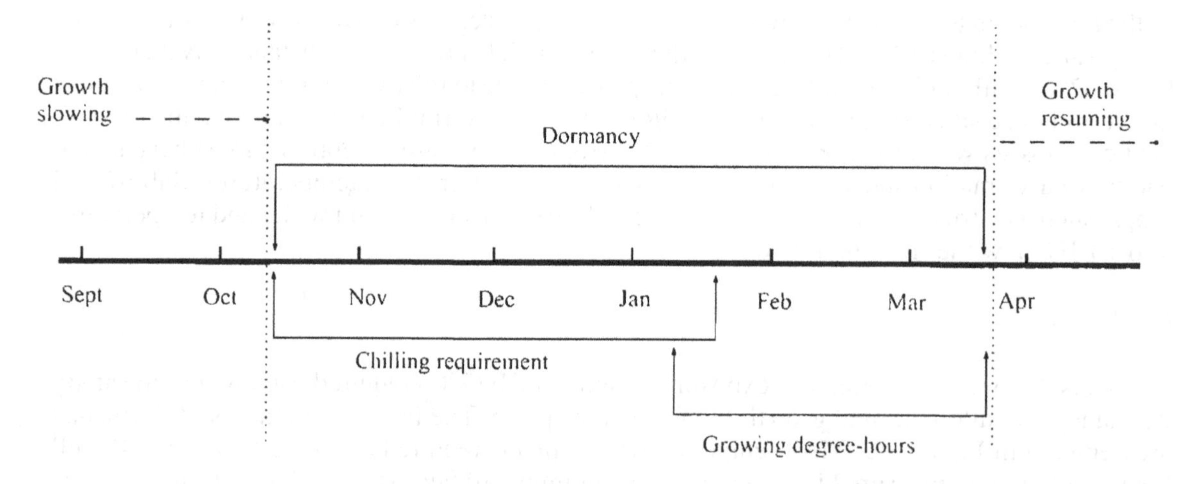

FIGURE 1.9. Typical timeline of dormancy in temperate fruit crops (Northern Hemisphere). Exposure to temperatures below 45°F satisfies dormancy and allows plants to respond to warm temperatures in late winter and eventually resume growth in spring.

for example. All fruit crops, except coffee and cacao, perform best in full sunlight rather than shade.

Rainfall and Humidity

Many crops that originated in humid, rainy climates perform well where these conditions are found. When grown in arid areas, irrigation must be provided during the growing season. High humidity and rainfall favor weed, disease, and insect outbreaks, and fruits grown in humid regions often require more pesticide applications to achieve the same yield and quality as fruits grown in arid climates. This is particularly true for species native to arid areas that have little natural pest resistance.

Temperature During Fruit Maturation

The flavor of a fruit is a function of the amounts and ratio of sugars and **organic acids** found in the **pulp.** Sugars increase and acids decrease as fruits ripen (Figure 1.10). Warm conditions during ripening favor sugar accumulation and organic acid degradation, rendering fruits sweeter and richer in flavor. Cooler than desirable temperatures do the opposite: they make fruits more watery and tart. This is one major reason why wines have "vintage" years and poor years. If the weather during late summer/early fall is too cool, then the wine will be acidic, dry, and perhaps low in alcohol because the grapes never achieved an optimal sugar level and/or sugar-acid ratio. Grape growers hope for sunny, warm days and cool nights during maturation to obtain maximum sugar content and the proper sugar/acid ratio. What is warm weather for grapes may be cool weather for pineapples, so no cardinal temperatures are applicable to all fruits.

Propagation

Most fruit crops are propagated by vegetative means to retain the exact fruit characteristics of the parent plant. Seed propagation generally results in highly variable fruit size, shape, color, and flavor and creates a management nightmare, since each seedling is genetically different from the others. People recognized this a few thousand years ago and began **grafting, budding,** or rooting **cuttings** of desirable fruit crops, rather than planting them by seed. Today, almost all tree fruits are grafted or budded, and most **small fruits** and some grapes are grown from cuttings. Specialized nurseries produce millions of new plants each year, generally by growing **rootstocks** for a year, then budding or grafting them with the desired **scion** cultivars the following year. Many tropical fruit crops are still propagated by seed, including cashew, coffee, oil palm, and coconut. They are either fairly uniform when

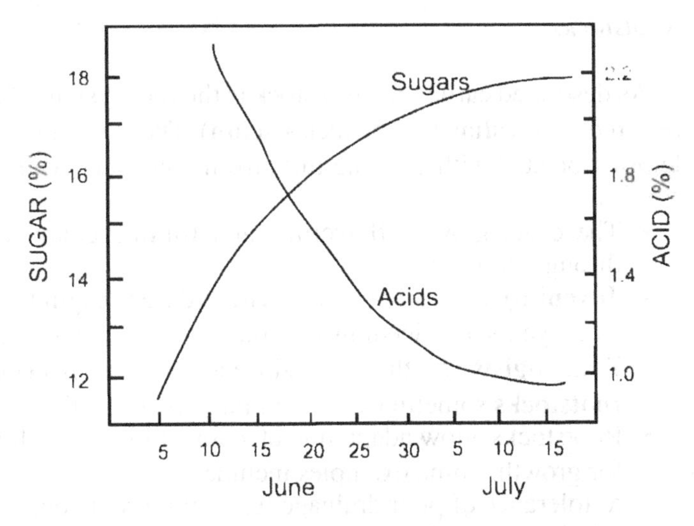

FIGURE 1.10. As fruits ripen, they accumulate sugars and lose organic acids, as shown. *Source:* Modified from Westwood, 1993.

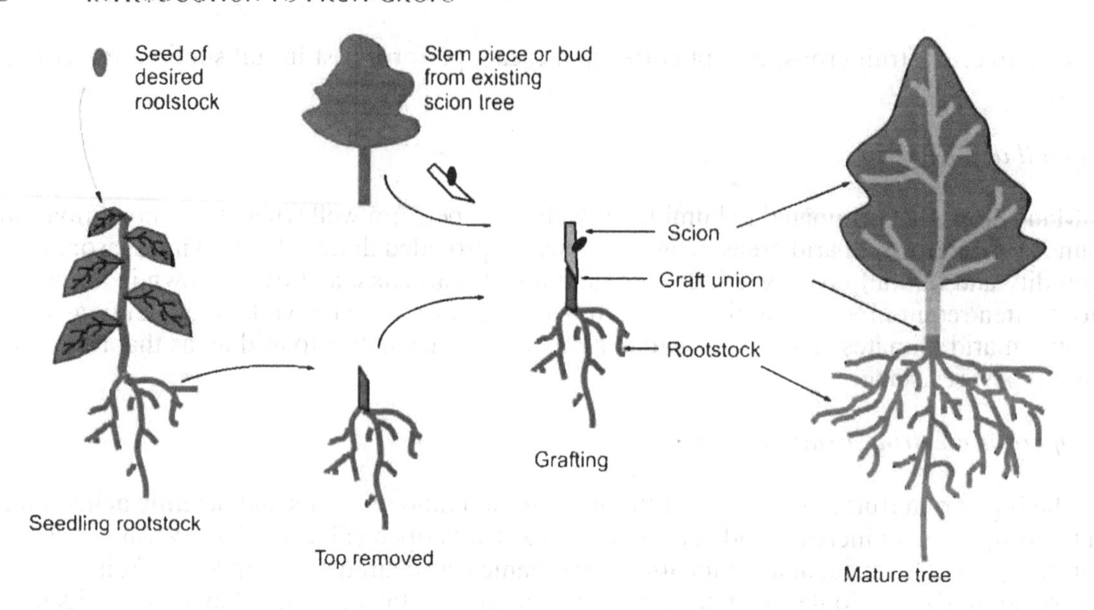

FIGURE 1.11. Basic process of grafting or budding a fruit tree. The rootstock is grown for about 1 year prior to grafting or budding. Shown here is a seedling rootstock, but some rootstocks themselves are vegetatively propagated. Once the rootstock is of sufficient size, one or more buds from the desired scion cultivar are joined with it via a number of techniques.

grown from seed or cannot be propagated easily by vegetative techniques.

The general process of budding or grafting is depicted in Figure 1.11. In this book, I give a brief description of the methods used to propagate a given species, but if you have not studied plant propagation recently, or at all, it would behoove you to refer to one of the texts in this section's list at the end of the chapter for general information. Several terms related to propagation are described in the glossary.

Rootstocks

As described earlier, the rootstock is the root system of a grafted tree. A two-part tree may be referred to as a **stion** (from **stock** + **scion**). The use of grafted trees overcomes many of the problems associated with growing trees from seed, such as the following:

- The exact growth, flowering, and fruiting characteristics of the cultivar are preserved through grafting.
- **Juvenility** is greatly reduced. Grafted trees begin fruiting at an early age, often when only 2 or 3 years old. Seedlings of some species may not fruit until they are 5 or more years old. **Pomologists** say that grafted trees are more **precocious** than seedling trees. **Dwarfing rootstocks** sometimes induce more precocity than nondwarfing or **seedling rootstocks.**
- Rootstocks allow adaptation of scion cultivars to climates and soils normally unfavorable for growth. Some examples include
  a. tolerance of poor drainage (e.g., plum rootstocks for peach in wet soils);
  b. tolerance to drought (e.g., rough lemon rootstock for sweet orange cultivation on the droughty sands of central Florida);
  c. tolerance of high pH or salinity (e.g., peach × almond hybrid rootstocks for high pH tolerance of peach);

d. improved coldhardiness and/or bloom delay of scions (e.g., trifoliate orange rootstock for sweet orange); and
e. tolerance to soil diseases and **nematodes** (e.g., 'Nemaguard' rootstock for root-knot nematode resistance in peach).

- Rootstocks provide tree size control in some species. This is most widely exploited in apple, where there is a range of tree size obtainable by using different dwarfing rootstocks. For example, a 'Red Delicious' tree on M.27 rootstock will be only 6 feet tall at maturity, but would be 20 or more feet tall if grafted onto an apple seedling rootstock. The availability of dwarfing rootstocks has allowed many innovations in tree fruit culture, such as more efficient planting designs and training systems, reduced pesticide use, and an earlier return on investment for orchardists.

Planting Design, Training, Pruning

Planting Design

To achieve high yield and the earliest and highest return on investment, plants must fill their allotted spaces as rapidly as possible and then be maintained within these spaces through annual training and pruning. Yield of all crops is positively related to the amount of sunlight absorbed by the **canopy** per acre, so we optimize canopy shape and plant spacing, leaving just enough room between rows to move equipment and labor. The design of a planting should consider this concept primarily, but other factors, such as pollinizer placement, row orientation, tree density (number of trees per acre), desired tree height, and training system, are also important.

With freestanding trees, we typically see rectangular planting schemes, where the distance between rows is wider than the distance between trees in a row. An orchardist may say that trees are planted 20 × 15 feet, meaning that rows of trees are 20 feet apart, and trees within the row occur at 15-foot intervals. Tree density would be the amount of square feet per acre (43,560) divided by the space allotted each tree (20 × 15 feet = 300 square feet), giving about 145 trees per acre in this example.

High-density orchards contain several hundred to a few thousand trees per acre, and are generally made possible by dwarfing rootstocks (Figure 1.12). In these orchards, individual trees lose their identities as they are trained into continuous fruiting surfaces that resemble long hedgerows. Dwarfing rootstocks keep trees to a manageable height and induce fruiting at an early age, generally during the second year. Since dwarf rootstocks are poorly anchored, high-density orchards are often supported by a trellis of one to four wires; the trellis also aids tree training. Vineyards are laid out very similar to high-density orchards, using a trellis to support the vines.

FIGURE 1.12. A high-density apple orchard with over 1,000 trees per acre.

Training System

The training system refers to the shape of the canopy, which in turn is controlled by pruning and positioning limbs. Again, the main motivation is to maximize light absorption and induce fruiting as soon as possible. There are many training systems for trees, but all stem from two basic forms: the **central leader** and the **open center** or vase (Figure 1.13). The central leader and all of its variations utilize one main central stem (the "leader") that extends from the trunk to the top of the canopy. At regular intervals along the leader, tiers of **scaffold** branches are trained to radiate outward from the leader. Each tier of scaffolds extends outward progressively less as you move from the ground up. As shown in Figure 1.13, top view, scaffold limbs are not placed directly above another one, as the upper scaffold would shade the lower one. The resulting canopy has a pyramidal or Christmas tree shape. This allows good light penetration to the lowest scaffolds, keeping them healthy and fruitful. Central-leader systems are useful for many species, particularly those with a strong tendency to grow upright, such as apple and pear (Figure 1.14).

The open-center system and its variations have scaffolds originating from a single point on the trunk, and no scaffolds oriented upright or in the middle of the canopy. Generally, about four limbs that are pointing in different directions (about 90° apart) are selected 1 to 3 feet above the soil. These limbs are then trained to grow upward and outward, branching repeatedly to fill one-quarter of a circular canopy. The canopy acquires a vase shape, as no structural limbs are allowed to grow in the center. This system allows good light penetration to all branches, as light comes in from the sides and through the center of

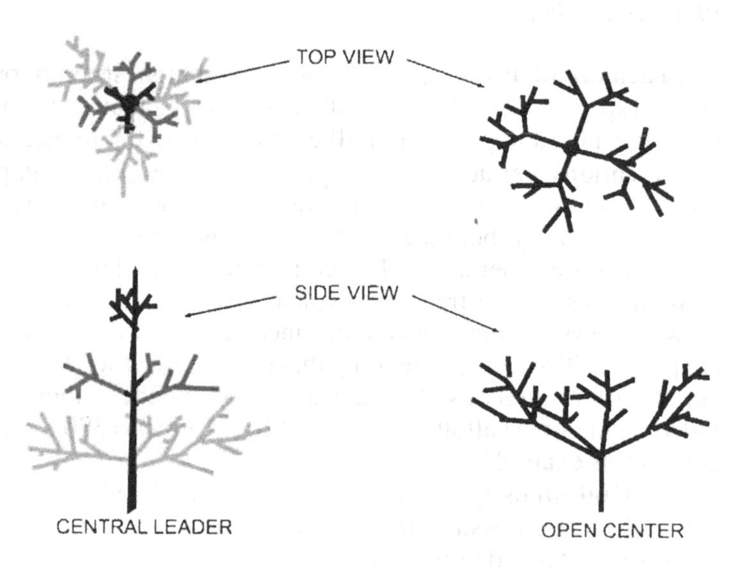

FIGURE 1.13. Stick figures of the two most popular tree-training systems. Most other training systems are modifications of these two basic forms.

FIGURE 1.14. Dead apple and peach trees reveal the basic framework of central-leader *(left)* and open-center *(right)* trees.

the canopy. It is used for trees that tend to produce rounded, dense canopies with no main leader naturally, such as peach, apricot, plum, and almond (Figure 1.14).

Pruning

Annual pruning is necessary for most fruit crops to keep them young, vigorous, and healthy. However, any pruning during the formative years of the plant extends the time it takes to fill its allotted space in the orchard. Thus, pruning is used sparingly when training young plants, but often practiced annually once the plant is mature.

The amount of annual pruning varies tremendously with species. For wine grapes, about 95 percent of the previous season's growth is removed every year, but for sweet cherry trees, only interfering branches and **water sprouts** are removed, when necessary. The severity and type of pruning depends on

- inherent vigor of the tree;
- anticipated regrowth response;
- fruit size or the number of fruiting sites needed for a full crop;
- the nature of the fruiting wood (i.e., spurs on 2-year-old or older wood or laterally on 1-year-old stems); and
- the training system.

If the cultivar is inherently vigorous, it will require more pruning to keep it in shape than would a slow-growing cultivar, or one grafted on a dwarfing rootstock. However, severe pruning invites strong, undesirable regrowth. Therefore, while a vigorous plant requires more pruning, it should not be pruned severely enough to stimulate unfruitful regrowth. **Summer pruning** is practiced when a single, dormant pruning is insufficient to maintain optimal growth control.

In large-fruited species such as peach, 100 pounds might be obtained from just a few hundred fruits, since each one weighs ¼ to ½ pound. But in the small-fruited cherry, several thousand fruits may be required to make the same total yield of 100 pounds. Thus, peach requires far fewer fruiting sites than cherry and can be pruned more severely without a significant effect on yield.

Recall that some species bear fruits on short, lateral branches called spurs, which may produce fruits for several years. Other species produce fruits only on elongated, 1-year-old or current season's shoots. In spur-bearing species, we want to encourage the spurs to remain fruitful, so light pruning is needed to keep good sunlight exposure, but severe pruning will remove spurs or result in spurs growing out into long, unfruitful shoots. On the other hand, the lateral-bearing species need to be pruned at least moderately to encourage formation of new shoots for next year's crop. This underscores the importance of proper pruning—it affects not only this year's crop but next year's crop as well.

Some training systems require specialized pruning to maintain tree form. For each crop, the text will give some details on these specific training systems, but here I want to make a more general point on the types of pruning cuts and their effects on regrowth.

Pruning cuts are of two basic types: **heading back** and **thinning out,** or just heading and thinning (not to be confused with fruit thinning). Heading back refers to when a branch is cut somewhere along its length, leaving some of it behind (Figure 1.15). Thinning out refers to when a branch is removed at its point of origin, leaving none of it behind. Pruning stimulates regrowth, regardless of the type of cut made, but the *location* of regrowth varies with the cut. Specifically, heading back causes a localized stimulus at the wound, such that regrowth occurs from buds just below the cut. Thinning out causes a more generalized stimulus throughout the tree canopy and does not stimulate regrowth at the cut.

Orchardists use **heading cuts** to induce branching at a specific point, say, where a tier of scaffolds should be positioned in a young, central-leader tree. They use **thinning cuts** where the canopy is too thick, or a branch is growing in the wrong orientation. Thinning misguided branches removes a problem and sends a stimulus to the remaining, properly oriented branches. Figure 1.16 shows how heading and thinning cuts are used to train a young tree to a central-leader system. Also shown in this figure is the proper time for pruning: late winter. Sometimes pruning is done in the summer to eliminate excess growth and improve light penetration into the canopy. Pruning in the autumn should be avoided since it can reduce coldhardiness.

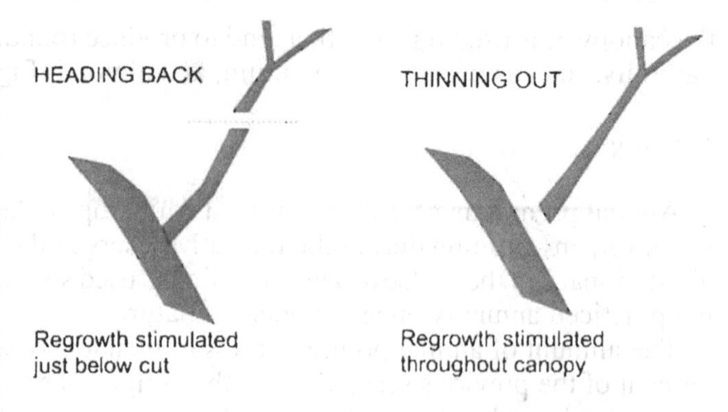

FIGURE 1.15. The two basic types of pruning cuts and associated regrowth response.

A Note About Backyard Fruit Growing

Some of the cultural practices detailed in this text may not be appropriate to backyard fruit culture. If you live in an area that happens to be one of the main production centers for a crop, then you are fairly safe in adapting the commercial practices to your own backyard. It stands to reason that I can offer only generalized advice, which may or may not be sufficient for you to be successful. In most areas of the world, particularly North America, information provided through government extension services is available, generally free of charge, on the Internet, or from the local extension agent or farm advisor. The following key land grant universities publish this type of information (web addresses current as of 2004):

Region	Land Grant University	Web Site
Northeast	Cornell	http://www.hort.cornell.edu/
	Penn State	http://tfpg.cas.psu.edu/ http://ssfruit.cas.psu.edu/
Mid-Atlantic	Virginia Tech	http://www.ento.vt.edu/Fruitfiles/VAFS.html
	North Carolina State	http://www.ces.ncsu.edu/depts/hort/hil/hfruitnew.html
	Regionwide	http://www.caf.wvu.edu/hearneysville/fruitloop.html
Southeast	University of Georgia	http://www.uga.edu/hort/
	University of Florida	http://edis.ifas.ufl.edu/DEPARTMENT_HORTICULTURAL_SCIENCES
Midwest	Michigan State	http://www.msue.msu.edu/fruit/
Intermountain West	Colorado State	http://hla.agsci.colostate.edu/
Southwest	Texas A&M	http://aggie-horticulture.tamu.edu/extension/fruit.html
Pacific Northwest	Washington State	http://treefruit.yakima.wsu.edu/
	Oregon State	http://oregonstate.edu/dept/hort/
California	University of California, Davis	http://fruitsandnuts.ucdavis.edu/

Pest Problems

For each crop, the major problems that occur in many regions or in the largest production region are highlighted. This section must be generalized because pests and diseases tend to be highly regional. Once again, refer to the sources in the previous list for specifics on pest management. Here, I present a brief overview of major pests of fruit crops and their management options.

Weeds

Although **weed** does not fit some people's definition of pest, the single greatest limitation to yield in agriculture is weeds, and more **herbicides** are used in the United States than all **insecticides** and **fungicides** combined. Fruit crops are intolerant of heavy weed infestations or turf grass growing next to the trunk (Figure 1.17). Three basic control strategies for weeds are herbicides, **mulches,** and **mechanical control** (tillage, mowing, burning). Commercially, growers generally keep a weed-free strip beneath crops using herbicides or mulch (Plate 1.6). Mulching is effective, but more labor intensive and costly than herbicides. Organic growers frequently use mulches because herbicides are not cleared for use in organic orchards. Tillage or soil cultivation is used often in arid climates in conjunction with flood irrigation, or with crops harvested from the ground (nuts). Soil cultivation can damage roots and tree trunks and increase erosion. In developing countries or on small organic farms, growers may weed plots with hand implements. In this text, nothing more will be said about weeds of individual crops, and it can be assumed that weeds are problems in the production of all fruit crops.

Insects and Mites

Literally hundreds of thousands of species of insects exist—more variety than is seen with any other life-form on the planet. In some fruit crops, up to 300 insects have been documented to cause damage to one or more parts of the plant. Fruits are particularly attractive to insects because they are sources of food and protection, as well as useful sites for raising the next generation (Plate 1.7). Generally, only a few cause economic injury to any given crop, and often one or

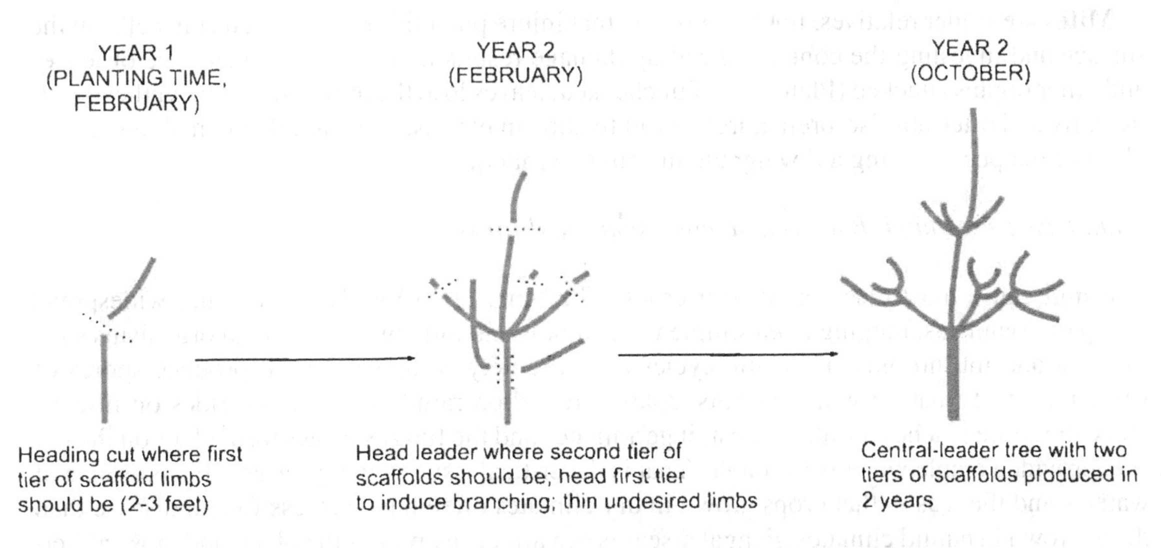
YEAR 1
(PLANTING TIME, FEBRUARY)
YEAR 2
(FEBRUARY)
YEAR 2
(OCTOBER)
Heading cut where first tier of scaffold limbs should be (2-3 feet)
Head leader where second tier of scaffolds should be; head first tier to induce branching; thin undesired limbs
Central-leader tree with two tiers of scaffolds produced in 2 years

FIGURE 1.16. Training of a single-stemmed, young tree to a central leader in 2 years, with heading and thinning cuts.

FIGURE 1.17. Three-year-old 'Empire' apple trees with different amounts of weed competition. *(left)* Tall fescue grass was allowed to grow up to the trunk. *(right)* Tall fescue was kept at least 4 feet away on all sides. *Source:* Photos courtesy of Mike Parker, North Carolina State University.

two represent the bulk of the outlay for insecticides or other forms of protection. The text focuses on these key pests, as a complete discourse on all species affecting the crop is prohibitively lengthy. Some of the most common insect and mite problems of fruit crops are listed in Table 1.2.

Mites are spider relatives, not true insects, that injure plant tissues by puncturing cells on the surface and ingesting the contents. Feeding damage results in a stippled appearance of leaves and other organs attacked (Plate 1.8). If unchecked, leaves lose the ability to photosynthesize efficiently and often abscise prematurely. Fruit feeding may cause undesirable blemishing or stippling of the peel, causing a downgrade in fruit external quality.

Plant Diseases: Fungi, Bacteria, Mycoplasmas, and Viruses

Fungi cause most diseases of fruit crops. The fungi are a highly diverse and widespread group of organisms, ranging from simple water molds and mildew to beneficial organisms such as yeast and mushrooms. Their life cycles are extremely complex, but all produce spores of some type that float on wind currents, splash around on raindrops, or hitch rides on insects. Once the spore reaches a suitable host, it germinates and the fungus grows through or on the tissue, provided ample water is available. That is the critical issue in many cases—the presence of water—and the reason that crops grown in dry climates often have far less fungal disease than those grown in humid climates. Fungal diseases can affect any part of the plant, and those affecting fruit directly are some of the worst problems facing fruit growers (Plate 1.9). Common diseases of fruit crops are listed in Table 1.3.

TABLE 1.2. Major insect and mite pests affecting fruit crops.

Insect	Crops affected*	Damage
Codling moth *(Cydia pomonella)*	Apple, pear, plum, walnut	Fruit feeding, fruit drop
Oriental fruit moth *(Grapholita molesta)*	Peach, plum, apricot, almond, apple	Shoot dieback, fruit feeding, fruit drop
Plum curculio *(Conotrachelus nenupar)*	Peach, plum, apple, cherry, blueberry	Surface scarring, catfacing, fruit feeding, fruit drop
Leaf rollers (e.g., *Platynoda, Argyrotaenia* spp.)	Apple, pear, peach, plum, grape, citrus, strawberry	Leaf and bud feeding, damage; webbing on fruit; fruit damage and subsequent rot
Scale insects (e.g., *Quadraspidiotus* spp.)	Most fruit crops	Fruit scarring, cosmetic damage; leaf feeding; limb and twig dieback, tree decline; honeydew secretion and sooty mold development
Stinkbugs and plant bugs (e.g., *Leptoglossus, Lygus* spp.)	Most fruit crops	Fruit catfacing, spotting
Aphids (e.g., *Aphis* spp.)	Most fruit crops	Leaf and shoot feeding, distortion; honeydew secretion and sooty mold development; virus transmission
Leaf miners (e.g., *Lithocolletis* spp.)	Citrus, apple	Leaf feeding by tunneling
Mites (e.g., *Tetranychus, Panonychus* spp.)	Most fruit crops	Leaf feeding, stippling, distortion; webbing at shoot tips

*Partial list of major crops affected.

Bacteria-related diseases of fruit crops are not as common or diverse as fungal diseases but can be more difficult to control. In fact, the only solution for some crops is to produce them in areas where bacterial growth and dissemination are disfavored. For example, pears were cultivated commercially in New York, Pennsylvania, Michigan, and other eastern states in the 1800s and early 1900s. The bacterial disease fire blight (Plate 1.9) wreaked havoc in these rainy, humid climates, and pear culture gradually moved to the arid Pacific Northwest where the disease does not develop as well. As in humans, bacterial diseases can be treated with antibiotics, but this is cost prohibitive for most commercial fruit growers and does not work as well. Quite often, the only solutions to bacterial diseases are to grow resistant cultivars or to remove and destroy infected plant parts as soon as the infection is noticed.

Mycoplasmas are somewhere between bacteria and **viruses** on the evolutionary scale of life. Diseases such as Pierce's disease of grape, lethal yellowing in palm, and pear decline are examples of the blight that mycoplasmas bring to the fruit world. Mycoplasmas colonize the water-conducting tissues of woody plants, and after multiplying for months or years, they eventually clog the **xylem** and kill or severely debilitate the plant. Insects that feed on xylem sap, such as leafhoppers and spittle bugs, carry them from plant to plant. It is not feasible to control the disease through insecticide spraying because it takes only one feeding event from one insect to

TABLE 1.3. Major fungal and bacterial diseases of fruit crops.

Disease	Crops affected*	Symptoms and damage
Fungal diseases—Leaves and stems		
Powdery mildew (*Podosphaera, Sphaerotheca,* and *Uncinula* spp.)	Apple, grape, strawberry, cherry, peach, plum	Distorted, stunted growth appears at shoot tips with white, powdery spore masses on both sides of leaves; weblike russeting or discoloration of fruit also occurs.
Leaf spots or scabs (e.g., *Mycospharella, Venturia*)	Apple, pear, peach, strawberry, many others	Circular, angular, or irregular blemishes or lesions appear on leaves. Spots often coalesce to form blotches if severe.
Fungal diseases—fruit		
Brown rot (*Monilinia* spp.)	Peach, apricot, plum, cherry, almond, quince	The blossom blight phase kills flowers at bloom; the fruit rot phase occurs within days of harvest. Brown, soft spots spread rapidly, producing powdery tan spores.
Gray mold or bunch rot (*Botrytis* spp.)	Grape, strawberry	Classic gray, velvety covering grows over the ripe fruit; fruit softens and then shrinks as a result.
Anthracnose (*Colletotrichum* spp.)	Banana, mango, avocado, papaya, pineapple	Small to large, brown or black, sunken lesions appear on fruit surface near harvest; lesions may coalesce in badly infected fruit. Lesions are usually dry and firm.
Fungal diseases—Trunk, crown, and roots		
Armillaria or oak root rot (e.g., *Armillaria*)	Apple, grape, peach, plum, cherry, apricot, walnut, citrus, many others	Wilting, decline, and/or dieback of the aboveground portion of the tree occurs; a conspicuous white mycelial mat forms between the wood and bark of affected trees, and clusters of mushrooms may grow at the base of the trunk.
Stem canker (e.g., *Leucostoma, Phomopsis*)	Apple, pear, peach, plum, cherry, apricot, almond, many others	Sunken, discolored, or rough areas form in bark; they are often round or elliptical in shape. Size is variable and may grow in size from year to year. Limb weakening or dieback occurs through girdling. Callus tissue may form at margins of large cankers.
Vascular wilt (e.g., *Verticillium, Fusarium*)	Peach, plum, apricot, cherry, blackberry, raspberry, almond, strawberry	Leaves wilt, become chlorotic, or turn brown, followed by shoot dieback. Often one limb or side of the plant is affected before the other(s). Yellow, red, or brown discoloration of vascular tissue is seen.

TABLE 1.3 *(continued)*

Disease	Crops affected*	Symptoms and damage
Phytophthora root/crown rot (*Phytophthora* spp.)	Apple, pear, peach, apricot, cherry, plum, citrus, others	Poor shoot growth, chlorotic leaves, and general lack of vigor are seen. Shoot dieback and tree collapse may occur after rainy periods.
Bacterial diseases		
Bacterial canker *(Pseudomonas syringae)*	Peach, plum, apricot, cherry, almond, walnut, others	Irregular, sunken areas form in bark; these vary in size. Often exudes amber gum from canker in spring. Tissue beneath cankers is discolored and often sour smelling. Twigs, limbs, or entire trees die back, but trees will sprout from rootstock since roots are alive.
Crown gall *(Agrobacterium tumefaciens)*	Over 200 species of woody plants	Galls or tumors from .25-6 inches form at the crown or on main roots. Tree may not exhibit foliar symptoms if galls are small; large galls may cause stunting and leaf chlorosis, particularly on young trees.
Fire blight *(Erwinia amylovora)*	Pear, apple, quince	Browning or blackening and withering of flower clusters or current season's shoots are seen. Shoots appear burned and curl at tips into a "shepherd's crook" shape. Entire limbs and trees can be killed by girdling.
Leaf scorch, scalds, declines *(Xylella fastidiosa)*	Peach, plum, grape, citrus; species in over 30 plant families	This bacteria-like organism, called a mycoplasma, grows in the xylem and restricts water and nutrient flow, often for years before the tree or vine succumbs.

*Partial list of major crops affected.

transmit the disease. Injections of antibiotics can slow the disease, but resistant cultivars, roguing, and quarantines are the only practical methods of control.

Viral diseases are similar to mycoplasmas in many respects. They are moved by insects in many cases. They may also move on pruning tools and some even in pollen, so they are a bit more mobile than the mycoplasmas. Once a plant is infected, it remains so for the rest of its life. As in human medicine, there is no cure for viral diseases. Viruses often cause striking symptoms, such as mosaic yellowing of leaves, unusual spotting patterns, or twisting and contortion of leaves and shoots, so it is easy to spot an infected plant and remove it from the site. However, some viral diseases produce no symptoms other than reduced growth or yield. Plants can live many years after being infected with a virus, sometimes without showing obvious symptoms, and thus be a source of infection for other plants for long periods of time. Roguing infected plants is often the only means of controlling viruses.

Nematodes

Nematodes are microscopic, nonsegmented roundworms that feed on roots of plants. Nematodes are one of the most diverse groups of organisms on the planet, with hundreds of thousands of species named (and still counting). Species in only two of 15 orders of nematodes actually parasitize plants, but that leaves more than enough to go around. Virtually every agricultural crop known to hummankind plays host to a dozen or so nematodes. The most common fruit pests tend to be in the root-knot *(Meloidogyne)*, ring *(Criconemella)*, dagger *(Xiphinema)*, root lesion *(Pratylenchus)*, and cyst *(Heterodera)* groups. Despite their diversity, nematodes cause similar injury symptoms because they simply feed on roots, reducing the ability of the root system to support the top of the plant. Stunting, wilting, chlorosis, and, in severe cases, toppling over are common symptoms. While they cannot be seen with the naked eye, the result of their feeding is often easily detected in the form of root galls, knots, lesions, or simply poor root development (Figure 1.18).

Soil fumigation prior to planting is often recommended in soils with high nematode populations. **Fumigants** and other **nematicides** are highly toxic chemicals that provide good control but are dangerous to work with and strictly regulated by federal, state, and local government agencies. Methyl bromide, the chief soil fumigant used in fruit crops, has been gradually phased out of production due to its ability to escape into the atmosphere and cause ozone depletion. Alternatively, tolerant rootstocks can be used for certain crops, for example, 'Nemaguard' rootstock for peach, which resists root-knot nematode feeding *(Meloidogyne incognita)*. Note that nematode-resistant rootstocks resist feeding by only a few species of nematodes at best, and no rootstock resists feeding by all nematodes.

Basic Approaches to Pest Control

Pesticides are not the only way to deal with pests. They are extremely effective and commonplace today, but agriculture was practiced for several thousand years before the development of synthetic pesticides. I divide approaches to pest control into five basic strategies, recognizing that elements of two or more strategies may be utilized by a given fruit grower.

1. *Let nature take its course.* Doing nothing is easy, so letting nature take its course is quite popular with backyard gardeners. In fact, there is solid scientific justification for doing nothing, since every *native* pest has some natural enemy, and sooner or later that enemy will bring the pest population back into balance, at least from Mother Nature's viewpoint. The real question is

FIGURE 1.18. *(left)* "Banana topple" disease, caused by *Radopholus* nematode feeding on roots. *(right)* Root death leaves plants poorly anchored, and they fall over easily in heavy rain or wind.

whether that natural balance occurs at a point where a commercial grower can make a profit. A commercial grower with the thinnest of profit margins can scarcely afford to throw out even one-tenth of the annual crop. This method is also used in **integrated pest management (IPM,** see strategy 5) when pest pressures are not severe enough to warrant action. The direct environmental impact of this method is obviously none, but, indirectly, if more land must be cleared for cultivation to compensate for lower yields, the impact of doing nothing can be severe.

2. *Cultural control.* This includes all nonchemical or biological means to control pests. Generally, these are not as rapidly effective as chemicals and are done well in advance of pest outbreaks. An example would be closely mowing the grass beneath orchard trees to remove stink bugs living there, just waiting to jump on the newly set fruits after bloom. Cultural controls help to relieve or reduce pest pressures and often reduce the number of chemical sprays required, but alone they infrequently reduce pests to insignificant levels. The environmental impact is generally minimal, but some cultural controls can damage the environment; for example, tilling weeds and leaving the soil exposed may accelerate erosion. Here are some important cultural controls in brief:

- *Sanitation:* One bad apple can indeed spoil the whole bunch, so locating and roguing out the bad ones, when feasible, is a good idea. A flower, fruit, leaf, or twig affected by a pest is often the source for further infection within the same tree or orchard. With sanitation, one is physically removing the pest organism and/or its means of propagating itself, which slows the rate of pest buildup. One example is pruning out fire blight–affected shoots in pears and apples to remove the bacteria that could otherwise affect another shoot within the tree. Disease organisms often **overwinter** in trees they affected the previous summer, so dormant pruning presents an opportunity to reduce the pest's potential numbers in the upcoming season.
- *Life cycle disruption:* If one can disrupt any point of a pest's life cycle, its population will collapse. For insects, this has been exploited commercially with the use of pheromone mating disruption. **Pheromones** are chemicals that allow insects to communicate, and certain pheromones allow male insects to find females during the mating period. Once the specific mating pheromones were identified, researchers reasoned that saturating an area with pheromones would prevent males from finding females. Commercially, this is done for such pests as grape berry moth and codling moth by attaching plastic ties to tree limbs that release pheromones. This works great for large orchards, or in any case where the mating takes place in the orchard. In small, irregularly shaped plantings, which have a lot of border area, adults can easily mate outside the orchard, after which the female can fly in to lay eggs. Nevertheless, this has reduced the need for chemical sprays for certain pests substantially and is a strategy with almost no nontarget effects.
- *Traps, baits, diversions:* Similar to the foolery involved in life cycle disruption, we can use pheromones or other attractants to lure insects to traps or diversions and kill them or at least steer them away from the crop. The insect is often lured by visual or chemical cues to a sticky trap where it lands and cannot escape. This has been very successful for controlling flies that produce maggots in apple, blueberry, and cherry. For apple maggot, decoys are made to resemble nice red fruit for laying eggs (Plate 1.10). The decoys are laced with insecticide, killing all flies that approach. They are also made of cornstarch so they break down naturally in the field. In some trials, pesticide application has been reduced by 90 percent using this method.
- *Resistant cultivars:* This is the ultimate in cultural pest control strategies, simply planting cultivars that are genetically resistant to the pest. Sounds easy and it has worked in the past for some key pests, but resistant cultivars are slow to develop and sometimes unavailable. Also, pest resistance is specific; years of breeding may go into conferring tolerance to one

species (or subspecies) of a pest, leaving several others available to affect the crop. Last, the pest often breaks down the resistance of the host if the two are left alone in the evolutionary landscape for a period of time, which makes breeding an ongoing effort.

3. *Biological control.* This involves introducing other organisms that prey on or parasitize the pest of the fruit crop. Biological control is extremely effective for specific pest problems but has yet to achieve the widespread use of most other pest control strategies. For example, introduction of the vedalia beetle to control cottony cushion scale in California citrus eliminated the need to spray for that pest. Since other insects may attack citrus, unfortunately, vedalia beetles are killed while spraying to control something other than cottony cushion scale. Another limitation is the need for at least some pest population to be present to support the biological control agent. This basal pest population may be above the threshold for economic loss. Also, the biocontrol agent may need other sources of food besides the pest to live, and we may not have these present or even know what they are in some cases.

4. *Chemical control.* Pesticides are much more responsive than cultural or biological controls, allowing a grower to react instantly to a crisis and quickly reduce pest populations. Unfortunately, many nontarget organisms are affected by chemical applications, and the potential exists for human and environmental damage from pesticide misuse. Chemicals have evolved over the years to be less persistent in the environment and have fewer nontarget effects, and some of the worst chemicals (e.g., DDT and EDB [ethylene dibromide]) were banned years ago. In some cases, availability of pesticides prompted growers to reduce cultural controls in favor of spraying chemicals every few weeks (days) to control pests without knowing whether they were present. This spray-by-calendar strategy is not common anymore due to the advent of integrated pest management.

5. *Integrated pest management (IPM).* As the name suggests, many methods of control are used in an integrated fashion to reduce crop losses. Cultural and biological controls are often used, and pesticides are applied only when other means fail to keep pests below a certain **threshold.** This threshold, determined by years of research, is the level of pest infestation that can be tolerated before economic losses occur. IPM generally reduces the effect of pesticides on the environment by reducing the number of spray applications, not eliminating sprays entirely.

Scouting is used in IPM to determine if thresholds have been reached. For example, a scout might sample 50 leaves in an orchard block and count the number of mites on the underside of each leaf. If an average of more than two mites per leaf is found, the grower would take action and spray, since research shows that the two-mites-per-leaf threshold is the point where economic losses begin to occur. Pheromone traps are often used in IPM to monitor pest populations (Figure 1.19). Unlike the apple maggot traps described earlier, they are not designed to control the insect population, but to serve as indicators of pest presence and to aid the grower in spray timing.

IPM demands greater technical knowledge; growers must be able to recognize all types of insects, know the different fungal signs and symptoms, and quantify their extent quickly. Alternatively, consultants or scouts are available for hire to do this work, or an extension agent may be assigned the task in commercially important fruit-producing regions.

Organic farming has received great attention over the past decade or so and is currently the fastest-growing segment of agriculture in the United States. Pest management in organic farming is really just a form of IPM that does not use synthetic chemicals as a control option. One common misconception is that "organic" means "not sprayed with chemicals." A certified organic farmer can use chemicals for pest control if they are naturally occurring or plant derived and are found on the USDA's "National List of Allowed Substances." In keeping with the theory of IPM, organic farmers exhaust all other approaches to pest control first and use sprays as a last resort. In some cases, such as with apples in humid climates, organic farmers may have to spray

FIGURE 1.19. University of Georgia entomologist Dan Horton places a pheromone trap in an apple tree to monitor insect populations and advise growers on spray timing.

more frequently than conventional farmers to get adequate pest control with natural substances. Organic fruit growers can afford higher crop losses because current prices are often double those paid for conventional produce. The upsurge in organic farming has prompted all fruit growers to rethink their approach to pest management and pesticide use.

Food Safety in Fruit Crops

In the United States, the Food and Drug Administration (FDA) has the responsibility for the enforcement of the Environmental Protection Agency (EPA) standards in the nation's food supply. They routinely test food destined for the consumer's kitchen for a variety of chemical and biological hazards. Some state agencies do essentially the same thing, adding another layer of safety assurance. Figure 1.20 shows an example of residue testing by the FDA in 2003. The domestic data are derived from fruit samples collected from 45 states and Puerto Rico. The 2003 data are somewhat atypical for domestic fruit, showing 2 percent violative samples as opposed to 1 percent in normal years. A violative sample is one containing even a trace amount of a pesticide not registered for that crop or an amount of legally registered pesticide above the crop tolerance. Most violative samples are those with traces of nonregistered chemicals, not those with high amounts of registered chemicals. Crop tolerances are set by the EPA at levels at least 100-fold below the level that caused no observable effects in lab animals that ate the pesticide every day of their lives. Imported fruit data are derived from fruit samples from 99 countries, with Mexico being the primary source. Imported fruits contained more violative samples in 2003 but also had a higher percentage of no detectable residues. Thus, 95 to 99 percent of the time, fruits consumed in the United States have either no residue or residues that fall well below EPA tolerance levels.

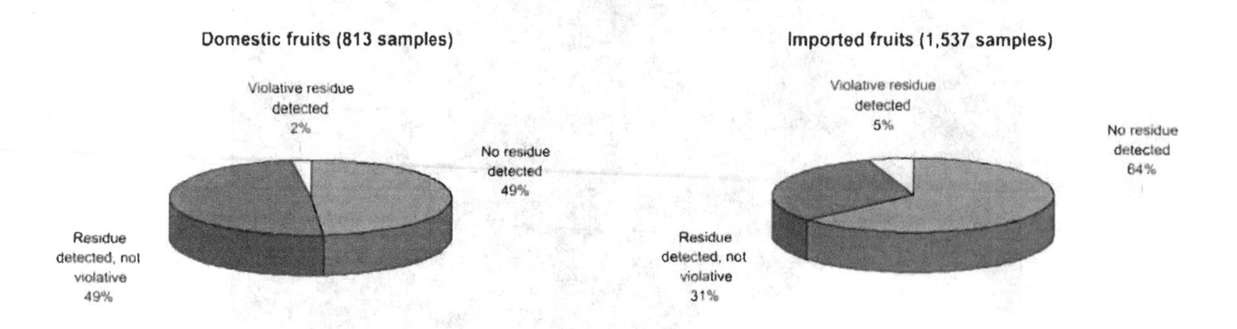

FIGURE 1.20. FDA pesticide residue test results from 2003 for *(left)* domestic fruits and *(right)* imported fruits (see text for description).

HARVEST, POSTHARVEST HANDLING

Harvesting is usually the most labor-intensive and expensive component of fruit growing, as it is usually accomplished by hand. Also, many workers are needed to sort, grade, and pack the harvested fruits. In this section, I present the basics of harvesting and handling common to most species.

As fruits reach maturity, one must decide *when* to harvest. Harvesting too early results in poor size and quality, and harvesting too late causes many of fruits to soften, bruise, or rot before reaching the consumer. Experience is key in this decision, but a few tools help commercial growers objectively determine harvest dates. Two of the most common tools are the **refractometer,** which measures the sugar content, and the firmness meter (or **penetrometer**), which quantifies the firmness of the flesh (Figure 1.21). As a fruit matures, sugars accumulate and the flesh softens. From research or past experience, growers know the ranges of sugar content and firmness that correspond to a given cultivar and its intended market. For example, apples intended to be stored for long periods are picked when more firm than are apples intended to be marketed immediately.

Fruit color is also important in determining maturity and strongly influences consumer acceptance. However, color may develop in the fruit skin well before or after the pulp reaches the optimal sugar content or firmness, so skin color should not be relied upon exclusively. The background color (or **ground color**) of the fruit is often better correlated with pulp characteristics than is the red or highlight color.

Other methods have been developed for specific crops. In fruits that store starch and gradually break this down into sugars, a starch test can be performed to assess how much breakdown has occurred. In cherry, the "fruit removal force," which is the tension required to pull the fruit off, is measured with a pull gauge. Fruit used for processing into juice or wine undergo more sophisticated measurements of sugar content, acid content, pH, sugar-acid ratio, and other chemical constituents before they are harvested.

As mentioned earlier, hand harvest is the norm, but mechanical harvesters have been developed for several crops. These devices shake, slap, or vibrate the plant to dislodge fruits and then collect fruits as they fall. Mechanical harvesters are most frequently used for nut crops and fruits intended for processing, since blemishes or bruises are unimportant or do not occur in these crops.

Fresh fruits are washed, graded, sorted, and packaged postharvest (Plate 1.11). Most fruits are graded by size and less often by color. They are packed in various containers: mesh and plastic bags, cardboard boxes, single-layer flats, and plastic "clamshells" for berries. Perishable fruits are generally shipped immediately after harvest and packing, but long-keeping fruits, such as apple and some pears, may be stored for months prior to packing and shipping.

Fruit storage temperature is dependent on species and cultivar. If possible, fruits are stored around 32°F, as they will last the longest at this temperature. Some fruits are susceptible to **chilling injury,** which manifests itself as internal breakdown, surface pitting or browning, or other disorders, after storage at low, nonfreezing temperatures. The best example of this is the rapid browning that occurs in bananas when refrigerated (Plate 1.12). Many tropical fruits cannot tolerate temperatures below 45°F without chilling injury. **Controlled-atmosphere storage** (called CA storage) is used for some fruits (mostly apple); the oxygen (O_2) level in the storage room is lowered and the carbon dioxide (CO_2) level raised to inhibit fruit respiration and subsequent breakdown. Some apples can be stored for one year using CA storage.

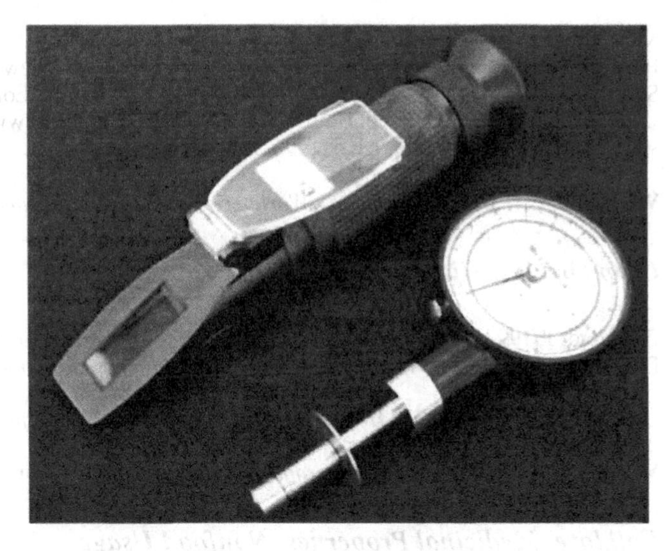

FIGURE 1.21. *(left)* A refractometer for measuring "soluble solids" or, basically, sugar content of fruit juice. *(right)* The penetrometer, or firmness meter, for monitoring crop development and determining when to harvest.

CONTRIBUTION TO DIET, FOOD USES

Fruits are an essential component of a healthy diet, being high in vitamins A and C, low in calories, and high in fiber. Nuts are packed with protein and nutrients but are generally high in fat and therefore calories. In this section, I have listed the nutrient composition of various fruits and nuts and provided the percentage of the recommended daily allowance (percent RDA) for each vitamin or nutrient. This is based on the FDA guidelines assuming a diet of 2,000 calories/day, as found on food labels in the United States. Also listed are the major food uses and the utilization statistics (i.e., percent fresh versus percent processed), largely based on USDA data.

BIBLIOGRAPHY

Taxonomy

Bailey, L.H. 1976. *Hortus third: A concise dictionary of plants cultivated in the United States and Canada.* New York: Macmillan.

Borror, D.J. 1960. *Dictionary of word roots and combining forms.* Mountain View, CA: Mayfield Publishing Company.

Downing, C. 1872. *Downing's encyclopaedia of fruits and fruit trees of America,* Parts I and II. New York: John Wiley and Son.

Facciola, S. 1990. *Cornucopia: A source book of edible plants.* Vista, CA: Kampong Publications.

Hedrick, U.P. 1922. *Cyclopedia of hardy fruits.* New York: Macmillan.

Lawrence, G.H.M. 1951. *Taxonomy of vascular plants.* New York: Macmillan.

Moore, J.N. and J.R. Ballington (eds.). 1995. *Genetic resources of temperate fruit and nut crops,* Volumes 1 and 2 (Acta Horticulturae 290). Belgium: International Society for Horticultural Science.

Morton, J.F. 1987. *Fruits of warm climates.* Miami, FL: Julia F. Morton, Publ.
Popenoe, W. 1927. *Manual of tropical and subtropical fruits.* New York: Macmillan.
Samson, J.A. 1986. *Tropical fruits,* Second edition. Essex, UK: Longman Science and Technical.
USDA. National Plant Germplasm Database. Available at http://www.ars-grin.gov/npgs/searchgrin.html.
Whealy, K. and S. Demuth (eds.). 1993. *Fruit, berry, and nut inventory,* Second edition. Decorah, IA: Seed Saver Publ.
Wiersema, J.H. and B. Leon. 1999. *World economic plants: A standard reference.* Boca Raton, FL: CRC Press.

Crop Origins, History

Diamond, J. 1997. *Guns, germs, and steel: The fates of human societies.* New York: W.W. Norton and Company.
Evans, L.T. 1998. *Feeding the ten billion: Plants and population growth.* Cambridge, UK: Cambridge University Press.
Upshall, W.H. (ed.). 1976. *History of fruit growing and handling in the United States and Canada 1860-1972.* Kelowna, BC, Canada: Regatta City Press.
Vavilov, N. 1926. *Origin and geography of cultivated plants.* Cambridge, UK: Cambridge University Press.

Folklore, Medicinal Properties, Nonfood Usage

Duke, J.A. 2001. *CRC handbook of nuts.* Boca Raton, FL: CRC Press.
Duke, J.A. and J.L. DuCellier. 1993. *CRC handbook of alternative cash crops.* Boca Raton, FL: CRC Press.
Lewis, W.H. and P.F. Elvin-Lewis. 1977. *Medical botany: Plants affecting man's health.* New York: John Wiley and Sons.
Morton, J.F. 1987. *Fruits of warm climates.* Miami, FL: Julia F. Morton, Publ.
Ritchason, J. 1995. *The little herb encyclopedia: The handbook of nature's remedies for a healthier life.* Pleasant Grove, UT: Woodland Health Books.

Botanical Description

Plant Morphology, Flowering, and Fruiting

Harris, J.G. and M.W. Harris. 1994. *Plant identification terminology: An illustrated glossary.* Spring Lake, UT: Spring Lake Publ.
Faust, M. 1989. *Physiology of temperate zone fruit trees.* New York: John Wiley and Sons.
Monselise, S.P. (ed.). 1986. *CRC handbook of fruit set and development.* Boca Raton, FL: CRC Press.
Nyeki, J. and M. Soltesz (eds.). 1996. *Floral biology of temperate zone fruit trees and small fruits.* Budapest, Hungary: Akademiai Kiado.
Sedgley, M. and A.P. Griffin. 1989. *Sexual reproduction of tree crops.* London: Academic Press.
Soule, J. 1985. *Glossary for horticultural crops.* New York: John Wiley and Sons.
Westwood, M.N. 1993. *Temperate zone pomology,* Third edition. Portland, OR: Timber Press.

General Culture

Soils and Climate

Barfield, B.J. and J.F. Gerber. 1979. *Modification of the aerial environment of crops* (Monograph No. 2). St. Joseph, MI: American Society of Agricultural Engineers.
Faust, M. 1989. *Physiology of temperate zone fruit trees.* New York: John Wiley and Sons.
Proebsting, E.L. and H.H. Mills. 1978. Low temperature resistance of developing flower buds of six deciduous fruit species. *J. Amer. Soc. Hort. Sci.* 103: 192-198.
Schaffer, B. and P.C. Andersen (eds.). *Handbook of environmental physiology of fruit crops,* Volumes 1 and 2. Boca Raton, FL: CRC Press.
U.S. Department of Commerce. National Climatic Data Center. Available at http://www.ncdc.noaa.gov/oa/ncdc.html.
Westwood, M.N. 1993. *Temperate zone pomology,* Third edition. Portland, OR: Timber Press.

Propagation

Garner, R.J. 1988. *The grafter's handbook.* Fifth edition. London: Cassell Publ., Ltd.

Garner, R.J. and S.A. Chaudhri. 1976. *The propagation of tropical fruit trees* (Horticultural Review No. 4). Farnham Royal, UK: CAB International.

Hartmann, H.T., D.E. Kester, and F.T. Davies. 1990. *Plant propagation: Principles and practices,* Fifth edition. Englewood Cliffs, NJ: Prentice-Hall.

Rom, R.C. and R.F. Carlson (eds.). 1987. *Rootstocks for fruit crops.* New York: Wiley Interscience.

Rootstocks

Rom, R.C. and R.F. Carlson (eds.). 1987. *Rootstocks for fruit crops.* New York. Wiley Interscience.

Tukey, H.B. 1964. *Dwarfed fruit trees.* Ithaca, NY: Cornell University Press.

Planting Design, Training, and Pruning

Baugher, T.A. and S. Singh (eds.). 2003. *Concise encyclopedia of temperate tree fruit.* Binghamton, NY: The Haworth Press.

Childers, N.F., J.R. Morris, and G.S. Sibbett. 1995. *Modern fruit science,* Tenth edition. Gainesville, FL: Norman F. Childers, Publ.

Galletta, G.J. and D.G. Himelrick (eds.). 1990. *Small fruit crop management.* Englewood Cliffs, NJ: Prentice-Hall.

Gilman, E.F. 1997. *An illustrated guide to pruning.* Albany, NY: Delmar Publ.

Jackson, D.I. 1986. *Temperate and subtropical fruit production.* Wellington, New Zealand: Butterworths of New Zealand.

Ryugo, K. 1988. *Fruit culture.* New York: John Wiley and Sons.

Teskey, B.J.E. and J.S. Shoemaker. 1978. *Tree fruit production.* Third edition. Westport, CT: AVI Publ.

Westwood, M.N. 1993. *Temperate zone pomology.* Third edition. Portland, OR: Timber Press.

Pests and Pest Control

The American Phytopathological Society has published a series of compendia on crop diseases, including apple and pear, blueberry and cranberry, citrus, grape, raspberry and blackberry, stone fruits (peach, plum, apricot, cherry), strawberry, and tropical fruits (banana, coconut, mango, pineapple, papaya, avocado). Full citations of compendia are given at the end of the relevant crop chapters.

The following Web resources are informative as well:

- On the issues of pesticides and food safety, Extoxnet, at http://ace.orst.edu/info/extoxnet, is a collaborative effort among extension specialists from several universities. The information is unbiased and science based.
- The USDA site offers information on organic food production at http://www.nal.usda .gov/afsci/ofp/.
- Photos of insects and diseases of many crop plants can be viewed at http://www.insect images.org and http://www.ipmimages.org/.
- The EPA site includes a page on pesticides at http://www.epa.gov/pesticides/.
- The actual data from the FDA's *Total Diet Study,* which monitors pesticide levels in food in the United States, can be found at http://www.cfsan.fda.gov/~comm/tds-toc.html.

Alford, D.V. 1984. *A colour atlas of fruit pests.* London: Wolfe Publ.

Avery, D.T. 1995. *Saving the planet with pesticides and plastic.* Indianapolis IN: Hudson Institute.

Croft, B.A. and S.C. Hoyt. 1983. *Integrated management of insect pests of pome and stone fruits.* New York: John Wiley and Sons.

Flint, M.L. 1998. *Pests of the garden and small farm: A grower's guide to using less pesticide.* Second edition (University of California Division of Agriculture and Natural Resources Publication 3332). Berkeley, CA: University of California Press.

Lind, K., G. Lafer, K. Schloffer, G. Innerhofer, and H. Meister. 2003. *Organic fruit growing.* Wallingford, UK: CABI Publ.

Ogawa, J.M. and H. English. 1991. *Diseases of temperate zone tree fruit and nut crops* (University of California Division of Agriculture and Natural Resources Publication 3345). Berkeley, CA: University of California Press.

Pena, J.E., J.L. Sharp, and M. Wyoski. 2002. *Tropical fruit pests and pollinators.* Wallingford, UK: CABI Publ.
Ploetz, R.C. (ed.). 2003. *Diseases of tropical fruit crops.* Wallingford, UK: CABI Publ.

Harvest, Postharvest Handling

Mitra, S.K. (ed.). 1997. *Postharvest physiology and storage of tropical and subtropical fruits.* Wallingford, UK: CABI Publ.
Salunkhe, D.K. and S.S. Kadam (eds.). 1995. *Handbook of fruit science and technology.* New York: Marcel Dekker.
Snowdon, A.L. 1990. *Color atlas of post-harvest diseases and disorders of fruits and vegetables.* Volume 1: *General introduction and fruits.* Boca Raton, FL: CRC Press.
Thompson, A.K. 2003. *Fruit and vegetables: Harvesting, handling, and storage.* Oxford, UK: Blackwell Publ.

Contribution to Diet, Food Uses

Hansen, R.G., B.W. Wyse, and A.N. Sorenson. 1979. *Nutritional quality index of foods.* Westport, CT: AVI Publ.
Lapedes, D.N. (ed.). 1977. *McGraw-Hill encyclopedia of food, agriculture, and nutrition.* New York: McGraw-Hill.
Morton, J.F. 1987. *Fruits of warm climates.* Miami, FL: Julia F. Morton, Publ.
Schneider, E. 1986. *Uncommon fruits and vegetables: A common sense guide.* New York: Harper and Row.
U.S. Census Bureau. 2001. *Statistical abstract of the United States.* Available at http://www.census.gov/statab/www/.

PLATE 1.1. Examples of ovary position in fruit crops: *(above left)* a superior ovary in thornless key lime flowers; *(left)* an inferior ovary of an Asian pear at petal fall; *(above right)* a perigynous ovary surrounded by the hypanthium in sweet cherry.

PLATE 1.2. *(right)* On the strawberry, the normal fruit at left has uniform achene set across the surface, whereas the fruit at right shows an area of poor set and subsequent underdevelopment. *(continued)*

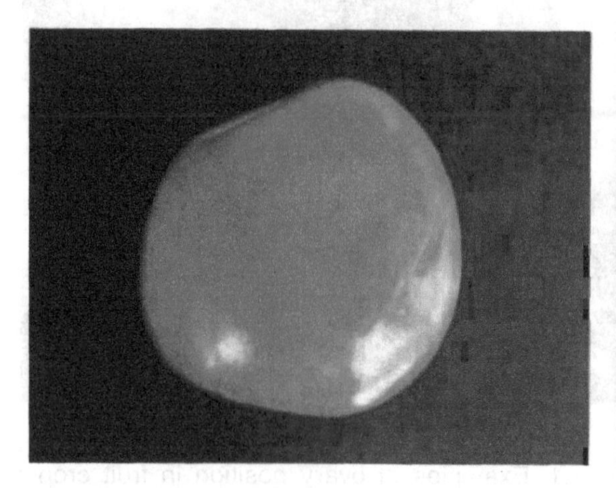

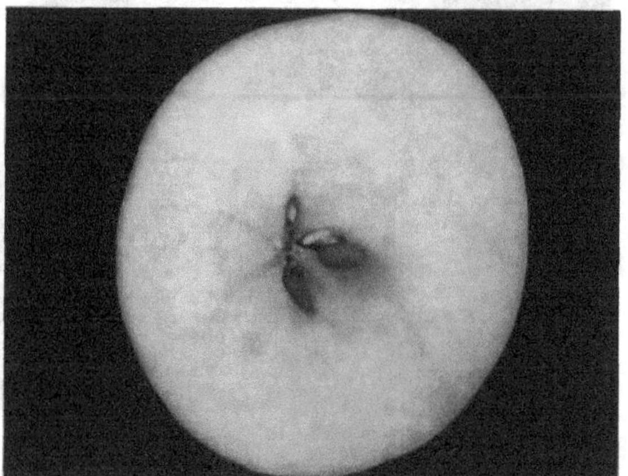

PLATE 1.2 *(continued)*. A lopsided apple *(left)* resulting from poor pollination and seed set in only three of the five locules *(right)*. The right side of the fruit was stimulated to grow more than the left because it was in closer proximity to the developing seeds.

PLATE 1.3. Fruit thinning is commonly practiced for large-fruited species, such as apple. In this case, three fruits are left, one per spur, spaced about 4 inches apart. Spacing is important; leaving three fruits on the same spur while removing all others would not yield the same increase in fruit size.

	Dormant	Silver tip	Green tip	Half-inch green	Tight cluster	First pink	First bloom	Full bloom	Post-bloom
°F	< –4	10	18	22	25	27	28	28	28.5
°C	< –20	–12	–7.5	–5.6	–3.9	–2.8	–2.3	–2.2	–1.9

PLATE 1.4. Change in killing temperature of apple flowers as they develop during late winter and early spring. *Source:* Modified from Proebsting and Mills, 1978, p. 192. The pictures depict a few of the stages of bud development.

PLATE 1.5. Peach flowers encased in ice during a frost event. Sprinkler irrigation is commonly used to protect fruit crops from frost damage in the temperate zone. The water releases heat as it freezes, keeping flower bud temperatures above the killing point of about 28°F.

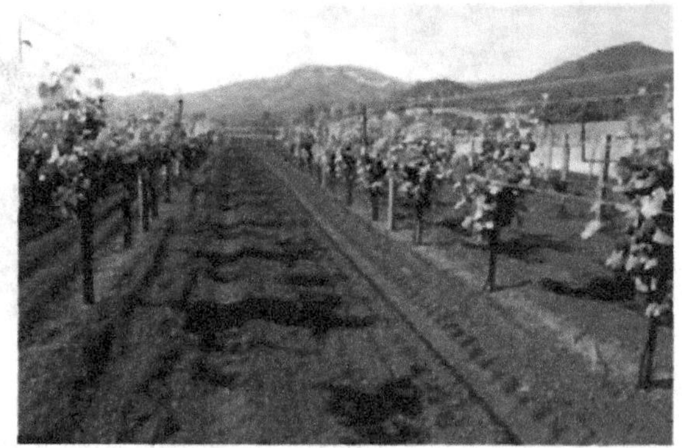

PLATE 1.6. Orchard floor management with *(above left)* herbicide strips beneath trees and grass row middles in apple, *(above right)* clean cultivation in grapes, and *(right)* mulch strips in an organic apple orchard.

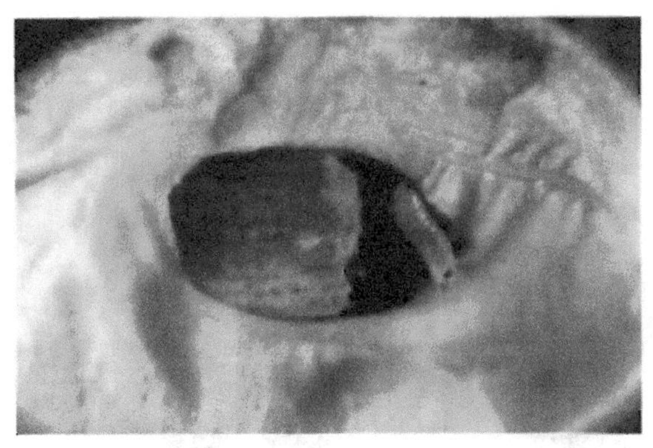

PLATE 1.7. The larva of the oriental fruit moth feeding on the seed of a developing peach fruit. Infestation this early generally causes fruit drop.

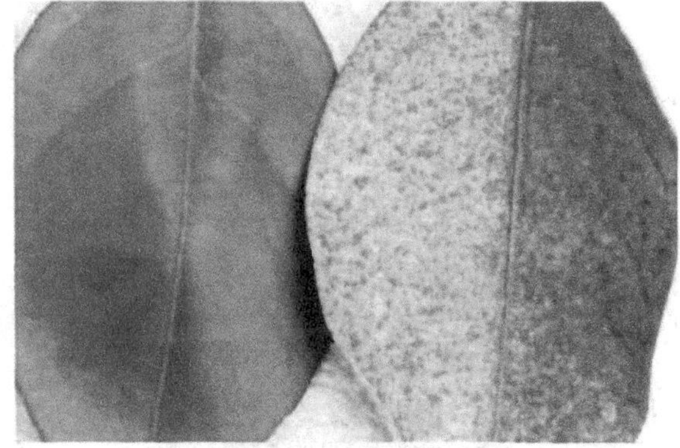

PLATE 1.8. *(at left)* A normal mandarin leaf and *(at right)* one damaged by mite feeding, showing the familiar stipple symptoms.

PLATE 1.9. *(left)* The brown rot fungus on ripe peaches. The tan dots are spore masses. Note the spread of the fungus between two adjacent fruits. *(right)* The bacterial disease fire blight on a pear shoot. Dying shoots turn black and exhibit a typical "shepherd's crook" shape.

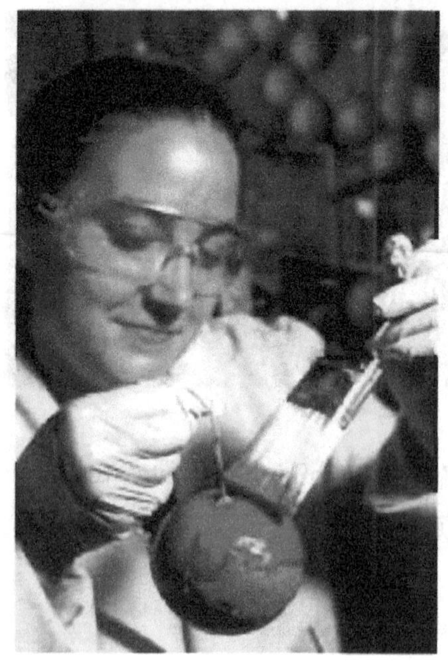

PLATE 1.10. Fruit decoys for control of apple maggot. *(left)* USDA entomologist Michael McGuire examines decoys in an apple tree. *(right)* Erica Bailey prepares a cornstarch-based apple decoy. The red color attracts the flies to the decoy for egg laying, and the insecticide within kills the adults. *Source:* Photos courtesy of USDA.

PLATE 1.11. Scene at a banana packinghouse where fruits are washed, culled, graded, and packed for distant shipment.

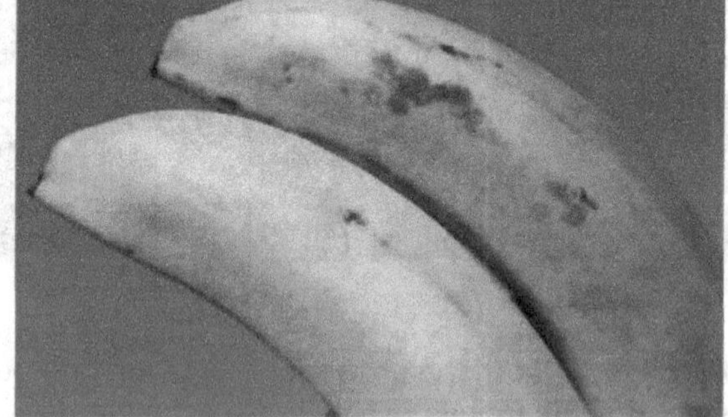

PLATE 1.12. *(at left)* A banana stored at room temperature and *(at right)* one stored at 40°F for 24 hours. Chilling injury is evident as a dull gray cast to the peel of the fruit at right.

PLATE 2.1. A single, mature almond tree showing the upright, tall, open-center habit.

PLATE 2.2. Almond flowers are similar to the "showy" flowers of peach but have whitish petals and are fragrant.

PLATE 2.3. Almond shells are the endocarp or pit of the fruit, containing a single seed or kernel.

PLATE 2.4. The shuck of almonds drying and splitting to reveal the nut, which is simply the pit of the fruit.

PLATE 3.1. 'Pacific Rose' apple, an example of some of the newer cultivars being introduced to the industry.

PLATE 3.2. A mature central-leader apple tree in a Pennsylvania orchard.

PLATE 3.3. *(left)* Apples have mixed buds that contain both leaves and flowers. The inflorescence is a cyme, where the central flower, or "king bloom," opens first and often produces the largest fruit at harvest. *(right)* The ovary is inferior, and the style is parted into five units, representing the five locules or seed cavities in the ovary.

PLATE 3.4. A branch of 'Manchurian' crab apple was grafted onto a 'Braeburn' tree at the end of a row to serve as a pollen source for other trees in the row. Crab apples will pollinate apples, and this prevents using space specifically for a pollinizer.

PLATE 3.5. An apple orchard near the town of Prosser in eastern Washington State. The climate of eastern Washington is particularly well suited to apple, being cool, temperate, and arid. The cheat grass growing on the hills above the orchard is the native vegetation.

PLATE 4.1. A large, old apricot tree with an open, spreading canopy. This tree is a remnant of an orchard planted in southern Utah in what is now Capitol Reef National Park. It received minimal care for many years but was still loaded with ripe fruits.

PLATE 4.2. Fruiting limb of apricot about one-half mature in May in California. The cordate leaves are distinct from leaves of other stone fruits. Note the predominantly spur bearing habit.

PLATE 4.3. Apricot flowers are borne laterally on spurs and tend to bloom early.

PLATE 4.4. Late-ripening 'Goldbar' apricots in July in British Columbia, Canada. Note that fruits achieved good size with light thinning due to the late ripening date.

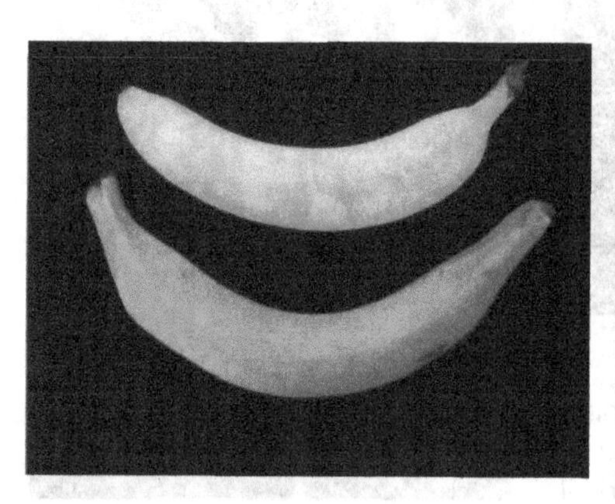

PLATE 5.1. Fruits of banana *(top)* and plantain *(bottom)* showing classic differences in size, shape, and color.

PLATE 5.2. A developing inflorescence of banana. Individual flowers are borne in double rows, spirally, along a large, terminal spike. A bract subtends each flower cluster. Fruits are negatively geotropic and curl upward after fruit set, giving the banana its familiar shape.

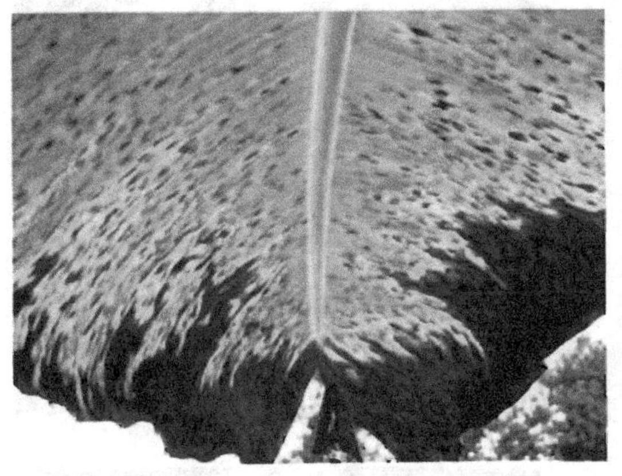

PLATE 5.3. Black sigatoka disease of banana is the greatest biotic threat to production in most areas. *(left)* The view from below a leaf infected with black sigatoka, showing characteristic marginal necrosis, interveinal spotting, and chlorosis. *(right)* Fungicides are applied from aircraft literally dozens of times per year to control sigatoka.

PLATE 6.1. Blackberries differ from raspberries by the position of the fruit abscission zone. In blackberry, the fruit plus the receptacle abscise from the plant at maturity. Raspberry drupelets abscise from the receptacle, which stays attached to the plant.

PLATE 6.2. Black raspberry floricanes die after the fruits mature. The primocanes surrounding the dead floricanes will grow vegetatively one year and fruit the next.

PLATE 6.3. Flowers of blackberry *(above left),* red raspberry *(above right),* and black raspberry *(right).*

PLATE 6.4. *(left).* A portion of a row of primocane raspberries that were mowed to the ground the previous winter. New primocanes are just emerging from the soil. *(right)* Four months later, the plants are beginning to bear fruits.

PLATE 7.1. Rabbiteye *(left)* and highbush *(right)* blueberries grown in northern Georgia. Note the more advanced development of the highbush, which has a much shorter maturation period than the rabbiteye.

PLATE 7.2. *(left)* A managed field of lowbush blueberries in Nova Scotia. *(right)* A closer view showing the creeping habit of lowbush blueberries. *Source:* Photos courtesy of D. Scott NeSmith, University of Georgia.

PLATE 7.4. A bumblebee pollinating blueberry flowers. Bumblebees are efficient pollinators of blueberry because they forage the face of the flower and sonicate the anthers properly to cause pollen release.

PLATE 7.3. Blueberry flowers are borne on stout racemes from axillary buds on 1-year-old wood. The inflorescences generally show a basipetal development sequence, as shown.

PLATE 7.5. Blueberries are "false" or epigynous berries since they derive from inferior ovaries. The calyx scar at the distal end of the fruit reveals that the ovary position is inferior.

PLATE 8.1. A new flush of growth on cacao is distinct due to the red coloration and drooping of foliage.

PLATE 8.2. Cacao is cauliflorous, bearing flowers and fruits on large branches and trunks. Flowers are borne from "cushions," which are old leaf scars.

PLATE 8.3. Fruits of the cacao plant can be yellow or red in color. The seeds are the economic product, termed "cocoa beans," and are surrounded by a fleshy, pleasant-tasting, edible aril.

PLATE 9.1. Cashew nuts still attached to the cashew apple (swollen peduncle) at a roadside fruit stand in Costa Rica.

PLATE 9.2. Cashew trees bearing a light crop in an orchard in southern Guatemala.

PLATE 9.3. Cashew flowers are either male or perfect. The style can be seen protruding from the two perfect flowers on the left, and it is absent from the other male flowers.

PLATE 9.4. In cashew, the nut develops first, followed by the apple. Here, nuts have reached their final size, while the "apple" (peduncle) remains small a few weeks after pollination.

PLATE 10.1. Sweet cherry *(left)* and sour cherry *(right)* in bloom. Flowers are essentially the same in structure, but they are borne more often on spurs in sweet cherry, rather than on 1-year-old lateral shoots, as in sour cherry.

PLATE 10.2. Ripe sweet cherries *(left)* borne on spurs and sour cherries *(right)* borne laterally on 1-year-old wood.

PLATE 11.1. 'Meiwa' kumquat, a cold-hardy citrus relative, is valued for its ornamental qualities as much as for its tart fruit.

PLATE 11.2. Two citrus fruits of minor importance are the pummelo, or shaddock *(left)*, and the citron *(right)*. Both have unusually thick albedo layers in the fruit (white layer around pulp).

PLATE 11.3. The 'Moro' blood orange contains dark red anthocyanin pigments in the pulp and peel.

PLATE 11.4. Tangerine hybrids are more common in the marketplace than pure tangerines. Those pictured are two of the most popular in the United States: *(left)* 'Murcott' is a tangor, or tangerine-orange hybrid, and *(right)* 'Minneola' is a tangelo, or tangerine-grapefruit hybrid.

PLATE 11.5. *(left)* The two major lime cultivars are 'Key' and 'Tahiti', the former being small, round, and seedy, and the latter, larger, oval, and seedless. *(right)* 'Lisbon' lemon is the most common lemon in the United States.

PLATE 11.6. Citrus flowers are borne in axillary cymes. *(above left)* A 'Meyer' lemon showing both bouquet bloom, profuse on older wood, and leafy bloom, on new growth at the tip. *(above right)* The king bloom of a 'Meyer' cyme shows typical flower morphology (the purple color is unique to lemons). *(left)* A 'Meyer' flower at petal fall showing a superior ovary on a raised nectary disc.

PLATE 11.7 Citrus fruits grown in a humid, tropical climate develop little orange or yellow pigmentation since temperatures remain high enough for chlorophyll synthesis and retention year-round.

PLATE 12.1. *(left)* Graceful coconut palms commonly grow on beaches, as their seeds are dispersed on ocean currents. *(right)* Coconut palm crowns retain about 30 leaves, with fruit clusters generally in each axil.

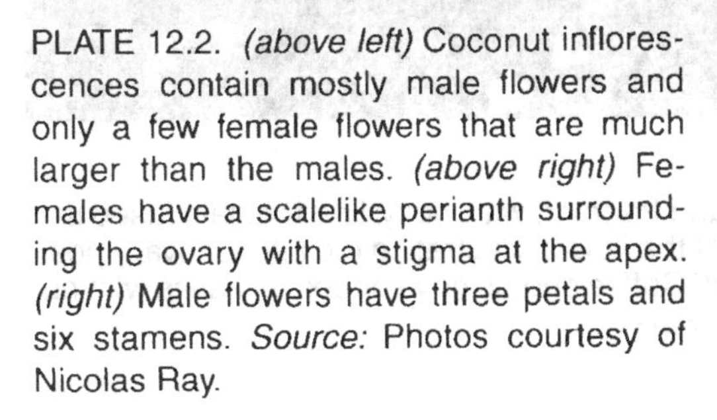

PLATE 12.2. *(above left)* Coconut inflorescences contain mostly male flowers and only a few female flowers that are much larger than the males. *(above right)* Females have a scalelike perianth surrounding the ovary with a stigma at the apex. *(right)* Male flowers have three petals and six stamens. *Source:* Photos courtesy of Nicolas Ray.

PLATE 12.3. Coconuts developing in a leaf axil of a dwarf golden cultivar.

PLATE 13.1. Coffee with ripe fruits growing in partial shade near Veracruz, Mexico.

PLATE 13.2. *(left)* Individual coffee flowers are white and fragrant, with waxy, linear petals fused in a narrow tube at their base. Note the fruitlet to the right of the flower, illustrating the inferior ovary position by virtue of the corolla scar on the distal end. *(right)* Coffee shoots have opposite leaves with numerous buds at each node, creating a profusion of flowers.

PLATE 13.3. *(top)* Ripe coffee fruits clustered at nodes. *(center)* The coffee fruit is an epigynous berry, containing two seeds, which are the coffee "beans." *(bottom)* Dried coffee in parchment, or still enclosed in the papery endocarp.

PLATE 14.1. Cranberry uprights flowering. *Source:* Photo courtesy of Teryl Roper, University of Wisconsin.

PLATE 14.2. Cranberries approaching full maturity in September in Cape Cod, Massachusetts. *Source:* Photo courtesy of Michael A. Dirr, University of Georgia.

PLATE 14.3. Cranberry bogs are constructed wetlands and represent the most sophisticated planting designs of all fruit crops. They are flooded for harvest and winter protection and have to be carefully engineered for water movement. *Source:* Photo courtesy of Teryl Roper, University of Wisconsin.

PLATE 14.4. Wet harvesting of cranberries involves flooding the bog, beating the fruits from the vines *(left),* and corralling the fruits in one corner where they are lifted by conveyor to trucks for transport *(right). Source:* Photos courtesy of Teryl Roper, University of Wisconsin.

PLATE 15.1. An oasis along the river Ziz in southern Morocco where dates, olives, and figs are grown. It is the only source of food or water for many miles.

PLATE 15.2. Date palms have persistent "boots," or leaf bases, and produce offshoots, or suckers, from the base of the trunk.

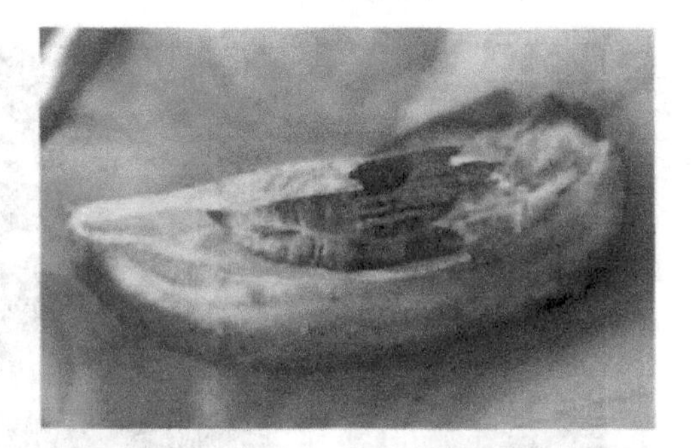

PLATE 15.3. 'Deglet Noor' dates from California: *(bottom)* at maturity, *(top)* showing the membranous endocarp surrounding the seed.

PLATE 15.4. *(left)* Young, developing fruits in the initial Hababouk or Kimri stages of development and *(right)* later in the Khalal stage when coloration occurs. *Source:* Photo at right courtesy of Gary A. Couvillon, University of Georgia.

PLATE 16.1. The three main species of grape: *(top left)* muscadine, *(top right)* American, and *(bottom)* European.

PLATE 16.2. White seedless grapes drying into raisins on the vine.

PLATE 16.3. *(top, left and right)* Flowering in grape occurs at the basal nodes of the current season's growth in all species. *(bottom left)* Individual flowers are tiny and have nonshowy petals fused at the apex into a structure called the calyptra. *(bottom right)* Muscadine inflorescences contain either pistillate flowers *(at left)* or perfect flowers *(at right)*, depending on the cultivar. Perfect-flowered muscadines have longer stamens, while the pistillate flowers have short, reflexed stamens.

PLATE 16.4. Winemaking is an art and a science. *(top)* A small-scale winery in Tuscany, with a grape press on the right, which extracts the juice, and fermentation vats on the left, where the juice is made into wine. *(center)* Large-scale fermentation vats in Sonoma, getting a test of their cooling capacity (note the frost formation around the middle portion of the vat). *(bottom)* These are 100-hectoliter oak casks for aging Chianti Classico wine in Italy.

PLATE 17.1. Nuts of 'Barcelona' *(at left)*, the major cultivar in the United States, and 'Ennis' *(at right)*, a minor cultivar with a larger nut size.

PLATE 17.2. In hazelnut, male flowers are borne laterally in catkins, and female flowers are borne terminally.

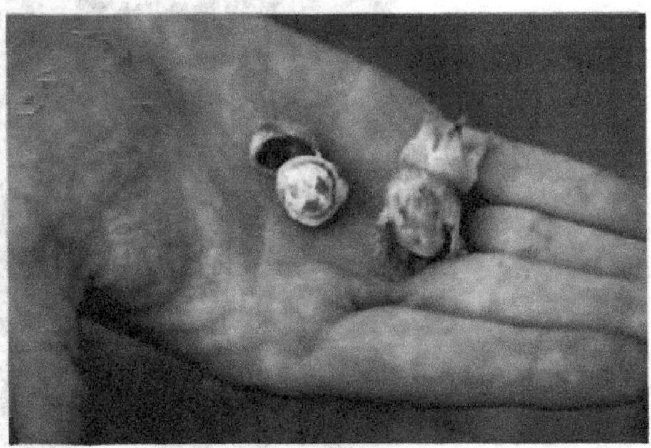

PLATE 17.3. *(left)* Hazelnuts are borne terminally in clusters of a few individuals. *(right)* At maturity, the nuts abscise from the plant but may remain enclosed in the leafy involucre.

PLATE 18.1. Macadamia foliage and fruits. Leaves are stiff with marginal spines, similar to holly (*Ilex* spp.) leaves.

PLATE 18.2. Racemose inflorescences of macadamia blooming in March in southern Florida.

PLATE 18.3. The fruit of *Macadamia* is a drupe, with the green hull surrounding the smooth, brown endocarp or shell.

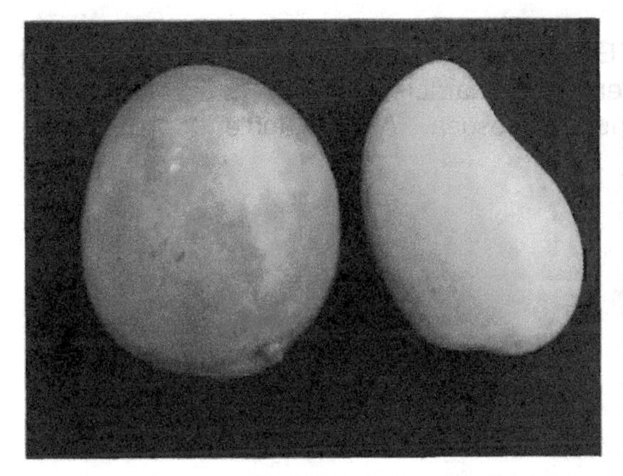

PLATE 19.1. Examples of the two types of mango cultivars: West Indian and Indo-Chinese. 'Kent' *(at left)* is a popular West Indian cultivar, and 'Ataulfo' *(at right)* is a popular Indo-Chinese cultivar.

PLATE 19.2. Mango flowers are produced terminally in panicles. Individual flowers are tiny, yellow, and mostly staminate.

PLATE 19.3. Fruit set is very low in mango; often only one fruit per panicle will develop from tens to hundreds of perfect flowers. Mangos are borne on the outside of the canopy, dangling on what appear to be long peduncles (panicle axes).

PLATE 19.4. Well-colored West Indian mangos at a roadside stand in Costa Rica. The fruit is a large drupe, as seen here, approaching the length of a small pineapple.

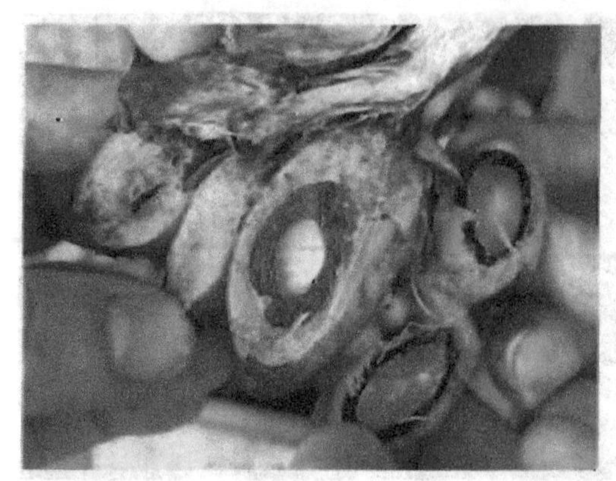

PLATE 20.1. Two breeding lines of oil palm illustrating different fruit characteristics, such as the relative thickness of mesocarp (yellow) and endocarp (black).

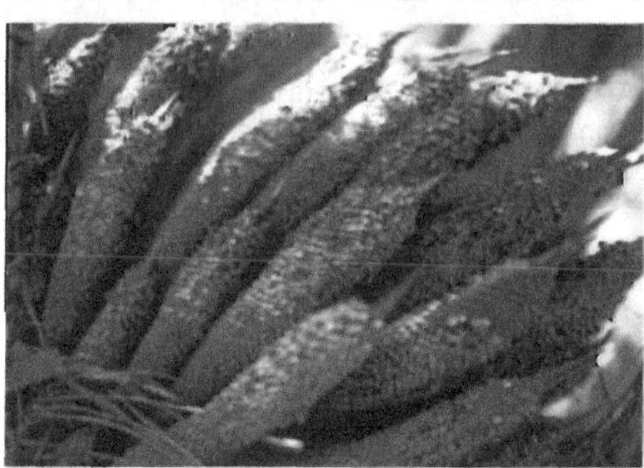

PLATE 20.2. Male *(left)* and female *(right)* inflorescences of oil palm. Both are a compound spadix. Note the beetle pollinators foraging on the male flowers at the tips of the spadix branches.

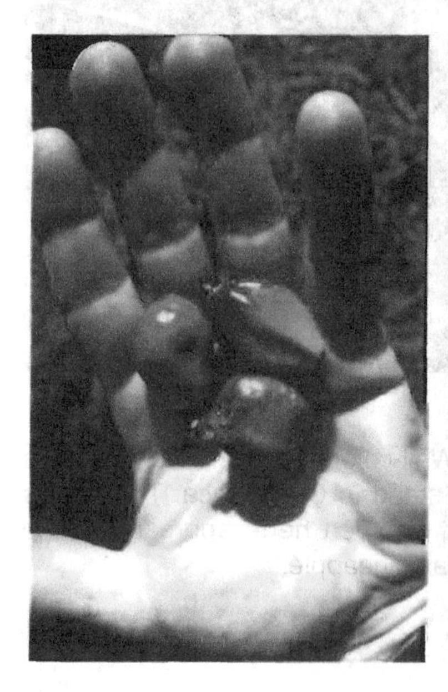

PLATE 20.3. *(left)* Individual oil palm fruits are drupes with an oily, fibrous mesocarp. *(right)* Bunches of fruits are borne axillary, with one ripening about every 2 weeks per tree.

PLATE 21.1. Olives are some of the longest-lived trees in the world. This tree, photographed near Agrigento, Sicily, is purportedly hundreds of years old.

PLATE 21.2. Olive flowers are produced on small panicles in the axils of 1-year-old wood.

PLATE 21.3. Olive fruits are drupes, green when immature and turning purple or black when fully mature. Table olives *(left)* are harvested when pale green or turning yellow, and oil olives *(right)* are harvested later when fully ripe.

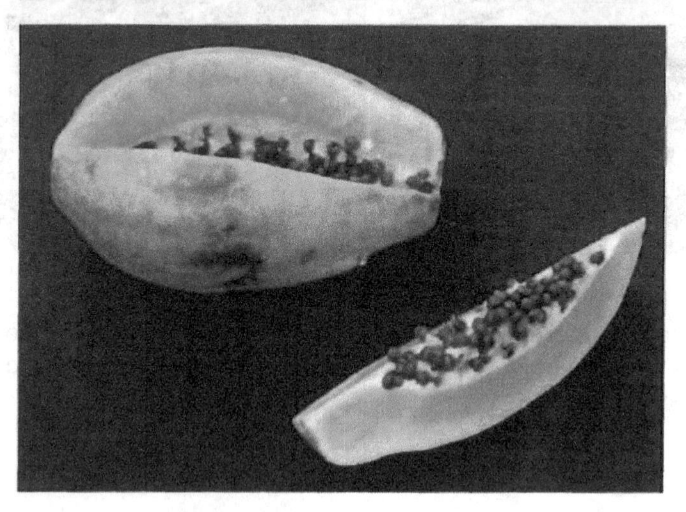

PLATE 22.1. Papaya fruits exhibit diverse morphology: *(top)* a large, rounded type from the Yucatan; *(center)* a cylindrical drupe from Costa Rica; *(bottom)* a small 'Caribbean Sunrise' fruit, one of several cultivars related to 'Solo', typical of Hawaii.

PLATE 22.2. A mature, bearing papaya resembles a tree but is actually a large, herbaceous plant.

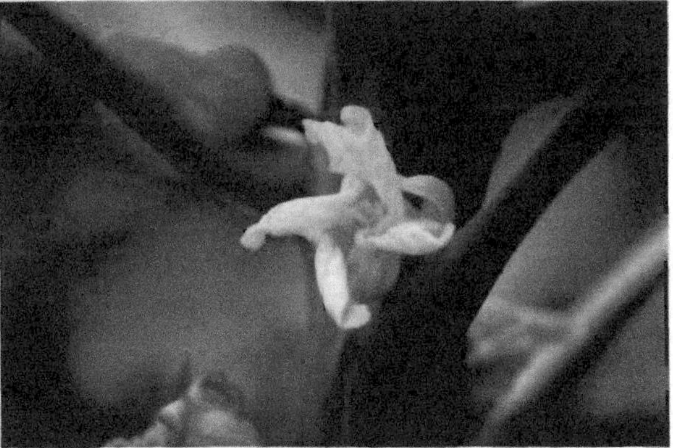

PLATE 22.3. *(left)* Male papaya flowers are distinguished by being smaller and borne on long stalks. *(right)* Female flowers are pear shaped when unopened and distinguished from bisexual flowers, which are cylindrical.

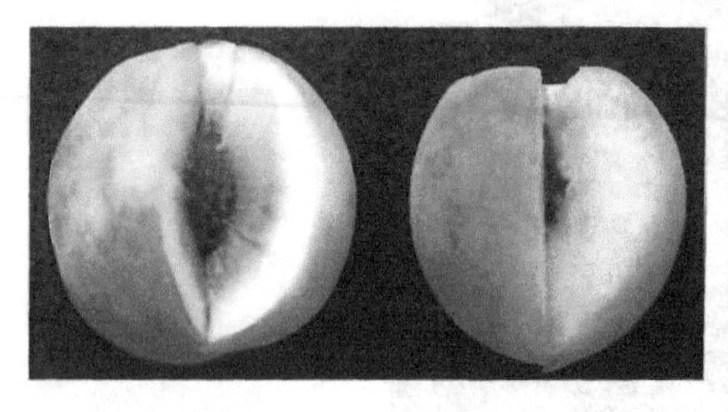

PLATE 23.1. 'September Snow' and 'Sweet September' white- and yellow-fleshed peaches. Both colors occur in peach and nectarine cultivars.

PLATE 23.2. The structure of peach flowers, showing the hypanthium or floral cup surrounding the ovary, which is glabrous in nectarine *(left)* and covered with hairs in peach *(right)*.

PLATE 23.3. Peach-packing operations. *(left)* Peaches are hydrocooled to reduce temperature and then subjected to culling by hand to remove defective fruits. The culls are placed on the top conveyor in this photo. Fruits are then defuzzed, brushed, waxed, and run through a computer-controlled packing line *(right)* that weighs each fruit and delivers it to a packing station with fruits of similar size.

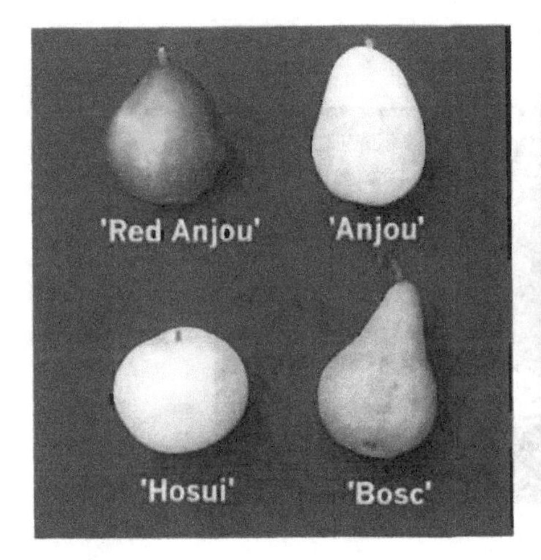

PLATE 24.1. Some major pear cultivars, showing a range of shapes and skin colors. 'Hosui' is an Asian pear; the others are European.

PLATE 24.2. Pear orchards in the Hood River Valley of Oregon. Pear culture in the United States is predominantly in the Pacific Northwest and California, a region having lower humidity and rainfall and less disease pressure.

PLATE 24.3. An old pear tree growing in the Yakima Valley of Washington State, showing the upright growth habit.

PLATE 24.4. Pear flowers are borne in corymbose inflorescences from terminal mixed buds.

PLATE 25.1. Nuts of 'Desirable' *(left)* and 'Wichita' *(right)*, two of the most popular cultivars in the United States.

PLATE 25.2. Pecans are monoecious. *(left)* Male inflorescences of pecan are catkins borne laterally on 1-year-old wood. *(right)* Female inflorescences are spikes borne terminally on the current season's growth.

PLATE 25.3. Fruit morphology in pecan. *(left)* Nuts are encased in a fleshy shuck, which splits at maturity to release the nut. *(right)* Pecan nuts contain two amber-colored, furrowed kernels, which are the cotyledons of the embryo.

PLATE 26.1. Pineapple plants cultivated on raised beds in the Dominican Republic.

PLATE 26.2. Pineapple flowering. Pineapples are induced to flower with ethylene-releasing compounds. *(left)* Red bracts subtend the inflorescence. *(right)* As the inflorescence develops, the rachis extends and small, tubular, red-purple flowers open acropetally over a period of weeks.

PLATE 26.3. *(left)* A large pineapple plantation in the rainshadow of the Talamanca range in southern Costa Rica. Note the bare areas in the distance being prepared for planting. *(right)* A Dole plantation on Oahu, Hawaii, shows pineapple as far as the eye can see. Pineapple is considered a "plantation crop" due to its large-scale production methods in many areas.

PLATE 27.1. A 'Kerman' pistachio tree, showing the typical open-center training and spreading crown.

PLATE 27.2. Male *(left)* and female *(right)* inflorescences of pistachio. *Source:* Photos courtesy of Louise Ferguson, University of California, Davis.

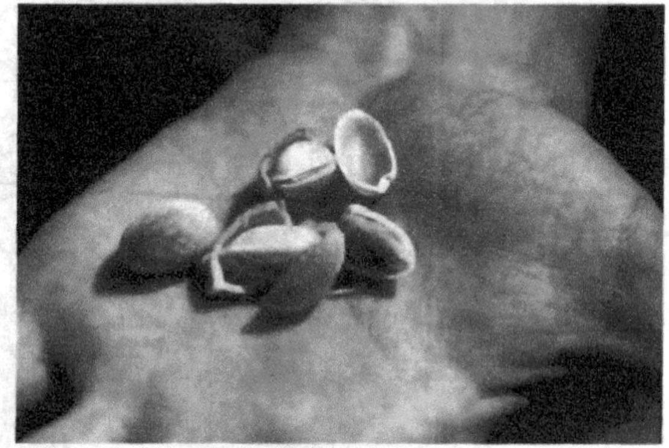

PLATE 27.3. Pistachio fruit morphology. *(left)* Fruits are borne in axillary clusters on 1-year-old wood. *(right)* Fruits are drupes, and the endocarp (shell) splits naturally prior to maturity.

PLATE 28.1. *(clockwise from top)* Japanese plum, apricot, and their hybrid, the pluot. Pluots are gaining popularity in the United States.

PLATE 28.2. European plums are particularly diverse in terms of size, color, and flavor. *(left)* 'Reine-Claude' or 'Greengage' plums, said to be among the best-tasting plums. *(right)* 'Stanley', a major prune cultivar.

PLATE 28.3. Flowers of European *(left)* and Japanese *(right)* plums are basically the same in structure but are more abundant in Japanese types.

PLATE 29.1. *(left)* Strawberry flowers are borne on dichasial cymes arising from the crown. Each flower contains hundreds of ovaries, borne on a raised, yellow receptacle that becomes the fruit at maturity. *(right)* All stages of floral and fruit development can be seen in strawberry in midspring.

PLATE 29.2. The true fruit of the strawberry is an achene, seen as numerous, tiny, ovate structures in depressions on the surface of the receptacle.

PLATE 29.3. Handpicking strawberries in Watsonville, California, in July. The cool, coastal climate of the Watsonville area allows strawberry harvest to continue for several months.

PLATE 30.1. A lateral-bearing walnut cultivar. Fruits are borne from lateral buds, not just terminals, as in older walnut cultivars and pecan.

PLATE 30.2. Walnuts are monoecious and similar to pecan with respect to flowering. *(left)* Catkins of eastern black walnut, produced laterally on 1-year-old wood. *(right)* Persian walnut female flowers produced terminally on the current season's growth.

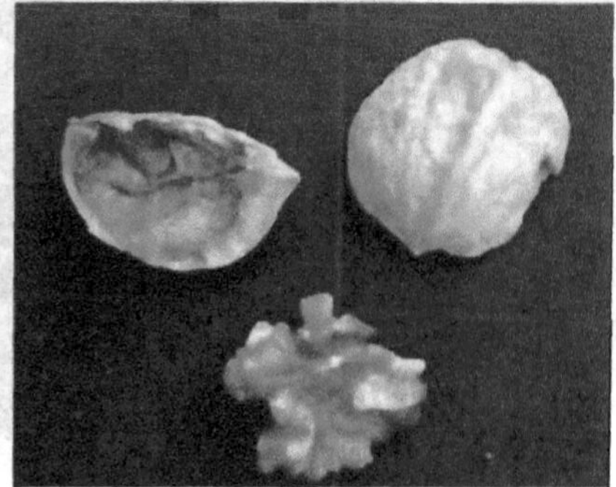

PLATE 30.3. Nuts of the 'Hartley' *(left, at left)* and 'Chandler' *(left, at right)* Persian walnut. *(right)* Walnut kernels are the cotyledons of the embryo, as in pecan, but are more deeply furrowed and rounded compared to pecan.

Chapter 2

Almond *(Prunus dulcis)*

TAXONOMY

Almonds are members of the Rosaceae (rose) family, subfamily Prunoideae, along with many other tree fruits, such as peaches, plums, cherries, and apricots. Within the genus *Prunus*, almond is most closely related to the peach, and the two crops share the subgenus *Amygdalus*. Peach and almond are generally cross- and graft compatible, and hybrids of the two are used as rootstocks.

Linnaeus first named the almond *Amygdalus communis* in the 1750s, which translates roughly as "the common Greek nut." A few years later, Miller classified almond as *Prunus dulcis*, the name used today, to indicate the sweet *(dulcis)* nature of the kernel of the cultivated form. However, *Prunus amygdalus*, proposed by Batsch shortly afterward, became the most widely accepted name of the almond, although some continued to use *A. communis* or *P. communis*. In the 1960s, botanical authorities revisited the subject and decided to use *Prunus dulcis* after all, leaving us with *Prunus dulcis* (Mill.) D.A. Webb.

The cultivated almond is placed within the section *Eumygdalus*, one of five sections that contain 20 to 25 species of near relatives. The bitter almond *(P. dulcis* var. *amara)*, which is used for its essential oil and as a flavoring, is the only close almond relative of commercial significance.

Cultivars

Dozens of almond cultivars are grown commercially around the world, with different centers of production having unique selections. Some areas of the Middle East continue to use seedlings, sometimes grafting existing trees over to superior local selections. Thus, thousands of cultivars or selections probably exist worldwide, but mainstays of production number fewer than ten in most regions. The top ten almond cultivars in California follow:

1. Nonpareil
2. Carmel
3. Mission (Texas)
4. Merced
5. Price Cluster
6. Ne Plus Ultra
7. Peerless
8. Thompson
9. Butte
10. Monterey

The major California cultivar is 'Nonpareil' (~60 percent acreage), with several of the other listed cultivars used as pollinizers, since almonds are largely self-incompatible. 'Nonpareil' was a chance seedling discovered by A. T. Hatch in 1879 in California. It has retained the number

Introduction to Fruit Crops

doi:10.1300/5547_02

one ranking largely due to its superior nut yield and quality. It has medium-sized, smooth, attractive kernels with thin shells, giving a very high shelling percentage (65-70 percent). However, the thin shell makes it more vulnerable to pests than are hard-shell cultivars such as 'Peerless'. 'Nonpareil' also tends to be a consistent bearer, giving yields of up to 3,000 pounds of kernels per acre.

ORIGIN, HISTORY OF CULTIVATION

Almond and related species are native to the Mediterranean climate region of the Middle East (Pakistan eastward to Syria and Turkey). The almond and its close relative, the peach, probably evolved from the same ancestral species in south-central Asia. The original population of peach-almond progenitor species was separated by the formation of mountain ranges in southern Asia millions of years ago. The resulting climate change allowed almonds to evolve in the arid western part of this region, while peaches evolved in the humid eastern areas of south-central China.

Almonds were domesticated at least by 3000 BC, and perhaps much earlier, since wild almonds have been unearthed in Greek archaeological sites dating to 8000 BC. In the wild state, most species of almonds are bitter and unpalatable, as they contain the cyanide-releasing compound amygdalin. Presence of amygdalin in almonds is controlled by a single dominant gene; thus, a simple mutation would produce nonbitter almonds, and by chance, early farmers found these trees occasionally. Trial and error combined with selection of the best-tasting almonds to plant new trees eventually led to the domestication of modern, sweet almonds.

The almond was spread along the shores of the Mediterranean in northern Africa and southern Europe by Egyptians, Greeks, and Romans. It was brought to California in the 1700s by Spanish padres who settled the mission at Santa Barbara. Larger plantings did not occur until the mid-1800s. Around the turn of the century, the industry started in California, due to development of superior cultivars in the late 1800s. Tariffs on almond imports were levied to protect the industry. From then until about 1960, the industry grew at a moderate pace, but acreage and production have increased severalfold since then, making California the clear world leader in almond production. In 2004, there were 550,000 acres of almonds in California, making it the most widely planted tree crop in the state.

FOLKLORE, MEDICINAL PROPERTIES, NONFOOD USAGE

Almonds are native to the Middle East and were probably abundant in ancient times, since they are mentioned several times in the Bible. The Hebrew name for the almond, *shakad,* means "hasty awakening," which probably derives from its prolific and fragrant bloom in late winter, ahead of most other orchard species. Aaron's rod was made of almond wood, and Jewish people often carry blooming almond branches to synagogues for festivals. Ancient pagans thought almonds symbolized virginity. Some even used almonds as an emblem of the Virgin Mary. While symbolizing virginity, almonds were also used as fertility charms and marriage blessings. Italians used to distribute almonds at weddings as tokens of fruitfulness. The almond was referred to as the "womb of the world," suggestive of its supposed birth-bringing powers. Reference to the flowering almond in older poetry often meant "hope," according to "the language of flowers." However, almond trees have a somewhat conflicting symbolism: giddiness, heedlessness, stupidity, indiscretion, and thoughtlessness. Almonds were left in King Tutankhamen's tomb to provide nourishment in his afterlife. Almond branches were used as divining rods in Tuscany, Italy.

Almonds have been used as a folk remedy for cancers, tumors, ulcers, corns, and calluses. Almonds were thought to prevent intoxication from drinking too much alcohol.

As with other members of *Prunus*, bitter almonds contain cyanogenic glycosides in seeds, bark, and leaves; if eaten in large quantities, they can cause convulsions and death. About 50 to 70 bitter almonds cause death in adults; seven to ten can cause death in children, while three can cause severe poisoning. The sweet almonds of commerce do not contain these compounds. Phloretin is an antibiotic-like compound found in bark and root extracts; in concentrated form, phloretin can kill certain bacteria.

The major nonfood usage of almond is for oil. Almond oil is highly valued for use in cosmetics and creams, and bitter almond oil is used as an essential oil. The oil is used to treat various forms of dermatitis.

PRODUCTION

World

Total production in 2004, according to FAO statistics, was 1,669,642 MT or 4 billion pounds. Almonds are produced commercially in 45 countries on 4.1 million acres. Production has increased 62 percent in the past decade. Worldwide, yield averages about 900 pounds/acre but can reach 3,500 to 4,500 pounds/acre in some countries. The top ten almond-producing countries (percent of world production) follow:

1. United States (49)
2. Syria (8)
3. Iran (7)
4. Spain (6)
5. Italy (5)
6. Morocco (4)
7. Tunisia (3)
8. Turkey (2)
9. Greece (2)
10. Algeria (2)

United States

Total production in 2004, according to USDA statistics, was 463,600 MT of shelled kernels or 1.0 billion pounds. Their industry was valued at $2.05 billion in 2004, an all-time high, ranging from $700 million to $2.05 billion over the past decade (Figure 2.1). Prices received for shelled kernels were $2.04 per pound in 2004; $0.86 to $2.48 per pound over the past decade. There are about 550,000 acres of almonds throughout the central valley of California. Of all fruit crops grown in California, only grapes are grown on greater acreage. Yields average 1,100 to 2,000 pounds shelled kernels per acre.

In 2002, the United States imported only 750 MT (1.6 million pounds) or the equivalent of 0.2 percent of production. Exports are typically over 50 percent of total production; 290,000 MT or 62 percent of production was exported in 2002. Exports have doubled since 1992.

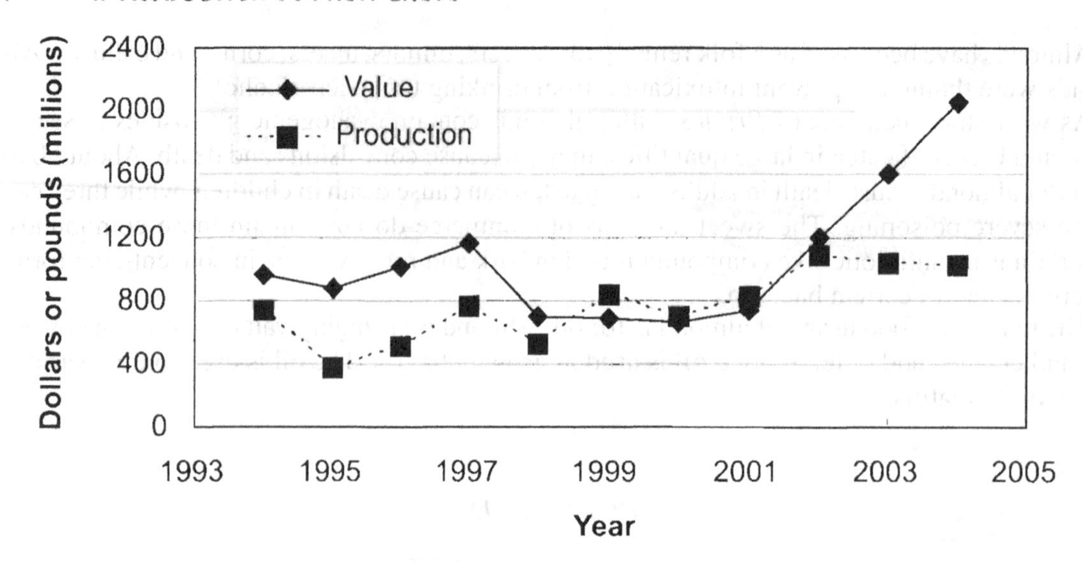

FIGURE 2.1. Ten-year trend of almond production and value in California. *Source:* USDA statistics.

BOTANICAL DESCRIPTION

Plant

Almonds grow on a small- to medium-sized tree with a spreading, open canopy, usually 10 to 15 feet in commercial orchards (Plate 2.1). Leaves are linear or slightly ovate, about 2 to 5 inches in length, with acute tips and finely serrate margins. Overall, leaves are smaller and less folded along the midrib than those of its close relative the peach.

Flowers

Almond flowers are nearly identical to peach and other *Prunus* flowers in structure, but light pink or white in color and fragrant (Plate 2.2). Flowers have five petals and sepals, and many elongated stamens; all of these appendages originate from a hypanthium or floral cup that surrounds the ovary. The ovary is perigynous. Flowers are borne laterally on spurs or short lateral branches, or sometimes laterally on long shoots, as in peach, particularly on younger trees. Flowering on basal spurs of 4- to 6-year-old wood gradually declines, since terminals fail to grow out when many flowers per spur set fruits; hence, older limbs become unfruitful and are pruned out.

Pollination

Almonds are self-incompatible and require cross-pollination. A few self-fertile cultivars (e.g., 'Halls Hardy') are available but are generally inferior in quality. Pollinators (honeybees) are absolutely essential, especially since cool, wet weather can occur at the relatively early blooming period.

Pollinizers are generally planted in separate rows so nuts can be shake harvested a row at a time without mixing different cultivars. Since pollination is so critical to fruit set and yield, in some cases two pollinizers are used, one blooming slightly ahead (but overlapping) of the main cultivar, and one slightly after.

Fruit

Almond's fruit is the nut. The entire fruit, including the hull, is a drupe; however, the hull dries and splits prior to harvest, revealing what appears to be the pit of the fruit (Plate 2.3). Botanically, this pit with the kernel inside fits the definition of a nut (dry, indehiscent fruit with a hard shell). Fruiting begins in 3- to 4-year-old trees, with maximal production in 6 to 10 years. Unlike its short-lived cousin the peach, almond trees can produce for 50+ years. Thinning is unnecessary; a high proportion of flowers must set fruits for normal yield.

GENERAL CULTURE

Soils and Climate

Almonds produce best on deep, loamy, well-drained soils but will tolerate poor soils, and even drought, during the latter portion of fruit development better than most tree crops. In intensive orchards, soils are managed similar to other stone fruits, but less-intensive plantings of the Mediterranean area occur on calcareous, rocky, and droughty soils. As with all *Prunus* species, almonds are intolerant of wet, poorly drained soil.

The almond is a true Mediterranean tree, requiring mild winters and long, rainless, hot summers with low humidity. Rainfall is deleterious anytime during growth; during bloom, it decreases bee activity and therefore fruit set; during fruit development, it causes fungal and bacterial diseases; prior to harvest, it can cause brown rot or *Rhizopus* rot of fruits as shucks split. It is no surprise that all of the almonds in the United States are grown in California. Almonds require only 300 to 500 chill hours. Almonds bloom in late February in California, and frost is a perennial problem. Open-flower hardiness is similar to that for other tree fruits, with death occurring below 28°F; young fruits may be killed by temperatures as high as 31°F. Frost is a serious limitation in most areas outside of California. Midwinter hardiness is not as limiting a factor as spring frost damage to blossoms. Almonds routinely survive temperatures in the single digits (°F) and even lower in some cases. 'Nonpareil' matures in late August/early September in California, after blooming in mid- to late February (6-7 months). 'Mission' requires about one additional month. Thus, growing season length needs to be at least 6 to 8 months for the major cultivars.

Propagation

Almonds are budded onto seedling rootstocks so that the integrity of the cultivar is kept intact. Propagation of trees by seed is possible and practiced in some regions of the Mediterranean. However, tree uniformity is poor in seedling orchards because almonds are self-incompatible and heterozygous as a species. Almonds root poorly from cuttings, as opposed to peach, which is easily rooted.

Rootstocks

Peach seedlings ('Lovell', 'Halford', 'Nemaguard', 'Nemared') are the primary rootstocks for almond. Trees on peach seedling stocks are more vigorous and bear earlier than those on almond seedling stocks. Peaches are slightly more tolerant of crown gall, verticillium, oak root rot, and *Phytophthora* than are almonds.

Almond seedlings are used as rootstocks in less-intensive orchards of the Mediterranean area, where soils are calcareous and irrigation is not available. These trees may tolerate drought better than trees on peach.

FIGURE 2.2. An almond orchard in California with clean cultivation to facilitate shake harvesting and flood irrigation.

Peach × almond hybrids (for example, GF 677, GF 557), are recent introductions but are now the dominant rootstock for peach and almond in France. These stocks have similar disease response as almond seedlings, but some have been produced with resistance to root-knot nematode and iron deficiency.

Planting Design, Training, Pruning

Trees are planted in rectangular or hexagonal arrangements, with separate rows of pollinizers and main cultivars, usually alternating with each other. Solid rows of pollinizers are used since trees are shake harvested, and this makes it easier to harvest without mixing cultivars. In some orchards, two different pollinizers are used, generally one blooming slightly before and the other slightly after the main cultivar. Spacing must be wide enough to permit movement of harvesting equipment, or about 18 feet. Spacings of 24 × 24 feet are common in California, yielding about 87 trees per acre (Figure 2.2).

Trees are trained to an open-center shape in the first year. Young trees are headed at 3 feet to allow adequate room for the trunk shaker attachment. Three scaffolds are selected initially, about 3 to 6 inches apart from each other on the trunk, spaced equidistant when viewed from above. Various degrees of pruning are practiced after the first year, with minimal pruning allowing the most rapid production of bearing area. In subsequent years, scaffolds are trained to fill their allotted spaces in the tree.

At maturity, pruning consists of water sprout removal, removal of dead and interfering branches, and limb thinning. Little pruning is needed because many fruiting points are needed for cropping, and vigor is low in mature trees. Since fruiting spurs live about 5 years, fruiting wood is renewed every 5 years, or 20 percent of the canopy is pruned back each year to allow new wood to grow and replace the old. In practice, trees are pruned every other year with no loss of productivity. Under proper management, trees should have many 15- to 20-inch-long, 1-year-old shoots. These shoots will branch, grow spurs, and become fruiting wood in subsequent years.

Pest Problems

Insects

Navel orangeworm. The larvae of this moth *(Amyelois transitella)* appear late in the season. Eggs are laid from April through October, in or on developing fruits. The larvae bore into and eat through the kernels. Several larvae may invade the same nut; this feature and the presence of webbing distinguish it from oriental fruit moth or peach twig borer damage. The pest overwinters in damaged nuts ("mummies") that remain in trees or on the ground, or in any intact fruits on the ground. Insecticide applications are poorly effective for control but can be applied in May and later in July when hulls split. Some natural enemies (wasps) exist but cannot control the pest completely. Early harvest is the best way to avoid damage. Shaking mummified nuts from trees in winter and removing all debris from beneath the tree eliminate the overwintering insect and greatly reduce infestation.

Scale insects. San Jose scale *(Quadraspidiotus perniciosus)* adults are small (<⅛ inch), round sedentary insects that appear as gray bumps on wood; they suck the sap from branches and twigs. Dormant oil sprayed 1 to 3 weeks before budbreak, combined with an insecticide if infestation is heavy, is one control measure. Additional sprays in May when crawlers emerge are sometimes required.

Oriental fruit moth. These small moths *(Grapholita molesta)* lay eggs in shoot tips early in the season, and larvae burrow downward a few inches; wilted and dead expanding leaves at the shoot tips indicate infestation. Later infestation results in fruit loss, as eggs laid on fruit produce ¼-inch, legged, pinkish larvae that burrow through the fruit. Although fruit infestation does not occur until 3 to 4 weeks before harvest, if at all, early season sprays can help reduce this pest. Insecticides at petal fall and shuck split control the first generation, and late May or early June sprays control later generations.

Peach twig borer. Larvae of this moth *(Anarsia lineatella)* cause damage to the fruits and kernels primarily and can kill shoot tips. The adult moth lays eggs on the green fruit, and the larvae bore into the hull and kernel to feed. The larva is ½ inch long with a dark-colored head and light and dark alternating bands along its body. The same dormant oil sprays used for scale in late winter before budbreak control this pest. Alternatively, an organic method involves sprays of the insecticide Bt *(Bacillus thuringiensis)* at first sign of pink color on the flowers up to the "popcorn" stage.

Mites. Pacific *(Tetranychus pacificus)*, strawberry *(T. turkestani)*, and two-spotted *(T. urticae)* mites are the most common spider mites in California. European red mite *(Panonychus ulmi)* also attacks almond. Mites feed on the lower leaf surface, causing "mite stipple" to the upper leaf surface. If severe, leaf photosynthesis is reduced and leaves can be shed. Dormant oil sprays with or without miticide before budbreak reduce initial populations. In California, mites must be monitored throughout the year, with miticide applications made when specific thresholds are met. Care is taken to avoid killing predators, such as avoiding pyrethroid and organophosphate insecticides. Preventing dusty conditions and water stress are important cultural controls for mites.

Diseases

Brown rot/blossom blight. The fungus *Monilinia laxa* causes flowers to rot. Infections during cool, wet springs can kill entire spurs or shoots. Overwintering spores are controlled by dormant oil sprays, and additional fungicide sprays at blossom time may be used in wet years.

Shot hole fungus. The fungus *Wilsonomyces carpophilus* destroys buds, blossoms, and young fruits in wet years. The disease name reflects the distinctive leaf and fruit symptoms of many small holes, where the leaf sacrifices small regions of tissue surrounding infection points that eventually abscise. Fungicide sprays at bloom and petal fall control this disease.

Verticillium wilt. The organism *Verticillium dahliae* kills roots of trees and is particularly prevalent in the western United States, and in fields previously planted to susceptible crops, such as tomato. Other root rot organisms can attack almond, but this is generally the worst problem. Control is achieved by using more-tolerant rootstocks (peach seedlings) or, primarily, by avoiding sites with a history of the disease. Preplant soil fumigation can be used but is dangerous and expensive.

Hull rot. At least three different types of fungi (*Rhizopus* or *Monilinia* spp.) can cause this problem, including the same fungus that attacks peach fruit and almond blossoms. The kernels are generally not affected, but leaves and twigs can be killed, causing a scorched appearance to the canopy if severe. Rapid harvest once shucks begin to split will lessen the problem. High humidity, rainfall, excessive nitrogen fertility, or irrigation close to harvest accentuates the problem. Generally control does not require sprays.

Bacterial canker and blast. Bacteria *(Pseudomonas syringae)* cause injury to twigs, branches, and large limbs. Branches/limbs are girdled in severe cases, and they die back quickly. Many "suckers" are produced from the base of the trunk on severely infected trees, since the lower trunk and roots are not killed. Young trees can be killed to the bud union. The bacteria produce a typical "sour sap" smell, similar to that of vinegar, and the wood turns brown-red. Blast is another form of infection on twigs and blossoms; they turn black and remain attached to the tree. This is nearly impossible to control with sprays, as with many bacterial diseases. The best defense is a healthy, well-maintained tree. Freezing injury, nematode weakening, other diseases, and drought or waterlogging all predispose trees to infection.

HARVEST, POSTHARVEST HANDLING

Maturity

The hull splits at maturity, and nuts physically separate from the tree at this point (Plate 2.4). Trees are harvested when hulls of fruit in the interior of the canopy are open, since these split last. The seed coat turns brown during the drying-out process of maturation. Delay in harvest increases risk of navel orangeworm infestation.

Harvest Method

In California, trees are harvested by mechanical tree shakers. Young trees may be damaged by shakers, so these are harvested by hand knocking in the first few years. Hand knocking is used in production regions that lack mechanization or are too hilly to accommodate shakers. Nuts are then left to dry on the ground for one to two weeks and then swept into windrows for harvesting. In situations where rain, higher humidity, or pest pressure necessitates action, nuts are shaken and taken immediately to a processing facility. A mechanical harvester then picks up the windrowed nuts, blowing off extraneous leaves and debris.

Postharvest Handling

Fruits may be dried and hulled immediately or stockpiled for fumigation against navel orangeworm after harvest. Nuts are dried by forced hot air until their moisture content reaches 5 to 7 percent. Nuts are then dehulled and shelled. The hulls are often sold for livestock feed. In-shell nuts can be stored in bins for weeks or months until final processing. Nuts are then shelled and sorted for size and appearance. Last, nuts are bleached for color improvement and then salted, roasted, and/or flavored before packaging.

Storage

Almonds can be stored for months either in the shells or shelled if dry, or for very long periods when frozen (years). Commercially, nuts for long-term storage are fumigated for navel orangeworm and kept at temperatures below 40°F.

CONTRIBUTION TO DIET

Almonds are the most widely used nut for confectionery items, such as candy bars, cakes, toppings, and so on. Much of the crop is roasted and flavored or salted and sold in cans; broken and small kernels go to confectionery processing. The highest prices are received for whole kernels. "A can a week, that's all we ask" was the old advertising slogan of the California almond growers.

About 99 percent of almonds in the United States are shelled prior to sale, and 1 percent are sold as in-shell nuts. Nutmeats can be sold as whole kernels or further processed by blanching or slicing. Per capita consumption of almonds has more than doubled over the past 25 years to 0.9 pound/year.

Dietary value per 100-gram edible portion:

Water (%)	5
Calories	578
Protein (%)	21.3
Fat (%)	50.6
Carbohydrates (%)	19.7
Crude Fiber (%)	11.8
	% of U.S. RDA (2,000-caloric diet)
Vitamin A	<1
Thiamin, B_1	16
Riboflavin, B_2	48
Niacin	20
Vitamin C	0
Calcium	25
Phosphorus	68
Iron	24
Sodium	<1
Potassium	21

BIBLIOGRAPHY

Duke, J.A. 2001. *Handbook of nuts*. Boca Raton, FL: CRC Press.

Kester, D.E. 1969. Almonds, pp. 302-314. In: R.A. Jaynes (ed.), *Handbook of North American nut trees*. Knoxville, TN: North American Nut Growers Assoc.

Kester, D.E. 1979. Almonds, pp. 148-162. In: R.A. Jaynes (ed.), *Nut tree culture in North America*. Hamden, CT: Northern Nut Growers Assoc.

Kester, D.E. and R.A. Asay. 1975. Almond breeding, pp. 382-419. In: J. Janick and J.N. Moore (eds.), *Advances in fruit breeding*. West Lafayette, IN: Purdue University Press.

Kester, D.E., T.M. Gradziel, and C. Grasselly. 1990. Almond *(Prunus)*, pp. 699-758. In: J.N. Moore and J.R. Ballington (eds.). *Genetic resources of temperate fruit and nut crops* (Acta Horticulturae 290). Belgium: International Society for Horticultural Science.

Kester, D.E. and C. Grasselly. 1987. Almond rootstocks, pp. 265-294. In: R.C. Rom and R.F. Carlson (eds.), *Rootstocks for fruit crops.* New York: John Wiley and Sons.

Kester, D.E. and W.C. Micke. 1984. The California almond industry. *Fruit Var. J.* 38:85-94.

Micke, W.C. (ed.). 1996. *Almond production manual* (University of California Division of Agricultural and Natural Resources Publication 3364). Berkeley, CA: University of California Press.

Rosengarten, F. 1984. *The book of edible nuts.* New York: Walker and Co.

Socias, R. (ed.). 2001. *Proceedings of the 3rd international symposium on pistachios and almonds* (Acta Horticulturae 591). Belgium: International Society for Horticultural Science.

Taylor, R.H. and G.L. Philp. 1925. *The almond in California* (University of California Agricultural Experiment Station Circular 184). Berkeley, CA: University of California Press.

Teviotdale, B.L., T.J. Michailides, and J.W. Pscheidt. 2002. *Compendium of nut crop diseases in temperate zones.* St. Paul, MN: American Phytopathological Society.

Watkins, R. 1979. Cherry, plum, peach, apricot, and almond: *Prunus* spp., pp. 242-247. In: N.W. Simmonds (ed.), *Evolution of crop plants.* London: Longman.

Weinbaum, S.A. and P. Spiegel-Roy. 1985. *Prunus dulcis*, pp. 139-146. In: A.H. Halevy (ed.), *CRC handbook of flowering*, Volume 4. Boca Raton, FL: CRC Press.

Woodroof, J.G. 1967. *Tree nuts: Production, processing, products*, Volume 1. Westport, CT: AVI Publ.

Chapter 3

Apple *(Malus domestica)*

TAXONOMY

The cultivated apple, *Malus domestica* Borkh., belongs to the Pomoideae subfamily of the Rosaceae, along with pear *(Pyrus* spp.), quince *(Cydonia oblonga),* loquat *(Eriobotrya japonica),* and medlar *(Mespilus germanica).* Subfamily Pomoideae contains 18 genera, all of which possess an unusual haploid chromosome number of 17, unlike other genera of the Rosaceae, which contain only 7 to 9 chromosomes. The genus *Malus* contains at least 15 primary species, although considerable debate still exists as to the exact number, which could be as high as 33. Other species of interest include the crab apples, represented by several species (e.g., *M. baccata, M. sargentii)* or interspecific hybrids (e.g., *Malus* ×*floribunda, Malus* ×*zumi*).

Linnaeus first named the apple *Pyrus malus,* but this was later replaced by *M. communis* and *M. pumila* in the 1800s. The latter two names are still used (incorrectly) today by some. In 1984, the name *Malus* ×*domestica* Borkh. was proposed to emphasize that the cultivated apple is an interspecific hybrid (denoted by the multiplication sign). This seems reasonable because (1) the cultivated apple does not exist in the wild, (2) its exact origin is unclear, and (3) biochemical evidence shows that the cultivated apple contains certain glucosides found only in *M. floribunda, M. zumi, M. sargenti,* and *M. sieboldii.* Despite the persuasive logic, *Malus* ×*domestica* has not been widely accepted to date.

Borkhausen is credited with the current name *M. domestica,* and he believed there were three species incorporated in the cultivated apple: *M. sylvestris* ("woodland apple"), *M. dasyphyllus* ("John's apple"), and *M. praecox.* Modern breeding programs have incorporated genes from at least *M.* ×*floribunda, M. micromalus, M.* ×*atrosanguinea, M. baccata jackii,* and *M. sargentii* into recent cultivars.

Cultivars

Throughout its history of cultivation, at least 10,000 apple cultivars were developed, many of which are now lost. This was due, in part, to the older practice of seed propagation of this heterozygous species. Commercially, about 100 cultivars currently are being grown, but only ten make up over 90 percent of U.S. production. Most cultivars have been selected from superior seedling trees or were sports of existing cultivars, rather than being produced through breeding. Notable cultivars produced through breeding include 'Gala' ('Kidd's Orange Red' × 'Golden Delicious'), 'Jonagold' ('Golden Delicious' × 'Jonathan'), 'Empire' ('McIntosh' × 'Delicious') and 'Fuji' ('Ralls Janet' × 'Delicious').

When discussing apple cultivars, it is important to note that almost every cultivar has a number of sports or strains—slight mutations that improve on the original. Many sports are redder than their parents; in the United States, redder is better in the marketplace. For example 'Red Max' is a red sport of 'McIntosh', and 'Redchief Delicious' is one of many red sports of 'Delicious'. 'UltraEarli Fuji' is a 'Fuji' sport that ripens a bit earlier, allowing growers to hit the 'Fuji'

Introduction to Fruit Crops

doi:10.1300/5547_03

market when prices may be higher. For many cultivars, a form is available that produces a more compact, precocious tree called a "spur type," since many more short, fruiting branches (spurs) are formed per unit of shoot length. These trees can be spaced closer together due to their compact form and will produce a smaller, more efficient tree on any given rootstock than will the nonspur cultivars.

Table 3.1 shows the top ten cultivars grown in the United States in 1997. Since 1997, it is likely that 'Fuji' and 'Gala' have increased in rank, while 'Red Delicious', 'Rome', and 'McIntosh' production has decreased. 'Braeburn' and 'Jonagold' had not entered the top ten by 1997 but are currently ranked sixth and seventh, respectively, in Washington State, the largest apple producer in the United States.

Nursery tree sales showed that 'Fuji' and 'Gala' were the most frequently ordered cultivars in the late 1980s and early 1990s. At that time, prices of $28 to $48 per box for 'Fuji' and $30 to $36 per box for 'Gala' (versus about $12 for 'Red Delicious') lured growers to plant these new cultivars, but prices have moderated now that many new orchards have come into bearing. 'Fuji' is a large-sized, late cultivar with long-keeping potential; 6 months in normal cold storage and 12 months in controlled-atmosphere storage. 'Gala' is a sweet summer apple, the first in the season to be harvested in Washington. Better colored sports ('Galaxy Gala', 'Scarlet Gala') are replacing the original. Drawbacks include smaller size and a short shelf life of only a few months, but great flavor keeps this apple in high demand. 'Braeburn', 'Empire', and 'Jonagold' were also popular nursery sellers in the early 1990s. 'Empire' (a New York State Experiment Station introduction) has now made its way into the top ten, surpassing the popular old processing cultivar 'York' in 1996. The most recent cultivars to appear are 'Cameo' and 'Pink Lady', both late-season, good-flavored, long-keeping apples. 'Cameo' was a chance seedling discovered in Washington in the mid-1980s, and 'Pink Lady' was introduced about the same time from Australia. 'Pacific Rose' is another up-and-coming cultivar (Plate 3.1).

Clearly, the trend is toward diversity and, to some extent, away from the Reds, Goldens, and Grannys, although these standard cultivars will probably remain at the top for quite some time. All of the top three cultivars can be stored for up to 12 months, whereas some of the newer cultivars cannot. Today's 'Red Delicious' strains derive from the original 'Delicious' cultivar,

TABLE 3.1. Top ten apple cultivars grown in the United States, 1997. Data are in thousands of boxes (42 pounds of apples/box).

Cultivar	Five-year average 1993-1997	1996	1997	% Change 1996-1997
Red Delicious	105,722	103,969	95,709	–8
Golden Delicious	36,710	35,712	35,725	0
Granny Smith	16,598	16,030	16,607	+3
Rome	15,978	13,624	14,720	+8
Fuji	10,648	12,692	16,205	+28
McIntosh	12,647	11,195	13,253	+19
Gala	7,036	8,149	9,870	+22
Jonathan	6,519	4,634	5,748	+24
Idared	4,899	3,987	5,097	+28
Empire	4,127	4,377	4,964	+14
All others	35,184	32,179	35,645	+11

which was a chance seedling found growing in Iowa in 1872. The higher-yielding, redder strains of 'Delicious' became widely planted in Washington in the early to middle part of the twentieth century, and by 1960, 'Red Delicious' was the dominant cultivar grown in the state. The cultivar is extremely well adapted to eastern Washington's arid, temperate climate. Its widespread appeal stems from outstanding quality attributes, long storage life, and promotion efforts of the Washington Apple Commission. Although interest in 'Red Delicious' has declined somewhat, this cultivar still represents 40 percent of Washington's apple crop.

'Golden Delicious', the second most popular apple in the United States, is probably the most widely grown cultivar in the world. It, too, has many strains, all derived from a single chance seedling found in West Virginia in 1914. 'Granny Smith' was discovered in Australia in 1868, predating both the Reds and Goldens, but was not widely planted in the United States until the 1970s. It provides a superfirm, tart alternative to the sweeter Delicious cultivars. A postharvest physiologist once commented that you could throw a Granny at a concrete wall and still not cause a bruise. 'Granny Smith' production is sometimes surpassed by that of 'Gala' and 'Fuji', so its ranking of third is tenuous today.

ORIGIN, HISTORY OF CULTIVATION

The center of diversity of the genus *Malus* is the east Turkey, southwest Russia region of Asia. Although a few *Malus* species are native to North America, they were never domesticated as a food crop on this continent. Apples were probably improved through selection over a period of thousands of years by early farmers. Domestication efforts were no doubt frustrating because planting apple seeds from good-quality fruit generally does not produce new trees with even remotely similar fruit. This stems from apples being a cross-pollinated and highly heterozygous species. Also, they do not root from cuttings the way grapes and olives would, and their complete domestication was not possible until grafting was invented, sometime in the first millennium BC. Once this was accomplished, Greeks and Romans selected superior types and spread them throughout Europe. Alexander the Great is credited with finding dwarfed apples in Asia Minor in 300 BC; those he brought back to Greece may well have been the progenitors of dwarfing rootstocks. The excellent keeping qualities of apples were discovered by at least 100 BC, as the Roman Varro provided written accounts of "fruit houses" for storing apples for winter. Throughout medieval times, several authors detailed fertility, water, site preferences, and pruning of apple trees.

Apples came to North America with the colonists in the 1600s, and the first apple orchard on this continent is said to have been located near Boston in 1625. From New England origins, apples moved west with pioneers, John Chapman (alias Johnny Appleseed), and missionaries during the 1700s and 1800s. In the 1900s, irrigation projects in Washington State began and allowed the development of the multi-billion-dollar fruit industry, of which the apple is the leading species. Today, apple production is growing most rapidly in California, and remaining steady or declining in the eastern states due to overproduction and greater disease and insect pressures. In the past 5 years, the industry has experienced somewhat poor economic times due to foreign competition, overproduction, and stagnant demand for apples. In fact, the apple industry received federal government assistance in 2001 and 2002 for losses in previous years, believed to total $1.7 billion. It is rare for growers of fruit crops to receive federal aid of this nature. A trend toward decreased production and increased price per pound since 1998 (see Figure 3.1) suggests that the industry is making the appropriate adjustments to return to profitability.

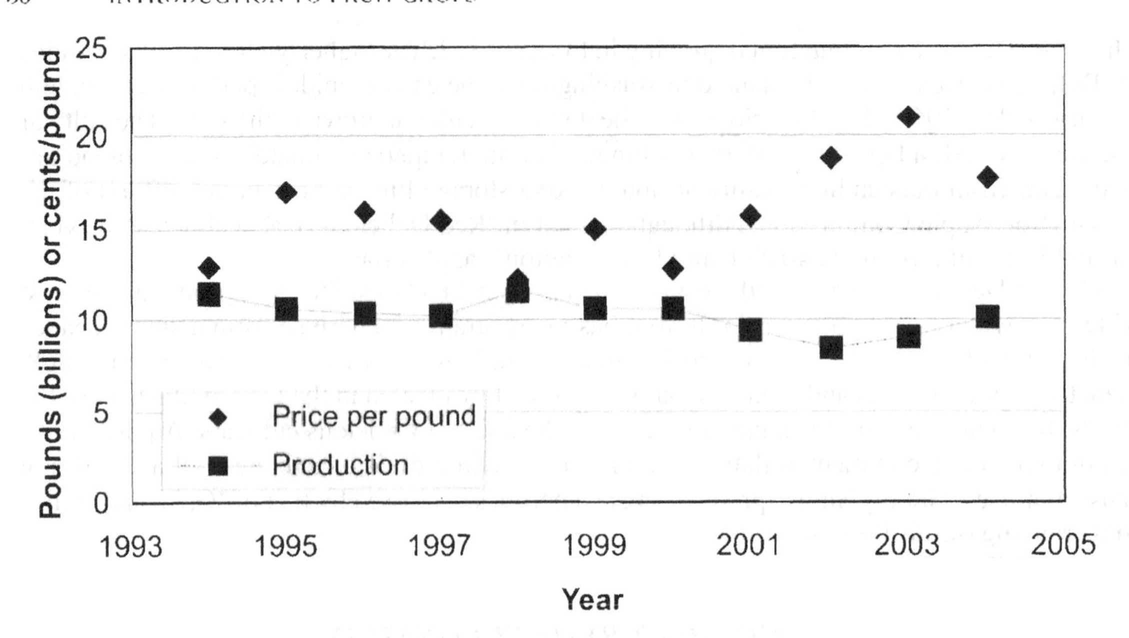

FIGURE 3.1. Ten-year trend of apple production and price in the United States. *Source:* USDA statistics. The 12 percent decline in production is offset by a 37 percent increase in price, resulting in a 20 percent increase in overall industry value.

FOLKLORE, MEDICINAL PROPERTIES, NONFOOD USAGE

Many myths and legends are associated with the apple. Apples were frequently used in Greek, Roman, Norse, and other mythologies as symbols of immortality or reincarnation. They are the food of the gods/goddesses or are given to humans by deities as rewards for various acts. The apple tree is also symbolic: King Arthur was taken by a goddess to Avalon, the "apple-land" of eternal life. Apple trees have often marked the "axis of the world" or the "center of life" in various tales, as in the Romanian folktale of the fairy Magdalina, who appeared sitting in a cosmic apple tree, "whose branches touched the sky, and whose roots reached into the bottom of the ocean."

As the apple fruit is five-locular, when cut in transverse section, it reveals the pentacle, a sacred sign in some cults and pagan beliefs. Apples were cut in this way at Gypsy weddings, each half eaten by the bride and groom. Because of connections with deities and magic, apples are said to confer some healing or telepathic powers. Fever could be cured if a holy name was written on an apple and eaten on three consecutive days. Abortion would occur if the words *sator arepo tenet opera rotas* (a palindromic charm) were inscribed on an apple before eating it. One's future spouse would be revealed in one's dream if an apple was eaten on Halloween, according to legend, which is the origin of bobbing-for-apples.

One of the most popular stories pertaining to apples is that of Adam and Eve, who ate the "forbidden fruit" of the tree of knowledge of good and evil in the center of the Garden of Eden. Actually, the account given in Genesis 2 and 3 never mentions what kind of fruit this tree produced, but numerous works of art commonly depict it as an apple. The Latin noun *Malus* has the dual meaning of either "apple" or "evil," which probably stems from this Bible story. Apples are symbolic of temptation, perhaps also as a result of this story. Apple blossoms are the state flowers of both Arkansas and Michigan.

An apple a day can indeed keep the doctor away, according to recent research on human health. The average apple contains about 5 grams of fiber, important in normal bowel function.

As recently discovered in many fruit crops, flavonoids may play an important role in preventing cancer. Research conducted by the Mayo Clinic, Cornell University, and the National Cancer Institute shows that the apple flavonoid quercetin can inhibit the growth of colon and prostate cancer cells, and eating quercetin-rich fruits may reduce the risk of lung cancer by 50 percent. Flavonoids may reduce the risk of heart disease and stroke as well. Eating two apples or drinking 12 ounces of apple juice daily was shown to reduce buildup of arterial plaque. The bark and roots of most *Malus* species contain phloretin, an antibiotic-like compound that acts on certain bacteria.

Apple juice has been implicated in chronic, nonspecific diarrhea found in infants and young children. Apple juice is high in fructose and sorbitol, which leads to "malabsorption" and diarrhea. Clinical studies have shown that elimination of apple juice consumption in certain children alleviates the condition. However, *M. diversifolia*, an American crab apple, was used against diarrhea by Native Americans of western Washington. They peeled the bark from the tree, soaked it in water, and then drank the decoction. Fruit decoctions made from the close apple relatives *Spiraea douglasii* and *S. tomentosa* were also used by North American Indians as powerful astringents against diarrhea.

Cyanogenic glycosides are poisonous compounds that yield hydrocyanic acid (HCN) upon hydrolysis. Amygdalin, the most common cyanogenic glycoside in members of the Rosaceae, is found in seeds of apple and pear, as well as in leaves, bark, and seeds of stone fruits, *Passiflora*, and *Sambucus*. HCN inhibits the key respiratory enzyme cytochrome oxidase, resulting in asphyxiation at the cellular level. Symptoms include spasms, coma, difficulty breathing, and, ultimately, death, as in the tale of the man who ate a cup of apple seeds and died. Amygdalin is also teratogenic (alters embryo growth) and may cause birth defects.

PRODUCTION

World

Total production in 2004, according to FAO statistics, was 59,059,142 MT or 130 billion pounds. Apples are produced commercially in 91 countries on about 13 million acres. World apple production has increased about 17 percent in the past decade. Average yields are 10,000 pounds/acre worldwide, but averages of over 40,000 pounds/acre are achieved in New Zealand, with some orchards producing double that amount. The top ten apple-producing countries (percent of world production) follow:

1. China (35)
2. United States (7)
3. Poland (4)
4. France (4)
5. Iran (4)
6. Turkey (4)
7. Italy (3)
8. Russia (3)
9. Germany (3)
10. India (2)

For many years, the United States led the world in apple production. Chinese production of apples has increased enormously in the past several years, and China surpassed the United

States for the first time in 1985. Just 7 years later (1992), Chinese production was double that of the United States and today is more than five times greater.

United States

Total production in 2004, according to USDA statistics, was 4.6 million MT or 10.1 billion pounds. Apples are grown commercially in 35 states, with the top ten states accounting for 93 percent of production (Table 3.2). The value of the industry is $1.76 billion. There are approximately 8,000 apple growers in the United States, who on average receive prices of $0.19 per pound. Apple acreage has decreased 17 percent over the past decade to 386,000. Yields average over 26,000 pounds/acre, more than double the world average. In recent years, production has declined about 12 percent in the United States (Figure 3.1), but prices have improved proportionately.

In 2002, the United States exported 12 percent of production. Leading countries of destination include (descending order) Mexico, Canada, Taiwan, Indonesia, Hong Kong, United Kingdom, Malaysia, and United Arab Emirates. U.S. apple exports have increased tremendously since 1988. NAFTA has had a positive impact on trade with Mexico. In 1990, Washington State exported about 700,000 cartons of apples to Mexico; they exported over 8 million cartons from the 1993 crop. The Mexican tariff on apples decreased from 18 percent to 16 percent, as of January 1, 1995, on a limited volume of shipments as a result of NAFTA. Export markets all over the world are experiencing growth, particularly in Europe. China and Japan imported fresh apples from the United States for the first time in 1994.

Imports total approximately 3 to 5 percent of domestic production. Leading countries exporting to the United States include (in descending order) Chile, New Zealand, Canada, and South Africa.

BOTANICAL DESCRIPTION

Plant

Apples grow on a small- to medium-sized tree with a spreading canopy. Apple trees can grow up to 30 feet in the wild, and generally 6 to 15 feet under cultivation (Plate 3.2). Tree size and shape are heavily dependent on rootstock and training system (see Planting, Design, Training,

TABLE 3.2. Leading apple states in the United States and recent production trends.

State	% of U.S. crop	Three-year trend (2000-2002)
Washington	62	–12%
New York	8	–35%
Michigan	6	–38%
California	5	–26%
Pennsylvania	5	–18%
Virginia	3	–22%
North Carolina	2	–16%
Oregon	2	–10%
West Virginia	1	+12%

and Pruning in the following section). Leaves are elliptical with serrate margins, and their color is dark green with light pubescence on the underside.

Flowers

Petals are white when open but have red-pink undersides when opening, hence the "pink" bloom stage. The ovary is epigynous or inferior, embedded in the floral cup or hypanthium (Plate 3.3). It contains five locules, with usually two ovules per locule. The inflorescence is a cyme of four to six flowers, with the center flower opening first; the central flower is often called the "king bloom" and has the potential to produce a larger fruit than the other flowers. Flowers are produced terminally from mixed buds (containing both leaves and flowers) on spurs or, to a lesser extent, on long shoots (see photos in Plate 1.4). Spurs form on 2-year-old and older wood and generally grow only a fraction of an inch each year. Floral initiation occurs the summer before the blossom period.

Pollination

Most cultivars require a pollinizer because they are commercially self-unfruitful. Cross-incompatibility is rare, and most cultivars that bloom at the same time and are not sports of each other will serve as pollinizers, including crab apples (Plate 3.4). Single apple trees produce some fruit when self-pollinated because most cultivars are not totally self-incompatible. A few cultivars are pollen-sterile (e.g., 'Jonagold', 'Winesap', and 'Mutsu'). Honeybees are the most effective pollinator and usually are supplied to the orchard at one-half to two hives per acre, depending on tree spacing and colony strength. Colonies may actually lose weight during bloom period because flowers are nectar poor, and bees chiefly are collecting pollen. Insecticide sprays are avoided during pollination. Poor pollination may result in misshapen fruits (Plate 1.2).

Fruit

Apple and related fruits belong to a special fruit type—the pome. The bulk of the fleshy edible portion derives from the hypanthium or floral cup, not the ovary. The fruit is five-locular, containing five seed cavities with generally two seeds each. Seeds are relatively small, black, and mildly poisonous. Fruiting begins 3 to 5 years after budding, although a few fruits may be produced in the second year. This varies with rootstock (dwarfing = more precocious) and cultural practices (excessive pruning = delay). Fruits should be removed from the upper part of the central leader of young trees to encourage growth and prevent leaning.

Fruits are usually thinned to one per spur, with spurs spaced 4 to 6 inches apart for attainment of marketable size (Plate 1.3). Fruits derived from the central flower or king bloom are preferable to retain since they have greater size potential. Apples are generally thinned with chemicals such as the insecticide Sevin or the synthetic auxins NAA and NAAm (see glossary). The materials are applied postbloom and cause abscission of developing fruitlets until they reach a size of about 1 inch in length. Chemical thinning is less expensive than hand thinning, but very dependent on weather during and shortly after chemical application.

GENERAL CULTURE

Soils and Climate

Deep, well-drained, loamy soils with a pH of 6 to 7 are best, but apples are grown on a wide variety of soils worldwide. An incredibly wide range of rootstocks are available for apple, allowing growers to adapt trees to many soil types. Apples are largely intolerant of poorly drained soils, as with all fruit crops.

Apples are adaptable to various climates but can be considered best adapted to the cool-temperate zone from about 35 to 50° latitude. They have a more northern range than many other tree fruits due to relatively late blooming and extreme coldhardiness. Generally, fruit quality is best in temperate climates with high light intensity, warm (not hot) days, and cool nights, hence the success of apple culture in eastern Washington State, which has a temperate, desertlike climate (Plate 3.5). The red skin color of many cultivars is stimulated by exposure to direct sunlight. However, overexposure of fruit to the sun can cause sunscald. High humidity and rainfall promote infection by apple scab, fire blight, and other diseases; therefore, apples grown in dry climates experience less disease pressure. Apples are relatively drought intolerant and require 20 to 24 inches of rainfall or irrigation distributed throughout the growing season.

Apples require about 1,000 to 1,600 hours of chilling (~45°F) to break dormancy, so winters cannot be too mild. However, some cultivars have been bred for mild climates (e.g., 'Anna', 'Dorsett Golden'), requiring only 500 hours. Wood and buds of dormant apple trees are hardy to about –40°F, but open flowers and young fruitlets are killed by brief exposure to 28°F or colder. Apples are among the latest blooming tree fruits and are therefore less frost prone than other species at any given location. Frost occasionally reduces the crop in the Pacific Northwest, since the arid climate is prone to radiative-type frosts through April.

On average, apples reach maturity about 120 to 150 days after bloom, with some cultivars maturing in as short as 70 days, and others as long as 180 days. Growing season length is generally not a problem, except in areas where late-season cultivars, such as 'Pink Lady', might not mature until November.

Propagation

Apples are T- or chip-budded in the nursery and sold as 1-year-old whips on 1.5- to 2-year-old rootstocks (Figure 3.2). Rootstocks are also produced vegetatively, generally by mound layering

FIGURE 3.2. Chip-budded apple trees growing in a nursery. Note the larger size of the rootstock, which is 1 year older than the scion.

(Figure 3.3). Young apple trees are often more expensive than those of other species due to the use of clonal rootstocks and the extra time required to produce trees. Seedling rootstocks are still used in some regions.

Rootstocks

Hundreds of years of selection and decades of research have led to more rootstock choices for apple than for practically all other major temperate tree fruits combined. There is a remarkable range of tree size control available among rootstocks. In addition, apple rootstocks vary greatly in coldhardiness and disease tolerance (Table 3.3).

Choice of rootstock is critical in apple because the rootstock controls the potential height and vigor of the tree, regardless of the cultivar used. A dwarf 'Red Delicious' tree is genetically identical to a standard-sized 'Red Delicious' tree *above the graft union*. Any cultivar can be dwarfed by using the appropriate rootstock. The dwarfing trait alone would not be as desirable if it were not combined with precocity, or the induction of fruit bearing at an early age. In Table 3.3, note that precocity is generally inversely correlated with degree of size reduction.

Apple rootstocks are generally given alphanumeric names, related to their origin, but not necessarily their degree of dwarfing. Several series of rootstocks are available, the most popular being the Malling and Malling-Merton. The single letter "M" stands for "Malling," after the East Malling research station in England where these rootstocks were introduced; "MM" designates "Malling-Merton," since these stocks were a joint venture between the East Malling station and an institute located in Merton, England. Other important series are Budagovsky ("Bud" or "B") from Russia; Polish ("P"), Ottawa ("O"), and Kentville ("KSC") from Canada; and Michigan ("MAC") and Oregon ("OAR") from the United States. The most common series are described in the following pages.

Apple Seedlings

Despite the plethora of clonal apple rootstocks, seedlings are the most common rootstocks worldwide, in countries where nonintensive orchards exist. Seed is obtained from processing

FIGURE 3.3. An apple rootstock stoolbed, where rootstocks are mound layered to retain the exact dwarfing characteristics.

TABLE 3.3. Some popular apple rootstocks and their characteristics.

Rootstock	Size reduction[a]	Susceptibility[b]				Precocity induction
		Crown rot	Fire blight	Cold damage	Burr-knots[b]	
M.27	25	1	3	3	2	High
Budagovsky 9	30	1	3	1	1	High
M.9	35	1	4	3	3	High
M.26	40	3	4	1	4	High
Mark	40	2	3	1	4	High
Ottawa 3	40	1	3	1	1	High
M.7	65	2	1	1	3	Medium
MM.106	75	4	2	2	3	Medium
M.2	80	2	2	1	2	Medium
M.4	80	1	2	2	2	Medium
MM.111	85	2.5	2	1	4	Low
Merton 793	95	2	—	2	2	Medium
Domestic seedling	100	2	2	1	1	Low
Maruba	110	1	1	—	1	—
Novole	110	1	1	2	1	Medium

Source: Modified from Westwood, 1993.

[a]Tree size as percentage of size trees on domestic seedling rootstock.
[b]Increasing susceptibility from 1 or 4.

plants; it is highly nonuniform, but inexpensive and easy to propagate. 'Delicious' is used most often, probably due to availability. Trees grafted on seedling rootstocks are very large (20-30 feet) and slow to bear.

East Malling Series

Dwarfing rootstocks for apple were common in Europe by the fifteenth century. Several selections of the extremely dwarf 'Paradise' (syn. French paradise) and less dwarfing 'Doucin' (syn. English paradise) were collected in 1912 at East Malling, England, by Sir Ronald Hatton. They were studied, categorized by degree of dwarfing, given the roman numerals I through XXIV, and subsequently distributed worldwide. The East Malling series stocks are now abbreviated with arabic numerals (e.g., M.1 . . . M.27). In the United States, the most common are M.26 (dwarf), M.7 (semidwarf), and M.9 (dwarf). Intensive plantings most often are on M.9. The M.27 rootstock is the most dwarfing, but trees are considered too small to be used commercially. M.9 is the most dwarfing of the commercially viable rootstocks; M.9 trees are easy to manage at a height of 6 feet. However, M.9 and the slightly larger M.26 are both susceptible to fire blight, a serious bacterial disease. The Bud 9, P.2, and O.3 rootstocks are as dwarfing as M.9

and have greater coldhardiness. However, M.9 still performs well in the cold winter climates of Washington, New York, and Michigan.

Almost all dwarf and semidwarf trees (i.e., those on M.9, M.26, and M.7) require staking because of poor anchorage and leaning due to heavy fruit loads borne in their early years. The use of these stocks is often combined with a trellis or tree stake in commercial orchards.

Once plant viruses were discovered, an effort was launched by the East Malling and Long Ashton research stations to remove the viruses from M series stocks. The resulting stocks, known as "EMLA," often are not as dwarfing as their virus-laden precursors. The program revealed that most viruses do not have debilitating effects in apple; in fact, they may contribute a desirable degree of dwarfing. As such, the EMLA series has not achieved the popularity of the M or MM series. A few rootstocks were cleansed of some, but not all, viruses; these carry a lowercase "a" after the name, for example, M.7a.

Merton-Malling (MM) Series (and Lesser-Known Merton-Immune [MI] Series)

In 1917, breeding programs were started to confer resistance of Malling series stocks to wooly apple aphid *(Eriosoma langerum),* a serious pest of bark and roots in England (massed together, they look like tufts of wool). The MI series was introduced in the 1930s, followed by the more successful MM series in 1952. Both series of rootstocks were crosses of Malling stocks (for dwarfing) with 'Northern Spy', a cultivar with resistance to wooly apple aphid. Although several MI stocks were released, only MI.793 is used to any extent, in South America, Australia, and New Zealand. Of the 15 MM stocks, MM.106 and MM.111 are the most popular in the United States, producing semivigorous trees (smaller than trees on seedling rootstocks) that lack the extreme dwarfing of the M series.

FIGURE 3.4. A tree topworked to a different cultivar in the orchard. The graft on the left side did not take, but it shows the type of wood and insertion point of the technique. The graft on the right took and grew out vigorously in 1 year because it had a large root system to support its growth.

Interstocks

To circumvent some of the problems of dwarfing rootstocks, three-part trees are used in some applications. Poor anchorage and wooly apple aphid problems can be avoided by using an MM series rootstock with an interstock of M.9 or B.9 grafted between the rootstock and scion. Coldhardiness of the trunk is sometimes a limitation to apple, and interstocks with greater hardiness than the M series stocks can be used to allow cultivation of dwarf trees in areas such as Russia and Poland. By default, trees that are topworked to new cultivars have an interstock (Figure 3.4).

Planting Design, Training, Pruning

The range of rootstocks available for apple permits a wide variety of orchard designs and tree-training systems. In most cases, trees are grown in rectangular arrangements or hedgerows. Conventional apple orchards on seedling rootstocks require spacings of 20 × 20 (110 trees/acre) to 40 × 40 feet (27 trees/acre). More intensive orchards (400+ trees/acre) are generally planted to form hedgerows, or solid rows of fruiting wood in which individual trees lose their identities (Figure 1.12). Spur-type scions, which are less vigorous, allow closer spacings than nonspur scions of the same cultivar/rootstock combination (Table 3.4).

The most common training system for apple is the central leader, in which one branch is allowed to grow vertically to form the main bole upon which scaffold limbs are spaced at selected intervals (Figure 1.14). The standard central leader consists of two or three tiers of scaffolds spaced 2 to 6 feet apart on the leader, depending on tree size. Variations include the slender spindle and spindlebush, which are compact, narrow forms with numerous short scaffolds (Figure 3.5). The central leader is sometimes removed from large, older trees to allow light into the center of the canopy; these trees resemble an open-center configuration.

More types of training systems have been used to form hedgerow apple plantings than for any other fruit crop. Trellised-hedgerow training is typically "palmette," which means consisting of two scaffold limbs tied to each of three to four wires spaced 18 inches apart vertically. In some field trials, researchers have found no difference in yield or quality between trees trained to intensive systems, such as palmette, Tatura, and Ebro, versus small central-leader trees that receive minimal pruning. Such elaborately trained, trellised orchards can be thousands of dollars per acre more expensive than simpler, single-wire trellis or staked systems. Minimal trellis or staked systems offer the advantages of high-density orchards at a lower cost and have gained popularity in recent years.

Pruning is typically done when trees are still dormant in late winter. Summer "tipping" is done on some red cultivars that color poorly due to genetic or climatic factors; this practice, done about 1 month prior to harvest, involves very light removal of shoots at the periphery of the canopy to allow light to penetrate deeper. The severity and type of pruning cuts depend on bearing habit, fruit size, inherent tree vigor, and training system thus making it difficult to generalize. Most apple trees bear fruits on spurs, so one objective in pruning is to retain spurs on 2-year-old and older wood, exposing them to light as much as possible. Since long 1-year-old shoots generally do not bear fruits, they are removed, unless needed to replace declining fruiting wood. Some of the diminutive central-leader systems have only one tier of permanent scaffolds at the base; the rest of the fruiting wood is pruned close to the leader and renewed periodically. Trees on dwarfing stocks require less pruning than those on standard rootstocks.

TABLE 3.4. Tree spacing in the orchard depends on rootstock and scion growth habit. Rootstocks are arranged in decreasing order of the tree size they induce.

	Between trees in row (feet)		Between rows (feet)	
Rootstock	Nonspur scion	Spur scion	Nonspur scion	Spur scion
Seedling	18	12	22	18
MM.106	16	10	20	16
MM.111	16	8	20	14
M.7	12	8	18	14
M.26	10	6-8	16	10-12
M.9	8	6-8	14	10-12

FIGURE 3.5. Intensive apple training. *(left)* A spindlebush apple tree is just a variation on the theme of central leader. *(right)* 'Gala' apples fruiting in their second year in the field trained to an intensive trellis system.

Pest Problems

Insects

Codling moth. Lasperyresia pomonella is one of the worst pests of apple worldwide. The moth (¾ inch long and of dark color) lays eggs singly on young fruits in midspring, and the larvae (pinkish-brown head with a white body, ¾ inch long) burrow through the fruits. The moth can produce multiple generations per year. Insecticides are applied at petal fall, then biweekly or as needed through the summer. Mating disruption with pheromones provides a nonchemical means of control.

Aphids. Aphids generally attack young leaves, shoot tips, and developing flowers or fruits. Rosy apple aphid *(Dysaphis plantaginea)* is one of the worst aphid pests, causing severe deformation of leaves and fruits. Wooly apple aphid *(Eriosoma lanigerium)* is unusual since it attacks the roots as well. These insects (1/32 to ⅛ inch long) come in various colors, from white to black, green, and red. They congregate in large groups usually and move slowly when disturbed. Their piercing, sucking mouth parts locate the sugar-conducting tissues of the tree, and they rob the young, growing leaves, twigs, and fruits of this food source. Dormant oil, combined with insecticide if aphids were a problem the previous year, is sprayed 1 to 3 weeks before budbreak. Sporadic infestations are easily controlled by spot spraying. Ladybugs are an excellent biological control agent for light aphid populations. The MM series rootstocks are resistant to wooly apple aphid.

Plum curculio. Curculio *(Conotrachelus nenuphar)* is generally not as problematic on apple as on stone fruits but can be a major fruit pest. Adults are gray to dark brown, ⅛- to ¼-inch-long snout beetles; grubs are small, yellow-white, and legless and burrow into the fruit's flesh, usually causing the fruit to abscise. Feeding damage by adults alone may cause catfacing (misshapen fruit). Insecticide sprays are applied at petal fall, then at 7- to 10-day intervals throughout spring. The insect is confined to the United States east of the Rocky Mountains.

Scale insects. Adults (*Quadraspidiotus* spp., others) are small (<⅛ inch), round or oval, sedentary insects that appear as bumps on twigs or branches; they suck the sap from branches and twigs. Dormant oil, combined with insecticide if infestation is heavy, is sprayed 1 to 3 weeks before budbreak.

Leaf miners. The larvae of these tiny moths (*Phyllonorycter* spp.) tunnel through between the upper and lower leaf surfaces and can destroy a large portion of the photosynthetic surface if unchecked. Leaves may drop early, and fruit may be indirectly affected by defoliation or leaf impairment. Natural enemies, such as wasps, can keep populations low, but they are often killed by the same sprays used against codling moth or other pests. Postbloom insecticide sprays target the first generation, but there may be two or three generations per year.

Apple maggot. This small, shiny fly *(Rhagoletis pomonella)* is largely a problem in eastern North America. The adults lay eggs just beneath the skin, and larvae feed on the flesh of the developing fruits, resulting in fruit drop. The use of insecticide sprays as fruits mature is being replaced with traps that lure the adults to lay eggs on red, sticky decoys.

Mites. Mites (*Panonychus, Tetranychus* spp., others) are tiny (<$\frac{1}{10}$ inch) spiders, not true insects. They congregate on the underside of leaves, feeding on the lower leaf surface; this causes a characteristic "mite stipple" or bronzed appearance to the upper leaf surface and, if severe, reduces leaf photosynthesis and ultimately yield. Mites may be white, red, or yellow and two-spotted, but they generally occur on leaf undersides and are worst in hot, dry weather. A dormant oil spray applied before budbreak reduces initial populations and is all that is needed in most years. In hot, dry summers, miticide may be required when visible injury occurs.

Diseases

Apple scab. The fungus *Venturia inaequalis* causes numerous, small (<$\frac{1}{8}$ inch), black lesions on the fruits and leaves and is one of the worst apple diseases in humid, cool, rainy areas of the world. Lesions may coalesce and become brown and corky later; fruit may become deformed and crack open if severe. Fungicide sprays from "silver tip" stage, when a silvery tip to the buds is visible after bud swell, through the petal fall stage can control scab. The fungus overwinters on dead leaves beneath the tree, so covering or removal of fallen leaves may reduce the disease somewhat. Resistant cultivars include 'Jonafree', 'Redfree', 'Macfree', 'Priscilla', 'Liberty', 'Britegold' and 'Trent', but, unfortunately, none of the major commercial cultivars.

Powdery mildew. The fungus *Podosphaera leucotricha* is a problem in drier areas of the world, such as the Pacific Northwest and Western Europe. Leaves, flowers, and fruits are affected and may abscise if the disease is severe. Its name comes from the white, feltlike patches of fungus that occur on the lower leaf surface and all over flowers and young fruits. Shoot growth of young trees is reduced, and fruits are cosmetically affected in most cases of flower infection. Fungicides applied as flower buds become distinct, but before they open, are effective in controlling mildew. Sprays may need to be repeated on highly susceptible cultivars until shoot growth ceases in summer. 'Delicious', 'Golden Delicious', 'Winesap', and 'York' are less susceptible than many other cultivars.

Fire blight. Erwinia amylovera causes this severe bacterial disease that infects some apple cultivars, particularly during warm, wet springs. In general, pears are more vulnerable than apples. The bacteria are carried by bees from tree to tree at bloom and can kill all or most of the flowers on a tree if severe. Highly susceptible cultivars also show twig and spur dieback or complete tree death. Flowers, twigs, and leaves often turn black and wilt; shoot tips will droop over, giving a distinctive "shepherd's crook" appearance. Resistant cultivars, including 'Delicious' and 'McIntosh', are the best means of control. Susceptible cultivars (e.g., 'Jonathan', 'Rome', 'Granny Smith') can be managed by spraying a copper-containing fungicide just as buds swell, but before they show green color at their tips. Sanitation is extremely important—infected shoots are pruned 6 inches below the blackened tissue and removed from the orchard. Pruning shears may be dipped in or sprayed with dilute bleach between the pruning of individual trees to avoid spreading the disease.

Fruit rots. Bitter rot, white rot, and black rot (caused by *Colletotrichum, Botryosphaeria* spp.) are common diseases of apple and pear in all countries. Rots can result in entire crop losses in a matter of days if weather prior to harvest is warm and wet. For apples in the southeastern United

States, these diseases are annual problems, whereas they are less prevalent in drier climates. Although leaves and stems can be affected, the primary concern is fruit lesions, which develop rapidly and can extend to the core, rendering the fruits useless and causing premature fruit drop. Control of all three rots is achieved by removal of mummified fruits, infected twigs, and limb cankers (through pruning) and the use of fungicides. Although rots occur near harvest, when fungicides are applied very early, from bud swell to bloom, they reduce the potential for late-season disease.

HARVEST, POSTHARVEST HANDLING

Maturity

Several methods are available for determining optimal harvest time. Days from full bloom is relatively constant from year to year and gives growers a rough estimate of picking date. Some cultivars, such as 'Gala', mature early and others, such as 'Fuji', very late. As maturity approaches, apples soften, so a fruit firmness meter can be used to measure the force required to puncture the flesh. Target values of firmness vary by cultivar and intended storage method, with firmer fruit reserved for long-term storage. Soluble solids can be measured with a refractometer, giving an indication of a fruit's sugar content, which increases as the apple ripens. Starch is broken down into sugars during ripening, and a quick iodine stain test on a half-cut apple can reveal how far this process has progressed. This test is used on stored apples to assess remaining shelf life. Apples are classified as climacteric fruits with respect to ripening.

Harvest Method

Apples must be picked by hand to avoid bruising and reduction of fresh-market quality grade (Figure 3.6). Fruits must be picked carefully to avoid damaging the spur, where next season's fruits will be borne.

Postharvest Handling

Standard packing-line operations—hydrocooling, washing, culling, waxing, sorting, and packing—are used for apples after harvest (Figure 3.7). Apples are packed most often in ⅘-bushel boxes (40 pounds), but polyethylene bags (holding 5-10 pounds) are also popular for retail marketing. Quality grade is based on size and appearance of skin; greater prices are obtained for larger fruits and those with minimal surface blemishes.

FIGURE 3.6. Apples being picked by hand and dumped gently into bulk bins for transport to the packinghouse.

Storage

The apple is one of the few fruits that can tolerate long-term storage without significant loss of quality. Controlled-atmosphere (CA) storage allows the marketing of apples

FIGURE 3.7. Apples working their way to a packing line via a water flume. Fruits are inspected to remove defects, waxed, brushed, graded for size, and packed into 42-pound cartons for shipment.

on a year-round basis. The storage room atmosphere is altered to retard respiration by reducing oxygen to 2 to 3 percent and raising CO_2 to about 1 percent. Firmer, less ripe fruits are placed in long-term (150-365 days) CA storage, while more mature fruits are sold directly or placed in short-term storage.

Several storage disorders affect apples. Chilling injury, or "low-temperature breakdown," occurs in some cultivars ('Cox', 'McIntosh', 'Jonathan') at 32 to 38°F, and core flush ("brown core") and soft scald also occur as low-temperature disorders. Inadequate nutrition (mostly lack of calcium), immaturity, and weather prior to harvest are often contributing factors to these disorders. Superficial scald occurs due to volatile products arising from the fruit itself (e.g., farnesene) that cause discoloration of the skin. Water core results from water-soaking of the flesh near the core due to abnormally rapid conversion of starch to sugar; this is common in 'Bramleys', 'Cox', 'Delicious', 'Jonathan', and 'Winesap' and accentuated by low calcium, heavy pruning, and hot summer weather. Bitter pit and cork spot ("corking") are calcium-related disorders affecting many cultivars, especially 'York'. This disorder is worse under drought stress and can be prevented by repeated calcium chloride ($CaCl_2$) sprays during the summer. Several postharvest diseases (caused by various species)—black mold *(Alternaria)*, gray mold *(Botrytis)*, blue mold *(Penicillium)*, brown rot *(Monolinia)*, and others common to fruit crops—occur routinely in storage unless packinghouse fungicide practices are used.

CONTRIBUTION TO DIET

Apples have a broad spectrum of food uses: pies and cakes, jams, sauces and juices, apple butter, dried apples, and much more. Apple juice has surpassed orange juice consumption by children in the United States. A medium-sized apple contains about 80 calories and is unusually high in fiber, having generally about 5 grams per fruit (mostly from pectin).

In 2004, U.S. consumers ate an average of 50.8 pounds of apples and processed apple products. Total consumption is up 27 percent since 1980, but fresh consumption has dropped

3 percent, canned consumption has dropped 13 percent, and juice consumption has increased 95 percent in the same period. Apple utilization is as follows:

Product	Percentage of Crop
Juice	51
Fresh	37
Canned	9-10
Dried	1
Frozen	1
Other (vinegar, wine, slices)	<1

Dietary value per 100-gram edible portion:

Water (%)	86
Calories	52
Protein (%)	0.3
Fat (%)	0.2
Carbohydrates (%)	13.8
Crude Fiber (%)	2.4

	% of U.S. RDA (2,000-calorie diet)
Vitamin A	1
Thiamin, B_1	1
Riboflavin, B_2	2
Niacin	<1
Vitamin C	8
Calcium	<1
Phosphorus	2
Iron	<1
Sodium	<1
Potassium	3

BIBLIOGRAPHY

Beach, S.A. 1905. *The apples of New York.* Volume 1. Albany, NY: J.B. Lyon Co.

Childers, N.F., J.R. Morris, and G. S. Sibbett. 1995. *Modern fruit science.* Tenth edition. Gainesville, FL: Norman F. Childers.

Ferree, D.C. and R.F. Carlson. 1987. Apple rootstocks, pp. 107-144. In: R.C. Rom and R.F. Carlson (eds.), *Rootstocks for fruit crops.* New York: John Wiley and Sons.

Ferree, D.C. and I.J. Warrington. 2003. *Apples: Production, botany and uses.* Wallingford, UK: CABI Publishing.

Forshey, C.G., D.C. Elfving, and R.L. Stebbins. 1992. *Training and pruning apple and pear trees.* Alexandria, VA: American Society of Horticultural Science.

Gur, A. 1985. Rosaceae—Deciduous fruit trees, pp. 355-389. In: A.H. Halevy (ed.), *CRC handbook of flowering.* Volume 1. Boca Raton, FL: CRC Press.

Jackson, J.E. 2003. *Biology of apples and pears.* Cambridge, UK: Cambridge University Press.

Jones, A.L. and H.S. Aldwinckle. 1990. *Compendium of apple and pear diseases.* St. Paul, MN: American Phytopathological Society.

Phillips, M. 1998. *The apple grower.* White River Junction, VT: Chelsea Green Publ.

Manhart, W. 1995. *Apples for the twenty-first century.* Portland, OR: North American Trees Co.

Smock, R.M. and A.M. Neubert. 1950. *Apples and apple products.* New York: Interscience Publ.

Teskey, B.J.E. and J.S. Shoemaker. 1978. *Tree fruit production.* Third edition. Westport, CT: AVI Publ.

Way, R.D., H.S. Aldwinckle, R.C. Lamb, A. Rejman, S. Sansavini, T. Shen, R. Watkins, M.N. Westwood, and Y. Yoshida. 1990. Apples *(Malus)*, pp. 1-62. In: J.N. Moore and J.R. Ballington (eds.), *Genetic resources of temperate fruit and nut crops* (Acta Horticulturae 290). Belgium: International Society for Horticultural Science.
Westwood, M.N. 1993. *Temperate zone pomology*. Portland, OR: Timber Press.

Chapter 4

Apricot *(Prunus armeniaca)*

TAXONOMY

Apricot, *Prunus armeniaca* L., is a member of the Rosaceae family, subfamily Prunoideae, along with peach and other stone fruits. The apricot is found in the *Prunophora* subgenus within *Prunus* along with plums. There are three to ten species in the *armeniaca* section of the *Prunophora*. Other well-known species very similar to *P. armeniaca* are *P. sibirica, P. mandshurica, P. mume*, and *P. ×dasycarpa* (a natural hybrid between *P. cerasifera* and *P. armeniaca*). Although this group presents great diversity in tree size, stress tolerance, bloom date, and fruit quality, very few cultivars of apricot are grown commercially throughout the world, and most of these derive straight from *P. armeniaca*. Cultivars also tend to be grown in only one region of a given country, and most would be virtually unknown outside of that region.

Recent hybrids produced by crossing plums and apricots are said to bear finer fruits than either parent. A "plumcot" is 50 percent plum, 50 percent apricot; an "aprium" is 75 percent apricot, 25 percent plum; and the most popular hybrid, the "pluot," is 75 percent plum, 25 percent apricot. Some pluots are marketed in stores as "dinosaur eggs," since they have an odd, mottled appearance to the skin. All plum-apricot hybrids have a fairly low chilling requirement (<600 hours) and thus are adapted to warmer regions. They are similar to Japanese plums in many respects.

The most common ornamental form of apricot is *Prunus mume*. Several cultivars, such as 'Kankobai', 'Kobai', and 'Koume', are grown as early blooming, small landscape trees. Some of these produce fruits that are rather dry and sour, yet edible, especially if processed. The bulk of Japanese apricot production is based on *P. mume*.

Cultivars

Commercially, 'Blenheim' (syn. 'Royal') and 'Tilton' are the major midseason California cultivars. 'Blenheim' accounted for 82 percent of production in 1950, which is similar to the level of production it still enjoys today. In Washington, 'Wenatchee Moorpark' is the main cultivar, followed by 'Tilton', 'Royal', and 'Perfection'. In the 1990s, the most requested cultivars from nurseries in the United States were the early ripening 'Earlicot' and the late ripening 'Autumn Royal', indicating a move toward extending the marketing season.

ORIGIN, HISTORY OF CULTIVATION

The center of diversity of the apricot is northeastern China near the Russian border (in the Great Wall area), *not* Armenia, as the name suggests. From there it spread west throughout Central Asia. Cultivation in China dates back 3,000 years, and movement to Armenia, and then to Europe from there, was slow. The Romans introduced apricots to Europe in 70-60 BC through

Introduction to Fruit Crops

doi:10.1300/5547_04

Greece and Italy. Apricots probably moved to the United States through English settlers on the East Coast and Spanish missionaries in California. For much of their history of cultivation, apricots were grown from seedlings, and few improved cultivars existed until the nineteenth century. Cultivars vary among countries, and in Turkey, Iran, Iraq, Afghanistan, Pakistan, and Syria, a great deal of the production is from seedling orchards.

Cultivation in the United States was confined to frost-free sites along the Pacific slope of California, due to early blooming but a relatively high chilling requirement, as well as fungal disease problems in humid climates. Now, most of the production in California is in the San Joaquin Valley.

FOLKLORE, MEDICINAL PROPERTIES, NONFOOD USAGE

The apricot flower is symbolic of doubt in the "language of flowers." Also, the apricot has been used to symbolize female genitalia, similar to the peach and other stone fruits.

As with all stone fruits, apricot leaves, flowers, and especially seeds and bark contain toxic compounds that generate cyanide, which is of course toxic or lethal in large doses. However, in the plant tissues, the cyanide concentration is low enough to be considered therapeutic, particularly for cancer (tumor) treatment, and has been used for this purpose since at least 25 BC. Apricot oil was used against tumors and ulcers in England in the 1600s. Apricot seeds contain the highest amounts of these cyanide-generating compounds, and the controversial cancer drug laetrile is derived from this source. Treatment is based on the theory that the apricot pit extract breaks down to release cyanide, but only when in contact with beta-glucuronidase, an enzyme active in tumor cells. The cyanide is released preferentially at tumor sites, killing cancerous cells. The efficacy of laetrile in treating cancer is supported by only a few reports of tumor regression and pain reduction. The National Cancer Institute in the United States claimed laetrile was an ineffective cancer treatment in 1980, but it is still legal in Mexico. Some patients still cross the border to seek laetrile therapy when other cancer treatments fail.

Phloretin is an antibiotic-like compound found in apricot bark and root extracts; in concentrated form, phloretin can kill certain bacteria.

PRODUCTION

World

Total production in 2004, according to FAO statistics, was 2,685,486 MT or 5.9 billion pounds. Apricots are produced commercially in 63 countries on about 988,000 acres. Production has been stable over the past decade. Yields average 5,980 pounds/acre, ranging from just a few thousand pounds to over 15,000 pounds/acre in some European countries. The top ten apricot-producing countries (percent of world production) follow:

1. Turkey (16)
2. Iran (10)
3. Italy (8)
4. France (6)
5. Pakistan (5)
6. Spain (5)
7. Ukraine (4)
8. Morocco (4)

9. United States (3)
10. China (3)

United States

Total production in 2004, according to USDA statistics, was 91,545 MT or 201 million pounds. Apricots are a very minor industry in the United States, valued at $35 million in 2004, ranging from $26 million to $48 million over the past decade. Prices are relatively low, at $0.19 per pound, which is typical of prices over the past decade. Apricots are produced commercially in three states, with California accounting for 94 percent of the crop. In 2004, California had 17,000 bearing acres of apricots, producing about 11,000 pounds/acre, for a value of around $29 million. Washington produces 5 percent of the crop, and growers there receive over $0.50 per pound, about threefold greater than prices paid in California, since most of the Washington crop is sold as fresh fruit. Utah produces less than 1 percent of the crop.

The United States exported 31 percent of production in 2002, mostly as dried fruit, with fresh fruit accounting for about one-third of exports. Exports have doubled since 1988. No import data are available, but small quantities of fresh fruit are imported during winter months from such Southern Hemisphere countries as Chile.

BOTANICAL DESCRIPTION

Plant

Apricots are small- to medium-sized trees with spreading canopies (Plate 4.1). They are generally kept under 12 feet in cultivation but are capable of reaching 45 feet in their native range. The 1-year-old wood and spurs are thin, twiggy, and shorter lived than those of other stone fruits. Leaves are elliptic to cordate, with acute to acuminate tips, about 3 inches wide, which is wider than the leaves of other stone fruits. Leaves have serrate margins and long, red-purple petioles (Plate 4.2).

Flower

Flowers are similar in morphology to those of peach, plum, and cherry trees. White flowers are borne solitary in leaf axils of 1-year-old wood or in leaf axils on short spurs, and they appear to be grouped in clusters (Plate 4.3). Flowers feature five sepals and petals, and many erect stamens, all of which emanate from the hypanthium or floral cup. Ovary position is perigynous.

Pollination

The major U.S. cultivars are self-fruitful and do not require a pollinizer; exceptions include 'Riland' and 'Perfection', which are self-incompatible. Honeybees are the major pollinator.

Fruit

The fruit of the apricot is a drupe, about 1.5 to 2.5 inches wide, with a prominent suture; its color ranges from yellow to orange with a red blush, having a light pubescent or a nearly glabrous surface (Plate 4.4). The pit is generally smooth and encloses a single seed. Flesh color is mostly orange, but a few white-fleshed cultivars exist. Trees are fairly precocious and begin fruiting in their second year, but substantial bearing does not begin until 3 to 5 years. Fruits are borne mostly on short spurs on mature, less vigorous trees but can also occur on long lateral shoots of vigorous trees. Fruits require 3 to 6 months for development, depending on cultivar, but the main harvest season is May 1 through July 15 in California. Apricots are thinned by hand, leaving one fruit for every 3 to 5 inches of shoot length.

GENERAL CULTURE

Soils and Climate

In their native range, apricots are found on dry, rocky hillsides. However, deep, fertile, well-drained soils are best for commercial production. Apricots are somewhat more tolerant of high soil pH and salinity than are other *Prunus* spp., although they are intolerant of waterlogging. As with the cultivation of most stone fruits, replanting a site where apricots, peaches, or plums were planted previously is discouraged because tree productivity and longevity is compromised on these soils.

Apricot culture is most successful in mild, Mediterranean climates where the danger of spring frost is limited and disease pressure is reduced. This is why commercial culture is restricted to Turkey, southern France, coastal Spain, Italy, and California. Rainfall and high humidity during the growing season, particularly at bloom or harvest, are serious limitations due to fungal diseases that kill the flowers and shoots or cause the fruits to rot. Cultivation of apricots in the eastern United States requires routine fungicide applications, and trees often succumb to bacterial diseases. Chilling requirements range from 400 to 1,000 hours. The heat requirement following chilling is very short, causing apricots to bloom early in almost any location. Apricots are considered frost sensitive due to their early bloom habit. Open flowers are killed by exposure to 28°F or lower. Coldhardiness during dormancy in U.S. cultivars is about the same or somewhat less than that for peach (–20 to –30°F), but very good (to –40°F!) in Asian cultivars with *P. sibirica* in the parentage. Several cultivars have been bred in Michigan and Ontario for adaptation to the cold Midwestern winters. Length of season is not a concern with apricot, since most ripen in late spring to midsummer (May-July), as they bloom early and are fairly small at maturity.

Propagation

Apricots are T- or chip-budded onto rootstocks during summer or fall usually, although June budding is practiced occasionally. Seedling trees are still grown in the Middle East, Asia, and parts of Europe.

Rootstocks

Apricot seedlings are the most popular stock worldwide. The cultivars used are important scion types: 'Blenheim' in California, 'Canino' in France, 'Hungarian Best' in Hungary, Czechoslovakia, and Romania. Seedlings are highly variable and give nonuniform stands, but stocks other than *P. armeniaca* are usually incompatible to some extent. Seedling rootstocks are well adapted to dry, light, or pebbly soil and yield vigorous trees. Peach seedling rootstocks, 'GF 305' in France and 'Lovell' and 'Nemaguard' in California, are used on acid-neutral soils where irrigation is available. Resistance to bacterial canker and verticillium root rot is improved in these rootstocks, but some cultivars exhibit delayed incompatibility.

Several rootstocks have been shown to dwarf apricot trees, including the near apricot relatives *P. sibirica* and *P. ×dasycarpa*. The sand cherry, *P. besseyi*, dwarfs apricot and induces precocity.

'Pixy', a form of damson plum *(P. insititia)*, and a few other clonal plum rootstocks will dwarf apricots as well. Problems stemming from delayed incompatibility and poor adaptation to local conditions have largely precluded the commercial use of dwarfing rootstocks.

Planting Design, Training, Pruning

Since apricots require good light exposure for fruit color development and do not require pollinizers, they can be planted in solid blocks, similar to peach tress. Spacings of about 20 to 24 feet between trees and rows are necessary for vigorous trees on apricot stocks, but closer spacings can be used for less vigorous rootstocks. The high-density systems used in Europe resemble the forms used for peach and other stone fruits (Figure 4.1).

Apricots are trained to a modification of the open-center system, called the "Winters" system; used in California, this tall, shallow-mantled, open-center tree shape improves fruit color and reduces unevenness of ripening within the tree, which allows for fewer pickings. Apricots can be pruned fairly heavily, since they bear too many fruits and are vigorous. They will bear fruits on spurs or long lateral shoots, and since spurs are short-lived, more severe pruning is needed to renew fruiting wood. Generally, most new growth and interfering wood is removed each year, exposing the spurs to maximal sunlight.

Pest Problems

Insects

Oriental fruit moth. These small moths *(Grapholita molesta)* lay eggs in shoot tips early in the season, and the resulting larvae burrow downward a few inches; wilted and dead expanding leaves at the shoot tips indicate infestation. Later infestation results in fruit loss, as eggs laid on fruits produce ¾-inch, legged, pinkish larvae that burrow through the fruit. Although fruit infes-

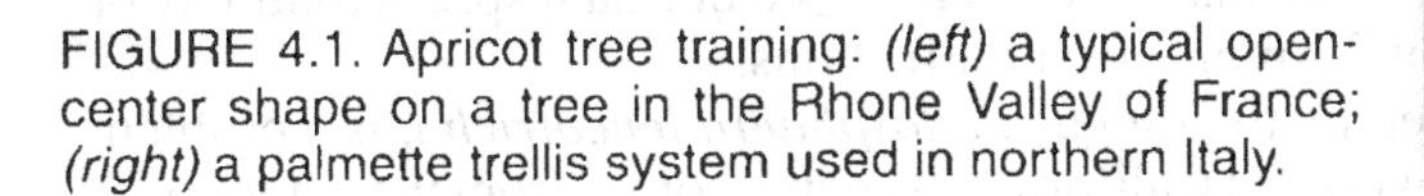

FIGURE 4.1. Apricot tree training: *(left)* a typical open-center shape on a tree in the Rhone Valley of France; *(right)* a palmette trellis system used in northern Italy.

tation does not occur until fruits are nearly half grown, early season sprays are critical for controlling this pest. Insecticides are applied at petal fall and shuck split.

Borers. Larvae of the peach tree borer *(Synanthedon exitosa)* cause damage to the lower trunk and scaffolds, sometimes killing young trees. The adult moth lays eggs on the tree, and the larvae bore into the bark to feed. Larvae are ½ to 1 inch long with a dark-colored head. Lower limbs of trees are sprayed thoroughly with insecticide in late summer when the adult moths are laying eggs. Peach twig borers *(Anarsia lineatella)* damage young shoots and fruits. Pheromone mating disruption is one means of control; another is the application of postbloom insecticides if populations are high. Pacific flathead borer *(Chrysobothris mali)* and shothole borer (*Scolytus rugulosus*) are occasional wood pests that can affect weak or diseased trees.

Leaf rollers. Fruit tree leaf roller *(Archips argyrospilus)* and oblique-banded leaf roller *(Choristoneura rosaceana)* can feed on leaves and young fruits. They roll marginal areas of leaves together with webbing. Fruit feeding causes scarring. Since they are caterpillars, they can be controlled by sprays of Bt or other insecticides, and they are vulnerable to a number of natural predators, which helps to keep populations in check as well.

Mites. Mites (*Panonychus, Tetranychus* spp., others) feed on the lower leaf surface, causing a characteristic "mite stipple" or bronzed appearance to the upper leaf surface. European red mite *(Panonychus ulmi),* Pacific spider mite *(Tetranychus pacificus),* and two-spotted mite *(T. urticae)* are economic threats to California orchards. A dormant oil spray before budbreak can control light populations, but miticides are sometimes required when webbing and visible injury occurs. Spot treatments in dusty areas can prevent more widespread infestations. Several natural predators exist, and these can provide a means of biological control. Care should be taken to monitor both pest and predator populations before sprays are applied.

Diseases

Brown rot/blossom blight. These diseases result from the same fungi *(Monilinia laxa, M. fructicola)* that cause flower death in spring and fruit rot near harvest. Apricots have the least resistance to blossom blight of all stone fruits; twigs and spurs may be killed. A brown, powdery mass of spores in concentric rings is visible around a soft lesion on the fruit. Fruits infested during a previous year's season will die, shrivel, turn black, and remain hanging on the tree; called "mummies," these desiccated fruits house spores for the following year's infection and should be removed from trees in winter. Fungicide applied as trees bloom controls blossom blight and will reduce brown rot later. During the summer, fungicide sprays may need to be applied at 7- to 10-day intervals up to a week before harvest. Resistant cultivars include 'Derby Royal', 'Moorpark', 'Nugget', 'Stella', 'Veecot', and a number of those grown in Italy.

Eutypa dieback. The opportunistic fungus *Eutypa lata* invades wounds on wood and can kill entire limbs very quickly in early summer. The fungus, which requires plentiful rainfall to invade wounded wood, is a severe limitation in rainy climates but can be managed in dry summer climates. In California, pruning 2 to 6 weeks prior to expected rainfall can control the disease.

Shot hole fungus. Caused by the fungus *Wilsonomyces carpophila,* this disease is characterized by small purple lesions on leaves and fruit; on leaves, the lesions abscise later in the season, causing the shot hole appearance. This disease infects and kills dormant buds in winter that will then exude gum, providing spores for infecting the next season's growth. Among the stone fruits, apricot is the most severely affected. A fungicide spray in autumn when trees go dormant controls the winter phase of infection. Another spray at fruit set in spring is often needed to control the leaf and fruit phases in summer. Trees that have not received dormant sprays routinely or have not been pruned regularly often suffer more severe infections.

Powdery mildew. This disease, caused by the fungus *Podosphaera leucotricha,* is a problem in drier areas, such as the Pacific Northwest and California. Its name comes from the white-gray, feltlike patches of fungus that occur on the lower leaf surface and all over flowers and young fruits. Leaves, flowers, and fruits are affected and may fall off if the infection is severe. Shoot growth of young trees is reduced, and fruits are cosmetically affected in most cases of flower infection. Fungicides are applied starting at shuck fall stage (~2-3 weeks postbloom). Sprays may need to be repeated until shoot growth ceases in summer on highly susceptible cultivars. All major cultivars are susceptible, with 'Perfection' being the most susceptible of them all.

Bacterial spot. This disease is caused by the bacterium *Xanthomonas campestris,* which exists in most parts of the world, except on the Pacific Coast. It is more prolific in humid, warm climates and in areas with sandy soils. The bacteria can infect leaves, twigs, or fruits and cause symptoms of infection that range from small black spots to large sunken black lesions. Sprays of copper or antibiotic materials are risky and often do not control the disease well. Several resistant cultivars were developed in Ontario and New Jersey, where the disease is severe, including 'Harcot', 'Harglow', and 'Jerseycot'. Moderate resistance is found in 'Newcastle', 'Stella', 'Superb', and 'Moorpark'.

Fungal and bacterial canker. These diseases are the result of pathogens *(Cytospora* spp., *Pseudomonas syringae)* that cause injury to twigs, branches, and large limbs. Branches/limbs become girdled in severe cases and experience rapid dieback. Many "suckers" are produced from the base of the trunk on severely infected trees because the lower trunk and roots are not killed by these pathogens. Canker, particularly the bacterial form, is difficult or impossible to control with sprays. Freezing injury, nematode weakening, other diseases, drought, and waterlogging all predispose trees to infection, so healthy trees are a good defense. Resistant cultivars include 'Haggith' and others grown in Asia.

Plum pox (syn. Sharka) virus. Symptoms of this discease include severe fruit malformation and sporadic occurrence of ring-shaped chlorotic spots on leaves. Fruit may have irregular spots or rings and be dry and flavorless. The virus spreads through infected bud wood and aphids. The disease, a serious problem in Europe, was introduced to North America in the mid-1990s. Quarantine and eradication measures have effectively reduced its spread outside of a few isolated areas, and the stone fruit industry in the United States has not been affected to date. Resistant cultivars include 'Earli Orange', 'Stella', and 'Harcot'.

HARVEST, POSTHARVEST HANDLING

Maturity

Apricots for fresh consumption are picked when firm but well before physiological maturity because they become too soft for shipping if allowed to approach ripening on the tree. It is said that proper flavor never develops in fruit picked prior to physiological maturity, for which firmness is a reliable indicator with apricots, as with plums. Gauging maturity using days from full bloom is a fairly reliable index, given the relatively invariable growing conditions in California. Apricots are classified as climacteric fruits with respect to ripening.

Harvest Method

Apricots for fresh consumption or processing are picked by hand and carefully handled. Trees are usually picked over two to three times each. Trunk shaking can be used for processed fruits, although apricots are said to be more susceptible to trunk damage than are the other stone fruits.

Postharvest Handling

Fresh apricots are shipped in shallow containers to prevent crushing/bruising. Dried apricots are harvested later (fully ripe) than those to be shipped fresh, so they are exposed to sulfur dioxide (SO_2) to combat postharvest diseases. The drying ratio is 5.5:1 (pounds fresh fruit : pound[s] dry fruit), and drying occurs through either natural processes, such as sun drying, or mechanical dehydration, as with prunes. Canned apricots are immersed in syrup so that 0.7 pound fresh equals 1 pound canned.

Storage

Apricots have an extremely short shelf-life of only 1 to 2 weeks at 32°F and 90 percent relative humidity. They are susceptible to all postharvest diseases (e.g., *Rhizopus* fruit rot) to which other stone fruits are susceptible. Apricots are not sensitive to chilling injury.

CONTRIBUTION TO DIET

Most of the U.S. crop is not sold fresh; drying and canning are popular options for apricots because they are so perishable. Cultivars that retain their color and flavor during drying, such as 'Royal' and 'Tilton', are best for this market. Dried apricots can be easily rehydrated and are particularly popular with backpackers. As with plums, drying apricots concentrates all nutrients severalfold. In 2004, utilization of apricots was as follows:

Product	Percentage of Crop
Dried	57
Canned and juices	23
Fresh	13
Frozen	5

Seed of Central Asian and Mediterranean apricots is generally "sweet," such that seed can be used as a substitute for almonds or crushed for almondlike cooking oil.

Per capita consumption was 0.92 pound in 2004 and has fluctuated between 0.9 and 1.6 pounds since 1980. Fresh and frozen consumption have remained largely unchanged, but canned consumption has dropped 43 percent since 1980, being compensated for by a 40 percent increase in dried consumption.

Dietary value per 100-gram edible portion:

Water (%)	86
Calories	48
Protein (%)	1.4
Fat (%)	0.4
Carbohydrates (%)	11.1
Crude Fiber (%)	2

	% of U.S. RDA (2,000-calorie diet)
Vitamin A	38
Thiamin, B_1	2
Riboflavin, B_2	2
Niacin	3

Vitamin C	17
Calcium	1
Phosphorus	3
Iron	2
Sodium	<1
Potassium	7

BIBLIOGRAPHY

Bailey, C.H. and L.F. Hough. 1975. Apricots. In: J. Janick and J.N. Moore (eds.), *Advances in fruit breeding*. West Lafayette, IN: Purdue University Press.

Crossa-Raynaud, P. and J.M. Audergon. 1987. Apricot rootstocks, pp. 295-320. In: R.C. Rom and R.F. Carlson (eds.), *Rootstocks for fruit crops*. New York: John Wiley and Sons.

Gu, M. 1988. Apricot cultivars in China, pp. 63-67. In: S.A. Paunovic (ed.), *Second international workshop on apricot culture and decline: Research and crop improvement* (Acta Horticulturae 209). Belgium: International Society for Horticultural Science.

Gulcan, R. 1988. Apricot cultivars in near east. pp. 49-54. In: S.A. Paunovic (ed.), *Second international workshop on apricot culture and decline: Research and crop improvement* (Acta Horticulturae 209). Belgium: International Society for Horticultural Science.

Gur, A. 1985. Rosaceae—Deciduous fruit trees, pp. 355-389. In: A.H. Halevy (ed.), *CRC handbook of flowering*, Volume 1. Boca Raton, FL: CRC Press.

Klement, Z. (ed.). 1986. *Eighth international symposium on apricot culture and decline* (Acta Horticulturae 192). Belgium: International Society for Horticultural Science.

Mehlenbacher, S.A., V. Cociu, and L.F. Hough. 1990. Apricots *(Prunus)*, pp. 63-108. In: J.N. Moore and J.R. Ballington (eds.), *Genetic resources of temperate fruit and nut crops* (Acta Horticulturae 290). Belgium: International Society for Horticultural Science.

Ogawa, J.M., E.I. Zehr, G.W. Bird, D.F. Ritchie, K. Uriu, and J.K. Uyemoto. 1995. *Compendium of stone fruit diseases*. St. Paul, MN: American Phytopathological Society.

Paunovic, S.A. 1988. Apricot germplasm, breeding, selection, cultivars, rootstocks, and environment, pp. 13-28. In: S.A. Paunovic (ed.), *Second international workshop on apricot culture and decline: Research and crop improvement* (Acta Horticulturae 209). Belgium: International Society for Horticultural Science.

Teskey, B.J.E. and J.S. Shoemaker. 1978. *Tree fruit production*. Third edition. Westport, CT: AVI Publ.

Watkins, R. 1979. Cherry, plum, peach, apricot, and almond; *Prunus* spp., pp. 242-247. In: N.W. Simmonds (ed.), *Evolution of crop plants*. London: Longman.

Chapter 5

Banana and Plantain (*Musa* spp.)

TAXONOMY

Bananas and plantains belong to the Musaceae, known simply as the banana family. It is a relatively small family with only five genera and at most 150 species. Some taxonomic treatments remove the genus *Heliconia* from the Musaceae, giving the 60 or so species of heliconias family status, reducing the Musaceae even more. A few species of *Heliconia* are grown as tropical ornamentals, along with *Strelitzia reginae*, the bird-of-paradise flower (small, orange-flowered plant); *Strelizia nicolai* (tall, white-flowered bird-of-paradise); and *Ravenala madagascarensis*, the traveler's palm. The family is distinguished by being composed of fairly large, often treelike, tropical herbaceous plants with flowers subtended by distinct bracts. Banana and plantain are unusual fruit crops since they are monocots, not dicots, as with most others. Taxonomically, this makes them more closely related to maize, wheat, and turf grasses than to apples or oranges.

There are 25 to 80 species in the genus *Musa*, depending on the taxonomist, but it is likely that the higher number stems from subspecies being given species status. *Musa* is important not only for fruit production; the genus has provided man with food, clothing, tools, and shelter since prior to recorded history. Manila hemp *(M. textilis)* is grown for fiber (not fruit) that is derived from its pseudostem. This fiber can be made into strong rope or abaca cloth.

The naming of banana and plantain and the distinction between the two types are issues mired in confusion. Linnaeus first named bananas *M. sapientum* and plantains *M. paradisiaca*, basing his separation on how the fruits were consumed, classifying sweet, soft fruits eaten raw as bananas and starchy fruits cooked prior to consumption as plantains. Today, bananas are still distinguished from plantains based largely on consumption of raw versus cooked fruits, but we know that they are not necessarily two different species (Plate 5.1). All banana and plantain cultivars derive from two main species:

1. *Musa acuminata* Colla (syn. *M. cavendishii* Lamb. ex Paxt., *M. chinensis* Sweet, *M. nana*, *M. zebrina* Van Houtee ex Planch.). This is the wild, edible banana that has undoubtedly been collected by humans for millennia. This species exists as either a seedless diploid ($2n = 22$) or a seedless triploid ($3n = 33$) in its native habitat of Malaysia. The fruits are sweet and soft at maturity because they convert most of their starch to sugars. Most dessert bananas derive directly from this species.
2. *Musa balbisiana* Colla. This seedy-fruited, unpalatable species from India and southeast Asia has been used as a parent in several cultivars, due to its disease resistance, drought tolerance, and general hardiness to stressful environmental factors. Its natural range does not overlap that of *M. acuminata*, but human dispersal allowed the two species to hybridize naturally thousands of years ago. Sterility in the hybrids makes their fruits seedless and therefore palatable. The fruits exhibit a slower conversion of starch to sugars at maturity; plantain cultivars generally have some parentage from this species.

Introduction to Fruit Crops

doi:10.1300/5547_05

Hybrids of *M. acuminata* and *M. balbisiana* are sometimes given the names *M. ×paradisiaca* L., *M. ×sapientum* L., or perhaps, most accurately, *M. acuminata* × *M. balbisiana* Colla. However, a shorthand method of distinguishing hybrids and accurately representing their parentage was developed in the 1950s and remains commonplace today. Each type is given a two- to four-letter designation consisting of A's, representing *acuminata,* and B's, representing *balbisiana.* For example, AA represents a diploid type derived only from *M. acuminata,* and AAB represents a triploid type with two-thirds *M. acuminata* and one-third *M. balbisiana* parentage. In general, the most important banana cultivars in the world are AAA, and plantains are mostly AAB, ABB, or BBB. A numerical rating system based on scoring 15 traits that differ between the two species allows parentage to be determined from plant morphology (Table 5.1). A cultivar is scored for each trait, earning a 1 if it matches the *M. acuminata* description, and a 5 for matching the *M. balbisiana* description. The scores are then totaled to determine the parentage (Table 5.2).

Cultivars

Banana

As mentioned earlier, virtually all dessert bananas sold to export markets have the AAA genome. 'Gros Michel' was formerly the most widely cultivated banana in the Western Hemisphere but was phased out due to susceptibility to Panama disease (fusarium wilt). It has pro-

TABLE 5.1. Traits scored in the system of Simmonds and Shepherd, 1955, that allow determination of the parentage of banana and plantain cultivars.

Trait	*M. acuminata*	*M. balbisiana*
Pseudostem	with brown and black blotches	without blotches
Petiolar canal	open and spreading, not clasping pseudostem	closed, clasping pseudostem
Peduncle	downy or hairy	glabrous
Pedicels	short	long
Ovules	in 2 rows in each of 3 locules	in 4 rows in each of 3 locules
Bract shoulder	more proximal	more distal
Bract curling	curl after opening	lift but do not curl after opening
Bract shape	lanceolate	broadly ovate
Bract apex	acute	obtuse
Bract color	outside—red, purple, or yellow inside—purple or yellow	outside—brownish purple inside—bright crimson
Color fading	inside bract color fades to yellow at base	inside bract color uniform, not fading
Bract scars	prominent	not prominent
Free tepal of male flower	corrugated below tip	rarely corrugated
Male flower color	creamy white	variably flushed with pink
Stigma color	orange or bright yellow	cream or pale yellow to pink

TABLE 5.2. Total scores for 15 traits presented in Table 5.1 and break points to determine cultivar parentage.

Genome	Score	Example cultivars	Food usage
AA or AAA	15-25	'Sucrier' (AA), 'Grand Nain' (AAA)	dessert banana
AAB	26-46	'Horn' or 'French Silk' (syn. 'Lady Finger' or 'Apple')	plantain dessert banana
AB	49	'Ney Poovan'	dessert banana
ABB	59-63	'Bluggoe', 'Pelipeta'	plantain
ABBB	67-69	'Klue Teparod'	plantain
BB or BBB	70-75	'Saba' (BBB)	plantain

Note: Scores represent the combined scales of Simmonds and Shepherd, 1955, and Silayoi and Chomchalow, 1987, as presented in Robinson, 1996.

duced several clones and has been used as the parent for newer cultivars. The Giant Cavendish subgroup within the AAA genome is now the most common group of cultivars. 'Mons Mari', 'Williams', 'Williams Hybrid', and 'Grand Nain' are the major cultivars. 'Grand Nain' is the most common banana imported into the United States; it has fairly large fruits with thick skin that withstand bruising. All are susceptible to race 4 of Panama disease, which currently is a problem only in isolated areas. Dwarf Cavendish (AAA genome) is a group of smaller-statured cultivars (4-7 feet) that bear medium-sized fruit with thin skin. These cultivars, which are grown in subtropical areas such as South Africa and Australia, resist windthrow better than Giant Cavendish types. Red-skinned variants are found in both Cavendish subgroups. 'Lady Finger' or 'Apple' bananas (AAB genome) are small (4-5 inches), very sweet fruits with thin skin (Figure 5.1). 'Sucrier', 'Senorita', or 'Lacatan' bananas (AA genome) are also small, sweet, thin-skinned fruits, but they are less popular due to lower yields and/or less disease resistance than is found with the triploids.

Plantain

In the AAB genome, plantains are one of two types: French or Horn. Each type has several variants, with French cultivars featuring persistent male flowers and Horn exhibiting rapid abscission of male flowers. In the ABB genome, 'Bluggoe' and its Moko disease–resistant variant 'Pelipita' are the main cultivars. 'Saba' is the leading cultivar in the BBB genome.

ORIGIN, HISTORY OF CULTIVATION

Edible *Musa* spp. originated in southeastern Asia, from India east and south to northern Australia. Early Filipinos probably spread the banana eastward to the Pacific Islands, including Hawaii, prior to recorded history. Westward, banana likely followed the major trade routes that transported other fruits, and it is known to have arrived in East Africa around 500 AD. Bananas were not carried to Europe until the tenth century, and Portuguese traders obtained it from West Africa, not Southeast Asia, during the age of discovery. Plants were taken from West Africa to the Canary Islands and South America in the sixteenth century, and spread throughout the Caribbean with settlement of the area in the sixteenth and seventeenth centuries. Bananas are now grown pantropically in 130 countries, more than any other fruit crop in the world.

FOLKLORE, MEDICINAL PROPERTIES, NONFOOD USAGE

Bananas are widespread in poor, tropical countries, and due to the abundance of folk remedies in those countries and/or lack of Western medicines, bananas proved useful for a number of medicinal applications. Many such uses are poorly documented, but reports do include descriptions of treatments for ailments of the skin, back, and blood; headaches, fever, and flu; both diarrhea and constipation; and several other problems, including gray hair and syphilis. Bananas contain moderate amounts of potassium, although the utility of bananas for restoring electrolytes or regulating blood pressure may be overstated. Some banana extracts have shown hypoglycemic activity in lab studies. Banana fruits contain the physiologically active compounds serotonin, norepinephrine, and dopamine.

FIGURE 5.1. 'Lady Finger' bananas at a roadside market in Ecuador. They are delightfully sweet and flavorful compared to the Giant Cavendish bananas sold throughout the world.

Banana leaves and stems contain large amounts of fertilizer elements and are often cut from old plants to serve as mulch and nutrition for the growing crop. Fertilizers are made from dried, chopped stems and leaves. Ash from burned leaves and stems is used as salt. Livestock often feed on banana culls and other plant parts. Fibers from the ground banana rachis can be mixed with wood fibers to make paper; this paper can be transformed into novelty items, such as notepads, envelopes, and wrapping paper, that can be produced and sold in local markets in countries such as Costa Rica.

Banana leaves are among the largest ones found on all plants. They can be used effectively as umbrellas, wrapping material for food or other loose items, plates or vessels for liquids, and roof thatch. In the animal world, tent bats exploit banana leaves by chewing along the midrib on the underside of the leaf, causing the leaf to fold, and thereby creating a dry tent for sleeping (hence their name).

PRODUCTION

World

Banana

Total production in 2004, according to FAO statistics, was 70,629,047 MT or 155 billion pounds. Banana ranks second among fruit crops in the world in terms of production, following first-place African oil palm. Many consider banana to be the first-place fruit crop, labeling oil

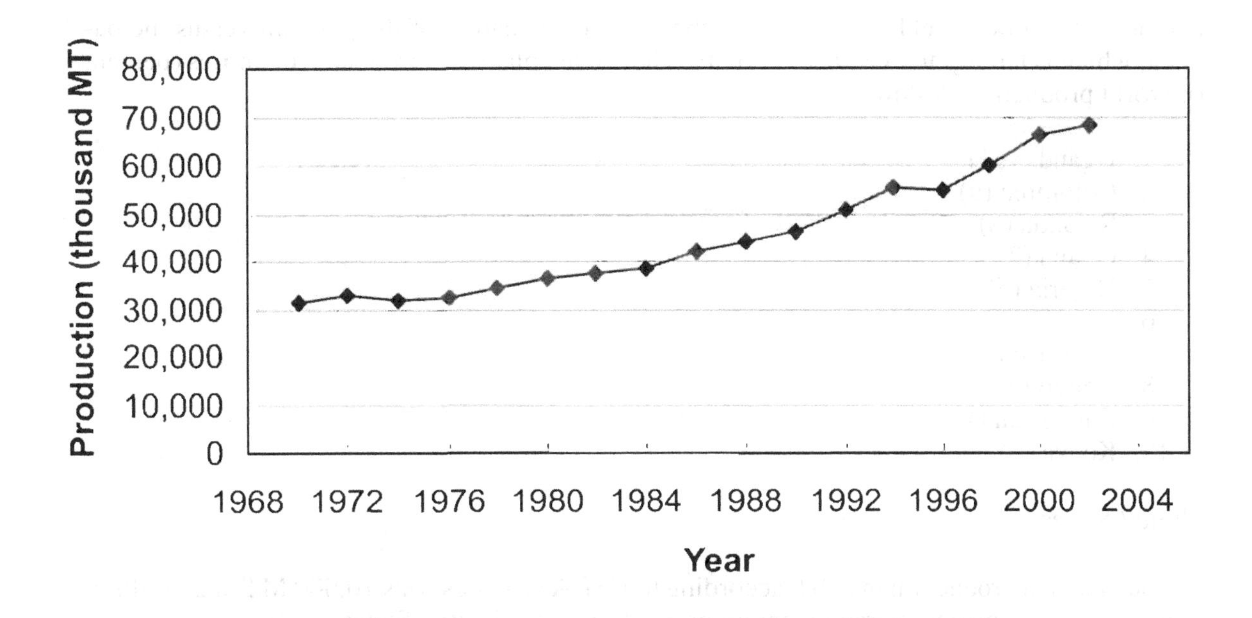

FIGURE 5.2. World banana production has more than doubled since 1970. *Source:* FAO statistics. Acreage has increased 66 percent since 1970, and yield per acre has increased 36 percent. Thus, most of the increase in production is due to the greater area cultivated rather than the intensity of cultivation. Over the entire period, the number of countries commercially producing bananas has remained constant at 130.

palm an "oil crop," not a fruit crop. Banana production has increased about 25 percent in the past decade (Figure 5.2). Bananas cover about 11 million acres, among the largest areas devoted to a single fruit crop. Worldwide, the average yield is about 14,000 pounds/acre, although yields in efficient plantations in Latin America are easily threefold higher. The top ten banana-producing countries (percent of world production) follow:

1. India (24)
2. Brazil (9)
3. China (9)
4. Ecuador (8)
5. Philippines (8)
6. Indonesia (6)
7. Mexico (3)
8. Costa Rica (3)
9. Thailand (2)
10. Colombia (2)

Plantain

Total production in 2004, according to FAO statistics, was 32,668,323 MT or 72 billion pounds. Plantains are grown as a staple food in 52 countries worldwide on about 12.8 million acres. Combined with bananas, acreage under cultivation of *Musa* spp. exceeds that of almost all other fruit crops. Plantain yields are much lower than those for bananas, around 5,600 pounds/ acre, although plantains are capable of yielding 35,000 to 40,000 pounds/acre under good management. The lower yield is reflective of the subsistence nature of the plantain versus the banana, which is largely an export commodity. The top ten plaintain-producing countries (percent of world production) follow:

1. Uganda (31)
2. Colombia (9)
3. Rwanda (8)
4. Ghana (7)
5. Nigeria (6)
6. Peru (5)
7. Cote d'Ivoire (4)
8. Congo (4)
9. Cameroon (4)
10. Kenya (3)

United States

Total banana production in 2004, according to USDA statistics, was 10,000 MT or 22 million pounds. There is no commercial plantain production in the United States.

All U.S. banana production is in Hawaii on 1,400 acres of land; this is smaller than a single average banana plantation in Latin America. The industry is valued at $9.2 million, and all fruits are sold fresh. Yields are slightly above the worldwide average at 15,700 pounds/acre in 2003, and ranging from 14,000 to 20,000 pounds/acre in the past decade. Prices received by growers in 2003 were $0.42 per pound, higher than the $0.35 to $0.40 per pound received over the past decade.

BOTANICAL DESCRIPTION

Plant

Both banana and plantain are large, herbaceous monocots, reaching 25 feet in some cultivars, but generally measuring 6 to 15 feet under cultivation (Figure 5.3). Plantains are often larger than bananas and have a "trunk" or pseudostem that is not a true stem, but only the clustered, cylindrical aggregation of leaf stalk bases. Banana tree leaves are among the largest found on all plants, expanding up to 9 feet long and 2 feet wide. Margins are entire and venation is pinnate; leaves tear along the veins in windy conditions, giving a feathered or tattered look. Each plant features 5 to 15 leaves, with 8 to 10 leaves considered the minimum for properly maturing a bunch of fruits. A total of up to 50 leaves may be produced by a shoot during its life cycle, but leaves are functional for only about 2 months. In the humid tropics, about one new leaf per week is produced. The perennial portion of the plant is the rhizome, often called a corm, which may weigh several pounds. It produces suckers, or vegetative shoots, that are thinned to two per plant: one "parent" sucker for fruiting and one "follower" to take the place of the parent after it fruits and dies back (Figure 5.4). It also produces roots and serves as a storage organ for the plant. The vegetative apex of the rhizome spontaneously initiates a reproductive meristem after approximately 40 leaves have been produced, usually 9 months after initiation of a sucker. A banana plant bears fruit 10 to 12 months after planting; plantains can take longer, roughly 14 to 19 months, particularly in cooler areas. See Table 5.1 for other plant traits that differ between banana and plantain.

FIGURE 5.3. A banana plantation in Costa Rica. Note the tram for transporting the fruit from the field to the packinghouse.

FIGURE 5.4. The perennial portion of the banana is a rhizome, entirely belowground. Shoots arise from the rhizome, fruit, then die or are cut back. This plant shows an old shoot that has been cut off *(right)*, a fruiting shoot *(center)*, and a "follower," or sucker, that will take the place of the fruiting shoot after it is harvested and cut back *(left)*.

Flowers

The inflorescence is a spike originating from the rhizome. Initially, it appears above the last leaves in an upright position and consists of only a large, purple, tapered bud. As the bud opens, the white, tubular, toothed flowers are revealed, clustered in whorled double rows along the stalk, each cluster covered by a thick, purple, hoodlike bract (Plate 5.2). The bract lifts from the first flower cluster in 3 to 10 days. Female flowers, with inferior ovaries, occupy the lower portion of the inflorescence , with neuter or hermaphrodite flowers in the center, and males at the top. Male flowers and bracts are shed immediately after opening, leaving the terminal portion of the rachis naked, except for the large, purple, fleshy bud at the tip that contains unopened male flowers (except Dwarf Cavendish bananas and French plantains, which have persistent male flowers). The inflorescence begins to droop down under its own weight shortly after opening; the flowers are negatively geotropic and turn upright during the first 10 weeks of growth.

Pollination

Bananas are male-sterile, and those of the Cavendish subgroups are female-sterile as well; fruits are set parthenocarpically. Traces of the undeveloped ovules appear as brown specks in the center of the fruits. Floral morphology suggests that wild bananas are bat pollinated in their native range, but no pollination is needed in cultivation.

Fruit

The banana is an epigynous berry; fruits are borne in "hands" of up to 20 fruits, with 5 to 20 hands per spike. Fruits appear as angled, slender, green "fingers" during growth, reaching harvest maturity in 90 to 120 days after flower opening (Figure 5.5). The terminal bud on the stalk may be removed if fruit set is high, to allow more complete filling of fruits (similar to thinning), since this organ continues to grow throughout fruit development. In commercial plantations, growers look for the "false hand," a hand of fruits with an abnormally small finger on the side (Figure 5.5). The rachis is cut above this point to prevent set of inferior fruits and to promote more uniform maturation of the remaining hands. A Giant Cavendish bunch can weigh over 100 pounds and contain 300 to 400 marketable fruits.

FIGURE 5.5. Young banana fruits developing sequentially from base to apex on the spike *(left)*. The "false hand" *(right)* is eventually produced, being distinguished by one or more misshapen fruitlets. It signals the point at which the developing bud should be cut off.

GENERAL CULTURE

Soils and Climate

Deep, well-drained, alluvial soils are best, but bananas and plantains can tolerate a wide variety of soil conditions. The water table should be at least 2 feet below the surface, and drainage or raised beds are necessary if waterlogging occurs. Drainage canals are often cut through plantations. The pH should be mildly acidic (~6.0), but liming is necessary if pH drops below 5.0, which encourages Panama disease (fusarium wilt). Bananas require heavy fertilization for adequate yield—200 to 500 pounds nitrogen (N)/acre and up to 500 to 600 pounds potassium (K)/acre (less for lower-yielding plantains). Bananas are considered "heavy feeders" of nutrients among fruit crops. Harvested shoots are cut down and left on the soil surface as mulch.

Both banana and plantain are adapted to hot, wet, tropical lowlands. However, in some parts of the world and in home gardens, cultivation may extend their range to 5,000 feet above sea level. Mean annual temperature should be 80°F, and minimum and maximum daily temperatures should remain within 72 to 90°F. Growth ceases when temperatures drop below about 56°F, and temperatures below 50°F can cause chilling injury to fruits. Suboptimal temperature greatly retards growth; bananas produce about one leaf per week under optimal conditions, but only one leaf per month if cool weather prevails. Frost kills plants to the ground, although the rhizome usually survives.

Plants require about 4 inches of rain per month, with dry seasons no longer than 3 months. Irrigation is supplied in arid areas or during dry seasons in the humid tropics. High humidity and rain decrease water requirement but greatly exacerbate foliar disease. Wind is often a problem in banana culture. Due to the large leaves, moderate winds (20-40 mph) can tear leaf blades or break petioles. Gale force to minimal hurricane winds of 40 to 80 mph will uproot plants bearing heavy fruit stalks and often flatten entire plantations.

Propagation

New plantings are established using rhizomes or pieces of rhizomes called "bits" or "eyes" (analogous to planting potatoes). Large rhizomes allow for earlier and higher yields. Rhizomes are cleaned with knives or brushes and commonly disinfested preplant by hot water dip (130°F), nematicide, and/or fungicide solutions. Rhizomes for propagation are dug from plants a few months old but not yet flowering. Entire suckers from existing stools may be removed and planted as well. Suckers with their first mature-type leaves are the most desirable size for transplanting. Tissue-cultured plantlets are sometimes used to start new plantations since they are completely pest free. They must be planted first in containers, usually black polyethylene bags, to obtain the size adequate for field planting (Figure 5.6).

Rootstocks

No rootstocks exist for either banana or plantain.

Planting Design, Training, Pruning

Spacing between plants varies proportionally to cultivar height at maturity, but generally 400 to 800 plants/acre are used. Plantations are laid out in hexagonal or bedded designs, the latter having double or triple rows and then alleys for machine traffic (Figure 5.7). Common spacings range from 6 × 10 to 12 × 15 feet. Complete coverage of the soil surface with a banana canopy not only maximizes yield but also reduces weed competition.

Stools are allowed to produce only two shoots at a time: a larger one for fruiting and a smaller sucker that will produce fruits about 6 to 8 months after the main stem is harvested (Figure 5.4). The life of a banana plantation is 25 years or longer, during which time the individual stools or planting sites may move slightly from their original positions, as lateral rhizome formation dictates. Latin Americans sometimes comment that the plants are "walking" over time.

In terms of training and pruning, almost nothing is required due to the unique growth habit of the banana and plantain. As mentioned earlier, the rachis below the last marketable hand of fruits is pruned off at time of fruit set. Propping or guying helps support most cultivars with heavy fruit stalks. Poles are used as braces, or plants are tied off or tied together with twine. This

FIGURE 5.6. Micropropagated banana plants provide pest-free propagation material for starting new plantings. They are more expensive than rhizomes or "bits" and require a nursery, as shown.

FIGURE 5.7. A double-row bed system for banana cultivation in southern Guatemala. Two rows close together facilitates tying off, which is done at fruit set to prevent toppling. Labor and equipment move in alternate, wider rows.

is commonly done at the same time that plastic bags are placed over developing fruit bunches. The bags, or bunch covers, prevent fruit damage by wind scarring, and because they are impregnated with insecticide, they also are effective against insects. Leaves (or portions thereof) that are severely infected with sigatoka fungus or otherwise damaged are pruned off periodically.

Pest Problems

Insects/Nematodes

Nematodes. Compared to their prevalence in other fruit crops, insect pests common to banana and plantain are relatively few. Nematodes, however, are a more serious problem. At least 22 species of nematodes are known to attack banana roots. The burrowing nematode *(Radophilus similis)* is the most severe pest, causing "banana topple" disease, or the uprooting of plants (Figure 1.18). Lesion nematodes (*Pratylenchus* spp.), the spiral nematode *(Helicotylenchus multicinctus)*, and root-knot nematodes (*Meloidogyne* spp.) are widespread banana pests. Control is achieved by nematicides applied to rhizomes at planting, and repeat applications as the pseudostem elongates. Alternatively, hot water dips (125-135°F for 15-20 minutes) will kill nematodes on rhizomes. Cultural control is achieved by plowing and exposing soil to sunlight for several weeks or up to 3 years prior to planting. Also, planting pangola grass *(Digitaria decumbens)* reduces the population of most species of nematodes.

Black weevil or banana stalk borer. Cosmopolites sordidus, the second most important banana pest, is a snout beetle, about ½ inch long, that enters at the base of the pseudostem to tunnel into the rhizome. Feeding by these beetles debilitates plants, causing toppling or outright death. Control is achieved by injecting insecticide into the pseudostems or by applying it around the base of the stem. There are no resistant cultivars.

Diseases

Panama disease or fusarium wilt. The soilborne fungus *Fusarium oxysporum* f. sp. *cubense* was responsible for the abandonment of 'Gros Michel' as the major banana cultivar, in favor of the more resistant Cavendish types. The fungus enters the rhizome through fresh injuries, then moves into the pseudostem, causing leaf yellowing and drooping from the base toward the tip.

Forty years ago, the disease severely reduced production in Central America, Colombia, Taiwan, and the Canary Islands. Control is accomplished by resistant cultivars. For banana, the major export cultivars of the Cavendish group are resistant to all but race 4 of the pathogen. Race 4 currently is isolated to subtropical areas of Asia. 'Lacatan', 'Monte Cristo', 'Datil', and 'Nino' are resistant to all races of the pathogen. For plantain, 'Maricongo', 'Enano', and 'Pelipita' are resistant to all races of the pathogen. Flooding fields for 6 months can eliminate the pathogen, or the use of a cover crop (nonhost) may reduce the inoculum sufficiently.

Sigatoka and black sigatoka. Sigatoka, also called yellow sigatoka, was described near the turn of the twentieth century and has been a major problem worldwide. The fungus *Mycosphaerella musicola* causes leaf spotting and subsequent plant debilitation, resulting in low yields, poor quality, and uneven ripening. Control is achieved through oil and fungicide sprays applied at approximately 3-week intervals throughout the year.

Black sigatoka, also called black leaf streak, is much worse than yellow sigatoka and has replaced the latter as the predominant crop disease affecting banana and plantain production in many areas of the world (Plate 5.3). The more virulent pathogen *M. fijiensis* var. *difformis* kills leaves outright and exposes bunches to sunburn. Fungicide and oil sprays have been only marginally effective at control, even at the application rate of 40 times per year (or every 10-12 days). Newer, more effective fungicides (ergosterol inhibitors) combined with field scouting have been successful in reducing the number of sprays to 10 to 25 per year. Black sigatoka has been a particular problem in Central America and South America, and now also in the Pacific Islands. Resistant cultivars are currently lacking.

Moko disease. This disease, brought on by the bacterium *Pseudomonas solanacearum*, causes internal decay and death of plants. It is a severe problem in the Western Hemisphere, especially in Colombia. Since it is a wilt disease, some of its symptoms mimic those of fusarium wilt. Bacterial ooze may seep from the male bud of the inflorescence. Control is achieved through eradication of infected plants and prevention of spread to other plants. The bacteria are carried on pruning and harvesting tools, so these should be disinfected frequently to limit spread. Depending on the strain of the pathogen, eradication of plants in a 15- to 30-foot radius around the infected plants may be necessary.

Banana bunchy top virus. This is the worst viral disease of banana and is a particular problem in areas where expensive eradication measures cannot be employed. It occurs in Southeast Asia, Australia, and Africa but has not yet become a problem in the Americas. The virus may be a single-stranded DNA or RNA (or both) and is transmitted in infected planting stock and by the ubiquitous banana aphid, *Pentalonia nigronervosa.* Infected plants are easily spotted due to the characteristic symptoms for which the disease is named; leaves at the top become upright, smaller, with chlorotic and wavy margins, and they will appear bunched together due to internode shortening. Alternating bands of yellow and green appear on leaves in what are referred to as "dot-dash" or "Morse code" patterns. Control is accomplished by killing infected plants and those in the surrounding area with herbicide. Accurate serological tests (ELISA and DNA probes) can detect the virus in planting material, and stringent phytosanitary programs have helped to reduce bunchy top vurus incidence in some areas.

HARVEST, POSTHARVEST HANDLING

Maturity

Fruits can be harvested when about 75 percent mature, as angles are becoming less prominent and fruits on upper hands are light green in color. At this stage, desiccated styles on tips of fruits can be easily rubbed off. This occurs at around 75 to 80 days after opening of the first hand.

Some managers manipulate the harvest date as per the direction of the buyer, and harvest may be delayed up to 100 to 110 days after opening of the first hand. Bananas and plantains are classified as climacteric fruits with respect to ripening.

Harvest Method

Entire bunches are cut from pseudostems by hand and carried on the shoulder or back to a nearby tramline for long-distance transport (Figure 5.8). The cutter leaves a portion of bare stalk as a handle for transporting the bunches to the packinghouse.

Postharvest Handling

Banana bunches are hung on tramways and pulled out of plantings manually or by machine; this expedites the process and limits handling of the fruits. Bananas should be kept out of light after harvest, since this hastens ripening and softening. For local consumption, hands are often left on stalks and sold to vendors, who then cut off hands/fingers per the customer's order. For export, hands are cut into units of 4 to 10 fingers, graded for both length and width, and carefully placed in plastic-lined 40-pound boxes (Plate 1.11; Figure 5.9). Prior to packaging, fruits are sometimes floated in water or dilute sodium hypochlorite solution to remove latex, which may cause black peel staining. Fruits are shipped by boat when green and ripened by exposure to ethylene gas (1,000 ppm for 24 hours) at their destination, in sealed "banana ripening rooms."

FIGURE 5.8. A harvest crew hangs a banana stalk on a tram wire and inserts padding between hands to prevent bruising in transit.

FIGURE 5.9. Bananas are measured to the nearest thirty-second of an inch as they are graded for export at the packinghouse. *(left)* This American buyer required a minimum width of 40 thirty-seconds of an inch (or 1¼ inches), measured with the device shown. *(right)* Bananas are packed green in 40-pound boxes for export and ripened by ethylene exposure at their destination.

Storage

Fruits can be stored for 2 to 4 weeks after harvest at temperatures no lower than 55°F, since fruits are susceptible to chilling injury (Plate 1.12). Postharvest diseases, particularly crown or finger rot (caused by *Botryodiplodia theobromae* or *Colletotrichum* spp.), cause economic losses occasionally but are controlled by packinghouse sanitation of fruit. Short transit times to market also lessen the impact of the diseases. Fruit freckle (caused by *Guignardia musae*) can cause unsightly skin spotting that results in unmarketable fruits, although it does not affect the pulp.

CONTRIBUTION TO DIET

In addition to fresh consumption, bananas and plantains are used in numerous ways. Banana puree is made into baby food and ice cream, as well as baked desserts. Flour can be derived from the dried fruits and used for pastries or mixed with other flours. Dried fruits of both banana and plantain are commonly made into chips by frying slices in oil and then salting them. In many countries, larger slices are deep-fried and eaten like french fries. Another dish, mofongo, combines fried green plantains with seasoned pork. As with most fruits, the fermented juices are made into beer and wine, most commonly in Africa. The young leaves and terminal inflorescence buds are also edible.

Per capita consumption of bananas is 25.8 pounds per year, higher than that for most other fruit crops. Consumption has risen about 24 percent since 1980.

Dietary value per 100-gram edible portion:

	Banana	Plantain
Water (%)	75	65
Calories	89	
Protein (%)	1.1	1.3
Fat (%)	0.3	0.4
Carbohydrates (%)	22.8	31.9
Crude Fiber (%)	2.6	2.3
	% of U.S. RDA (2,000-calorie diet)	
Vitamin A	1	22
Thiamin, B_1	2	4
Riboflavin, B_2	4	3
Niacin	3	3
Vitamin C	14	31
Calcium	<1	<1
Phosphorus	3	5
Iron	1	3
Sodium	<1	<1
Potassium	10	14

BIBLIOGRAPHY

Duke, J.A. and J.L. duCellier. 1993. *CRC handbook of alternative cash crops*. Boca Raton, FL: CRC Press.
Gowen, S. (ed.). 1995. *Bananas and plantains*. London: Chapman and Hall.
Lessard, W.O. 1992. *The complete book of bananas*. W.O. Lessard.
Morton, J.F. 1987. *Fruits of warm climates*. Miami, FL: Julia F. Morton.

Nakasone, H.Y. and R.E. Paull. 1998. *Tropical fruits.* Wallingford, UK: CAB International.
Ploetz, R.C., G.A. Zentmyer, W.T. Nishijima, K.G. Rohrbach, and H.D. Ohr (eds.). 1994. *Compendium of tropical fruit diseases.* St. Paul, MN: American Phytopathological Society Press.
Robinson, J.C. 1996. *Bananas and plantains.* Wallingford, UK: CAB International.
Simmonds, N.W. and K. Shepherd 1955. The taxonomy and origins of the cultivated bananas. *Journal of the Linnaean Society, London, Botany* 55:302-312.
Stover, R.H. and N.W. Simmonds. 1987. *Bananas,* Third edition. Essex, UK: Longman Science and Technical.
Sturrock, D. 1959. *Fruits for southern Florida.* Stuart, FL: Southeastern Printing Co.
Von Loesecke, H.W. 1950. *Bananas,* Second revised edition. New York: Interscience Publ., Inc.

Chapter 6

Blackberry and Raspberry (*Rubus* spp.)

TAXONOMY

Blackberries and raspberries, often termed "brambles," are a diverse group of species and hybrids in the genus *Rubus*. They are members of the Rosaceae family, closely related to strawberry in the subfamily Rosoideae. *Rubus* is one of the most diverse genera of flowering plants in the world, consisting of 12 subgenera, some with hundreds of species. The geographic distribution ranges from the Arctic Circle (Arctic berry) to the tropics (Mysore raspberry), on every continent except Antarctica. Three subgenera contain edible fruiting plants:

1. *Eubatus*—Blackberries and dewberries (hundreds of species). Dewberries are basically smaller, trailing, low-chill blackberries native mostly to the southeastern United States.
2. *Idaeobatus*—Red and black raspberries (~200 species).
3. *Cylactis*—Arctic berries (*R. arcticus* and *R. stellatus*). The few named cultivars that exist are interspecific hybrids between these two species. They are grown on a limited scale in northern Sweden and Finland, where an aromatic liquor is made from the berries.

Blackberries (*Eubatus*) as a group are far more complex than are raspberries in terms of growth habit, number of species cultivated, and genetic background. Blackberries have three growth forms: erect, semierect, or trailing (Figure 6.1). The semierect types are often hybrids of the erect and trailing. Red raspberries are either erect or trailing, and black raspberries are strictly erect in growth form. Erect blackberries were probably derived from such species as *Rubus allegheniensis*, *Rubus argutus*, and *Rubus frondosus*. This is a fairly cold-hardy, vigorous, thorny group. Many trailing cultivars are grown commercially in Oregon and Washington, and they are derived from species native to that area—*Rubus ursinus*, *Rubus macropetalous*, and *Rubus loganobaccus*, to name a few. They are generally less cold hardy than the erect species.

Another major distinction in blackberries is thorny versus thornless cultivars. The thornless condition is most desirable, since blackberry thorns are larger and more stiff than those of raspberry. Thornless types have been developed for both trailing and erect growth forms. Many thornless cultivars were selected as chance mutations that resulted in lack of thorns; the genes for making thorns were deleted or turned off in the outermost layer of stem tissue, and these "one in a million" plants were propagated by stem cuttings or tip layers to maintain the thornless condition. However, the genes for thorns are still carried in internal tissues, yielding a partial mutation known as a periclinal chimera. Thus, if these cultivars are allowed to produce root suckers or are propagated by root cuttings, they turn into thorny shrubs (e.g., 'Thornless Evergreen', 'Thornless Logan') (Figure 6.2) because new shoots arise from inner, not outer, tissues of the roots, where the thorn genes persist. Some newer thornless cultivars were developed by breeding a thornless European cultivar with those native to North America, and these are not chimeras, but true-breeding, completely thornless types (e.g., 'Black Satin', 'Hull Thornless', 'Smoothstem').

Introduction to Fruit Crops

doi:10.1300/5547_06

FIGURE 6.1. Extremes of growth habits in *Rubus:* erect *(right)* and trailing *(left)*. For trailing types, floricanes are trained to a trellis and primocanes are left to trail on the ground. Semierect types *(not shown)* are intermediate and often grown on a trellis.

FIGURE 6.2. Thorny and thornless canes from the same 'Thornless Boysenberry' plant. The cultivar is a periclinal chimera of the original 'Boysenberry' and retains genes for thorniness in the roots. The top cane arose from the roots, and the bottom cane from the crown.

Three species are recognized as having the greatest horticultural importance:

1. *Blackberry*—*Rubus* spp. is the best approximation to a scientific name, considering that blackberry may be the most taxonomically complex of any fruit crop. *Rubus ursinus* Cham. & Schlect. is native to the Pacific Northwest and belongs to the *Ursini* section of *Eubatus*. It has been useful in producing commercial cultivars grown in that region. *Rubus ursinus* is sort of a take-all designation, often encompassing blackberry cultivars derived from several species, even when *R. ursinus* (proper) was not a parent(!). To add to the confusion, *R. ursinus* is thought to be an ecotype of a single aggregate species that also includes *R. macropetalous* and *R. vitifolius*. In Europe, six species in the *Moriferi* section have been used to produce cultivars of local importance; they are referred to as the aggregate species *R. fruticosus* L. agg. In North America, seven species of the *Moriferi* section are involved in eastern North American wild types and cultivars, but, again, none is

R. ursinus. Dewberries may belong to *R. trivialis*, *R. canadensis*, *R. hispidus*, or *R. baileyanus*, depending on location and source. *Rubus trivialis* is native in the southeastern United States. *Rubus glaucus* (Andean blackberry) is a South American species that has formed the basis of an out-of-season fruit industry centered in Bogota, Colombia. This species produces large, cosmetically appealing fruit year-round, unlike other *Rubus* species. It is thought to be a natural hybrid between black raspberry and *R. bogotensis*.
2. *Red raspberry—R. idaeus* L. The European subspecies of this group is designated *R. idaeus* ssp. *vulgatus* Arrhen., whereas the North American red raspberry is termed *R. idaeus* ssp. *strigosus* Michx. or, more simply, *R. idaeus* (European) and *R. strigosus* (North American).
3. *Black raspberry—R. occidentalis* L. is fairly straightforward, being a good species of its own, with a range which overlaps that of *R. strigosus*, but extending further to the south.

The distinction between blackberries and raspberries revolves around fruit characteristics. All bramble fruits are aggregate fruits, which means they are formed by the aggregation of several smaller fruits, called drupelets in *Rubus*. The drupelets are all attached to a structure called the receptacle, which is the fibrous central core of the fruit. In raspberries, the receptacle remains with the plant when fruits are picked, creating the hollow appearance of the harvested fruits. In blackberries, the drupelets remain attached to the receptacle, which comes off with the fruits when picked (Plate 6.1). Hence, since they contain some receptacle tissue in the center, blackberries have a bit more crunch and dietary fiber than raspberries. A second distinction: raspberry drupelets have fine hairs and adhere to one another, whereas blackberry drupelets are hairless and smooth.

Red raspberries differ from black raspberries by being more cold tolerant, higher yielding, less disease prone, and more suited to commercial markets than black raspberries, which are mostly grown in home gardens or on small retail farms. Purple raspberries are hybrids between red and black raspberries and thus retain many of the characteristics of black raspberries. Because fruit quality is inferior in purple raspberries, there is no commercial production, but a fair amount of backyard production.

Some of the most important commercially grown brambles are actually blackberry–red raspberry hybrids. Examples include 'Boysenberry', 'Loganberry', and 'Youngberry'. The fruit flavor is unique, but culture and management for these hybrids resemble the practices for blackberry more closely than for raspberry. Major *Rubus* hybrids include the following:

- *'Loganberry'*—A cross between *R. vitifolius* and *R. idaeus*, this blackberry-raspberry hybrid more closely resembles a blackberry. The plant bears large, reddish-purple fruits, has a low yield, and exhibits vigorous growth. This hybrid was first produced by Judge J. H. Logan of Santa Cruz, California, in 1883, most likely through crossbreeding between 'Aughinbaugh' blackberry and 'Red Antwerp' red raspberry, as these were two of the three *Rubus* cultivars planted in his yard that year.
- *'Youngberry'*—A cross between 'Phenomenal berry' and 'Austin Mayes' dewberry, this hybrid is sweeter than loganberry but cold sensitive.
- *'Phenomenal berry'*—This second-generation cross between 'Aughinbaugh' blackberry and 'Cuthbert' raspberry, produced by Luther Burbank in 1905, is also called 'Burbank's Logan'. As the latter name suggests, it is similar to 'Loganberry' but with larger, lighter-colored fruit.
- *'Boysenberry'*—This cross between loganberry and dewberry was discovered by Ralf Boysen in California in 1920. It is preferred over 'Youngberry' and 'Loganberry' for its larger, earlier, more acid-tasting fruit.

- *'Olallieberry'*—This hybrid is a cross between 'Black Logan' and 'Youngberry'. It is one of the parents of 'Marion', which is the most widely grown blackberry hybrid in the Pacific Northwest today.
- *'Tayberry'*—This hybrid was produced by a cross between 'Aurora' blackberry and an unnamed tetraploid raspberry selected at Hort Research Institute in Scotland. It is longer, 50 percent larger, less downy, and brighter in appearance than 'Loganberry'.

Cultivars

Blackberries

Cultivars are classified by growth habit—trailing (prostrate), erect, or semierect—and also as thorny or thornless. In the Pacific Northwest, trailing types, such as 'Thornless Evergreen', 'Marion', and 'Kotata', cover 98 percent of acreage. 'Marion' accounts for 50 percent of the acreage; growers in Oregon call them "Marionberries," not blackberries. 'Thornless Evergreen' is a thornless mutant of the European species *R. laciniatus*. The hybrids 'Boysenberry', 'Loganberry', 'Olallieberry', and 'Youngberry' are grown in significant quantities in California and the Pacific Northwest. In the southeastern United States, thorny, erect types dominate. Thorny types include 'Cherokee', 'Comanche', 'Cheyenne'. 'Shawnee', and 'Choctaw' (all 'Brazos' × 'Darrow' crosses). Thornless cultivars include 'Hull', 'Chester', 'Navaho', and 'Arapaho'.

Red Raspberries

In the Pacific Northwest, 'Meeker' is grown on about 60 percent of the acreage. 'Willamette' is grown on 30 percent, and a few others are grown on the remaining 10 percent. The popularity of the leading cultivars stems from ease of machine harvest and their relatively high yield potential. As with blackberry, cultivars are distinguished by growth habit, either trailing or erect. Primocane- or autumn-fruiting raspberries are a unique group in that they can produce fruits in their first year and up to two crops per year (see Botanical Description section). 'Heritage', 'Amity', 'Redwing', and 'Autumn Bliss' are the major primocane-fruiting types.

Black Raspberries

Few cultivars of this species are grown. 'Munger' is grown on greater than 90 percent of the acreage in Oregon. 'Allen', 'Blackhawk', 'Bristol', and 'Jewel' are grown on the remaining commercial acreage and are also grown by home gardeners. 'Royalty' and 'Brandywine', crosses between black and red raspberries, are purple raspberries planted mainly by home gardeners.

ORIGIN, HISTORY OF CULTIVATION

Blackberry

Blackberries are native to Asia, Europe, North America, Australia, Africa, and South America and have the most widespread geographic origin of any fruit crop. It follows that blackberries grown in specific regions are largely derived from species indigenous to those regions, and no single species dominates world production. Blackberries have been used in Europe for over 2,000 years; they are consumed as food, employed for medicinal purposes, and planted in hedgerows to keep out intruders. In the United States, *R. allegheniensis, R. argutus, R. cunei-*

folius, and *R. canadensis* have been important in developing "northern" blackberry cultivars, including thornless types (cultivars popular in the western United States also). In the southeastern United States, *R. trivialis* has been used to confer low-chill tolerance and disease resistance in such cultivars as 'Brazos'. *Rubus lacinatus* ("cut leaf" or "evergreen"), the first domesticated species in Europe, was imported into the Pacific Northwest in 1860, where it produced one of the main cultivars for that region, 'Thornless Evergreen'. *Rubus ursinus* is native to the Pacific Northwest and has been important in the development of trailing cultivars currently grown in that region.

Red Raspberry

Red raspberries are indigenous to Asia, Europe, and North America (see Taxonomy section), although the specific epithet *(idaeus)* denotes Mount Ida, in the Caucasus Mountains of Eastern Europe. Fruits were gathered from the wild by the people of Troy around the time of Christ. Records of domestication appear in the fourth-century writings of Palladius, a Roman agriculturist; add to this the seeds discovered at ancient Roman forts in Britain, and it seems probable that the Romans spread red raspberry cultivation throughout Europe. The British popularized and improved raspberry cultivation throughout the Middle Ages and were exporting the plants to New York by 1771. Throughout the 1800s, red raspberries of European origin were grown in North America more extensively than were varieties derived from the native *(R. strigosus)* red raspberry. However, the superior hardiness of *R. strigosus* was recognized over time, and its selections or hybrids with European types (e.g., 'Brandywine', 'Marlboro') became more popular toward 1900. A successful breeding program in Washington in the 1950s and 1960s produced several cultivars that form the basis of the Pacific Northwest industry today.

Black Raspberry

Black raspberries are indigenous to only North America, where they are most abundant in the East, exclusive of the Gulf states, but also found in the West, along with the related species *R. leucodermis*. Domestication appears to have been delayed until the 1800s, due to the popularity of red raspberries and an abundant supply of wild fruits. In 1850, H. H. Doolittle discovered tip layerage as an efficient propagation method and released 'Doolittle', the first such cultivar. A rival of Doolittle's released his own version, 'Doomore', shortly thereafter, but both were superseded by 'Ohio Everbearer'. By 1880, at least 17 named cultivars existed, and several thousand acres of land were under cultivation in New York. Today, black raspberries are the least important of the brambles, enjoying very little commercial production compared to blackberries or red raspberries. They make fine garden plants in the mid-Atlantic region of the eastern United States, the Midwest, and the Pacific Northwest.

FOLKLORE, MEDICINAL PROPERTIES, NONFOOD USAGE

The word *bramble* means "envy" in the "language of flowers"; blackberries and dewberries are symbolic of remorse.

On both sides of the Atlantic Ocean, brambles were used medicinally hundreds of years ago. European blackberry juice was used to treat infections of the mouth and eyes until the sixteenth century. In the Pacific Northwest, the powdered bark of salmonberry *(R. spectabilis)* was used for toothache relief. A tea made from the leaves of *Rubus macropetalus* in western Washington is said to aid digestion. Thimbleberry *(R. odoratus)* and blackberry are being studied to determine the presence of certain tannins that may be useful in developing anticancer drugs; the roots and stems of these plants are peeled and boiled to produce a liquid that will arrest vomiting upon drinking it. Blackberry and raspberry root decoctions are used to remedy dysentery.

Blackberries and raspberries contain relatively high quantities of ellagic acid, which has a wide range of documented functions in humans: anticarcinogen/antimutagen, inhibition of HIV binding to cells, inhibition of blood clotting, and free radical scavenging. See Chapter 29, "Strawberry," for more information on ellagic acid.

PRODUCTION

World

Total production in 2004, according to FAO statistics, was 389,061 MT or 856 million pounds. *The world production figures here are for raspberries only,* as the FAO does not keep separate statistics on blackberries. Raspberries are produced in 38 countries worldwide, on about 195,000 acres. Production has increased 28 percent in the past decade, largely due to a 24 percent increase in acreage, as yield has remained relatively constant, at about 4,400 pounds/acre, ranging from less than 1,500 to about 9,000 pounds/acre.

Blackberries are lumped with other similar fruits, such as mulberries (*Morus* spp.), myrtle berries (*Myrtus* spp.), and huckleberries (*Vaccinium* or *Gaylussacia* spp.). Assuming that blackberries account for half of the fruits in the FAO's supercategory of "Berries NES," a crude approximation of total production would be about 337,000 MT (741 million pounds) of blackberries worldwide.

The top ten raspberry-producing countries (percent of world production) follow:

1. Russia (28)
2. Serbia and Montenegro (20)
3. United States (13)
4. Poland (11)
5. Germany (5)
6. Ukraine (5)
7. Canada (3)
8. Hungary (3)
9. United Kingdom (2)
10. France (2)

United States

Total production in 2004, according to USDA statistics, was 85,454 MT or 188 million pounds. Virtually all of the commercial production of brambles occurs in Oregon, California, and Washington. The total value of the industry is about $236 million/year, produced on just under 22,000 acres. The industry value has doubled within the past 4 years, largely due to higher prices received by growers. Prices range from $0.59 per pound for processed fruit, up to about $1.50 per pound for fresh fruit, although black raspberry prices were over $2.00 per pound in 2004, regardless of utilization. Generally, 90 percent or more of the crop is processed. Table 6.1 presents a detailed breakdown by commodity.

Foreign trade data are unavailable. Small quantities of fresh berries are produced in the off-season and imported from Chile and other Latin American countries. In the tropics, berries are grown at elevations of 4,000 to 7,000 feet for export to North America in winter.

TABLE 6.1. Production data for blackberries and raspberries in the United States.

Crop	Production (million pounds)	Yield (pounds/ acre)	Area (acres)	Value (million $)	Leading states
Blackberry and hybrids (all)	52.3	—	7,310	39.6	1. Oregon
Marion	28.3	6,580	4,300	21.1	2. California
Evergreen	5.8	6,440	900	3.6	
Boysen	5.2	5,470	950	6.0	
Logan	0.2	2,830	60	0.1	
Other	12.8	11,600	1,100	8.8	
Red raspberry	133	9,520	14,100	191	1. California 2. Washington
Black raspberry	2.2	2,200	1,000	5.0	1. Oregon

Source: USDA statistics, 2004.

BOTANICAL DESCRIPTION

Plant

Blackberries and raspberries are erect, semierect, or trailing, generally thorny shrubs that produce renewal shoots called "canes" from the ground. These perennial plants are composed of biennial canes that grow vegetatively for one year, initiate flower buds in late summer, fruit the following summer, and then die (Plate 6.2). The first-year canes are called "primocanes," and the second-year flowering canes are called "floricanes." Raspberry thorns are finer and more flexible than the thorns on blackberries. Within raspberries, the black and purple varieties have more-prominent thorns than the red variety. Leaves are palmately compound with three to five leaflets, the middle one being the largest. Leaf margins are finely serrate.

Red raspberries are unique in that some primocane-fruiting (syn. autumn-fruiting) cultivars produce fruits in their first summer, which allows the crop to be treated more or less as an annual. The upper portion of the primocane produces fruits and dies back, leaving lower buds to overwinter and produce fruits the following season. Primocane-fruiting types can produce fruits in both spring and autumn. Currently, blackberry germplasm with the primocane-fruiting trait is being developed by breeders in the southeastern United States, suggesting that blackberry culture may become greatly simplified in the future.

Flower

White to pink flowers (1-inch diameter) are borne terminally on several-flowered racemes, cymes, or corymbose inflorescences on the current season's growth (Plate 6.3). For dewberries, raspberries, and some wild blackberries, inflorescences are cymose, and some flowers are borne individually in axils of leaves on fruiting laterals. Blackberry flowers generally have larger petals than raspberry flowers. The flowers are initiated in late summer in biennial types, and early to midsummer in primocane-fruiting types. The gynoecium consists of 60 to 100 ovaries, each of which develops into a drupelet, and there are 60 to 90 stamens, five sepals, and five petals. All species produce copious nectar and attract bees.

Pollination

Most cultivars of blackberries, black raspberries, and red raspberries are self-fruitful and do not require pollinizers. However, dewberries are self-incompatible and must be interplanted for good fruit set. Honeybees are naturally attracted to brambles, and wind also aids pollination.

Fruit

In all brambles, the fruit is an aggregate of drupelets, which are simply small drupes, so each tiny bump has the same structure as a plum, cherry, or apricot—a thin skin with flesh beneath and a pit in the center containing the seed. The drupelets of blackberries lack the fine hairs on the fruit surface that raspberries have and thus appear shiny and smooth. Blackberries retain the receptacle within the fruits at harvest, whereas raspberry fruits detach from the receptacle and contain no nonovarian tissue.

Fruiting begins in the second year of the planting and can continue for more than 10 years if properly managed. Primocane- or autumn-fruiting types produce two crops per year if not mowed. Fruit development occurs rapidly, taking only 30 to 50 days for most raspberries and 40 to 70 days for blackberries.

GENERAL CULTURE

Soils and Climate

Brambles are suited to a wide range of soil types, from sandy to clay loams, as long as they are provided with good drainage and a pH of 5 to 7. Depth is not important because plants develop shallow root systems, although blackberries generally are more deeply rooted, and therefore more drought tolerant, than are raspberries. Irrigation is helpful in most climates, since rooting depth tends to be shallow, and yield is increased with supplemental water. Brambles are susceptible to soilborne diseases and several viruses. Soils previously planted to solanaceous crops (pepper, tomato, eggplant, potato) in the past 5 years are avoided, since these plants can harbor a fungus that kills raspberry roots.

Climatically, blackberries have a more warm-temperate adaptation than do raspberries. Black and red raspberries generally require cooler summers than do blackberries, and they are poorly adapted to the southern United States and to hot, arid climates. Midwinter coldhardiness can be a great limitation in growing blackberries in the northern United States, and growth is also limited in areas outside the moderate, maritime climates of Oregon and Washington. Tender cultivars can die at temperatures ranging from 0 to 10°F, and the hardiest will tolerate temperatures only as low as about –10°F. Erect cultivars have greater hardiness than the trailing, and thorny cultivars are generally more hardy than the thornless. New York represents about the northern limit to blackberry cultivation in the eastern United States. Trailing types can be cultivated further north if the dormant primocanes are covered with heavy mulch during the winter to protect against freezing. Raspberries are more cold hardy than are blackberries, with the red being hardier than the black or purple raspberries. Red raspberry plants can tolerate temperatures to –20°F, whereas black and purple raspberries are injured at temperatures between –5 and –10°F. Frost damage is generally not a problem for brambles because flowering occurs terminally on the current season's growth, and later in spring than for most other fruit crops. Emerging buds are more cold hardy than open flowers. Blackberries generally have lower chilling requirements, ranging from 200 to 800 hours, to break dormancy than do raspberries, which have chilling requirements ranging from 800 to 1,600 hours. Heat stress on ripening fruits from direct exposure to sunlight can cause sunscald; in affected plants, one to several drupelets lose color and shrivel.

Propagation

Unlike many fruit crops, brambles are easily propagated by a number of techniques. In fact, most self-propagate quite well and can be invasive or even weedy. Certain types lend themselves to one or two main propagation techniques, as described in the following paragraphs.

Root Cuttings

Using root cuttings in propagation is the fastest and most economical method to start a new planting. The basic procedure is to cut roughly pencil-size roots into pieces 4 to 6 inches long and then plant them 2 to 3 inches deep, about 1 to 2 feet apart, in rows. Root cuttings are dug in January/February and planted in March/April. Some species have up to 20 shoot buds per inch of root, but not all of them will grow into canes.

Some notable exceptions to successful root-cutting propagation are the trailing blackberries as well as black and purple raspberries, which do not produce adventitious (or enough) buds on roots and therefore cannot be propagated this way. They are tip-layered. Thornless mutants of trailing cultivars ('Thornless Evergreen', 'Thornless Logan', and 'Thornless Youngberry') are periclinal chimeras that will produce thorny shoots if propagated by root cuttings because the genetic mutation for thornlessness does not exist in adventitious buds on roots. In some of these mutants, however, the chimera has ingressed into the inner tissues, making root cuttings possible. Thornless, erect blackberries can be propagated by root cuttings because they are produced through crossbreeding with thornless species and are not partial mutations.

Suckering

Propagating through suckering is basically the same as using root cuttings, except the adventitious buds on roots have already produced canes. Suckers are detached from the parent plant, leaving some roots, and then transplanted. This method shares many of the same drawbacks found with root-cutting propagation.

Tip Layering

This technique is used for black raspberries, hybrids, trailing types, and thornless mutants in lieu of root cuttings. Most brambles have a natural tendency to root where shoot tips from long canes bend down and touch the ground. This can be facilitated by propagators. In late summer, shoot tips on long primocanes are buried in shallow holes. Lateral roots form on the buried stems, and the following spring, the layered plants are removed from the original cane, dug, and transplanted (Figure 6.3).

Leafy Stem Cuttings

This method can be used for any bramble as they are easily rooted. Semihardwood cuttings containing one to a few leaves are removed in spring to midsummer and placed in well-drained media. Intermittent mist is provided to prevent drying out, and rooting hormone may be used to improve rooting success. Cuttings root in a 2 to 3 weeks.

FIGURE 6.3. Tip layering in *Rubus*. The tip of a cane rooted and produced two shoots where it touched the ground; the original cane has died back.

Tissue Culture

This technique is used to only a limited extent, especially when virus-free material is difficult to obtain by other means. The expense of tissue culture plant propagation holds pace with the cost of buying individual plants, but it is greater than the costs associated with propagation through root cuttings or suckers. Another drawback is that plants are generally tiny (1 inch tall) when produced from tissue culture and therefore require close supervision to ensure their successful establishment. It is often used by nurseries to ensure disease-free conditions of stock plants.

Rootstocks

No rootstocks exist for blackberries and raspberries.

Planting Design, Training, Pruning

Most erect *Rubus* species are grown in hedgerows, spaced 8 to 16 feet apart (commonly 12 feet), to allow room for maneuvering equipment along the rows (see Figure 6.1). Plantings are often started by root cuttings or layers planted at 2- to 3-foot intervals. Plant density has little influence on yield after the first season because hedgerows quickly fill in space between the original plants. Hedges are kept to heights of 4 to 6 feet, with widths of 30 inches on top, and canes arising more than 1 foot from the center of the row are removed to facilitate mechanical-harvesting equipment. Mechanical harvesters shake fruits from plants and catch them as they fall; wide hedgerow bases allow more fruits to fall to the ground than do narrow ones. Hand-harvested hedges are allowed to spread to 3 feet in width; wider growth prevents picking from the center of the hedge.

Trailing types must be trained on a trellis, usually two or three wires spaced 18 inches apart, with the top at 5 to 6 feet. Primocanes are trained by spiraling them around the wires and spacing them as evenly as possible. An alternative practice is to bundle them and tie them to the trellis at regular intervals (Figure 6.4). This is done after harvest or in winter, when dead fruiting canes are pruned out.

An innovative trellis system, developed by researchers at Virginia Tech University in the 1980s, and known as the shift trellis, capitalizes on the negative geotropic growth habit of the

FIGURE 6.4. *(left)* A simple trellis system where red raspberry canes are bundled and tied to a trellis wire. This photo was taken in late summer after harvest when canes were dying back. *(right)* Primocane raspberries are trained along strings tied to a trellis wire since the weight of the fruits produced terminally cannot be supported by the cane itself.

fruiting laterals of brambles (i.e., lateral buds on floricanes grow toward the sky regardless of cane orientation). Canes are tied to the trellis, which is placed parallel to the ground during flowering. After fruit set, the trellis is pivoted to the vertical position, so the fruits can be easily picked by moving down only one side of the trellis row.

Alternate-year production is a common method of simplifying bramble culture. Canes are pruned to the ground every other year and produce a crop in alternate years, which amounts to about 85 percent of a single season's crop produced by traditional methods. However, costs of production are substantially reduced, and disease/insect buildups do not occur. In areas with a long growing season, such as the southeast, it may be possible to have some yield every year on early cultivars that are pruned to 12 inches directly after harvest.

In primocane raspberry culture, as mentioned earlier, primocane- or autumn-fruiting raspberries produce fruits terminally on the current season's canes. This allows growers to prune all canes to the ground each year without a yield penalty, greatly reducing disease and insect problems and simplifying cultural practices (Plate 6.4). All that is needed is a simple support system, as canes top-heavy with fruit tend to bend over.

For erect cultivars not cultivated via alternate-year production, three main pruning practices are required each year: (1) topping, (2) dead floricane removal, and (3) primocane thinning. Summer topping induces branching on vigorous, long-growing primocanes and maintains proper height for harvest. Blackberries, black raspberries, and purple raspberries are generally more vigorous and need to be topped to induce lateral branching in midsummer. Canes are topped at a height of 3 to 5 feet, and laterals induced by topping are shortened to 12 to 18 inches during winter-pruning operations. Floricanes die after fruit are harvested, so dead floricane removal is done to ease cultural operations and reduce the possibility of disease and insect buildup, since pests may overwinter in dead canes. Last, cane thinning is required because more primocanes often are produced each year than are needed. Canes are thinned to about five per linear foot of row to stop overcrowding. The first canes to appear in summer are often removed as a head start on primocane thinning. Thorny canes arising on thornless cultivars as root suckers are selectively removed to maintain the thornless condition.

Trailing and some semierect cultivars require a two- or three-wire trellis for support. Their basic pruning needs differ only in the lack of need for cane topping. Primocanes may grow over 10 feet in 1 year, but they grow along the ground and do not interfere with floricane growth. They can be shortened to 6 to 8 feet during the winter, when floricane removal and cane thinning are practiced. Floricane removal is actually facilitated by the trailing habit, which allows for easy pruning of the floricanes without becoming entangled in or damaging the primocanes. Thus, the winter pruning/training operation involves these steps: (1) dead floricane removal, (2) selection of eight to ten vigorous primocanes canes per plant, (3) shortening primocanes to 6 to 8 feet in length, and (4) bundling and/or tying them to the trellis.

Pest Problems

Insects

Raspberry crown borer. Pennisetia marginata is a widespread and destructive insect, and likely one of the only insects that requires control in bramble plantings. The adult is a moth that lays eggs on leaf undersides in late summer; the resulting larvae migrate to the base of canes to overwinter at the soil line. In the spring, they bore into the canes, causing wilting, poor vigor, and laying over or breaking of canes. Symptoms appear when the fruit is about half grown. Removal of alternate hosts, such as wild brambles, will reduce infestations. Mowing or pruning may not completely eliminate the insect population because the larvae are found on or near the soil surface. Insecticide drenches in early spring may be the only recourse if infestation is severe.

Rednecked cane borer. Agrilus ruficollis is a serious pest of brambles in the eastern United States. Similar to the crown borer, this insect lays eggs that hatch into larvae which then burrow into the cane. Larvae feed on and girdle primocanes and cause galls to form, usually near ground level. Infested canes are unproductive and may die. Pruning out galled canes is the most effective means of control, but the insect has a number of natural enemies that limit populations in some cases. Insecticide can be sprayed onto the bases of the canes at first bloom and perhaps one to three additional times afterward for severe infestations.

Strawberry bud weevil (clipper). Anthonomus signatus is a dark red-brown weevil or "snout beetle," about ⅒ inch long, that is primarily a pest of strawberry, but also a threat to blackberry and raspberry, and even blueberry. This insect lays its eggs in a developing flower and then girdles the flower stem so that the flower appears broken, wilted, and dangling. The larvae feed in the wilted flower bud. The insect overwinters in the dead leaves and mulch beneath the plants. Late-blooming brambles are damaged less because they develop after the egg-laying cycle. Postbloom insecticides on early blooming types may be necessary when one weevil per 40 feet of row or one dangling flower bud per 2 feet of row is seen.

Mites (Tetranychus spp., *Phyllocoptes gracilis).* Spider mites feed on the lower leaf surface and cause stippling of leaves. The most common mites are the two-spotted ones, and these mites proliferate in hot dry weather. Raspberry bud mite and redberry mite attack buds and developing fruits of brambles, causing misshapen, oddly colored, or abnormal fruits. A dormant oil spray before budbreak, combined with sulfur or lime sulfur if infestation has been severe, should control most mites. Pruning of canes after harvest and a shift to alternate-year production reduces the need for sprays. Red raspberries are the most susceptible of the brambles.

Diseases

Anthracnose. This widespread fungal disease, caused by *Elsinoe veneta,* must be controlled in humid climates for profitable production. The disease causes the most damage on trailing blackberries and on black and purple raspberries; it causes the least severe damage on red raspberries. Symptoms appear on canes, leaves, and fruits, but cane damage is the greatest threat. Infected canes develop sunken areas or spots, which may enlarge and girdle the canes. Partially girdled canes are low in vigor and yield. Defoliation can result from severe leaf infections, the symptoms of which begin as small spots that later become shot holes. Fruit infections cause the greatest losses on trailing blackberries, resulting in reduced fruit size and a woody texture to the fruit. Control is accomplished mostly through sanitation of floricanes after harvest, air circulation within the planting, reduction of wild brambles from around the planting, and, if severe, early season fungicide applications to reduce overwintering inoculum. Fungicides are applied as plants come out of dormancy, measured from when a green tip is seen on buds until they are about a half-inch long. If weather is wet during the growing season, additional applications throughout the spring may be required. Alternate-year production reduces the problem. 'Cherokee', 'Cheyenne', and 'Shawnee' are moderately resistant blackberries; 'Wilamette', 'Heritage', and 'Autumn Bliss' are resistant red raspberries. Most thornless blackberry cultivars are resistant as well.

Rosette or double blossom. Cercosporella rubi causes this serious disease of blackberries, which occurs less often on red or black raspberries. This disease poses a problem along an extensive range, from Maryland south to Florida and west to Texas in the eastern United States. Symptoms are visually striking, due to the fungus's growth regulator–type effect, changing the entire appearance of the plant. Numerous, leafy sprouts ("witches'-brooms") develop small, distorted, bronzed leaves at budbreak. Flowers may be green and leaflike or abnormally shaped; fruits will not develop from these flowers, but the flowers do serve to spread the disease, as insects visiting the infected flowers will leave covered with whitish spore masses that are deposited on contact with other flowers. The fungus grows between bud scales and surrounds the young tissues within buds the year prior to symptom development, which makes spray control of these insects difficult. Control is best accomplished by sanitation of infected shoots and dead canes, removal of wild plants from nearby plantings, and cultivation of resistant cultivars (e.g., 'Gem', 'Humble'). Isolating plants from wild brambles that serve as sources of infection is also important. Use of disease-free plants or plants started from root cuttings is necessary to avoid "planting" the problem into the site. Mowing plants down after harvest is an extremely effective means of eliminating the problem. Short of this, infected shoots showing the witches'-broom appearance should be cut out immediately and burned or buried. Fungicides can be applied, but this is often less effective because the timing of infection is variable and sometimes unpredictable.

Crown gall and cane gall. This same disease also affects major tree fruits, but the problems caused by the bacteria *Agrobacterium tumefaciens* and *A. rubi* can be particularly severe for brambles, debilitating the canes by interfering with water and nutrient flow. Galls form at or just under the soil line. Achieving control over the problem is difficult or impossible once the plants are infested. The bacteria survive for several years in the soil, so sites previously planted to tree fruits or grapes should be avoided in bramble cultivation. Planting stock should be inspected, but some galls may only be pinhead sized in the early stages and therefore difficult to spot. 'Willamette' and 'Canby' show some resistance to crown gall.

Botrytis fruit rot. Botrytis cineria is a common fungus among brambles that causes fruit rot of several species. Fruit rot can be severe if berries are harvested wet, rain occurs during harvest, or fruits are injured by other means and then become wet. The fungus creates a fuzzy, gray mass around ripening fruits, causing rot, particularly in cool, wet weather. It is a problem in all fruit-growing regions, and probably the main fungal disease of brambles. Infection actually begins during the blossom stage and is most damaging to ripe fruit. The same fungus can also damage the canes, although outright killing of primocanes is rare. Cane botrytis is more common in red raspberry than in other brambles. Fungicides are applied as new growth begins and continued throughout bloom and harvest, if weather is cool and wet. Widely spaced plants and weed control will improve air flow and light exposure in the leaf canopy, thus reducing disease incidence. Alternate-year production does not reduce the cane botrytis problem entirely because the fungus is omnipresent.

Root rot. This disease is caused by members of these genera: *Verticillium, Phytophthora, Pithium,* and *Rhizoctonia.* The verticillium fungus infects a broad range of plants and is persistent in soil for many years. It causes root death, followed by cane wilting and dieback. The phytophthora fungus requires constant high soil water status to move among plants and is most often seen in low spots where water collects. Verticillium can be avoided by planting in soils not planted with tomato, pepper, eggplant, or tomato culture for several years because these plants harbor the fungus. Other root diseases can be avoided by using raised beds, improving soil drainage at planting, and avoiding low spots. The soil surface can be drenched with fungicides, but good site selection is generally all that is required. 'Latham' red raspberry is said to be resistant to phytophthora, but little genetic resistance to root rot exists.

Septoria leaf and cane spot. This disease, caused by the fungus *Septoria rubi,* affects blackberries primarily, not raspberries. It is a problem mostly in the southeastern United States and the Pacific Northwest. On leaves, a characteristic "frog eye" spotting occurs—spots have white centers and brown/purple margins; on canes, spots are more elongated. The fungus can cause premature defoliation and reduced vigor. As with anthracnose, good sanitation measures and practices that increase air flow and light penetration into the canopy reduce this disease. The same fungicides used for anthracnose are applied in the dormant season prior to budbreak to combat infection.

Viruses. Over 30 virus diseases of brambles have been recorded worldwide, with about half of these making their home in North America. Many are carried from plant to plant by aphids. Viruses often cause stunting and deformation of foliage and stems, as well as mottled yellowing or spotting of leaves, symptoms that mimic nutrient deficiencies or herbicide injury. The only way to control virus diseases is to prevent their entry into the planting and to remove promptly any infected plants once detected. Spraying to kill the aphids that transmit the viruses is often futile, and resistant cultivars are often not available. Purchasing virus-free (certified) nursery stock and eliminating wild hosts of the viruses, especially wild blackberries or raspberries, from areas around the planting can reduce the incidence.

HARVEST, POSTHARVEST HANDLING

Maturity

Berries are mature when they have completely developed their characteristic color and are easily detached from the plant. All brambles require frequent pickings over a period of a few weeks. Black raspberries tend to produce small fruit clusters that may ripen over a fairly short period. Primocane- or autumn-fruiting red raspberries ripen from midsummer until first frost in some cases, so pickings at 2- to 3-day intervals may occur for several weeks. Brambles are classified as nonclimacteric fruits with respect to ripening.

Harvest Method

Large operations use over-the-row mechanical harvesters almost exclusively, since hiring laborers for handpicking is expensive and a willing workforce often is unavailable (85 percent of Oregon acreage is mechanically harvested). The machine moves at 1 mph, harvesting 1 acre/hour at 10-foot row spacings—work that would require 80 to 85 pickers. Machine-harvested berries are superior in quality to hand-harvested fruits, since ease of abscission is probably the best single indicator of maturity. Hand-harvested fruits are picked directly into 12-pint flats that are usually hung around the neck or strapped to the waist so that both of the picker's hands are free. Pickers are generally paid by the flat (Figure 6.5).

FIGURE 6.5. *(left)* Handpicking raspberries in Latin America and *(right)* a mechanical harvester used in large plantings in Oregon.

Postharvest Handling, Storage

Brambles are extremely perishable, lasting only 2 to 3 days at temperatures of around 32 to 45°F. They are not sensitive to chilling injury. Fruits intended for fresh market are easily crushed and must be picked into shallow flats—typically 12-pint flats or half-pint or full-pint plastic "clamshell" containers that can be marketed directly without further handling. Since most of the fruits are processed, some injury during harvest is of no consequence. As with all fruit crops, lack of storage potential is one of the main barriers to fresh fruit production.

CONTRIBUTION TO DIET

Almost all brambles are processed; perhaps 10 percent of the crop is sold fresh. Among the products using bramble fruits are the following:

Product	Percentage of Crop
Preserves, jam, jelly	40
Bakery products	25
Individually quick-frozen berries	18
Juices, extracts	7
Ice cream, yogurt	5
Canned berries	5

Per capita consumption is .08 pound/year for blackberry and .22 pound/year for raspberries, among the lowest of any fruit crop in the United States.

Average nutritional values for blackberries and raspberries per 100-gram edible portion:

	Blackberry	Red Raspberry	Black Raspberry
Water (%)	88	86	81
Calories	43	52	74
Protein (%)	1.4	1.2	1.5
Fat (%)	0.5	.6	1.6
Carbohydrates (%)	9.6	11.9	15.7
Crude Fiber (%)	5.3	6.5	6.8
	% of U.S. RDA (2,000-calorie diet)		
Vitamin A	4	<1	<1
Thiamin, B_1	1	2	1
Riboflavin, B_2	2	2	4
Niacin	3	3	2
Vitamin C	35	44	40
Calcium	3	2	4
Phosphorus	3	4	5
Iron	3	4	5
Sodium	<1	<1	<1
Potassium	5	4	5

BIBLIOGRAPHY

Bowling, B.L. 2000. *The berry grower's companion.* Portland, OR: Timber Press.

Crandall, P.C. 1995. *Bramble production: The management and marketing of raspberries and blackberries.* Binghamton, NY: Food Products Press.

Crandall, P.C. and H.A. Daubeny. 1990. Raspberry management, pp. 157-213. In: G.J. Galleta and D.G. Himelrick (eds.), *Small fruit crop management.* Englewood Cliffs, NJ: Prentice-Hall, Inc.

Ellis, M.A., R.H. Converse, R.N. Williams, B. Williamson. 1991. *Compendium of raspberry and blackberry diseases and insects.* St. Paul, MN: American Phytopathological Society Press.

Jennings, D.L. 1988. *Raspberries and blackberries: Their breeding, diseases, and growth.* London: Academic Press.

Jennings, D.L., H.A. Daubeny, and J.N. Moore. 1990. Blackberries and raspberries, pp. 329-390. In: J.N. Moore and J.R. Ballington (eds.), *Genetic resources of temperate fruit and nut crops* (Acta Horticulturae 290). Belgium: International Society for Horticultural Science.

La Vine, P.D. 1982. *Growing boysenberries and Olallie blackberries* (University of California Division of Agricultural Sciences Leaflet 2441). Berkeley, CA: University of California Press.

Moore, J.N. and J.D. Caldwell. 1985. Rubus, pp. 226-238. In: A.H. Halevy (ed.), *CRC handbook of flowering,* Volume 4. Boca Raton, FL: CRC Press.

Moore, J.N. and R.M. Skirvin. 1990. Blackberry management, pp. 214-244. In: G.J. Galleta and D.G. Himelrick (eds.), *Small fruit crop management.* Englewood Cliffs, NJ: Prentice-Hall, Inc.

Pritts, M.P. (ed.). 1989. *The bramble production guide: Northeast region* (Agricultural Engineering Series Publication). Ithaca, NY: Cornell University.

Shoemaker, J.S. 1978. *Small fruit culture,* Fifth edition. Westport, CT: AVI Publ.

Stiles, H.D. 1990. *Shift trellises for better management of brambles* (Rubus *cvs*) (Virginia Agricultural Experiment Station Bulletin 95-2). Blacksburg, VA: Virginia Agricultural Experiment Station.

Turner, D. and K. Muir. 1985. *The handbook of soft fruit growing.* London: Croom Helm.

Chapter 7

Blueberry (*Vaccinium* spp.)

TAXONOMY

Blueberries are members of the Ericaceae (or heath) family, genus *Vaccinium*, subgenus *Cyanococcus*. The genus is very diverse, containing 150 to 450 species, mostly found in the tropics at high elevations, but also in temperate and boreal regions. Most species are shrubs, similar to the blueberry, but, again, a diverse range of growth forms, from epiphytes to trees, exists. The Ericaceae family contains several important ornamentals: rhododendrons and azaleas *(Rhododendron)*, mountain laurel *(Kalmia latifolia)*, heather *(Calluna)*, heath *(Erica)*, and leatherleaf *(Leucothoe)*. Other species of *Vaccinium* grown for fruit include the following:

- Cranberry *(V. macrocarpon)*. The most economically important *Vaccinium* in the world, cranberry is grown primarily in the northeastern United States. The plant is a creeping, rhizomatous, bog species that grows wild in swampy areas.
- Huckleberry (*V. ovatum*—evergreen; *V. parvifolium*—red; *V. ovalifolium*—tall mountain; *V. deliciosum*—low-growing mountain). Prized as wildlife species, huckleberry plants provide an edible landscape element and are also used as ground covers. Huckleberry is also classified in the genus *Gaylussacia*.
- Lignonberry *(V. vitis-idaea)*. A creeping evergreen from Northern Europe, this plant produces small, cranberrylike fruits.
- Bilberry *(V. myrtillus)*. A rhizomatous shrub native to Northern Europe, Russia, Scandinavia, and the Pacific Northwest, this plant produces aromatic, purple berries that are collected from the wild and used in jellies, jams, and wine.

Three commercially important blueberry species are recognized, along with two interspecific hybrids:

1. Northern highbush blueberry (*V. corymbosum* L.). The main cultivated species of blueberry, its native range is sunny, acidic, swampy areas of eastern North America, from Nova Scotia west to Wisconsin, and south to northern Georgia. Commercial cultivars of highbush blueberry may also have been products of hybridizations with *V. australe*, especially in southern regions; also with *V. lamarckii* and *V. brittonii* in the north; and with *V. arkansanum*, *V. simulatum*, and *V. marianum* in the south. Michigan, the leader in highbush blueberry production in the United States, grows over 20 cultivars, with 'Jersey', 'Bluecrop', 'Elliot', and 'Rubel' making up the top four. In a 1992 nursery survey, the top-selling cultivars were 'Duke', 'Sierra', 'Nelson', 'Bluegold', 'Toro', and 'Sunrise'. 'Weymouth' is grown widely in New Jersey.

2. Rabbiteye blueberry (*V. ashei* Reade). This species is native to the river bottoms and swampy, acid soils of southern Georgia and Alabama to northern Florida. Larger and more vigorous than highbush, the rabbiteye blueberry has a lower chilling requirement (earlier bloom) and generally a longer period from flowering to maturity (Plate 7.1). Rabbiteye fruits have a somewhat thicker skin, a larger "button" or sepal scar at the distal end, as well as more (and larger) seeds. 'Tifblue', 'Woodard', 'Climax', 'Delite', and 'Brightblue' are the mainstays of the industry, but many cultivars are grown commercially. Production of this species is largely con-

doi:10.1300/5547_07

fined to the southeastern United States on about 8,000 acres. Georgia is currently the largest producer of rabbiteye blueberries, with 6,000 acres in the southeastern part of the state allocated for their growth.

3. Lowbush blueberry (*V. angustifolium* Ait., and *V. myrtilloides* Michx., primarily, possibly with other coexisting species, such as *V. britonii* and *V. lamarckii*). *Vaccinium myrtilloides* is the predominant species in recently established fields, but *V. angustifolium* is most abundant in older plantings and is the lowbush blueberry of commerce. As the name suggests, plants are creeping shrubs, about 1 foot tall or shorter, and the fruits are smaller and lighter blue than those produced by other species. Over 1,000 selections have been evaluated and made available by the Agriculture Canada breeding program in Nova Scotia, but wild populations still predominate in commercial plantings. Selections do not form rhizomes as well as wild plants, and the low rate of bush spread is one of the most limiting factors in lowbush blueberry culture. 'Brunswick' is a purported self-fertile cultivar available from some nurseries.

Lowbush blueberry cultivation is basically the management of native stands of wild blueberries (Plate 7.2). Large areas with a high density of native blueberries may be fertilized or sprayed to control weeds and insects, provided with bees for pollination during bloom, and harvested with hand rakes or, to a lesser extent, mechanical harvesters. Rakes are drawn across the upright, fruiting shoots, and fruits are "combed" off the stems and dumped into small containers. In some cases, harvest is the only cultural practice performed. Fields are cropped in alternate years and burned or mowed every other year to control weeds, insects, and fungal diseases; burning sometimes injures rhizomes and promotes oxidation of the organic matter at the soil surface. Mowing has become more popular due to high fuel costs and loss of productivity in older fields. The entire lowbush blueberry crop is processed (and usually frozen, for use in baked goods, such as pies, muffins).

4. Southern highbush blueberry (*V. corymbosum* hybrids with *V. darrowi, V. ashei,* and other southern *Vaccinium* species). Breeders have combined the characteristics of northern highbush and southern highbush blueberries to yield cultivars that are adapted to the deep south, with a shorter fruit development period and very early maturity. In April to early May, growers in Florida and Georgia receive prices of up to several dollars per pound for these cultivars, whereas later-season blueberries are worth a dollar per pound or less. Southern highbush plants are similar in most ways to northern highbush plants, but they have low winter chilling requirements (only 200-700 hours at 45°F) and can be grown into southern Florida. 'Avonblue', 'Flordablue', 'Sharpblue', and 'Georgiagem' are the main cultivars.

5. Half-high highbush blueberry (*V. corymbosum* × *V. angustifolium*). These plants are the recent products of the Minnesota and Michigan breeding programs. The bushes are short-statured (2-4 feet), cold hardy, and similar to highbush plants in fruit characteristics. They are designed to be adapted to the extreme winters and snow loads of the northern continental United States. Cultivars include 'Northland', 'Northsky', and 'Northblue'.

ORIGIN, HISTORY OF CULTIVATION

Origins are described previously.

Cultivated blueberries were domesticated only in the twentieth century but probably were collected from the wild for thousands of years in North America. Prior to 1900, superior wild bushes were known to be cultivated, supplementing the wild harvested berries. Plantsmen and plant breeders, notably Frederick V. Coville, selected and bred large-fruited cultivars after the turn of the century that form the foundation of today's modern blueberry industry. Today, 75 to 100 named cultivars exist, and new cultivars are being produced annually. The present-day industry in the United States is worth about $300 million per year and is concentrated in the northeastern states. In the southeastern United States, rabbiteye blueberries have become an important crop, having been domesticated only in the past 50 years or so. Over the past two decades, interest in blueberry as a crop has increased in other countries and in the western states of the United States.

FOLKLORE, MEDICINAL PROPERTIES, NONFOOD USAGE

The bilberry is symbolic of treachery in the "language of flowers." Blueberries are such a recent crop relative to others that folklore surrounding their cultivation and usage is scant. However, the blueberry's medicinal properties were discovered hundreds of years ago, and several recent studies have shed light on its hitherto unknown medicinal values.

Fruit and root bark decoctions of *V. arboreum* (sparkleberry; native to southeastern United States) were used to treat dysentery. *Arctostaphylos uva-ursi* (kinnikinnick or bearberry), an evergreen ground cover related to blueberry, was used by North American Indians, the Chinese, and Europeans for kidney/urinary tract ailments. The leaves contain arbutin, a chemical that acts as a diuretic and produces an antiseptic effect in the urinary mucous membrane. Blueberries also produce arbutin and may reduce the incidence of urinary tract infections. Kinnikinnick is supposedly useful for everything from controlling bed-wetting to kidney stones, and even in the treatment of gonorrhea. In females, a tea made from kinnikinick can be used as a bath or douche after childbirth to aid healing and prevent infections, though drinking this tea during pregnancy should be avoided, as it may stimulate the uterus to contract. Indians used a tea derived from the fruits or leaves to control weight, and a similar tea can be purchased in health food stores today for this purpose. Fruits are bland and mealy but can be used for preserves. Eastern North American Indians mixed the leaves of this plant with tobacco to create the smoking mixture "kinnikinnik," from which the common name is derived.

Anthocyanins, responsible for the blue color of blueberries, function as antioxidants and anti-inflammatories in mammals. Although anthocyanins are more or less ubiquitous in horticultural crops, blueberries contain more than any other fruit or vegetable. Antioxidants are known to reduce the incidence of heart disease and cancer in humans. Recent research with rats and mice has shown blueberries to protect against aging-related brain damage and even to reduce the effects of Alzheimer's disease on brain tissue. Blueberries may reverse short-term memory loss as well.

The bilberry *(V. myrtillus)* is a creeping shrub that produces aromatic, highly colored berries that are collected from the wild in Europe to make jelly, jam, and wine. Bilberry has many purported medicinal properties, for example, improvement of blood circulation, treatment of dysentery and diarrhea, and treatment of urinary problems, such as kidney stones. Herb decoctions are said to pass through the stomach, without affecting it, going to work directly on the small intestine. Such decoctions have proven valuable for the treatment of dyspepsia and diarrhea in infants, and they are also purported to prevent smoker's cough.

Well-documented effects include the improvement of vision; reduction of eye fatigue, irritation, nearsightedness, and night blindness; and restraint of cataract development. The story of its discovery as a visual aid is interesting: British Air Force pilots who often ate bilberry jam on their bread were far more accurate in the bombing of German targets during World War II. As many of these raids were made at night, scientists began to research bilberry for the sharpening of night vision and for the treatment of eye disorders in general. With generous quantities of anthocyanins, bilberries promote the strengthening of capillaries and improved peripheral circulation. Possibly, this improved blood flow in the eye allows for better vision and prevents eye problems. More recently, anthocyanins in blueberries have been studied for similar effects on vision, including macular degeneration in the elderly.

PRODUCTION

World

Total production in 2004, according to FAO statistics, was 238,620 MT or 525 million pounds. Blueberries are produced commercially in 16 countries worldwide on about 120,000 acres. Worldwide average yields have increased almost 50 percent in the past 10 years to just over 4,300 pounds/acre. Acreage has also increased by 31 percent, causing production to double in the past 10 years. The top ten blueberry-producing countries (percent of world production) follow:

1. United States (50)
2. Canada (34)
3. Poland (7)
4. Ukraine (2)
5. Netherlands (2)
6. Romania (1)
7. Lithuania (1)
8. New Zealand (<1)
9. Italy (<1)
10. France (<1)

United States

Total production in 2004, according to USDA statistics, was 124,945 MT or 275 million pounds. Production has more than doubled in the United States over the past decade, yielding an industry worth $295 million, with $276 million from cultivated plantings and $19 million from wild plantings in Maine. Prices for cultivated blueberries average $1.21 per pound, relatively high compared to those for other fresh fruits, and range from $0.81 to $1.36 per pound for most states. Florida is the exception, with prices received ranging from $4.50 to $6.40 per pound over the past 3 years, due to the ability of Florida growers to produce the first fresh-market berries of the season, which receive extremely high prices. Processed fruits receive only $0.67 to $0.95 per pound, depending on state, with an average of $0.80 per pound. In 2004, cultivated blueberries covered 44,430 acres, with yields ranging from about 1,400 to 10,000 pounds/acre; the nationwide average is 5,120 pounds/acre.

Wild lowbush blueberries are harvested from about 30,000 acres annually, although total acreage is estimated at 50,000 to 60,000 acres. Fields are often burned or mowed after harvest to rejuvenate plantings, such that production occurs about every other year from any given field. The 2004 crop in Maine was only 46 million pounds, the lowest since 1991; it is more typically 60 million to 80 million pounds. The price received by growers was $0.41 per pound in 2004, about average for the decade. Since greater than 99 percent of the crop is processed, prices are lower than those received for cultivated blueberries.

The leading blueberry states (percent of U.S. production) in 2004 follow:

1. Michigan (29)
2. Maine (17; typically 20-24)
3. New Jersey (14)
4. Oregon (12)
5. North Carolina (8)

Other states producing blueberries include Alabama, Arkansas, Florida, Georgia, Indiana, New York, and Washington.

BOTANICAL DESCRIPTION

Plant

- *Highbush (northern and southern).* These erect, deciduous shrubs grow to 12 feet in the wild and 4 to 7 feet in cultivation. Shrubs are composed of canes that arise from the crown or roots; canes reach maximal production in about 4 years and begin to decline thereafter. Leaves are small (1-2 inches in length), ovate or elliptic, with entire margins and acute tips (Figure 7.1).
- *Rabbiteye.* These erect, deciduous shrubs grow to 20 feet in the wild and 4 to 10 feet in cultivation. Rabbiteye canes arise from the crown or roots, as with the highbush blueberry, but they have a longer productive life of about 7 years. Leaves are similar to those of the highbush blueberry.
- *Lowbush.* These plants are low-growing (<2 feet, usually 8 inches), creeping, rhizomatous shrubs (Plate 7.2). The main portion of the plant is the rhizome or underground stem, 1 to 3 inches below the surface, which produces upright shoots along its length. The uprights grow vegetatively for 1 year and then fruit in subsequent years. Leaves are smaller than those of the highbush or rabbiteye blueberry and have mildly serrate margins.

Flowers

For all species, white or cream flowers (1-16, usually 7-10) are borne on short racemes (1-2 inches) on the upper portion of 1-year-old wood. Flowers are urn-shaped and inverted, on very short pedicels (nearly sessile), with inferior ovaries (Plate 7.3). Flower (inflorescence) buds are noticeably larger and more conical shaped than vegetative buds, except in lowbush blueberry, which sports spherical floral buds. Lowbush and northern highbush blueberries are photoperiodic with respect to flowering, being short-day plants with critical photoperiods of 16 hours, which is unusual for a woody fruit crop. However, virtually no production regions have daylengths greater than 16 hours, and thus natural photoperiod does not play a meaningful role in flowering.

FIGURE 7.1. *(left)* Mature rabbiteye blueberry bushes. *(right)* New canes emerge from the crown and roots of a mature bush.

Pollination

Most northern highbush blueberries are self-fruitful, but for some cultivars, higher fruit set and larger fruits can be produced through cross-pollination. Southern highbush cultivars, in contrast, are only partially self-fruitful and are often provided with pollinizers in commercial production. Rabbiteye and lowbush cultivars are highly self-incompatible and must have a pollinizer to set fruits. Bees are necessary for pollination even in self-fruitful types because flowers are inverted and the pollen falls out of the flower without landing on the stigma.

Bumblebees (*Bumbus* spp.) and wild bees (e.g., the southeastern blueberry bee *Harbropoda laboriosa*) pollinate flowers naturally, whereas honeybees *(Apis mellifera)* are reportedly less effective (Plate 7.4) due to the differential ability of bumblebees and honeybees to sonicate or "buzz" flowers and stimulate pollen release. The frequency of the sound generated by the wings of bumblebees is such that anthers dehisce and release pollen; honeybees flap their wings at a different frequency and may also work the base of flowers already slit open by carpenter bees, retrieving only nectar and not performing pollination.

Fruit

In all cases, the fruit is an epigynous or "false" berry. This means that the fruit is berrylike but derived from an inferior ovary, unlike true berries that derive from superior ovaries. The "button" on the far end of the fruit is actually the calyx scar—the former point of attachment of the sepals (Plate 7.5). All species require a high degree of fruit set for a full crop (60-80 percent); no thinning is practiced.

- *Northern and southern highbush.* Berries are blue-black in color and possess good to excellent fruit quality. These cultivars feature the shortest period from flowering to maturity of all blueberries (45-75 days).
- *Rabbiteye.* Berries are blue-black in color and possess good fruit quality. Fruits mature in about 90 days following bloom, so, despite blooming earlier than highbush blueberries, rabbiteye fruits ripen later and have somewhat tougher skins, larger buttons, and larger seeds than highbush fruits, but these characteristics are hard to distinguish.
- *Lowbush.* Berries are black to bright blue in color, with fair to poor fruit quality, as compared to other cultivated types; virtually all commercially grown lowbush blueberries are processed for muffins and cakes. Lowbush berries have an intermediate fruit maturation period of 70 to 90 days.

GENERAL CULTURE

Soils and Climate

Unlike most fruit crops, blueberries require light, acid soils (pH 4.5-5.2) with high organic matter (20-50 percent) for best growth. Also, they cannot tolerate nitrate nitrogen and are often fertilized with ammonium sulphate. In areas with high soil pH, peat incorporation at planting is recommended, along with an appropriate amount of sulfur to reduce pH. Irrigation water with high pH (~7) is avoided on mineral soils.

Clearing pine forest land to plant blueberries, although expensive, is recommended as the best means of ensuring the appropriate soil type for blueberry plantings (Figure 7.2). Mulching is generally beneficial, especially on heavy clay soils where roots will not penetrate more than a few inches. Bushes are obviously very susceptible to drought in this case, and mulching reduces

FIGURE 7.2. Harvesting a light crop from 2-year-old blueberries in south Georgia. The area was formerly a pine flatwoods ecosystem, which has a sandy, low-pH soil rich in organic matter suited to blueberries. *Source:* Photo courtesy of Gerard Krewer, University of Georgia.

weed competition and evaporation. Rabbiteye shrubs are generally more tolerant of drought and heat than northern highbush shrubs.

All species of *Vaccinium* are relatively tolerant of short-term flooding, which is uncommon for fruit crops. However, raised beds are used where the water table is less than 20 inches below the surface during the growing season.

Various species and hybrids of blueberry are adapted to the humid, temperate climate of the eastern United States, with lowbush being adapted to the most northern areas (southeastern Canada to New England), northern highbush planted from New England and Michigan to North Carolina, and rabbiteye planted from North Carolina to Florida and westward to eastern Texas. Southern highbush shrubs are best adapted to the Gulf Coast and south Georgia, extending south to Immokalee, Florida. Although some blueberries are being grown in California, Oregon, and Washington, (areas with higher pH soils), the soil is heavily amended with organic matter to permit cultivation. Soil more than climate has limited the westward expansion of blueberry culture.

A large constraint on the distribution of various blueberry types in the eastern United States is winter chilling requirement. The average requirements follow:

Northern highbush: 800-1,100 hours
Rabbiteye: 350-800 hours
Southern highbush: 200-700 hours
Lowbush: >1,000 hours

Coldhardiness is generally not a limitation to growing blueberries, provided the appropriate species is selected, as mentioned earlier. In northern highbush, canes are killed at –5 to –40°F, depending on acclimation, and flower buds are killed at approximately –10 to –20°F when dormant. In rabbiteye, canes are killed at –5 to –20°F, depending on acclimation, and dormant flower buds are also killed in this range. The lowbush cultivars are probably the hardiest of the blueberry species, although great variation exists among native populations, with less tolerance in plants grown in maritime areas (Nova Scotia) and the greatest tolerance in those grown in continental areas (Minnesota). All blueberries bloom relatively early and often experience frost damage. Open flowers of all types are killed at 28°F. Southern highbush shrubs are particularly frost prone because they have very short winter chilling requirements.

Humidity is not a problem for blueberry culture because the plants have relatively few disease and insect problems.

Propagation

Hardwood and softwood cuttings are the most popular methods. Hardwood cuttings (pencil-sized 1-year-old shoots) are taken in January and February. Softwood cuttings are taken at the end of the first growth flush (around June). Cuttings of both types are usually grown in pots prior to field planting because the survival rate of cuttings transplanted directly to the field is poor.

Rootstocks

Blueberries have no rootstocks.

Planting Design, Training, Pruning

Northern highbush cultivars are grown in hedgerow configurations, usually spaced 4 × 9 to 10 feet (~1,000 bushes/acre); 10-foot row middles accommodate mechanical harvesters better. Rabbiteye blueberries have inherently more vigor and are spaced 5 to 8 × 12 to 14 feet, with wider spacings for pick-your-own plantings (Figure 7.3) and closer spacings for plantings to be mechanically harvested. Hedge height is maintained at 6 to 7 feet for mechanical harvest (Figure 7.4).

Lowbush blueberries are typically not planted, but native populations of variable density are managed. Well-managed plantings cover the ground completely. Lowbush blueberries are pruned by burning or mowing in alternate years. Shoots are burned to a height of 1 to 2 inches. Charred shoots usually rot by the time of next harvest and therefore do not interfere with raking.

Pruning of highbush and rabbiteye blueberries is similar, involving removal of weak wood and older, unproductive canes. Small-diameter wood tends to produce smaller berries. In highbush cultivars, canes decline in productivity after about 4 years, so removal of one or two of the oldest canes each year maintains vigor. In rabbiteye cultivars, canes are productive for about 7 years, so bushes with several canes have the oldest ones removed each year, with only one or two new, vigorous canes left to replace them. Flower buds are removed from young plants to enhance vegetative growth in early years.

For mechanical harvest, canes should be maintained within a strip no wider than 12 inches; bushes narrowed to 9 inches at the base have higher berry recovery from mechanical harvest, but lower yields per acre because the bushes are smaller.

FIGURE 7.3. Painstakingly slow hand harvest of southern highbush blueberries in south Georgia. The berries are intended for fresh-market sales and cannot be mechanically harvested. Workers are paid by the bucket.

FIGURE 7.4. *(above)* An over-the-row mechanical harvester slaps the bushes with vibrating arms and collects the fruits on conveyors below. *(right)* The inside view, showing the vibrating arms and pivoting plates that, respectively, shake and catch the fruits.

Pest Problems

Insects

Blueberry maggot. Rhagoletis mendax is the major fruit pest of blueberry. The adult is a small fly that lays an egg under the skin of a developing fruit. The tiny larva feeds within the fruit, often undetected, but when picked the berries appear mushy and leak juice easily. Insecticides timed to the season of adult egg laying effectively control the pest; this season typically begins mid-May in the South to late June in the Northeast and extends until about 1 week before harvest.

Cranberry fruitworm. Similar to the blueberry maggot, this adult insect *(Acrobasis vacinii)* lays its eggs on the calyx end of small, green berries, and since bloom periods are often extended in blueberry, egg laying can occur for several weeks. The larvae that develop can eat several fruits, leaving behind a brown frass, along with some webbing. Infested fruits turn blue prematurely and shrivel and may have pin-sized entry holes near the fruit stem. Control is achieved by insecticide application.

Plum curculio. Conotrachelus nenuphar adults are gray to dark brown, ⅛- to ¼-inch-long snout beetles; small, yellow-white, legless grubs burrow into the fruit flesh, usually causing the fruits to fall off prematurely. Feeding damage by adults alone may cause misshapen fruits—D-shaped brown depressions in the fruit surface. Insecticide sprays timed at postpollination, when fruits are ¼ inch in diameter, usually control this pest. The insect is confined to the United States east of the Rocky Mountains.

Blueberry bud mite. Found throughout the eastern United States, these tiny (<$\frac{1}{100}$ inch) pests *(Acalitus vacinii)* inhabit the leaf and flower buds, feeding on them before they emerge. Affected tissues have a roughened, blistered, red discoloration, and berries may have red "pimples" on them. Annual pruning of old canes reduces the mite population and is often all that is needed for control. If the infestation is severe, miticides are sprayed directly after harvest, and again 2 to 4 weeks later, to kill the mites before they reenter developing buds.

Diseases

Mummy berry. Caused by *Monilinia vaccinii-corymbosi*, this is probably the most widespread threat to blueberry throughout the United States, and sometimes mummy berry is the only fungal disease that requires sprays for control. The disease kills leaves, shoots, and flowers in spring, producing spores on these dead tissues that will later infect the fruits. Infected fruits remain pink, rather than turning blue, and drop to the ground, where they house the fungus over the winter. This disease has two phases: a primary infection of expanding shoots and inflorescences, called "shoot blight," and a secondary infection of the flowers, resulting in mummified fruits at harvest. The disease results in production losses of 8 to 10 percent in average years, and up to 50 percent for plantings where no control measures are implemented. Sanitation is important in reducing the incidence of the disease, as apothecia produced on leaf litter below the plants release spores that are responsible for the primary infection. Mulch removal or covering with new mulch is a good cultural control. If a planting has a history of the disease, two fungicide sprays are recommended: one just after budbreak and the second at full bloom.

Anthracnose. Colletotrichum gleosporioides is a broad-host fungus that attacks the ripening fruits of many species, from blueberry to mango. The fungus overwinters on twigs and infects fruits anytime it rains during the season. Fruit infections are latent; they do not appear until fruits are mature. The disease is easily identified by masses of salmon-colored spores produced on decaying fruits. Rabbiteye blueberries are far less susceptible than are highbush blueberries, with 'Murphy', 'Morrow', and 'Reveille' being the least susceptible highbush cultivars. Fungicide sprays at 7- to 10-day intervals, starting at bloom and following through to harvest, may be necessary in cases of severe infestation.

Stem blight and cankers. Several fungi produce cankers on or cause outright death to blueberry canes, including godronia canker *(Fusicoccum putrefaciens)*, gleosporium canker *(Gleosporium minus)*, botryosphaeria stem blight *(Botryosphaeria dothidea, B. corticus)*, and phomopsis twig blight *(Phomopsis vaccinii)*. These fungi generally do not kill entire bushes, affecting only individual canes. These pathogens are major problems in North Carolina, with godronia canker being a problem in northern areas, where it is responsible for cankers on and the death of 1- and 2-year-old wood. Rabbiteye blueberries are more resistant to these diseases than are the highbush blueberries.

Botrytis blight or fruit rot. The common fungus *Botrytis cineria* causes the decay of many small, soft fruits just prior to harvest and in storage. It can attack any tender blueberry tissue. Commonly, blossoms are attacked under cool, wet conditions, along with young twigs and fruits. Early infections lead to problems with fruit rot at harvest as well. Both rabbiteye and highbush blueberries are susceptible. If a planting has a history of the disease, two fungicide sprays are recommended: one just after budbreak and the second at full bloom.

HARVEST, POSTHARVEST HANDLING

Maturity

Blue color is currently the most often used indicator of maturity. Berries turn from green to pink and then gradually to a full blue color, at which time they are ripe. Ripe berries will remain attached for several days or weeks, and sugars continue to accumulate. Blueberries are classified as climacteric fruits with respect to ripening.

Harvest Method

Bushes are picked over several times by hand for fresh sales (Figure 7.3) and are mostly machine harvested for processed fruits. Since berries ripen over a period of weeks, more than one pass with the harvester is necessary. Once per week over 3 weeks is normal.

Mechanical harvesters are of two types: over-the-row (Figure 7.4) or hand-held vibrator and catch frame harvesters. Both result in loss of berries on the ground, and berry quality is generally inferior to that of hand-harvested fruits. However, one over-the-row harvester can cover up to 1acre/hour, replacing over 100 hand pickers. Economics may dictate mechanical harvesting, despite the drawbacks.

Lowbush blueberries are harvested by hand rakes, although some mechanical harvesters have been developed (Figure 7.5).

Postharvest Handling

Berries for fresh-market sales are picked into retail containers, usually flats containing 12 1-pint containers, to reduce handling damage. Berries harvested by mechanical means or picked into bulk containers are sometimes run through a "wet line:" trash is removed, and the berries are dipped in fungicide (in cold water), culled by hand, packaged in 1-pint containers, covered with plastic, and placed in flats.

Processed berries are often frozen after harvest by the individual quick freeze method (IQF) or dehydrated about sixfold by the "explosion puff" process.

Storage

Blueberries are fairly perishable, having a shelf life of only 2 weeks when stored at 32°F and 90 to 95 percent humidity. Blueberries are not sensitive to chilling injury. Fruit rots, such as botrytis, alternaria, phomopsis, and anthracnose, are major storage problems.

CONTRIBUTION TO DIET

Processed blueberries are used mostly in baked goods (e.g., muffins, pies, cakes). Blueberry juice is a recent innovation, but by itself it is unpopular. Blueberry juice is too dark and thick for

FIGURE 7.5. New and old ways of harvesting lowbush blueberries. *(left)* Hand rakes comb the fruits off the bushes and into buckets. *(right)* A mechanical rake mounted to a tractor speeds the process. *Source:* Photos courtesy of D. Scott NeSmith, University of Georgia.

drinking, so it is diluted 50 percent, before adding sugar and citric acid back in to taste. This juice is then mixed with cranberry or another fruit juice (e.g., some Ocean Spray products) for a more palatable fruit beverage.

Per capita consumption of blueberries is 0.78 pound/year, two-thirds of which is consumed as fresh fruits and the remainder frozen. Fresh consumption has increased 250 percent since 1980. Frozen consumption has increased since 1980 as well but peaked in the early 1990s, at almost 0.5 pound/year, and has declined slightly, perhaps due to greater availability of fresh fruits.

Dietary value per 100-gram edible portion:

Water (%)	84
Calories	57
Protein (%)	0.7
Fat (%)	0.3
Carbohydrates (%)	14.5
Crude Fiber (%)	2.4
	% U.S. RDA (2,000-calorie diet)
Vitamin A	1
Thiamin, B_1	2
Riboflavin, B_2	2
Niacin	2
Vitamin C	16
Calcium	<1
Phosphorus	2
Iron	2
Sodium	<1
Potassium	2

BIBLIOGRAPHY

Austin, M.E. 1978. Rabbiteye blueberries. *Fruit Var. J.* 33:51-53.

Bowling, B.L. 2000. *The berry grower's companion.* Portland, OR: Timber Press.

Brightwell, W.T. 1971. *Rabbiteye blueberries* (University of Georgia Agricultural Experiment Station Research Bulletin 100). Athens, GA: University of Georgia.

Caruso, F.L. and D.C. Ramsdell. 1995. *Compendium of blueberry and cranberry diseases.* St. Paul, MN: American Phytopathological Society Press.

Eck, P. 1988. *Blueberry science.* New Brunswick, NJ: Rutgers University Press.

Eck, P. and N.F. Childers (eds.). 1966. *Blueberry culture.* New Brunswick, NJ: Rutgers University Press.

Eck, P., R.E. Gough, I.V. Hall, and J.M. Spiers. 1990. Blueberry management, pp. 273-333. In: G.J. Galleta and D.G. Himelrick (eds.), *Small fruit crop management.* Englewood Cliffs, NJ: Prentice-Hall.

Galletta, G.J. 1975. Blueberries and cranberries, pp. 154-196. In: J. Janick and J.N. Moore (eds.). *Advances in fruit breeding.* West Lafayette, IN: Purdue University Press.

Gough, R.E. 1994. *The highbush blueberry and its management.* Binghamton, NY: The Haworth Press.

Hancock, J.F. and A.D. Draper. 1989. Blueberry culture in North America. *HortScience* 24:551-556.

Hanson, E. and J.F. Hancock. 1990. Highbush blueberry cultivars and production trends. *Fruit Var.* J. 44:77-81.

Kalt, W. and D. Dufour. 1997. Health functionality of blueberries. *HortTechnology* 7:216-221.

Luby, J.J., J.R. Ballington, A.D. Draper, K. Pliszka, and M.E. Austin. 1990. Blueberries and cranberries (*Vaccinium*), pp. 391-456. In: J.N. Moore and J.R. Ballington (eds.), *Genetic resources of temperate fruit and nut crops* (Acta Horticulturae 290). Belgium: International Society for Horticultural Science.

Lyrene, P.M. 1990. Low-chill highbush blueberries. *Fruit Var. J.* 44:82-86.

Mainland, C.M. 1985. *Vaccinium*, pp. 451-455. In: A.H. Halevy (ed.), *CRC handbook of flowering*, Volume 4. Boca Raton, FL: CRC Press.

Milholland, R.D. and J.R. Meyer. 1984. *Disease and arthropod pests of blueberries* (North Carolina Agricultural Research Service Bulletin 468). Raleigh, NC: North Carolina State University.

Scott, D.H., A.D. Draper, and G.M. Darrow. 1978. *Commercial blueberry growing* (USDA Farmer's Bulletin 2254). Washington, DC: U.S. Government Printing Office.
Shoemaker, J.S. 1978. *Small fruit culture,* Fifth edition. Westport, CT: AVI Publ.
Smagula, J. 1990. Changes in the lowbush blueberry industry. *Fruit. Var. J.* 44:72-76.
Spiers, J. 1990. Rabbiteye blueberry. *Fruit Var. J.* 44:68-71.
Trehane, J. 2004. *Blueberries, cranberries, and other* Vacciniums. Portland, OR: Timber Press.
Turner, D. and K. Muir. 1985. *The handbook of soft fruit growing.* London: Croom Helm.
Vander Kloet, S.P. 1983. The taxonomy of *Vaccinium* section *Cyanococcus:* A summation. *Can. J. Bot.* 61:256-266.
Vander Kloet, S.P. 1988. *The genus* Vaccinium *in North America* (Agriculture Canada Publication 1828). Ottawa: Agriculture Canada.

Chapter 8

Cacao (*Theobroma cacao*)

TAXONOMY

Theobroma cacao L. is a member of the Sterculiaceae (or sterculia) family, a medium-sized family with 50 genera and 750 species. More recent DNA evidence reclassifies the genus under the Malvaceae family, although this has yet to become widely accepted. The genus *Theobroma* contains about 22 species, with the near relatives *T. bicolor* and *T. angustifolium* being used as a substitute or adulterant of cacao or in breeding programs. Related species are relatively few and minor, with some native to the southern United States *(Ayenia, Melochia)*. Another beverage crop, *Cola nitida,* or African cola nut, was once a component of Coca-Cola and is probably the best recognized relative. Other ornamentals include bottle tree *(Brachychiton)*, dombeya *(Dombeya)*, and Chinese parasol tree *(Firmiana)*.

The name of the genus *Theobroma* translates as "food of the gods" when broken down into its Greek roots. This derives from the Mayan word *cacao,* which also meant "food of the gods" in their language. It is unclear who rearranged vowels in the common name to obtain the term *cocoa,* but those associated with managing the crop generally refer to the plant as "cacao" and the food product derived from the plant's seeds as "cocoa."

Cultivars

Cultivars fall into three categories, one of two botanical varieties or their hybrids:

1. Criollo *(T. cacao* var. *cacao)*. The word *criollo* means "native." This form is distributed from southern Mexico to South America, north and west of the Andes. It may be native to this region or was naturalized there in pre-Columbian times. Criollo fruits are oblong to ovoid in shape, tapering to a point, and have five or ten longitudinal ridges. The seeds have yellowish white cotyledons. Three forms are recognized within this group: *pentagonum,* with five prominent ridges, often considered the best quality; *leiocarpum* with five shallower furrows and a more obtuse tip; and *lacandonense,* a rare form from the Lacandon region of Mexico, with acute tips and ten furrows.
2. Forastero *(T. cacao* var. *sphaerocarpum)*. The word *forastero* means "foreign." This form was introduced to Mesoamerica from the Amazon basin. The fruits are ellipsoid to round, lacking a pointed tip, and may be furrowed but have a smooth surface otherwise. The cotyledons are violet. Forasteros are higher yielding and more vigorous than criollos, but they are considered to have inferior quality.
3. Trinitario (hybrids of criollo and forastero forms). These hybrids, which originated in Trinidad, are sometimes classified as a subgroup of the forasteros. Since they are hybrids, they are highly variable from seed, unless the seed is derived from known crosses. The seed quality is intermediate between that of the criollos and the forasteros, as are other characteristics.

Introduction to Fruit Crops

doi:10.1300/5547_08

About 80 to 90 percent of cacao production is based on the forastero form, due to its superior yield, vigor, and disease resistance. In chocolate making, the lower quality of the seed has not hindered its use because the production of milk chocolate, the predominant form of chocolate consumed today, does not require the best-quality seed. 'Amelonado' is the major West African cultivar, and the predominant type grown worldwide. A major advantage of 'Amelonado' is its self-pollinating behavior, which in turn gives rise to homozygosity and trueness to type. 'Amelonado' is also resistant to a major cacao disease, black pod, and more tolerant of cocoa shield bug. Breeding programs have produced crosses of 'Amelonado' with trinitarios or other selections, such as the upper Amazon types. 'Comum' and 'Para' are forasteros grown in Brazil, and the forastero 'Matina' is grown in Costa Rica. Breeding programs have produced a number of selections with alphanumeric names, such as the T series from the Tafo Experiment Station in West Africa, or the Imperial College Selections (ICS) from Trinidad. Trinitario types are not grown in great quantities commercially, but they have been important in breeding. Criollo types are grown on a small scale for their high-quality chocolate potential, but they are low yielding, slow growing, and more susceptible to diseases. The original criollos, such as 'Java Criollo', have been supplanted by trinitario hybrids that have nearly the same seed quality as criollos but are somewhat more vigorous and higher yielding.

ORIGIN, HISTORY OF CULTIVATION

Cacao is thought to be native either to the upper reaches of the Amazon basin exclusively or to a wider region stretching northward into present-day Chiapas, Mexico. Whether cacao is truly native in Mesoamerica or was taken there in 1450 to 2200 BC by the Valdivia Indians, who were known to have dispersed other crop plants, is still debated. Regardless, the Aztec and Maya were cultivating cacao in Mesoamerica long before European contact and are credited with the domestication of the criollo form. They made a drink called *chocolatl* (also *xocolatl*) by grinding roasted cacao seeds with maize, vanilla, chili, annatto, and other spices. *Chocolatl,* said to have the consistency of honey and a bitter flavor, was a drink prized by all but consumed mostly by the elite. Cortez is credited with bringing cacao back to Spain in the early 1500s, although Columbus had recorded seeing it earlier during his fourth voyage. The Spaniards substituted sugar for the spices but retained the maize and vanilla, and also the custom of it being a drink for the wealthy class. Once sweetened with sugar, cocoa became a popular beverage throughout Europe, at about the same time that coffee and tea were catching on. This stimulated production to spread throughout Central and South American countries and the Caribbean, and trade between mother countries and their American colonies increased gradually. During the eighteenth century, cacao eventually spread throughout the world via tropical colonies settled by the Dutch, French, Portuguese, British, and Spanish. It did not arrive in West Africa until about 1879, but today this region produces most of the world's crop.

Cocoa was consumed only as a beverage until about 1828, when the cocoa press was invented. This allowed cocoa powder to be separated from cocoa butter, which together with sugar are the basic ingredients of solid chocolate. The popularity of the beverage declined as solid chocolate became a major commercial product. A second innovation occurred in 1876; Daniel Peter of Switzerland blended condensed milk with chocolate to produce the first milk chocolate, which eventually became the preferred form of solid chocolate. Milton Hershey of Pennsylvania created the Hershey Chocolate Company in 1894 based on German innovations with chocolate. Hershey Kisses, first manufactured in 1907, are still popular today, and less than 100 years later, this company has grown into a $2 billion/year giant of the chocolate trade. In 2001, the global cocoa confectionery industry was valued at $73 billion, up about 20 percent from 5 years earlier.

FOLKLORE, MEDICINAL PROPERTIES, NONFOOD USAGE

Cacao has few reported nonfood uses, perhaps because the seeds and fruits were such prized foods. Cacao seeds were used as currency by the Aztec and Maya; in fact, cacao's first Latin name, *Amygdalae pecuniariae*, means "money almond." The pod waste from seed processing is used as fertilizer or livestock feed. The "shell" or seedcoat, removed at the beginning of chocolate processing, is used as a peat substitute in soilless potting media.

Medicinal uses are relatively few compared to other tropical crops. Most notable is the application of cocoa butter to soothe minor skin irritation, including burns, chapped lips, sores, wrinkles, and wounds. It is also reported to be a folk remedy for cough, fever, malaria, rheumatism, and snakebite. For centuries, chocolate was, and still is, considered an aphrodisiac.

Several myths about chocolate and health problems exist, but many are unfounded. Allergic properties may be overstated; food allergies to cocoa per se are poorly documented, and of far lower incidence than allergies to milk, eggs, and nuts, which are often used in making chocolate products. The contribution of chocolate to tooth decay may also be overrated because the tannins and other chemicals in chocolate may inhibit acid formation (acids derived from oral fermentation processes cause tooth decay). No statistical correlation exists between chocolate consumption and obesity, despite the common perception of chocolate causing weight gain. The avoidance of chocolate by diabetic people is commonplace but also may be unwarranted. On a glycemic index scale, with glucose valued at 100, a chocolate bar scores only 49, which is lower than the glycemic index of soft drinks, bananas, and most breads and breakfast cereals, but about even with canned baked beans. The evidence linking normal chocolate consumption to triggering headaches and migraines is scant, although large amounts of chocolate might contain enough theobromine, caffeine, and other chemicals to cause adverse health effects, possibly headache. Finally, no link exists between chocolate and acne in teenagers, according to the American Dietetic Association, the American Academy of Dermatology, and the Mayo Clinic.

Chocolate is said to be one of the most craved foods in the world, particularly by women experiencing premenstrual syndrome. Much research has been devoted to uncovering a physiological basis for the craving. Chocolate contains three stimulant compounds—theobromine, caffeine, and phenylethylamine—but at levels too low to have a significant effect on nervous system function and thus mood. Tests on lab rats show that high carbohydrate intake may trigger increased synthesis of serotonin by the brain, which in turn enhances mood, creating a better overall feeling. However, chocolate contains a significant amount of protein, which counters this reaction, so this is unlikely. Rats also produce more endorphins when fed chocolate, and endorphins have similar biological activity to morphine, an opiate painkiller. Likewise, rats given drugs that decrease endorphins do not eat as much chocolate. Thus, endorphins are more likely to be involved in the craving response than are other chemicals. White chocolate contains the sugars and fats of normal brown chocolate, but none of the pharmacological constituents (these are all in the brown cocoa powder). Thus, it is possible to separate the sensory experience of eating chocolate from the chemicals contained in it by administering either white chocolate, brown chocolate, cocoa powder capsules, or various combinations thereof. These studies showed that both white and brown chocolate satisfied cravings, but cocoa powder taken as a capsule did not. This suggests that the source of the craving lies in the pleasurable experience of eating chocolate, a sweet, fatty food that tastes good, resulting in the release of endorphins that reinforce the behavior.

PRODUCTION

World

Total production in 2004, according to FAO statistics, was 3,302,441 MT or 7.3 billion pounds of cocoa beans (the seeds of the fruits). Cacao is produced in 63 countries worldwide on over 17 million acres. Production is 24 percent higher than a decade ago. Worldwide average yield is only 419 pounds/acre, among the lowest of any fruit crop. Low yields are partially due to the fact that only the seeds of the fruits are used, which account for only a fraction of fruit fresh weight. Yields can reach 900 pounds/acre under good management. Growers received about $0.35 to $0.45 cents per pound in 2001 for dried cocoa beans, about half the amount currently paid at commodities exchanges in New York and London (Figure 8.1). The top ten cacao-producing countries (percent of world production) follow:

1. Cote d'Ivoire (30)
2. Ghana (22)
3. Indonesia (13)
4. Nigeria (12)
5. Brazil (5)
6. Cameroon (4)
7. Ecuador (3)
8. Colombia (2)
9. Mexico (1)
10. Dominican Republic (1)

Some major changes in cacao production have occurred over the past decade or so (Figure 8.2). In Brazil, the witches'-broom disease devastated trees in Bahia state, the major cacao-producing region, reducing industry output by more than half over the past decade. Consequently, Brazil dropped from third to fifth in world production. Malaysian production has declined six-

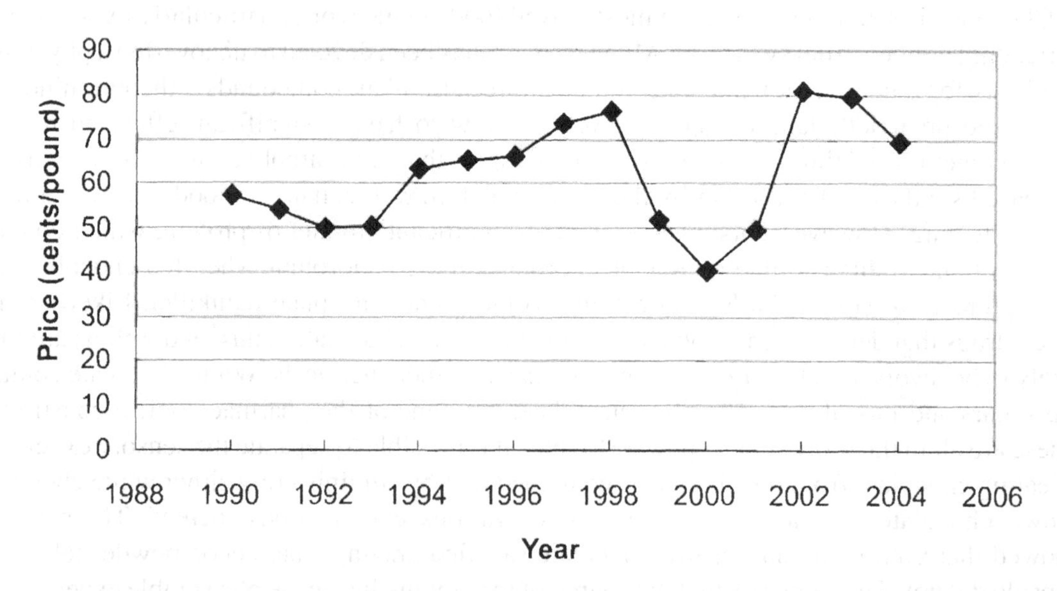

FIGURE 8.1. Prices paid for cacao at international commodities markets in London and New York. An abrupt decrease in 1998 brought prices close to the costs of production for a 3-year period. *Source:* International Cocoa Organization.

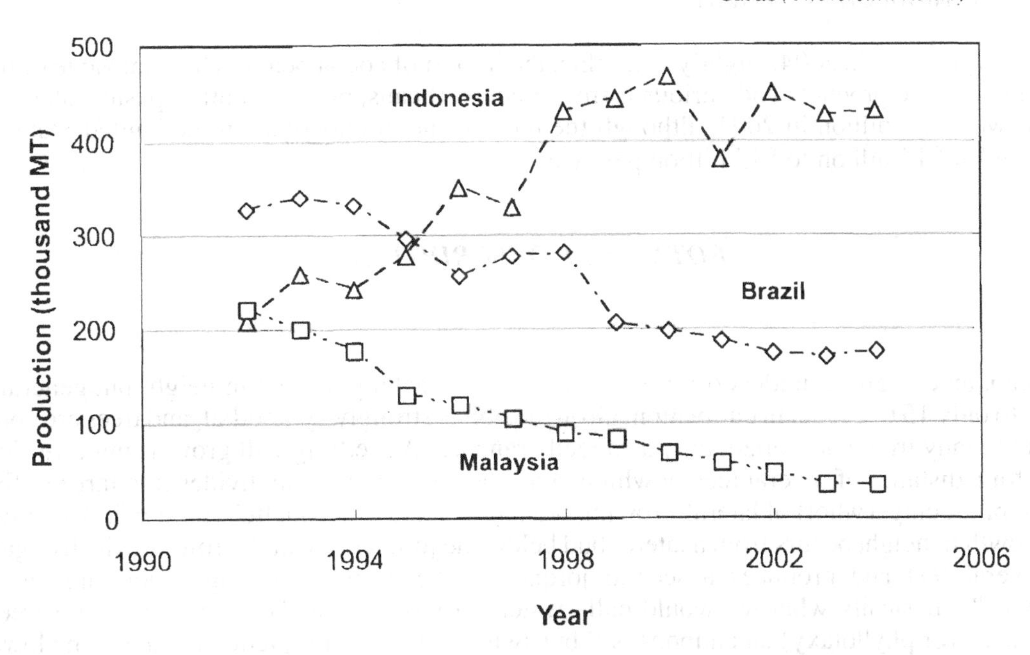

FIGURE 8.2. Changes in cacao production precipitated by different causes. In Brazil, disease reduced the crop; in Malaysia, growers switched to growing the more lucrative African oil palm. In Indonesia, a successful government program targeting small landholders has brought about a 3,000 percent increase in production in 20 years, offsetting reductions in Brazil and Malaysia. *Source:* FAO statistics.

fold as growers have diversified into oil palm, which is more profitable; Malaysia now leads the world in oil palm production, which is currently the country's top fruit crop. The rise in Indonesian cacao production is a case study of a successful government program in agriculture. In the 1970s, the government obtained high-yielding germplasm from Brazil and made plants, capital, and technology for growing cacao available to small landholders. By 1982, the country was producing 17,000 tons on under 100,000 acres. The Indonesian Cocoa Association was formed, extending technical help to small growers and promoting industry expansion. By 2002, Indonesian production had risen 30-fold, while acreage increased only 14-fold, as yield more than doubled in the same period. Today, small landholders produce over 80 percent of Indonesia's cacao crop, receive prices higher than the world average, and enjoy among the highest yields in the world. Contrast this to Cote d'Ivoire, the world's leading cacao producer, where 87 percent of the crop is controlled by foreign multinational companies and growers have been known to smuggle cocoa beans across the border to Ghana to receive better prices.

United States

No U.S. production data are available, although it is probably grown in Hawaii in small quantities.

Foreign Trade

The United States is one of the largest consumers of cocoa beans in the world. In 2003, the United States imported 852 million pounds, or 12 percent of the world's crop. Imports decreased 11 percent between 2001 and 2003, largely due to reduced yields from the leading cacao producers, Cote d'Ivoire, Indonesia, and Ghana. The United States imported 995 million pounds of chocolate products in 2004, slightly more than the import of cocoa beans. The total value of imported chocolate products and various forms of cocoa (beans, powder, butter, paste, cake, and shells) was $2.6 billion in 2004, although the retail value of chocolate in the United States is estimated at $14 billion to $15 billion per year.

BOTANICAL DESCRIPTION

Plant

Cacao, an evergreen, understory tree, is capable of reaching 45 feet in height but generally grows to only 15 to 25 feet in cultivation. Growth habit is strongly sympodial, and trees are composed of many irregular, horizontally arranged branches. A seedling will grow a single vertical shoot for a distance of several feet, at which point the apical meristem divides into three to five pieces, producing a whorl of lateral-growing branches called a "jorquette" (Figure 8.3). Continued growth in height occurs from a lateral bud below the jorquette, which grows vertically again for several feet and produces a second jorquette. The upright-growing shoots are called "chupons"—basically what we would call "water sprouts" or "suckers" on other fruit trees. Phyllotaxis (or phyllotaxy) on chupons is $3/8$ but switches to $1/2$ on jorquette branches. The lower jorquettes may shade out, while upper ones continue to branch, producing plagiotropic shoots that eventually become the fruiting wood of the tree.

Cacao exhibits a flushing-type growth habit, with two to four growth flushes per year. Leaves are short-lived for a tropical evergreen and generally drop after two flushes, as nutrients from old leaves are extracted to feed the current flush. Leaves average 4 to 8 inches long and 2 to 5 inches wide and are glabrous, with entire margins and acute to acuminate tips. Leaves on new flushes are pink to red and hang down vertically (Plate 8.1). The red anthocyanin pigments and orientation are thought to protect new leaves from high light stress until they harden, at which time they become green and more or less horizontally oriented. Leaves have pulvini at the end of each petiole that can regulate orientation throughout the day in response to light intensity—vertical in strong sunlight and horizontal in shade. Thus, cacao is extremely well adapted to shade and most often grown with an overstory tree crop, similar to shade coffee (see Chapter 13).

FIGURE 8.3. A young cacao tree growing in a botanical garden, showing the sympodial growth habit resulting in production of a jorquette about 3 feet above the ground. Note the fruit on the trunk and profuse flowering along the jorquette branches on the right.

Flowers

Flowers are small (½-¾ inch) and white to pink in color, borne in compressed dichasial cymes of a few to several individuals. Cacao is cauliflorous, with flowers arising on old wood at the site of a former leaf, called a "cushion" (Plate 8.2). Flowers are perfect, with one whorl of five functional stamens and five nonfunctional staminodes protruding beyond the reflexed petals. The ovary is superior, with five locules, and 20 to 60 ovules with axile placentation. Flowering begins 3 to 4 years after planting. Flowering is strongest in the rainy season, but at least some flowers are produced year-round in all but the most extreme wet/dry tropical areas.

Pollination

Compatibility is highly variable in cacao, with some types being highly self-incompatible and cross-incompatible with some other cultivars as well. Self-incompatibility is higher in criollo and trinitario types, but 'Amelonado', the major forastero cultivar, is self-compatible. Thus, plantings of 'Amelonado' are self-fruitful, but other types, particularly trinitarios, require cross-pollination. Since many areas use seed propagation, plantings are composed of genetically distinct plants, facilitating cross-pollination.

Cacao is insect pollinated, as pollen is heavy, sticky, and not frequently wind-borne. Cacao has a unique and diverse group of pollinating insects compared to other fruit crops. Midges of the genus *Forcipomyia* are the primary pollinators, but those from *Dasyheolea* and *Stylobezzia* are also involved, as are some species of aphids and thrips. Midges feed on the staminodes, rubbing their bristly backs against the anthers and stigmas and thereby collecting pollen. Midges had to be introduced to nonnative areas to achieve successful cultivation of cacao. Midges breed in banana stems, and intercropping with bananas has been shown to increase cacao fruit set in some areas. Fruit set is generally low, largely due to the profusion of flowers and brief effective pollination period (<24 hours); about 2 to 3 percent of flowers set fruits, but higher fruit set can cause crop load stress and is therefore undesirable. Pollination research has shown that a high percentage of ovules must be fertilized (50-75 percent or more) to allow fruit set.

Fruit

The fruit, erroneously called a "pod," has a hard outer covering that encloses seeds which are lined up in rows. However, the fruit is fleshy at maturity and indehiscent and thus not a pod: it is best described as a pepolike berry (some texts refer to the fruit as a drupe). Fruits are 4 to 12 inches long, less than half as wide, and are borne singly or in small groups on the main stems and older wood (Plate 8.3). The outer portion of the fruit is leathery, usually ridged or lobed, and about ¼ to ½ inch thick. The bulk of the fruit volume is occupied by the seeds, arranged in five longitudinal rows along a central axis. Each seed is enclosed in a fleshy, mucilaginous, sweet aril, termed the "pulp," which is important for development of chocolate flavor, as well as for seed dispersal. In nature, fruits are visited by monkeys and other mammals that chew through the fruit wall, eat the pulp, and discard the seeds. Seeds have either white (criollo) or purple (forastero) cotyledons. Fruits are either red or yellow at maturity, with the red ones showing color immediately and darkening later, and the yellow ones initially appearing light green. Fruits require 5 to 6 months to mature, and they mature year-round, although distinct peaks of production may occur in wet/dry tropical areas, such as West Africa, where flowering is strongest in the rainy season, and thus fruit maturation is highest in the dry season. Cacao thins itself via a process known as "cherelle wilt" (young fruits are called cherelles). The process is similar to that of "June drop" in temperate fruit trees, except that cacao fruits shrivel and mummify on the trees because fruits do not abscise. Self-thinning occurs during the first 3 months as the fruits grow rapidly and compete for resources. Fruiting begins 3 to 4 years after planting, and full crops can be expected at about 10 years.

GENERAL CULTURE

Soils and Climate

Cacao is similar to most other tree crops, growing best on deep, well-drained, loamy soils with a slightly acidic pH. Roots proliferate in the top layer of soil, which is commonly highly organic and covered with litter. The litter layer, underlain by a crumbling layer of decomposing organic matter, is considered important for cacao production. Complete or partial clearing of tropical rain forests is often practiced to find sites with appropriate soils. Cacao is cultivated on over 17 million acres worldwide, more acreage than is devoted to cultivation of most other fruit crops, so relatively large areas of rain forest have been sacrificed for this crop.

Cacao is native to and cultivated in hot, wet tropical lowlands in a belt between 10°N and 10°S latitude. Only a quarter of cacao production occurs outside of this region, in areas that have unusually well-distributed rainfall given their latitude. West Africa, the main production region, has a bimodal rainfall distribution, with short dry seasons in between. Rainfall is 45 to 70 inches per year, and temperatures do not drop below 65°F. High humidity prevails year-round. This is the ideal climate for cacao, but if provided with irrigation or a high water table, cacao can be cultivated in areas with more pronounced dry seasons. Cacao is intolerant of prolonged low or high temperatures, but brief periods of temperatures between 32 and 50°F are not harmful. Frost is lethal, and monthly averages below 65°F reduce growth, as do temperatures above 90°F. Moderate to high winds cause leaf tearing and wind damage to fruits. Thus, the ideal temperature regime is found beneath a canopy of shade trees in the rainy, humid tropics.

As with coffee, cacao is grown normally with a shade tree overstory. The shade canopy provides protection from high light intensity, preserves the litter layer on the soil surface, acts as a windbreak, and provides nutrients to the cacao trees. In West Africa, native forests are thinned by selective felling of trees and the cacao planted beneath them, but in other areas, the land is cleared entirely and shade trees are planted a few years in advance of the cacao. The latter method is more expensive but allows greater regulation of light levels. Nitrogen-fixing leguminous trees, such as those in the genera *Leucaena, Gliricidia, Erythrina,* and *Albizzia,* are generally used. Nonlegume cash crops are also employed (e.g., banana and plantain, rubber, coconut). Optimal yield of unfertilized cacao occurs at a level of 50 percent shade, but fertilized trees yield more at any given light intensity and have an optimum level of 75 percent. Lighter shade has also reduced the incidence of black pod disease, a severe limitation to yield. However, light shade results in drought stress in areas with a long dry season, particularly on shallow soils. Young trees are less tolerant of full sunlight than older trees, which have more extensive root systems and a higher degree of self-shading.

Propagation

Cacao differs from most fruit crops in that seed propagation is the norm. Seeds are carefully selected from superior types that self pollinate or are derived from "seed gardens." A seed garden is an isolated plot used to produce seeds only for planting stock. The self-incompatible nature of many cacao cultivars is exploited in seed gardens; two self-incompatible cultivars will produce only hybrid seeds without having to bag or emasculate flowers. As long as the seed garden is 200 yards or more from the nearest cacao planting, the hybrid seeds produced are uniform and of known parentage. The parent plants in seed gardens are often propagated vegetatively to ensure uniformity. If only one parent is self-incompatible, it is used as the female of the cross and is planted more densely than the male parent, whose fruits are discarded. Flowers are sometimes hand pollinated to ensure a particular cross and to control the time of fruit ripening, and thus the availability of seeds for propagation. All other fruits are removed from the trees when hand pollinated, which prevents off-type seed production and reduces the amount of fruit loss from cherelle wilt (natural thinning).

When desired uniformity cannot be obtained from seed propagation, rooted cuttings are used as the main form of vegetative propagation. The highly heterozygous trinitario cacaos and parents used in seed gardens are propagated this way. Semihardwood cuttings taken from lateral-growing "jorquette" branches root at rates of 50 percent or more when treated with rooting hormone. Cuttings root under mist or in humid enclosures in a few weeks and are grown in a nursery for 1 to 4 months longer before planting in the field. Budding and grafting are possible but infrequently used.

Vegetative propagation is problematic compared to seed propagation for several reasons:

1. It is more expensive and difficult to root cuttings, bud, or graft than it is simply to plant seeds.
2. Cuttings from jorquettes branch at ground level, creating problems for harvest, weed control, and movement of field labor. They must be pruned to a single stem shortly after rooting to obtain a satisfactorily high crotch.
3. All cultivars do not root well, and the major cultivar, 'Amelonado', roots poorly.
4. Cuttings root poorly during the dry season, which would be an optimal time for producing material of transplantable size by the beginning of the rainy season.
5. Suckering may occur on budded or grafted material, such that the rootstock overtakes the scion later.

It is easily understood why seedling propagation is the preferred method.

Rootstocks

Common seedlings are used in the rare cases when cacao is propagated by budding or grafting. Cultivars, such as 'Amelonado', that are sensitive to swollen shoot virus serve as rootstocks for indexing material suspected to carry the virus.

Planting Design, Training, Pruning

Cacao plantings are developed either by thinning existing forest or clear-cutting and planting shade trees in advance of the cacao. Depending on tree size, shade trees are planted 20 to 60 feet apart, and forests are thinned to leave 2 to 3 very large trees and 15 to 20 smaller trees per acre. Cacao plants are spaced 8 to 16 feet apart beneath the shade in either case. Young plantings from clear-cuts are often intercropped with such staples as maize or cassava, to make use of the land, or less often are sown with cover crops, to provide soil cover in the first few years. Shade is too dense to grow crops between cacao plants after a few years. Cacao seeds can be planted "at the stake," as is commonly done in West Africa, which means planting two or more seeds at regularly spaced sites in the planting to ensure establishment. In other areas, cacao seedlings are grown in a nursery for several months, until they are about 2 feet tall and well rooted, and then transplanted during the wet season.

Very little pruning or training is required in cacao production. The seedlings have a naturally sympodial habit and generally do not branch below a height of 3 feet, which would be problematic. Occasionally seedlings produce two chupons (erect stems), in which case one is thinned.

Young rooted cuttings produce jorquettes (lateral branches) too low to the ground and are pruned back to stimulate a second chupon. Mature trees require occasional pruning to remove suckers from the base, low interfering branches, as well as dead and diseased wood. If black pod disease is severe, pruning may open the canopy to allow better spray penetration and/or to create an unfavorable microclimate for disease spread.

Pest Problems

Insects

Mirids and capsids. A variety of these sap-sucking insects damage cacao plants worldwide. All damage cacao stems and pods with their piercing, sucking mouthparts that are designed to extract phloem sap. Feeding sites turn into small black spots that can be hotbeds for infection by opportunistic fungi. Damage ranges from twig dieback and leaf wilting to death of entire trees; in some areas, direct damage to pods is the major problem. Each region has its own unique species: *Sahlbergella singularis* and *Distantiella theobroma* in Africa; *Monalonion* spp. in Central America and South America; *Helopeltis* and *Pseudodoniella* spp. in Southeast Asia. All are winged insects about ¼ inch long with prominent antennae, often as long or longer than the body. Multiple applications of insecticides, timed to the more humid periods of the year when populations are high, are often needed for successful control of these pests.

Shield bugs. These insects posses long feeding stylets that penetrate the walls of young pods all the way to the seeds. Feeding on developing seeds causes young pods to turn black and abort or to stop growing and turn yellow. Two species of *Antiteuchus* occur on Central and South American cacao, causing not only feeding damage but also an increase in the incidence of watery pod rot *(Moniliophthora roreri)*, up to 77 percent. *Bathycoelia thalassina*, the green shield bug, is the major pest in African cacao. Natural predators and parasites have been shown to keep populations down to less than 10 percent, but spraying for mirids eliminates these natural controls. Minimizing mirid sprays reduces shield bug damage. 'Amelonado' suffers less severe damage because the bugs feed on only cacao, and this cultivar goes through a period of very low fruit production once per year, providing a break in the insect's life cycle. Other cultivars, such as the Amazon selections, produce fruits more evenly throughout the year and thus experience greater damage from this pest.

Cocoa moth or pod borer. Conopomorpha cramerella, a serious pest of cacao, is currently confined to Southeast Asia and the Pacific Islands. Adults are ¼-inch moths that lay eggs in furrows of fruits, that are 2 or more inches long. The larvae bore through the pod wall and feed on the placental and aril tissue between the seeds, leaving frass in its wake. The external appearance of the fruit is unchanged. Up to 40 larvae per fruit have been recorded, in which case the seeds cannot be salvaged; with lighter and/or later infestations, some of the seeds can be used. The pest has several alternate hosts, most notably the tropical fruit rambutan *(Nephelium lappaceum)*. Since the insects remain inside the fruits most of the time, insecticide sprays are futile, unless timed and targeted to adult egg-laying or mating periods. Pheromone-based traps and mating confusion have been studied, but the results of field trials are mixed. Release of *Trichogrammatoidea*, a parasitic wasp, has been effective in some regions but is an expensive option. The main control at present, called "rampassen," is the removal of all fruits from trees after the main harvest period, which breaks the insect's life cycle. The method has produced variable results, probably due to the wide array of alternate hosts. Cultivars with thicker fruit walls (e.g., 'Djati Roenggo') may be more resistant than those with thinner walls (criollos).

Diseases

Black pod. This is the most serious disease of cacao worldwide, causing at least 10 percent losses annually, and making cacao an unprofitable crop altogether in some regions. Black pod was thought to be caused exclusively by *Phytophthora palmivora*, but *P. megakarya*, *P. capsici*, and *P. citrophthora* are now known to be involved as well. The most important damage occurs on the fruits (pods), but the disease also attacks stems, young shoots, flower cushions, and even roots. The stem canker phase of the disease can provide inocula for fruit infection. Fruit infection begins as a small brown lesion that quickly spreads to consume the entire fruit in periods of rainy, humid weather. In young fruits, the seeds are still in contact with the fruit wall and thus are susceptible to infection about 2 weeks after the external lesions form. However, in mature fruits, a gap develops between the seeds and fruit wall, allowing seeds to be salvaged from blackened fruits. Copper-based fungicides are the mainstay of control but have to be applied frequently, particularly in rainy seasons. Plants in areas with extended dry seasons have a lower incidence of black pod, and growers in these areas may spray only a few times per year, whereas growers in wet areas spray a dozen or more times per year. Biological control via application of antagonistic bacteria (*Pseudomonas flourescens*) was shown to be as effective as fungicides in some trials. Pruning to open the canopy and removal or pruning of shade trees can decrease the incidence of the disease. Cultivars or breeding lines showing some resistance to black pod include SCA 6, SCA 12, K82, 'Catongo', and 'Amelonado', whereas criollo and Amazon hybrids are more susceptible. However, breeding is complicated by the number of species, and perhaps races, of these pathogens.

Witches'-broom. This serious disease, caused by the fungus *Crinipellis perniciosa*, is confined to the Americas and is responsible for the recent decline of the Brazilian cacao industry. The fungus infects rapidly expanding tissues, notably the buds, which turn into the "brooms." It also affects the growing fruits. Expansion of tissues ceases, and tissues become thickened and compressed. Small fruits become distorted and die, and larger fruits develop hardened necrotic lesions or black speckling. Seeds can be salvaged from fruits infected close to maturity. Losses can reach 70 percent but average around 30 percent. The fungus produces mushrooms on dead wood several months after the initial infection, and these mushrooms release airborne spores. Thus, removal and burning of infected stems, bark, and fruits is a main cultural control. Frequent spraying has proven uneconomical. Long-term breeding of new cultivars with resistance derived from SCA 6, SCA 12, Selicia 1 and 55, and some ICS lines will be needed to control the disease. Some success in breeding for witches'-broom resistance was reported from Brazil in 2002. Brazilian acreage has rebounded to levels near those of a decade ago, but total production and yield remain 35 percent below levels reported prior to the witches'-broom crisis.

Swollen shoot virus. This disease afflicts cacao trees in western Africa, primarily Ghana, where over 180 million trees were cut down in an unsuccessful attempt to control it. The virus is transmitted by mealybugs (*Planococcoides* spp.), which in turn are tended by ants, making it difficult to control the spread of the disease. The virus has several strains, the most severe being 1A (syn. Near Jauben), which can kill seedlings within months and mature trees in about 2 years. In the first year of infection, yield can be reduced up to 50 percent. Symptoms include swellings on shoots and roots and red vein banding patterns in leaves. Removal of infected trees by cutting below the soil surface can control isolated outbreaks, particularly if surrounding trees are removed and regrowth is prevented. However, this has not stopped large outbreaks. Programs targeting the vector have been ineffective as well. Upper Amazon types and trinitario hybrids have shown some resistance and are being used in breeding programs. Cacao biotechnology has advanced to the point at which genetic transformation for virus resistance may be possible (as with the genetic modification of papaya for virus resistance).

HARVEST, POSTHARVEST HANDLING

Maturity

Fruits change color when mature; the red ones darken and develop an orangish cast to the furrows, and the green pods change to yellow. Ripe fruits can be left on the trees for 2 to 3 weeks after their color change without a loss of quality, but fruits left longer on trees are prone to vivipary (premature seed germination), disease, and rodent attack. Fruits require 5 to 6 months to mature, and some mature year-round, with growth peaks that coincide with warmer periods. Cacao is classified as a nonclimacteric fruit with respect to ripening.

Harvest Method

Fruits do not abscise at maturity and must be cut off by hand with a knife or machete. The workers are careful not to damage the bark or "cushion" where the fruits are attached, as more fruits will be borne on these sites. Fruits are harvested ideally at 1- to 2-week intervals, although in practice this may be longer, depending on season and farm size. The first step in processing is fermentation, which requires about 200 pounds of seeds. Small farms may require well over a month for enough fruits to ripen to yield this amount.

Postharvest Handling

The initial processing of the seeds of the fruits into cocoa powder and chocolate consists of producing cocoa "beans" (fermented and dried seeds), which is generally done on or near the farm. Cocoa beans are then exported from producing countries and traded on world commodity markets. The second stage is far more complex and generally done in chocolate factories in the importing countries. Figure 8.4 outlines the main steps in processing cacao fruits.

First, the fruits are cut or cracked open and the seeds are removed by hand. Delaying opening for 3 to 4 days reduces the number of off-grade purple beans, so temporary fruit storage is a common practice. Next, the placenta is discarded, and the pulp-covered seeds are placed in piles or in large wooden boxes for fermentation. Anywhere from a few hundred pounds to a couple tons of seeds are fermented in a heap or box. The fermentation step is critical to cocoa quality, since it is during this process that the color changes to chocolate brown and many of the flavor components of chocolate are synthesized. Bacteria and yeasts on the pulp break down sugars and produce alcohol, liquefying the pulp, which drains away. The temperature rises to about 120°F during the first 2 days of the process, killing the embryo and preventing the seed from germinating. Mixing at this stage aerates the beans, accelerates bacterial activity, and ensures uniform fermentation. Within the bean, membranes break down, enzymatic reactions hydrolyze proteins and anthocyanins, and polyphenols are oxidized. Fermentation takes about 5 to 6 days for forastero types and 2 to 3 days for criollo types. Fermented beans are moved to drying tables or yards to reduce water content from 55 percent to 6 to 7 percent. Sun drying of cocoa beans, similar to drying coffee beans, takes a week or more, and in some areas artificial dryers are used to speed the process. Beans with less than 6 percent water are too brittle, and those with 8 percent or more water can develop mold while in storage. Finally, the beans are cleaned of debris, bagged, and prepared for export.

Chocolate manufacture begins with dried beans, which are cleaned and roasted. As with coffee beans, roasting cacao beans is a delicate process, and the second point at which color and flavor develop. Roasting is done at 210 to 250°F, from as little as 10 minutes up to 2 hours, depending on bean type and the type of chocolate to be made. Underroasted beans still have a fruity flavor and lack characteristic aromas, and overroasting increases bitterness and drives off aro-

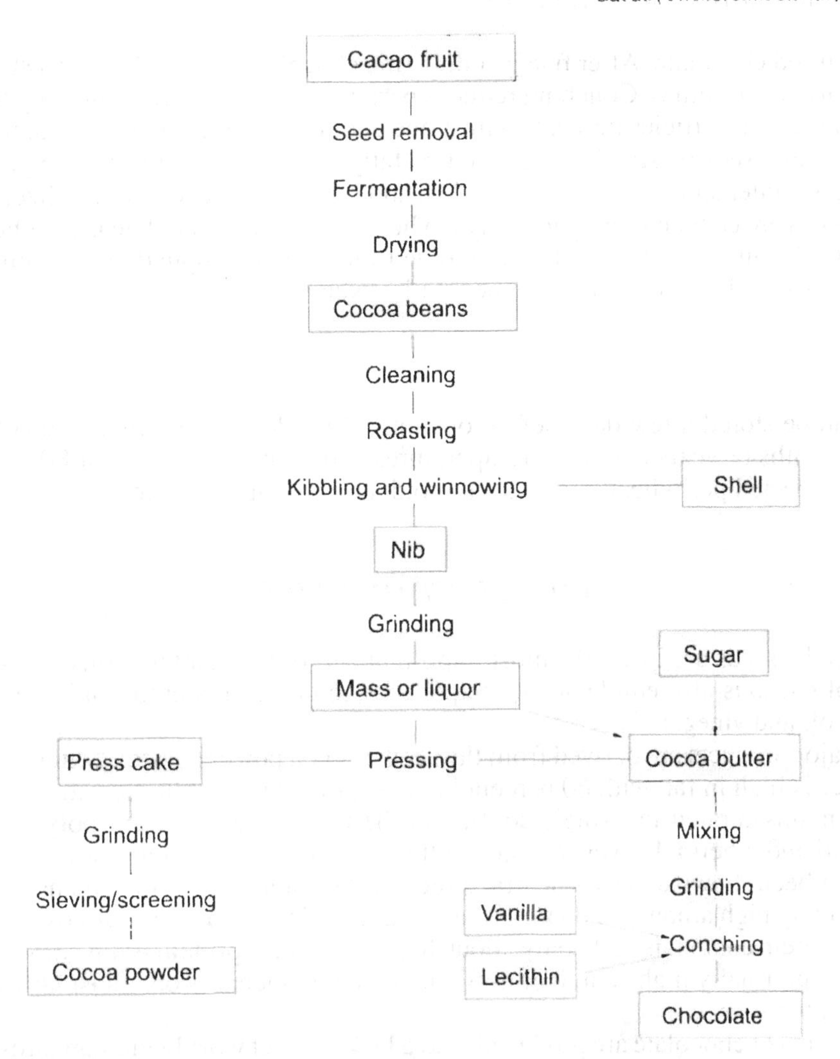

FIGURE 8.4. Schematic representation of the major processes involved in making cocoa powder, cocoa butter, and chocolate from the seeds of the cacao fruit. *Source:* Modified from Wood, 1975.

mas. Criollo beans, used for making high-quality chocolate, are roasted at lower temperatures than are forastero beans. Next, the beans are broken or kibbled to remove the seedcoats ("shells") from the cotyledons ("nibs"). The broken seeds are passed over a current of air to blow out the shells (winnowing), and the cleaned nibs of various types of beans may be blended at this point to obtain different quality grades. The nibs are ground by stones at a temperature of 120 to 155°F to produce cocoa "mass" or "liquor," which is basically cocoa particles suspended in cocoa butter. The mass is liquid at this temperature and contains about 55 to 58 percent fat. The cocoa particles are separated from the cocoa butter by pressing, yielding cocoa press cakes with varying fat contents. Normal cocoa powder is obtained by pressing until 22 percent fat remains, but cocoa powder with fat contents down to 10 percent or less can be obtained by continued pressing and solvent extraction. The press cake is ground, sieved, and screened to yield a fine powder—cocoa powder. Chocolate is created by combining the pressed cocoa butter with various quantities of mass, sugar, milk (for milk chocolate), emulsifiers, and perhaps other ingredients, depending on the type of chocolate. Once mixed together, the liquid chocolate is ground again to reduce particle size from about 100 microns down to 18 microns, yielding a smooth-textured chocolate. After fine grinding, the chocolate is "conched" at 180°F by mixing slowly in various machines. Conching reduces particle size even more, simply by the friction of the sugar and cocoa particles moving against one another, making a smooth, homogenous mixture with higher viscosity. Acidity is reduced, as fatty acids are volatilized during conching. Additional cocoa butter and vanilla may be added, and lecithin is added as a stabilizer. The world's best chocolates are conched for up to 5 days, whereas low-cost chocolate is conched only a few hours, if at all. Finally, the chocolate is molded into bars or applied as a coating over nuts, caramels, fruits, and so on, to produce the retail product.

Storage

Fruits can be stored a few days before opening, if needed. Once dried, cocoa beans can be stored for months or years at normal temperatures and relative humidity of 80 percent or less. Beans are inspected periodically for insect, mold, and rodent infestation.

CONTRIBUTION TO DIET

Although the seeds are by far the most important part of the plant in terms of food usage, the fruit pulp of cacao is also edible and quite pleasing in taste. This pulp can be made into jelly, juice, alcohol, and vinegar.

Three major products are derived from the seeds: cocoa powder, cocoa butter, and chocolate. Cocoa butter is high in fat, with 60 percent being saturated fats (palmitic and stearic acids), 35 percent monounsaturated fats (oleic acid), and the remaining 5 percent polyunsaturated fats (linoleic acid and others). However, cocoa butter contains no cholesterol or harmful trans fatty acids and has been shown to have a neutral effect on cholesterol levels in humans. Cocoa powder has an unusually high amount of dietary fiber for a fruit-derived product, almost 30 percent. Although the protein content is high, only about 40 percent of the protein is digestible. Cocoa powder is also an unusually high source of nutrient elements, such as iron, phosphorus, potassium, copper, and others.

Several forms of chocolate are produced, and a lack of strict worldwide standards has allowed the development of many different styles and brands. In the broadest sense, there are three major types based on ingredients used: white chocolate includes only cocoa butter, sugar, milk, and vanilla or other flavorings; dark chocolate contains cocoa liquor, cocoa butter, sugar, and various flavorings; and milk chocolate differs from dark only by the addition of milk. In the United States, regulations define six types of chocolate: milk, skim milk, buttermilk mixed dairy, sweet, semisweet, and bittersweet.

The melting point of cocoa butter is between room temperature and body temperature, and thus it remains solid until consumed. The melt-in-your-mouth sensation is said to be responsible for some of the pleasure associated with eating chocolate. Over 300 volatile compounds contribute to the unique aroma and taste of cocoa.

In 2002, worldwide per capita consumption of cocoa was just over 1 pound, although consumption in Europe approached 4 pounds, and in the Americas, about 2.5 pounds. Ironically, per capita consumption is lowest in the areas of major production—Africa and Asia—being well below 0.5 pound. About 60 percent of the world's cocoa is consumed in Europe and the United States. For many years, per capita consumption of chocolate in Switzerland was the highest of any country at about 22 pounds, but the Swiss were eclipsed in 2005 by the Irish, who consumed 24.6 pounds per person. In the United States, per capita chocolate consumption is often quoted to be 12 pounds per year, but 2004 import and consumption data from the USDA show about 9 to 10 pounds per year.

About one-half of the chocolate consumed in the United States is in products also containing candy, fruit, nut, or granola, mostly in the form of bars. The other half includes pure chocolate, bonbons, pralines, and other confectionery products. The highest chocolate consumption rates occur among teenagers and young adults in the United States and, surprisingly, are 10 to 20 percent higher among males than females. The highest consumption rates of chocolate snacks actually occur among individuals with lower body mass indices, not among the overweight or obese. About 82 percent of chocolate is consumed as between-meal snacks, with the next highest percentage being consumed at lunch.

Dietary value per 100-gram edible portion:

	Cocoa Powder	Milk Chocolate	Plain Chocolate (sweetened)
Water (%)	3	2	0.5
Calories	229	535	505
Protein (%)	19.6	7.6	3.9
Fat (%)	13.7	29.7	34.2
Carbohydrates (%)	54.3	59.4	59.6
Crude Fiber (%)	33.2	3.4	5.5
	% of U.S. RDA (2,000-calorie diet)		
Vitamin A	0	4	0
Thiamin, B_1	5	8	1
Riboflavin, B_2	14	8	14
Niacin	11	2	3
Vitamin C	0	0	0
Calcium	13	19	2
Phosphorus	105	30	21
Iron	77	13	15
Sodium	<1	3	<1
Potassium	44	11	8

BIBLIOGRAPHY

Alvim, P.T. 1985. *Theobroma cacao*. pp. 357-365. In: A.H. Halevy (ed.), *CRC handbook of flowering*. Volume 5. Boca Raton, FL: CRC Press.

Chatt, E.M. 1953. *Cocoa*. New York: Interscience Publ, Inc.

Duke, J.A. and J.L. duCellier. 1993. *CRC handbook of alternative cash crops*, pp. 448-452. Boca Raton, FL: CRC Press.

Hansen, M. 1983. Chocolate (*Cacao*), pp. 81-83. In: D.H. Janzen (ed.), *Costa Rican natural history*. Chicago, IL: University of Chicago Press.

Keane, P.J. and C.A.J. Putter (eds.). 1992. *Cocoa pest and disease management in southeast Asia and Australia* (FAO Plant Production and Protection Paper 112). Rome: FAO.

Kennedy, A.J. 1995. *Cacao, Theobroma cacao* (Sterculiaceae). pp. 472-475. In: J. Smartt and N.W. Simmonds (eds.), *Evolution of crop plants*. London: Longman.

Knight, I. (ed.). 1999. *Chocolate and cocoa*. Oxford, UK: Blackwell Science.

Lass, R.A. and G.A.R. Wood (eds.). 1985. *Cocoa production: Present constraints and priorities for research* (World Bank Technical Paper No. 39). Washington, DC: The World Bank.

Ogata, N. 2003. Domestication and distribution of the chocolate tree (*Theobroma cacao* L.) in Mesoamerica, pp. 415-438. In: A. Gomez-Pompa, M.F. Allen, S.L. Fedick, J.J Jimenez-Osornio (eds.), *The lowland Maya area: Three millenia at the human-wildland interface*. Binghamton, NY: Food Products Press.

Olaya, C.I. 1991. *Frutas de America*, pp. 38-51. Barcelona, Spain: Editorial Norma.

West, J.A. 1992. A brief history and botany of cacao, pp. 105-121. In: N. Foster and L.S. Cordell (eds.), *Chilies to chocolate: Food the Americas gave the world.* Tucson, AZ: University of Arizona Press.
Willson, K.C. 1999. *Coffee, cocoa, and tea.* Wallingford, UK: CABI Publ.
Wood, G.A.R. 1975. *Cocoa.* Third edition. London: Longman Group Ltd.
Wood, G.A.R. and R.A. Lass. 2001. *Cocoa,* Fourth edition. Oxford, UK: Blackwell Publ.

Chapter 9

Cashew *(Anacardium occidentale)*

TAXONOMY

The cashew, *Anacardium occidentale* L., is a member of the Anacardiaceae family, allied with mango, pistachio, poison ivy, and poison oak. The family contains 73 genera and about 600 species. *Anacardium* contains eight species, native to tropical America, of which the cashew is by far the most important economically. *Anacardium excelsum,* known as espavel, a tall, riparian tree of the tropical dry forest, produces a nut similar to the cashew but is not cultivated.

Members of the Anacardiaceae are known for having resinous bark and often caustic oils in leaves, bark, and fruits. Several species cause some form of dermatitis in humans. It is therefore ironic that two of the most delectable nuts and one of the world's major fruit crops come from this family. The cashew industry, in particular, had to overcome severe limitations imposed by caustic oils in the nut shell. Today, the caustic substance that made plant domestication difficult is a valued by-product of cashew nut production.

Cultivars

Cashew is one of the few fruit crops normally grown from seed, and few improved cultivars exist, at least in commercial production. Yellow and red apple forms exist naturally and do not appear to hybridize readily, but each is genetically variable in its own right and is not recognized as a cultivar. Several cultivars have been selected in India (e.g., the BPP series), and these show exploitable variation in kernel oil and shell oil content. In Brazil, an ongoing breeding program in Ceara (northeast) has produced dozens of dwarf clones, some yielding twice as much as seedling trees, with a higher percent kernel (38 percent more) and good cracking characteristics (<2 percent broken kernels). Cashew is easily grafted, and future release of these cultivars may greatly improve productivity and profitability of cashew production.

ORIGIN, HISTORY OF CULTIVATION

Cashew is native to northeastern Brazil, in the area between the Atlantic rain forest and the Amazon rain forest. The vegetation type of the region is dry forest, savannah woodland, or thorn scrub, and it includes the almost desertlike Caatinga. Cashew is sometimes referred to as a rain forest species, and cashew nuts are found in products that have a rain forest–friendly label or connotation. Although the trees will grow in tropical wet forests, they rarely produce many nuts, and production is far greater in areas with distinct wet and dry seasons, such as in its native range in Brazil, India, and East Africa.

The Portuguese introduced cashew to the west coast of India and East Africa in the sixteenth century, shortly after its discovery in 1578. It was planted in India initially to reduce erosion, with the exploitation of the tree for its edible products, the nut and cashew apple, occurring

doi:10.1300/5547_09

much later. The trees were well adapted to the region and became naturalized. Trees also became naturalized in Central America and the Caribbean Islands. Nut domestication predated the arrival of Europeans to Brazil, although international nut trade did not occur until the 1920s. Native South Americans discovered that roasting nuts in fire would remove the caustic shell oil, allowing the nuts to be cracked and consumed without any ill effects. The roasting practice was either not known or not appreciated outside the cashew's native range, and, as a result, the cashew apple was the first product consumed, with the nut being discarded. Natives also knew of many medicinal uses for the apple juice, bark, and caustic shell oil that were later exploited by Europeans.

India, which developed more refined methods for removing the caustic shell oil, is credited with developing the modern nut industry. India led the world in cashew production for many years, until just recently, when production in Vietnam surged about threefold over a few years. Cashew's native country Brazil ranks in the top five for world cashew nut production, and virtually all cashew apples and juice products come from this country. Cashew nut production surpassed that of almond in 1999, and thus cashew now claims the title of number one nut crop in the world.

FOLKLORE, MEDICINAL PROPERTIES, NONFOOD USAGE

Medicinal uses of cashew bark, leaves, and apple juice are plentiful and were well known prior to recorded history in the native region of Brazil. Bark teas were used for diarrhea, and the caustic shell oil was used to treat skin infections, warts, worms, and botfly larvae beneath the skin. Teas and fruit juices are known to have antimicrobial, anti-inflammatory, astringent, diuretic, hypoglycemic, and other medicinal properties. The active principles are thought to be tannins, anacardic acid, and cardol. The red apples have a higher tannin content than the yellow. Modern uses of shell oil and fruit juice include facial peels, scalp conditioners, and shampoos. Clinical studies have documented the anti-inflammatory properties of tannins and the antimicrobial properties of anacardic acid against several species, including *Escherichia coli* and *Helicobacter pylori*. Leaf extracts produce hypoglycemic activity in rodents and a reduction in artificially induced diabetes. Cashew apples contain up to five times the amount of vitamin C as is found in citrus and strawberries, as well as higher amounts of some minerals than are found in other fruits.

The caustic shell oil, termed "cashew nut shell liquid" or CNSL is sandwiched in a honeycomb layer of tissue between the two walls of the nut shell. Industrial uses include automobile brakes, adhesives, paints and varnishes, insecticides, electrical insulation, and antimicrobials. CNSL is highly caustic, causing moderate to severe skin irritation. When cashew wood is burned or its nuts roasted, contact with or breathing of the fumes can cause skin and eye irritation, inflammation, and poisoning. In addition to CNSL, resins and gums from fruit stems or bark are used as a varnish for books, wood, and flooring to protect them from damage caused by ants and other home-invading insects.

PRODUCTION

World

Total production in 2004, according to FAO statistics, was 2,082,101 MT or 4.6 billion pounds. Cashew is produced commercially in 32 countries. World production has doubled since 1994, with most countries experiencing substantial increases, particularly Vietnam (Figure 9.1).

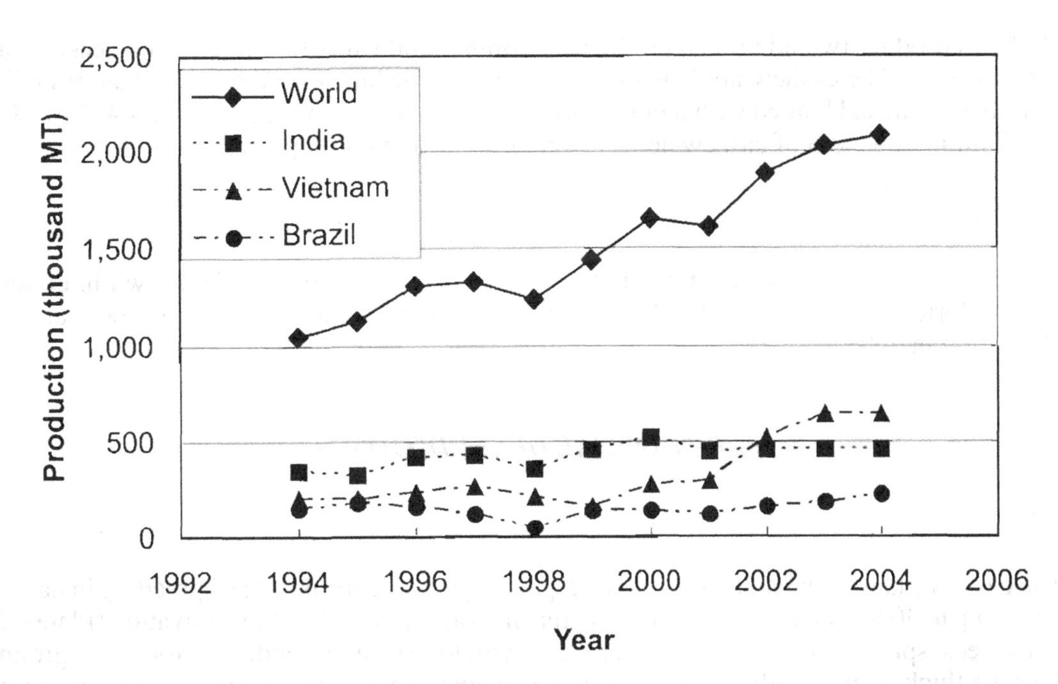

FIGURE 9.1. World cashew production has doubled in the past decade, in part due to a tripling of production in Vietnam. Data from 2004 show Vietnam with 39 percent more production than India, the perennial giant of the world cashew trade. *Source:* FAO statistics.

India, which pioneered the modern processing of nuts, had been consistently the world's leading producer for decades prior to 2002 but was recently surpassed by Vietnam. Cashew is now the leading tree nut crop in the world in terms of production, catching up with and then surging 30 percent beyond almond production in a period of only 5 years.

Cashews occupy just over 7.5 million acres of land area in the world, which is extremely high given the level of production. This reflects the low intensity of production in most areas; many nuts are harvested from wild or naturalized stands of trees. Average yields worldwide are about 600 pounds/acre. In its native range of Brazil, yields are only 290 pounds/acre. In Vietnam, yields are the highest of the top five countries, at just under 2,200 pounds/acre, reflecting the intensive management of the crop in that country. Thus, Vietnam produces about threefold more cashews than Brazil on only 38 percent of the land area. The adoption of high-yielding dwarf cultivars of cashew could easily reduce land area by 50 percent, while allowing increased production to meet rising world demand. The top ten cashew-producing countries (percent of world production) follow:

1. Vietnam (31)
2. India (22)
3. Nigeria (9)
4. Brazil (11)
5. Tanzania (6)
6. Indonesia (4)
7. Guinea-Bissau (4)
8. Cote d'Ivoire (4)
9. Mozambique (3)
10. Benin (2)

Cashews produce two additional products of commercial value from their fruits: cashew apples and CNSL. The cashew apple is the juicy, swollen peduncle of the fruit (Plate 9.1). The juice is astringent and loaded with tannin, particularly in red cashew apples. In 2004, 1,671,010 MT (3.7 billion pounds) of cashew apples were produced, with 96 percent from Brazil.

United States

Cashews are not produced commercially in the United States but can be grown in extreme southern Florida and in Hawaii. The United States imported 102,000 MT of cashews valued at $398 million in 2003.

BOTANICAL DESCRIPTION

Plant

The cashew plant is a small to medium tree, generally single trunked and spreading in habit; it can grow up to 40 feet in height but generally reaches only 10 to 20 feet in cultivation (Plate 9.2). In older trees, spread may be greater than height, with lower limbs bending to touch the ground. Leaves are thick, prominently veined, oval to spatulate in shape, with blunt tips and entire margins. New foliage contains red anthocyanin pigments.

Flowers

Flowering is similar to that seen in cashew's close relative the mango: both male and perfect flowers are borne in the same inflorescence (polygamous). Individual flowers are ¼ inch across, with crimson petals, often striped longitudinally and reflexed. They are borne terminally on panicles, generally at the beginning of the dry season (Plate 9.3). Flowering may occur over several weeks, and it is not uncommon to have ripening fruits and flowers on the tree at the same time.

Pollination

Trees are at least partially self-fruitful, as lone trees can bear many fruits. One study of pollination biology showed no difference in pollen tube growth between self- and cross-pollinated flowers, yet final yield was higher when flowers were cross pollinated. In practice, cashews are often grown from seed, and cross-pollination in orchards must occur to a high degree. Another study showed no indication of self-incompatibility, but a low percentage of fruit set (1-18 percent). Fruit set is highest in flowers that open first, suggesting some type of apical dominance with respect to fruit set, as is found in cashew's other close relative the pistachio. Fruit set is similar for female and perfect flowers. Various insects, even flies and ants, provide pollination.

Fruit

The true botanical fruit is a nut, about 1 inch long, and shaped like a small boxing glove; it hangs below a fleshy, swollen peduncle called the cashew apple or pseudofruit. The cashew apple resembles a pear in shape and size and is juicy, fibrous, and astringent tasting. It has thin skin of either yellow or light red color and yellow flesh. Fruits are borne singly or in small clusters, and mature in 60 to 90 days. The nut develops first, followed by the rapid swelling of the cashew apple in the last few weeks of fruit maturation (Plate 9.4).

The nut shell has an inner and outer wall, separated by a honeycomb-like tissue infused with caustic oil. Cracking the nuts while they are fresh causes oil to contaminate the kernel, so oil is removed by burning or extraction prior to cracking (see Postharvest Handling section). The nuts are about 22 to 30 percent kernel, and cashew kernels are difficult to extract whole compared to those of other tree nuts.

GENERAL CULTURE

Soils and Climate

Cashews are said to be tolerant of sandy, poor soils where many other crops will not thrive. Soil pH is generally on the acidic side, around 4.5 to 6.5. Trees are strongly taprooted and drought tolerant, if soils are deep, and can grow in areas receiving only 30 to 50 inches of rain per year. Cashews are especially intolerant of poor soil drainage.

Climatically, cashews prefer hot, tropical lowlands (<3,000 feet elevation) with a distinct dry season. High rainfall and humidity favor diseases that destroy the flowers and reduce fruit set. Cashews have no cold tolerance whatsoever, requiring protection from cold even in extreme southern Florida.

Propagation

Propagation is most often by seeds that are planted directly in the field where the trees are to be sited. Improved cashews are propagated by grafting, layering, or cuttings. Fruit production occurs in 4 to 5 years when trees are planted from seed, and 2 to 3 years when vegetatively propagated.

Rootstocks

No specific cashew rootstocks exist, but cashew seedlings can be used for rootstocks for grafted trees.

Planting Design, Training, Pruning

Cashews are planted at various densities, depending on the intensity of production, amount of rainfall, and other factors. New plantings are often established with 20 to 35 feet between trees and rows. Higher densities are possible but require selective tree thinning as canopies enlarge. In arid Tanzania, research showed that when about 50 to 60 percent of the ground area was covered by cashew tree canopy, yield per acre was maximized. This is a relatively low amount of light interception compared to intensive orchards and probably reflects the need for trees to exploit large soil volumes in arid areas.

Cashews form open, spreading canopies naturally, and very little information on pruning exists.

Pest Problems

Insects

Thrips. These tiny insects (*Rhynthothrips revenis, Selenothrips* spp.) feed on the surfaces of leaves, causing a silvering or bronzing symptom. Damage is sporadic—some areas of Africa spray multiple times per year to avoid defoliation, and other areas do not even list thrips as a major pest of cashews. Insecticides are applied if leaf silvering is severe. Trees derived from Indian seed sources tend to be more susceptible than those from African sources.

Stem and root borers. The larvae of these beetles *(Plocaederus ferrugineus, Apate terebrans, Analeptes trifasciata)* are wood pests of cashew, girdling limbs or killing entire trees. *Apate terebrans* is a broad-host species, being found on citrus, coffee, and guava as well. Adults lay eggs in bark crevices or girdle branches to provide dead wood for rearing their young. The larvae tunnel through the sapwood or bark, debilitating the conducting tissues of the tree. Affected trees are cut down in severe cases, and the bark is peeled to reveal (and kill) the larvae. *Apate terebrans* larvae are physically crushed by inserting a flexible wire into the hole. Girdled branches that house larvae are removed and burned. In India, an insecticide dust is applied to the soil at the base of the trunk prior to the rainy season, when insects are prevalent.

Tea mosquito. The adult and nymph forms of this pest *(Helopeltis antonii)* feed on developing nuts and inflorescence branches with piercing, sucking mouthparts. Brown feeding lesions can be large enough to cause dieback of new shoots and sometimes invite secondary pathogen infections. The insects multiply during the rainy season, and populations generally peak at the beginning of the dry season, when flowering occurs. One to three insecticide applications are timed just prior to and following flowering. This pest occurs primarily in India, whereas *H. anacardii* and *H. schoutedenii* occur in Africa. A similar piercing, sucking insect, *Anoplocnemis curvipes*, occurs in Africa as well.

Diseases

Anthracnose. The same fungus *(Colletotrichum gleosporioides)* attacks many tropical crops and is the most serious fungal disease of mango, a cashew relative. New leaves and flower panicles are attacked in wet weather, causing defoliation and drop of flowers and developing fruits. Bordeaux mixture or other copper-based fungicides are used for control.

Powdery mildew. Powdery mildew, caused by the fungus *Oidium anacardii*, is a serious disease in African production regions, and in Brazil as well. The fungus earns its name by covering expanding leaves and panicles with whitish gray, powdery mycelia, causing leaf deformation, defoliation, and flower and nut drop. It is most common in dry weather, which prevails in many production regions at the time of flowering. Fungicides are used for control.

HARVEST, POSTHARVEST HANDLING

Maturity

The cashew apple and nut abscise from trees naturally when ripe. Maturation occurs over a period of several weeks during the dry season. The fruit is classified as nonclimacteric with respect to ripening.

Harvest Method

Nuts are collected from the ground by hand. Frequent passes though the planting must be made if apples are to be utilized, as they are highly perishable. Rain at harvest may increase rot and stimulate nut germination.

Postharvest Handling

The presence of caustic CNSL in the shells makes cashew processing more difficult and hazardous than for other nut crops. After harvest, the nuts are dried in the sun or in simple tray driers and stored for processing later. Dried nuts can be stored for about 2 years at room temperature after reaching water contents of 5 to 10 percent. Nuts are rehydrated partially by soaking them or storing them in a high-humidity environment because moisture facilitates extraction of whole kernels and CNSL. Nuts are separated by size before roasting to ensure uniformity of the roasting process.

Nuts are roasted in one of two ways: heating in pans or drums over fire or in hot oil. Pan heating is common to small-scale operations. About 2 pounds of nuts are placed in a shallow, broad pan (similar to a wok) that is heated over a fire. The CNSL exuded from the shells ignites, and the nuts are stirred while they flame and smoke for about 2 minutes. The smoke and flames are dowsed with water, and the nuts are removed from the pan to dry. Drum roasting is a semimechanized form of pan roasting. Nuts are fed into the high end of a large, metal, rotating drum that is tilted from horizontal. The nuts catch fire during their transit along the drum and are sprayed with water at the low end where they exit the drum. The hot oil method is the most sophisticated and generally reserved for larger processing facilities. Nuts are immersed in a tank of CNSL heated to 375°F, and CNSL is driven out of the shell, increasing the volume of hot oil in the tank. About 50 percent of the oil is removed in 1 to 4 minutes, which is enough to allow safe cracking of nuts. Roasting in hot oil is the only method that captures CNSL as a by-product of cashew production.

After roasting, nuts are shelled either by hand or in machines. Hand shelling uses wooden mallets to crack nuts, which is very slow and tedious (only a few nuts per minute). However, cracking employs a large number of people in poor, rural areas where cashews are grown. Machine cracking can increase efficiency to a few dozen to over a ton of nuts per hour but is used in only sophisticated operations. Machines use either knives to open shells or centrifugal force to throw nuts against a hard surface. Hand cracking yields high numbers of whole kernels, but whole kernel yields from machines vary from 20 to 75 percent. Kernels are then dried in the sun or ovens at 160°F until the moisture content is about 3 percent. The papery seedcoat must be removed from the nut, usually by hand, at rates of up to about 25 pounds/day. Nuts are graded by hand and separated into whole kernels and variously sized pieces of broken kernels. Nuts are vacuum packed or packed in carbon dioxide for export.

Cashew apples can be juiced easily because they do not contain seeds; however, juicing must be done immediately at harvest. Apples yield about 60 to 70 percent of their weight as juice. Poor flavors, due to tannins and oils, can be removed by steam treatment, cooking in brine, or adding gelatin to juice, with subsequent filtration (tannins are precipitated by gelatin, as in wine). Juice can be concentrated and frozen or treated with sulfur dioxide and stored at room temperature for several weeks. Juice has properties that allow it to be fermented to wine easily; *feni* is a distilled beverage of about 40 percent alcohol content that is made from cashew wine distillation.

Storage

Vacuum packed, roasted nuts can be stored for up to 1 year, and packing them using carbon dioxide extends life an additional year.

CONTRIBUTION TO DIET

The cashew apple may be consumed fresh, but it contains high quantities of tannins that yield a bitter taste and cause dry mouth. It is more often cooked, partially dried, or candied, as is common in the Dominican Republic and India.. The wine made from the juice is said to be the finest made from tropical fruits.

Per capita consumption of cashew nuts is about 0.8 pound/year (based on import and population data), making it the second most consumed tree nut (almond is first).

Dietary value per 100-gram edible portion:

	Cashew Nut	Cashew Apple
Water (%)	5	86
Calories	553	50-75 (?)
Protein (%)	18.2	0.1
Fat (%)	43.8	0.3
Carbohydrates (%)	30.2	9-15
Crude Fiber (%)	3.3	0.8
	% of U.S. RDA (2,000-calorie diet)	
Vitamin A	0	0
Thiamin, B_1	28	2
Riboflavin, B_2	3	15
Niacin	28	2
Vitamin C (%)	<1	200-600
Calcium	4	<1
Phosphorus	85	2
Iron	37	2
Sodium	<1	<1
Potassium	19	<1 (?)

BIBLIOGRAPHY

Azam-Ali, S.H. and E.C. Judge. 2001. *Small-scale cashew nut processing*. Rome: FAO.

Bhaskara Rao, E.V.V. and H.H. Khan (eds.). 1984. *Cashew research and development*. Kerala, India: Indian Society of Plantation Crops.

Duke, J.A. 2001. *Handbook of nuts*. Boca Raton, FL: CRC Press.

Van Eijnatten, C.L.M. 1985. *Anacardium occidentale*, pp. 15-17. In: A.H. Halevy (ed.), *CRC handbook of flowering*. Volume 5. Boca Raton, FL: CRC Press.

Morton, J.F. 1987. *Fruits of warm climates*. Miami, FL: Julia F. Morton.

Ohler, J.G. 1979. *Cashew* (Communication of the Department of Agricultural Research, No. 71). Amsterdam: Koniklijk Instituut voor de Tropen.

Olaya, C.I. 1991. *Frutas de America*, pp. 38-51. Barcelona, Spain: Editorial Norma.

Popenoe, J. 1969. Coconut and cashew, pp. 315-320. In: R.A. Jaynes (ed.), *Handbook of North American nut trees*. Knoxville, TN: North American Nut Growers Association.

Rosengarten, F. 1984. *The book of edible nuts*. New York: Walker and Co.

Russell, D.C. 1969. *Cashew nut processing* (FAO Agricultural Services Bulletin No. 6). Rome: FAO.

Sturrock, D. 1959. *Fruits for southern Florida*. Stuart, FL: Southeastern Printing Co.

Thomson, P.H. 1979. Jojoba and cashew, pp. 203-210. In: R.A. Jaynes (ed.), *Nut tree culture in North America*. Hamden, CT: Northern Nut Growers Association.

Topper, C.P. 2002. *Issues and constraints related to the development of cashew nuts from five selected African countries* (International Trade Centre, Common Fund for Commodities Report of Project No. INT/W3/69). Geneva, Switzerland: International Trade Centre.

Wardowski, W.F. and M.J. Ahrens. 1990. Cashew apple and nut, pp. 66-87. In: S. Nagy, P.E. Shaw, and W.F. Wardowski (eds.), *Fruits of tropical and subtropical origin*. Lake Alfred, FL: Florida Science Source.

Woodroof, J.G. 1967. *Tree nuts: Production, processing, products*. Volume 1. Westport, CT: AVI Publ.

Chapter 10

Cherry *(Prunus avium, Prunus cerasus)*

TAXONOMY

Cherries are members of the Rosaceae family, subfamily Prunoideae. They occupy the *Cerasus* subgenus within *Prunus*, being fairly distinct from their stone fruit relatives: plums, apricots, peaches, and almonds. *Prunus avium* L. is the sweet cherry, and *Prunus cerasus* L. the sour, pie, or tart cherry. The *Cerasus* subgenus is subdivided into seven sections, each with two to ten species; sweet and sour cherries are in the same section.

As a group, cherries are relatively diverse and broadly distributed around the world, being found in Asia, Europe, and North America. In addition to the main species just mentioned, *P. fruticosa* (ground cherry) and *P. pseudocerasus* (Chinese cherry) are minor fruit species grown in Russia and China. While sweet cherries are pure *P. avium*, the sour cherry group may include hybrids between *P. avium* and *P. cerasus* (referred to as "Duke cherries"), ground cherry, and hybrids of ground cherry with *P. cerasus*. Duke cherries resemble sour cherries more than sweet cherries. Both sweet and sour cherries can be grafted onto the same rootstocks. However, none of the other stone fruits are cross- or graft compatible with cherries.

There are probably as many ornamental cultivars of cherry grown in the United States as there are fruiting cultivars since all cherries flower profusely in spring. 'Yoshino' *(Prunus ×yedoensis)* is perhaps the most common and widely planted landscape tree. Other common landscape trees include *P. serrulata* ('Kwanzan', 'Shirotae'), *P. sargentii* (Sargent cherry), *P. serrula* (birch bark cherry), and *P. subhirtella* (autumn-flowering cherry).

Sweet Cherry Cultivars

Fewer than 100 sweet cherry cultivars are grown in the major production regions around the world today. Thousands of cultivars may have existed prior to the 1800s, when cherries were grown largely from seed. When Romans dispersed cherries throughout Europe, cultivars of local importance were selected. Breeders collected the best of these and produced contemporary cultivars directly, or after a modest amount of genetic improvement.

Cultivars are sometimes categorized into the "Heart" (syn. "Guigne" in France, "Gean" in England) and "Bigarreau" groups. The former are heart-shaped, softer fruits, while the latter are round, firm, crisp fruits. Both groups contain white- and red-fleshed cultivars as well as dark- and light-skinned cultivars. Many of the top cultivars today are in the Bigarreau group, since firmness is an important postharvest criterion.

'Bing', 'Napoleon' (syn. 'Royal Ann'), 'Ranier', and 'Lambert' are the most important cultivars in North America. 'Bing' is the number one cultivar, used almost exclusively for fresh-market sales. 'Bing' and 'Lambert' were selected from chance seedlings in the 1800s. 'Bing' is a large, dark, firm cherry that ships well. However, it tends to crack if rain occurs near harvest and therefore does poorly in rainy summer climates. 'Napoleon' is the main cultivar for maraschino cherry production. It is picked early, prior to color development, and has white flesh, a desirable

Introduction to Fruit Crops

doi:10.1300/5547_10

characteristic for maraschino cherries. It is also canned. 'Lambert' is used for canning and fresh-market sales. 'Ranier', a USDA introduction from Washington, has become a major fresh-market cultivar in the past 20 years. It is light red with a yellowish background color, in contrast to the normally deep red to black cultivars used for fresh market sales.

Sweet cherry cultivars exhibit a gametophytic type of self-incompatibility, which also results in a certain degree of cross-incompatibility. When incompatible pollen reaches the stigmatic surface, the pollen tube forms but cannot complete its growth through the style, resulting in a lack of fertilization. The genetics of self-incompatibility in sweet cherry are well known, and cultivars are divided into about 12 groups based on cross-compatibility. Hence, cross-pollination and choice of pollinizer are critical in sweet cherry culture. Common pollinizers for 'Bing' are 'Early Burlat', 'Black Tartarian', and 'Van'. The few existing self-compatible cultivars, such as 'Stella' and 'Lapins', are of poorer quality than 'Bing' and the other cultivars that form the basis of the industry.

Sour Cherry Cultivars

Sour cherries are also divided into two groups: Amarelles and Morellos. Amarelles are upright, vigorous trees, with pale-colored fruits or with reddish fruits that contain a light-colored or clear juice, and with low acid content ('Montmorency', 'Kentish', 'Early Richmond'). Morellos are small, bushy, compact trees, with dark red fruits that are more spherical in shape, have higher acid content, and contain red-colored juice ('Stockton', 'Vladmir', 'North Star'). 'Montmorency' is by far the main sour cherry cultivar grown in the United States and Canada, accounting for 99 percent of all production. In 1915, there were 270 sour cherry cultivars listed in Hedrick's *The Cherries of New York*, but superior yield and quality in 'Montmorency' caused growers to abandon other cultivars. In Europe, 'Schattenmorelle' is a major cultivar in the Morello group. Only Hungary, Russia, and Romania have industries based on five or more cultivars.

ORIGIN, HISTORY OF CULTIVATION

Sweet Cherry

This species originated in the area between the Black and Caspian Seas of Asia Minor. Birds may have carried it to Europe prior to human civilization. Cultivation probably began with Greeks and was perpetuated by Romans, who consumed it as an essential part of the legionnaire's diet (this led to sweet cherry's spread throughout Europe). Trees were planted along roadsides and were valued for their timber as well as their fruits. Sweet cherries came to the United States with English colonists in 1629 and later were introduced to California by Spanish missionaries. In the 1800s, sweet cherries were moved west by pioneers and fur traders to their major sites of production in Washington, Oregon, and California. Cultivars selected at that time still form the base of the industry today.

Sour Cherry

Good evidence suggests that *P. cerasus*, a tetraploid, arose from a natural cross between *P. avium* and *P. fruticosa* (ground cherry). Unreduced ($2n$) pollen from sweet cherry would have allowed successful reproduction between the two species. The geographic ranges of the two species overlap in northern Iran and Turkmenistan, which is the center of origin of sour cherry. From there, sour cherry followed a similar course to Europe as did sweet cherry and ultimately came to North America with English settlers. It is more tolerant of the humid, rainy eastern conditions and therefore proliferated there more than sweet cherries did, and it is still cultivated primarily in that region today. Low monetary returns make sour cherry a less attractive investment than sweet cherry. Thus, it has been planted in western states only to a limited extent. Michigan, the leading producer, grows sour cherries along the eastern shore of Lake Michigan, where the moderating influence of the lake on winter and spring temperatures is beneficial to production.

FOLKLORE, MEDICINAL PROPERTIES, NONFOOD USAGE

The cherry has been associated with virginity from ancient to modern times, a link that probably arose from the appearance of the red-colored fruit with an enclosed seed, which came to symbolize the uterus. Maya, the virgin mother of Buddha, was offered fruit and general support by a holy cherry tree while she was pregnant. In European folklore, one's future spouse would be revealed by repeating the following chant while counting cherry pits: "This year, next year, sometime, never." In Danish folklore, cherry trees housed forest demons. A good crop of cherries was ensured by having the first ripe fruits eaten by a woman shortly after her first child is born. Cherries are symbolic of both good education and deception according to the "language of flowers."

As in all stone fruits, cherry leaves, flowers, and especially the seeds and bark contain toxic compounds that generate cyanide, which is of course toxic or lethal in large doses. However, in plant tissues, cyanide is low enough in concentration to be considered therapeutic, particularly for cancer (tumor) treatment, and has been used for this purpose since at least 25 BC. *Prunus serotina*, the wild black cherry of eastern North America, is the most dangerous species in the Rose family in terms of consumption. Children have been poisoned by chewing twigs, eating seeds, and drinking tea made from the leaves. All classes of livestock have been killed from eating the leaves. Keracyanin, from sweet cherry fruits, has been shown to maintain normal levels of uric acid in people suffering from gout.

Ojibwa Indians treated coughs and colds with tea made from the bark of wild cherry (*P. serotina* and *P. virginiana*). Cherry stalk tea is a diuretic and is thought to cure kidney diseases. Tests on rats and mice indicate that a compound in cherry stalk extract (Novicardin) may reduce inflammation caused by caustic materials if administered 30 to 60 minutes before the inflammation is induced. This is consistent with the known anti-inflammatory activities of flavonoids, which are similar to compounds found in cherry stalks. Phloretin is an antibiotic-like compound found in bark and root extracts; in concentrated form, phloretin can kill certain bacteria.

PRODUCTION

World

Sweet Cherry

Total production in 2004, according to FAO statistics, was 1,896,522 MT or 4.2 billion pounds. Sweet cherries are produced commercially in 65 countries on over 926,000 acres. Acreage has increased 22 percent over the past decade. Worldwide yield averages just over 4,500 pounds/acre and has remained constant or decreased slightly in recent years. Large decreases in yield have occurred in central and eastern European countries, as well as in the United States and some Middle Eastern countries, masking an increase in yield in most production regions. Production data show a fair amount of year-to-year variation but an overall increase of 20 percent in the past decade. The top ten sweet cherry–producing countries (percent of world production) follow:

1. Turkey (13)
2. Iran (12)
3. United States (12)
4. Ukraine (7)
5. Germany (6)
6. Italy (5)
7. Romania (5)
8. Russia (5)
9. Spain (4)
10. France (3)

Sour Cherry

Total production in 2004, according to FAO statistics, was 1,035,650 MT or 2.3 billion pounds. Sour cherries are grown in 27 countries worldwide on 613,000 acres. World production has increased 10 percent in the past decade, and, as with sweet cherry, sour cherry production shows a fair amount of annual variation. Acreage has increased about 15 percent since 1994, and this is responsible for the increase in production, since yield has been stable at about 3,700 pounds/acre. Yields can reach over 6,000 pounds/acre in the most productive countries. The top ten sour cherry–producing countries (percent of world production) follow:

1. Russia (20)
2. Poland (18)
3. Turkey (14)
4. United States (11)
5. Serbia and Montenegro (10)
6. Germany (8)
7. Iran (5)
8. Hungary (4)
9. Czech Republic (2)
10. Georgia (1)

United States

Sweet Cherry

Total production in 2004, according to USDA statistics, was 253,854 MT or 564 million pounds. The industry value is $436 million. Prices received by growers are just under $1.00 per pound for fresh fruit and $0.30 per pound for processed, higher than those for most other fruit crops. Prices have increased steadily from around $0.40 per pound in the mid-1980s to current values. Sweet cherries are produced commercially in nine states on about 78,000 acres.

Sour Cherry

Total production in 2004, according to USDA statistics, was 96,800 MT or 213 million pounds, valued at $70.8 million. More typical are the production figures over the past decade: 125,000 to 165,000 MT, valued at $40 million to $55 million (Figure 10.1). Returns to growers

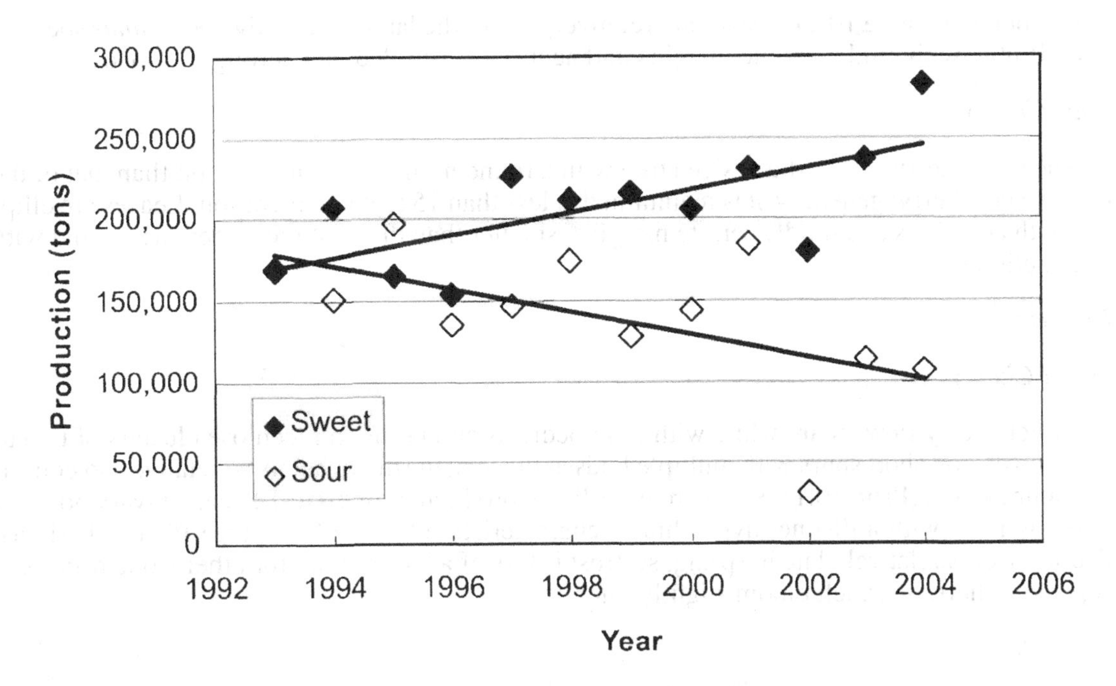

FIGURE 10.1. Sweet and sour cherry production in the United States are on different trajectories. Approximately equal in 1993, sweet cherry production has increased at an average rate of 6,800 tons per year, while sour cherry production has declined at a similar rate. *Source:* USDA statistics.

have improved to $0.33 to $0.45 per pound as a result of lower production in recent years; previously, returns were only $0.10 to $0.20 per pound. Higher prices have more than offset declining production. Sour cherries are produced commercially in eight states on about 37,000 acres. The top cherry-producing states (percent of U.S. production) follow:

Sweet Cherry
1. Washington (47)
2. California (26)
3. Oregon (15)

Sour Cherry
1. Michigan (70)
2. Utah (10)
3. Washington (8)

The United States exports about 15 to 20 percent of its production fresh and 5 to 10 percent canned. Imports are only about 5 percent of domestic production. Fresh exports and imports have increased several-fold since 1984, with imports outpacing exports. Increased imports of fresh cherries are due in part to out-of-season production from Southern Hemisphere countries, such as Chile.

BOTANICAL DESCRIPTION

Plant

Sweet Cherry

The sweet cherry is a vigorous tree with strong apical control with an erect-pyramidal canopy shape that is capable of reaching 50 feet. In cultivation, sweet cherries are maintained at 12 to 15 feet in height (Figure 10.2). Leaves are relatively large (the largest of cultivated *Prunus* species) and elliptic, with mildly serrate margins and acute tips, petioled, and strongly veined.

Sour Cherry

The sour cherry is a medium-sized tree with a rounder, more spreading habit than that of the erect sweet cherry; generally, it is maintained at less than 15 feet in cultivation. Leaves are elliptic, with acute tips and mildly serrate margins, smaller than those of the sweet cherry, and with long petioles.

Flowers

Sweet Cherry

Sweet cherry flowers are white with long pedicels and borne in racemose clusters of two to five flowers, on short spurs with multiple buds at the tips; the distal bud is vegetative and continues spur growth (Plate 10.1). Spurs are long-lived, producing for 10 to 12 years. Ovary position is perigynous with a distinct hypanthium, characteristic of stone fruits (see Plate 1.1, right). Bloom occurs relatively late in spring, so frost is less of a hazard than for other stone fruits, except sour cherries, which bloom slightly later.

Sour Cherry

Individual flowers are the same as for sweet cherry. Sour cherry inflorescence buds usually produce two to four flowers, with long pedicels, as in sweet cherry. However, many are borne laterally on 1-year-old wood, not exclusively on spurs, as in sweet cherry. Spurs are more short-lived on sour than on sweet cherry, gradually declining in productivity over 3 to 5 years. Sour cherries are the latest blooming of the stone fruits and therefore would be less frost prone than sweet cherries. However, frost is still a major growth limitation for sour cherry production in Michigan.

Pollination

Sweet Cherry

Pollination is absolutely essential for production, since sweet cherries are self-incompatible and need a high degree of fruit set (25-50 percent) for a commercial crop. A few cultivars, such

FIGURE 10.2. *(left)* Young 'Bing' sweet cherry trees on semidwarfing MxM rootstock *(background)* or Gisela dwarf stock *(foreground)*. *(right)* A mature 'Montmorency' sour cherry orchard in Michigan.

as 'Stella' and 'Lapins', are self-fruitful. In addition to self-incompatibility, sweet cherry has a high degree of cross-incompatibility, which is well understood. For example, the major U.S. sweet cherry cultivar, 'Bing', is cross-incompatible with 'Emperor Francis' and 'Kristin', meaning that these cultivars will not pollinate one another because they are in the same incompatibility group. About 12 incompatibility groups have been identified for sweet cherry.

Pollinizers are set every third tree in every third row, or at a ratio of 8-9:1. Since sweet cherries are handpicked, these arrangements present no imposition to harvest. Alternate-row arrangements, grafting limbs of other cultivars onto trees, and the use of bouquets are alternatives to distributing pollinizers throughout the orchard. Honeybees are the main pollinator.

Sour Cherry

Sour cherries are self-fertile and therefore require no pollinizers.

Fruit

Sweet Cherry

The sweet cherry fruit is a drupe, ranging in size from ½ inch to 1¼ inches, round or heart-shaped, and glabrous, with a long pedicel attached (Plate 10.2). The pit is generally smooth and encloses a single seed. The skin color is generally deep red or purple (often referred to as "black"), yellow, or white (rarely). Yellow fruits often have a red cheek. The flesh color varies from white to dark red. Fruits are borne on short spurs that arise from older wood. Sweet cherries require only about 2 to 3 months for fruit development.

Thinning is unnecessary for fruit size development, and since a high proportion of flowers must set for a crop, this is not practiced. Maximum yields are obtained beginning in the fifth or sixth year after budding, and trees are productive for 25 to 30 years, although the trees live much longer. Dwarfing rootstocks can reduce the nonbearing period to 3 years.

Sour Cherry

The sour cherry is also a drupe with a generally smooth pit that encloses a single seed. Sour cherries generally have lower sugar and higher organic acid contents than do sweet cherries, giving them their distinct flavor. Major cultivars are generally bright red in color and exhibit less color variation than the sweet cherries. Fruiting begins earlier for sour cherry than for sweet cherry, after 3 to 4 years. Productive life is shorter, however, lasting only 20 to 25 years. Fruit size ranges from ½ inch to almost 1¼ inches. Fruits are borne on short spurs that arise from older wood as well as laterally on 1-year-old wood. The proportions vary with cultivar: 'Montmorency' produces about two-thirds of fruits laterally on longer wood, and one-third on spurs. Hungarian cultivars produce 70 to 90 percent of fruits on spurs. Sour cherries require only about 2 to 3 months for fruit development. No thinning is required.

GENERAL CULTURE

Soils and Climate

Deep, well-drained, gravelly to sandy loam soils are best. Cherries develop extensive root systems, so water-holding capacity is not as important as aeration. Flooded or wet, heavy soils slow growth and reduce productivity.

Sweet cherry culture is most successful in cooler, drier climates where the danger of spring frost is limited, disease pressure is reduced, and rainfall does not occur during harvest. This is why commercial culture is primarily limited to eastern Washington State, Oregon, and California. Some crack-resistant cultivars that are used in backyard or small-scale production perform well in the mid-Atlantic and northeastern states. Diseases, high temperatures, and high humidity in the southern states make sweet cherry cultivation very difficult. Intense heat during floral initiation and development causes doubling of ovaries, which results in spurred or doubled fruits the following year.

Sour cherries do best in cooler, humid climates. Sour cherries can be grown in the western states, but, commercially, fruit size is compromised in arid climates, so eastern areas dominate production.

Both sweet and sour cherries have relatively high chilling requirements, about 1,000 to 1,500 hours. This precludes production in warmer climates. Attempts to produce low-chill cherry cultivars have met little success. Cherries are the last of the stone fruits to bloom in a given location, with sweet blooming before sour. Coldhardiness in cherries is better than in peach or apricot, but not as great as in European plum or apple. Fully dormant buds are killed at about –20°F. Cherries produce well in Northern Michigan, Ontario, and British Columbia, as well as in northern Europe and England; they are among the hardiest of stone fruits. Sour cherries are more cold tolerant than are sweet cherries. Cherries are the earliest maturing of the stone fruits, with most cultivars maturing in 2 to 3 months after bloom. Length of growing season is not a concern.

Rainfall and high humidity during the growing season, particularly at bloom or harvest, is a serious limitation due to fungal diseases that kill the flowers and shoots or rot the fruits. Rainfall during harvest is especially harmful, causing a condition known as fruit cracking. For this reason, most sweet cherries are grown in dry climates. Brown rot is also extremely damaging to cherries and is exacerbated by high humidity and rainfall near harvest.

Propagation

Cherries are T-budded onto rootstocks during late summer and forced to grow the next season.

Rootstocks

Until the 1990s, there were primarily two types of rootstocks for cherries: 'Mazzard', a wild selection of sweet cherry, producing a vigorous large tree, and 'Mahaleb', a separate species *(Prunus mahaleb)* that does best in well-drained soils. Mahaleb was selected in France and is good for planting in calcareous (higher pH), droughty soils. In the United States, most sweet cherries are grown on Mazzard and most sour cherries on Mahaleb. Both are propagated in nurseries from seed, but clonal selections of both also exist. Hybrids of the two, designated "M×M," are used commercially to a limited extent. For example, M×M 14 is semidwarfing, reducing size by 30 percent compared to Mazzard seedlings. 'Stockton Morello', a type of sour cherry, is used to control tree size and induce precocity and may be the best choice for heavier, wetter soils. In England, F12/1 is a vigorous *P. avium* selection often used for cherry, and 'Colt' *(P. avium × P. pseudocerasus)* is used as a semidwarfing stock. 'Colt' has poor hardiness and a tendency to produce suckers.

A new series of dwarfing rootstocks for cherry, known as "Gisela" or the Geissen series, first developed in Germany, are now available to growers (Figure 10.2, left). These stocks could do for cherries what the Malling series did for apple, that is, allow production from dwarf trees with high yield efficiency and precocity. Gisela stocks have been more readily adopted for sweet cherry, since it is more vigorous than sour cherry and less precocious. Also, sour cherries are shake harvested and require less regulation of tree size than do hand-harvested sweet cherries. Four of the several original stocks have been adopted in the United States (height and yield efficiency compared to a tree on Mazzard):

Rootstock	Height (%)	Yield Efficiency (%)
Gisela 5	45	210
Gisela 7	50	160
Gisela 12	60	170
Gisela 6	70-80	125

Planting Design, Training, Pruning

Sweet Cherry

Sweet cherries are capable of growing very large, yielding low numbers of trees per acre. On standard rootstocks, tree densities are around 100 trees/acre, but the Gisela rootstocks have allowed densities of hundreds of trees per acre. Filler trees may be used to increase returns in early years; trees may be planted on diagonals, 20 to 25 feet apart, and then alternate diagonal rows are removed, yielding trees 30 to 35 feet apart in straight rows. Pollinizers can be dispersed throughout the planting, for example, on every third tree in every third row (or a ratio of 8:1), or they can be planted in solid rows, alternating with the main cultivar. For fresh-market production, dispersed placement presents no problem because the fruits are hand harvested.

Sweet cherries are generally trained using the central-leader or modified central-leader system because the growth habit is naturally upright. It is difficult to induce branching and obtain wide-angled scaffolds on young trees, as with pears. High-density systems, such as hedgerow, French axe, or candelabra systems, are used to a limited extent with dwarfing stocks. The Spanish bush system, where several scaffolds originating from near the soil surface have grown out, fruited briefly, and been removed or headed back, has been used to a limited extent. Older orchards typically use the modified central-leader system, in which the central leader is removed to allow light into the center of the tree canopy.

Trees require the least pruning of any of the tree fruits because fruits are borne on long-lived spurs and many fruiting points are needed for full production. Some dormant heading is necessary during formative and noncropping years to encourage branching, since young trees produce long, succulent, vertical scaffolds. Otherwise, dead or interfering branches are removed routinely, and only about 10 percent of the fruiting wood is renewed each year.

Sour Cherry

Sour cherry trees do not attain the size of sweet cherries and are generally pruned more to stimulate new shoot production. Hence, closer spacings are used: 18 × 24 feet and rectangular or up to 25 feet and square. The shake and catch harvesters are relatively large and preclude the use of high-density plantings.

Sour cherries are trained to modified central-leader or open-center systems. The trees have more or less spherical canopies at maturity, with fruiting heaviest at the periphery. Some selective limb thinning may be necessary to maintain adequate light and spur development inside trees, as well as to prevent the peripheral fruiting area from becoming too thin. Young trees are headed higher than are other stone fruits to allow ample room for shaker attachment to the trunks. Branches interfering with harvest equipment are removed. Selection of scaffolds targets those with wide angles of attachment to the trunk and several inches of space between adjacent scaffolds. Sour cherries require more pruning than sweet cherries due to their tendency to produce lateral flower buds exclusively on short shoots. The so-called "blind wood" in sour cherries results from 1-year-old wood devoid of lateral vegetative buds; in subsequent years, the plants cannot produce new shoots or flowers. Blind-wood formation is accentuated if extension growth of lateral shoots is minimal, or less than 6 inches/year.

Pest Problems

Insects

Black cherry aphid. These shiny, black, small (1/8 inch long) insects *(Myzus cerasi)* suck sap from newly emerging foliage. Sweet cherries are more affected than sour. Leaves and stems are crinkled and deformed; sooty mold fungus can grow on the honeydew excretions of the aphids. Insecticides prior to bloom or just after petal fall are used for control.

Cherry fruit fly, black cherry fruit fly. Adult flies (*Rhagoletis* spp.) emerge from the soil in late spring. Females lay eggs on fruits near harvest, and the resulting maggots (white, 1/4 inch long) feed on the flesh. A problem in the northern United States and Canada, black cherry fruit fly prefers sour cherry over sweet cherry. One or two insecticide sprays during the month prior to harvest are used for control.

Plum curculio. Conotrachelus nenuphar adults are gray to dark brown snout beetles, 1/8 to 1/4 inch long; grubs are small, yellow-white, and legless and burrow into the fruit flesh, usually causing the fruit to fall off. Feeding damage by adults alone may cause "catfacing" (misshapen fruit) or D-shaped brown depressions in the fruit surface. Insecticide sprays at petal fall and then again 7 to 10 days later provide control. The insect is confined to the United States east of the Rocky Mountains.

Scale insects. Adults (*Quadraspidiotus* spp., others) are small (<1/8 inch long), round or oval, sedentary insects that appear as bumps on twigs or branches; they suck the sap from branches and twigs. Dormant oil is sprayed 1 to 3 weeks before budbreak, and combined with insecticide if scale infestation is severe.

Oriental fruit moth. These small moths *(Grapholita molesta)* lay eggs in shoot tips early in the season; the resulting larvae burrow downward a few inches. Wilted and dead expanding leaves at the shoot tips indicate infestation. Later infestation results in fruit loss, as eggs laid on the fruits produce 3/4-inch, legged, pinkish larvae that burrow through the fruits. Although fruit infestation does not occur until fruits are about half grown, early season sprays (petal fall, shuck split) are critical for controlling this pest.

Mites. As in other fruit crops, mites (*Panonychus, Tetranychus* spp., others) feed on the lower leaf surface, causing a characteristic "mite stipple" damage to leaves. Mites may be white, red, or yellow and two-spotted. They generally appear on leaf undersides and are worst in hot, dry weather. Dormant oil is sprayed before budbreak, reducing initial populations and eliminating mites for the season. In hot, dry summers, miticide may be required when visible injury occurs.

Diseases

Brown rot/blossom blight. The same fungi *(Monilinia laxa, M. fructicola)* cause flowers to rot after blooming and fruits to rot near harvest. *Monilinia laxa* is known as European brown rot, and *M. fructicola* as American brown rot. A brown, powdery mass of spores in concentric rings is visible around a soft lesion on the fruit. Flowers affected with blossom blight turn brown, and spurs or twigs can be killed. Fruits infested in the previous year die, shrivel, turn black, and hang on the tree. These "mummies" house spores for next year's infection and are thus removed from trees in winter. Fungicides applied at first bloom control blossom blight, and preharvest sprays, with multiple applications in rainy years, control fruit rot.

Cherry leaf spot. One of the most serious diseases of sour cherry, cherry leaf spot is caused by the same fungus (*Blumeriella jaapii,* syn. *Coccomyces hiemalis*) that also causes leaf spotting and defoliation on sweet cherry. Defoliated trees produce poor-quality fruits, are less winter hardy, and bloom and fruit less in subsequent years. Fungicides can control the disease, starting at petal fall stage and continuing at 10-day intervals if leaves continue to yellow and drop. A postharvest application (2 to 3 weeks after) may be needed if trees are defoliating too soon in late summer. The disease is more difficult to control in sour cherry than in sweet cherry.

Powdery mildew. The fungus *Podosphaera leucotricha* is a problem in drier areas, such as the Pacific Northwest and California. Leaves, flowers, and fruits are affected and may fall off if the disease is severe. Its name comes from the white, feltlike patches of fungus that occur on the lower leaf surface and all over flowers and young fruits. Shoot growth of young trees is reduced, and fruits are cosmetically affected in most cases of flower infection. Fungicides are applied, starting about 2 to 3 weeks postbloom, and may need to be repeated until shoot growth ceases in summer on highly susceptible cultivars. Major cultivars of sweet cherry are susceptible, as is 'Montmorency' sour cherry.

Perennial canker. Caused by *Cytospora, Leucostoma,* and *Valsa* spp., this is an important trunk and limb disease in the northern United States. The fungus causes elliptical depressions, called "cankers," in the wood; they often have brown gum surrounding them and blackened tissue beneath and around them. Cankers may grow larger over several years and form concentric rings of wound tissue around them. Twigs, limbs, or even trees are eventually girdled and killed. Fungicides are relatively ineffective, so cultural control is the primary option. The fungus enters through wounds and affects weak trees most often. Thus, avoiding winter injury, limb breakage, sunscald, insect wounds, and large pruning wounds is important. Broken limbs, those with large slow-healing pruning cuts, and those with visible cankers are pruned out. The bark can be surgically removed about 1 inch around the canker on the trunk or a large limb in summer, when wounds heal fastest.

Bacterial canker. The bacteria *Pseudomonas syringae* pv. *morsprunorum* causes injury to twigs, branches, and large limbs. Branches and limbs are girdled in severe cases, and they die back quickly. Many "suckers" are produced from the base of the trunk on severely infected trees because the lower trunk and roots are not killed. This is nearly impossible to control with sprays, as with many bacterial diseases. Freezing injury, nematode weakening, other diseases, and drought or waterlogging all predispose trees to infection, so healthy trees are a good defense. There is little resistance to this disease in sweet or sour cherry, but M×M rootstocks have some resistance, and some European sweet cultivars have moderate resistance.

Viruses. Cherries suffer from more virus diseases than any other stone fruit. Various symptoms include fruit discoloration; foliage, fruit, and stem deformation; reduced yields and tree vigor; union separation; and tree death. Common viruses include the following:

Cherry leaf roll
Cherry raspberry leaf
Green ring mottle
Prune dwarf virus (sour cherry yellows)
Prunus ringspot (necrotic ringspot)
Rusty mottle
Spur cherry
Yellow bud mosaic

The dwarfing and precocity characteristics of 'Stockton Morello' rootstock seem to have been induced by rusty mottle, crinkle leaf, fasciation, and perhaps other viruses. Most viral diseases are graft transmissible, and, therefore, care in selecting virus-free bud wood and using seed from virus-free mother trees for rootstocks are essential for control.

HARVEST, POSTHARVEST HANDLING

Maturity

Traditionally, color change and soluble solids content are used to signal maturity. Soluble solids may approach 23 percent in ripe sweet cherries. However, "fruit removal force" has been used more recently as a predictor of maturity and is more reliable. This measure is based on the progressive abscission of the fruits from the pedicels, starting about 2 weeks before maturity. It is measured by a pull gauge, which pulls the fruit from the pedicel and registers the force required to remove the fruit. Both sweet and sour cherries are classified as nonclimacteric fruits with respect to ripening.

Harvest Method

Sweet cherries for fresh consumption are harvested by hand, usually leaving the pedicels intact. They are harvested at firm-mature stage to reduce bruising. Sweet cherries intended for processing are hand harvested also, but without pedicels.

Sour cherries intended for processing are shaken from trees when ripe. Two machines are required: a trunk shaker, which grips the trunk and shakes the tree, and a catch frame, which catches and funnels shaken fruits onto a conveyor for collection. An entire tree is harvested in a matter of seconds (Figure 10.3). Ethephon, an ethylene-releasing compound, is applied about 2

FIGURE 10.3. Shake harvesting of sour cherries. *(above, left)* The shaker attached to the trunk ready to shake. *(above, right)* The tarp rolling up onto a conveyor belt. *(left)* Cherries dumped into cold water at the end of the conveyor.

weeks prior to harvest to reduce fruit removal force, and to increase the percentage of fruits harvested. Tree trunks are often damaged by shaking equipment, if done incorrectly. Trees that experience rapid growth are more prone to trunk injury.

Postharvest Handling

Both sweet and sour cherries have an extremely short shelf life and must be handled gently to reduce bruising and oxidation. Sour cherries for processing are dumped into cold water immediately following harvest. They are then transported to processing plants, where they are washed, destemmed, pitted, and packed for freezing, all within hours after harvest.

Sweet cherries are hydrocooled or dumped into cold water by pickers and packed in shallow flats, after first being sorted based on color and size, with usually the largest being about 15/16 inch or larger. They are transported to market quickly, often by air freight for exported fruits.

Storage

Shelf life of fresh cherries is only a few days at room temperature and up to 3 weeks at 32°F. Sweet cherries can be stored longer than sour cherries. Sour cherries, canned or frozen, can be stored for several months. Fresh cherries are subject to the same postharvest diseases (brown rot, gray mold, blue mold, *Rhizopus, Alternaria,* etc.) as other stone fruits. Cherries are not sensitive to chilling injury.

CONTRIBUTION TO DIET

Maraschino cherries are made mostly from sweet cherries, but a small proportion of sour cherries are brined for this purpose. Cherries with clear flesh are picked slightly early, sometimes are decolorized with sulfur dioxide (SO_2), and then steeped in marasca, a liqueur distilled from the fermented juice of wild cherries. Sour cherries are primarily processed into pie fillings. Sour cherry growers prefer the term *pie* or *tart* over *sour,* since the latter connotes bad flavor.

Per capita consumption of cherries in the United States was 1.9 pounds/year in 2004. Fresh consumption (largely sweet cherries) was about 1 pound/year, increasing 45 percent since 1980. Frozen and canned consumption (mostly sour cherries) was 0.9 pound/year in 2004, with canned consumption down about 50 percent and frozen consumption up 40 percent since 1980.

Cherry utilization (percent of crop) is as follows:

Sweet Cherry	Sour Cherry
Fresh (50-72)	Frozen, pie filling (52-70)
Brined, maraschino (18-30)	Canned, juice, jam, etc. (25-38)
Frozen (8)	Brined, maraschino (5-6)
	Canned, juice, wine, brandy, etc. (5)
	Fresh (1-2)

Dietary value per 100-gram edible portion:

	Sweet Cherry	Sour Cherry
Water (%)	82	86
Calories	63	50
Protein (%)	1.1	1.0
Fat (%)	0.2	0.3
Carbohydrates (%)	16	12.2
Crude Fiber (%)	2.1	1.6

	% U.S. RDA (2,000-calorie diet)	
Vitamin A	1	26
Thiamin, B_1	2	2
Riboflavin, B_2	2	2
Niacin	<1	2
Vitamin C	12	17
Calcium	1	2
Phosphorus	3	2
Iron	2	2
Sodium	0	<1
Potassium	6	5

BIBLIOGRAPHY

Childers, N.F., J.R. Morris, and G.S. Sibbett. 1995. *Modern fruit science,* Tenth edition. Gainesville, FL: Norman F. Childers.

Fogle, H.W. 1975. Cherries, pp. 348-366. In: J. Janick and J.N. Moore (eds.), *Advances in fruit breeding.* West Lafayette, IN: Purdue University Press.

Gur, A. 1985. Rosaceae—Deciduous fruit trees, pp. 355-389. In: A.H. Halevy (ed.), *CRC handbook of flowering,* Volume 1. Boca Raton, FL: CRC Press.

Hedrick, U.P. 1915. *The Cherries of New York.* Albany, NY: J.B. Lyon Co.

Iezzoni, A. 1988. 'Montmorency' sour cherry. *Fruit Var. J.* 42:74-75.

Iezzoni, A., H. Schmidt, and A. Albertini. 1990. Cherries *(Prunus),* pp. 109-174. In: J.N. Moore and J.R. Ballington (eds.), *Genetic resources of temperate fruit and nut crops* (Acta Horticulturae 290). Belgium: International Society for Horticultural Science.

Marshall, R.E. 1954. *Cherries and cherry production.* New York: Interscience Publ.

Ogawa, J.M., E.I. Zehr, G.W. Bird, D.F. Ritchie, K. Uriu, and J.K. Uyemoto. 1995. *Compendium of stone fruit diseases.* St. Paul, MN: American Phytopathological Society.

Perry, R.L. 1987. Cherry rootstocks, pp. 217-264. In: R.C. Rom and R.F. Carlson (eds.), *Rootstocks for fruit crops.* New York: John Wiley and Sons.

Roper, T.R. and C.R. Rom. 1990. 'Bing' sweet cherry. *Fruit Var. J.* 44:106-108.

Teskey, B.J.E. and J.S. Shoemaker. 1978. *Tree fruit production,* Third edition. Westport, CT: AVI Publ.

Watkins, R. 1979. Cherry, plum, peach, apricot, and almond; *Prunus* spp., pp. 242-247. In: N.W. Simmonds (ed.), *Evolution of crop plants.* London: Longman.

Way, R.D. 1974. *Cherry varieties in New York State* (New York Food and Life Sciences Bulletin 37). Ithaca, NY: New York State Agricultural Experiment Station.

Webster, A.D. and N.E. Looney. 1996. *Cherries: Crop physiology, production, and uses.* Wallingford, UK: CAB International.

Chapter 11

Citrus Fruits (*Citrus* spp.)

TAXONOMY

The genus *Citrus* belongs to the Rutaceae or Rue family, subfamily Aurantoideae. This family contains 150 genera and 1,600 species distributed throughout the world. Rutaceous plants are often aromatic by virtue of scented oils in leaves, flowers, and fruits. Holding a leaf to a light source reveals numerous pellucid dots, which are the oil glands. Another unique feature of the family is a raised nectary disc subtending the ovary in flowers. Ornamental plants include common rue *(Ruta)*, prickly ash *(Zanthoxylum)*, orange jessamine *(Murraya)*, and some of the near citrus relatives, such as trifoliate orange *(Poncirus)*. The Aurantoideae subfamily is further distinguished by having a unique fruit type—the hesperidium, which is basically a berrylike fruit with a leathery rind. There are many edible species, most within the genus *Citrus*, but some distantly related ones, such as white sapote *(Casimiroa edulis)* and wampee *(Clausena lansium)*. While *Citrus* is by far the most economically important genus, two other genera contain species important in citriculture:

1. *Fortunella* spp. (kumquats). Originally classified with citrus, kumquats were then moved to their own genus, named after Robert Fortune, who introduced kumquats to Europe. They are evergreen shrubs or small trees (8-15 feet) native to southern China, but they can be grown around the world into subtropical areas (Plate 11.1). Unlike citrus fruits, the peel of the kumquat fruit is edible and usually sweeter than the pulp. Fruits are eaten out of the hand and often used as a decoration in gift packs of citrus fruits for holiday trade. Kumquats are grown commercially in China and Japan. They are exceptionally high in calcium, potassium, and vitamins A and C, as are most citrus fruits. Four major cultivars were given species status by Swingle, a noted citrus taxonomist:
   - 'Nagami'—*F. margarita* Swing. Also called oval or long kumquats, the fruits are longer (1-1½ inches) than they are wide, with a thin, yellow-orange peel.
   - 'Meiwa'—*F. crassifolia* Swing. These large, round kumquats are possibly a hybrid between 'Nagami' and 'Marumi'. The fruits are round, with a thick, orange-yellow peel, and often seedless.
   - 'Hong Kong' or 'Hong Kong Wild'—*F. hindsii* Swing. These fruits are orange or scarlet when ripe and have many seeds.
   - 'Marumi'—*F. japonica* Swing. (syn. *Citrus madurensis* Lour.). Also called round kumquats, these fruits have a thin, golden yellow peel that surrounds aromatic and spicy pulp. These are said to be the most cold-hardy variety of the kumquats.
2. *Poncirus trifoliata* (L.) Raf. (trifoliate orange). This plant is important as a rootstock for citrus, especially in Japan, although the fruit is scarcely edible. It is used as a male parent in production of citrange (sweet orange × trifoliate orange) and citrumelo (grapefruit × trifoliate orange) rootstocks, and as an ornamental. It has a deciduous habit in cooler areas and can tolerate more freezing than any other citrus relative. It is native to northern China and grown as far north as Philadelphia in the eastern United States (Figure 11.1).

Introduction to Fruit Crops

doi:10.1300/5547_11

Taxonomic treatment of *Citrus* is confusing, at best, and misleading, at worst. There is considerable debate over the number of species within *Citrus,* with estimates varying between 1 and 162 species, depending on the taxonomist. Swingle recognized 16 species, Tanaka 162, and Hodgson 36. Some "lumpers" argue that all citrus fruits belong to one large species, as they are widely graft and cross-compatible. On the other extreme, Tanaka gave almost every cultivar or variant a species name, which is clearly in error, since some cultivars are known hybrids. The greatest amount of disagreement occurs in the naming of tangerines, which are the most diverse citrus fruits, containing more commercially important hybrids than lemons, limes, oranges, or grapefruit. More recent work with specific chemicals and DNA suggests that there are (or were) probably four basic species of *Citrus:*

FIGURE 11.1. The trifoliate orange is the most cold-hardy citrus relative. The 'Flying Dragon' cultivar shown here is scarcely edible but has serpentine stems and recurved thorns, making it an attractive ornamental.

1. *C. halimii*—Native to southern Thailand and west Malaysia, this may have been the possible progenitor species for *Poncirus* and *Fortunella.*
2. *C. medica*—The citron may be the progenitor species for all lemons and limes.
3. *C. reticulata*—The tangerine or mandarin may be the ancestral form of all oranges and tangerines, or it may be the progeny of a now extinct ancestor.
4. *C. maxima* (syn. *C. grandis*)—The pummelo or shaddock is likely one of the parents of grapefruit.

The four-species scheme has not become widely accepted, and some species names, right or wrong, are entrenched in the literature. As a compromise, I prefer the classification of Swingle for the five commercially important fruit crops (sweet orange, tangerine, grapefruit, lemon, and lime); each is recognized as a single species. Other types used for their fruits on a minor scale, as rootstocks, or as ornamentals are commonly given species status as well. The major fruit crops of commerce are as follows:

1. *C. sinensis* (L.) Osb.—Sweet oranges. This is a widely accepted name for this crop, containing four groups of cultivars: common oranges, blood oranges, navel oranges, and acidless oranges (see the following Cultivars section). The term *orange* is used rather loosely, sometimes for fruits that look like oranges but are not *C. sinensis*. Examples include 'Temple' and 'Page' oranges (tangerine hybrids), satsuma orange (a cold-hardy variant of tangerine), and trifoliate orange *(Poncirus trifoliata).*
2. *C. reticulata* Blanco—Tangerine, mandarin, or satsuma. Due to the success of breeding with these types (as they are often monoembryonic), many cultivars and hybrids have been produced or formed naturally, with some erroneously being given species status. I prefer to use *C. reticulata* for all tangerines, but other species names sometimes given in the literature include *C. unshiu* (satsuma), *C. deliciosa* (willowleaf), *C. reshni* ('Cleopatra'), *C. nobilis* ('King'), and *C. temple* ('Temple').

3. *C. paradisi* Macfad.—Grapefruit. This is clearly not a true species, but its economic importance today has granted species status that even a "lumper" could not deny. Grapefruit is thought to be a hybrid of pummelo and sweet orange that occurred naturally somewhere in the Caribbean between the time of Columbus's voyages and its introduction to Florida in 1809.
4. *C. limon* Burm. f.—Lemons. I would lump the rough lemon *(C. jambhiri)*, sweet lemon *(C. limetta)*, and Volkamer lemon *(C. volkameriana)* as variants within this species.
5. *C. aurantifolia* L.—Limes. The literature distinguishes the two main cultivars—'Key' and 'Tahiti'—as separate species, with the latter labeled *C. latifolia* Tanaka or *Citrus ×tahiti* Campbell. I would lump these two, along with the 'Rangpur' lime (*C. limonia)* and sweet lime *(C. limettioides)*, under *C. aurantifolia*.

Three others given species status and worthy of mention include the following:

1. *C. grandis* (L.) Osb. or *C. maxima* (Burm.) Merr.—Pummelo or shaddock. This species originates from Southeast Asia, where it is as common as grapefruit is in the United States. It is much larger and thicker peeled than grapefruit but said to have a milder flavor (Plate 11.2). There is little argument about its nomenclature.
2. *C. aurantium* L.—Sour orange. This is allied with limes by some but is a very important rootstock and ornamental. Since it often appears in the literature, it is convenient to keep this species name. Cultivars and variants include 'Bittersweet', 'Oklawaha', 'Vermillion-Globe', 'Paraguay', 'Trabut', *C. aurantium* var. *myrtifolia* (myrtle), 'Bergamot', daidai (Japanese), 'Leaf of Chinnoto', and *C. taiwanica* Tanaka.
3. *C. medica* L.—Citron. This lemonlike fruit may be the progenitor species of modern lemons and limes. The peel is very thick, and the white, spongy portion of the peel is edible. Rarely seen in the United States, it is used in Mediterranean countries, particularly Israel. 'Etrog' citron is used in the Jewish feast of the Tabernacles. 'Buddha's Hand' or 'Fingered Citron' is a striking fruit sometimes used in a religious context; outgrowths of peel tissue appear as fingers (Plate 11.2).

Hybrids

Several hybrids among *Citrus* species, and between *Citrus* and *Poncirus* or *Fortunella*, have been produced either naturally or through controlled breeding. Different prefixes and suffixes are used to denote the parents of such hybrids:

1. Trifoliate orange—The prefix "citr-" is used to denote parentage from *Poncirus trifoliata*. Due to the parent's coldhardiness, many hybrids were made to try to improve the hardiness of commercial citrus fruits.
   - Citrange (sweet orange × trifoliate orange)
   - Citrumelo (grapefruit × trifoliate orange)
   - Citradia (sour orange × trifoliate orange)
   - Citrangequat (citrange × kumquat)
2. Kumquat—The suffix "-quat" is used.
   - Limequat (kumquat × lime)
   - Citrangequat (kumquat × citrange)
3. Sweet orange—The root "ange," the suffix "-or," or, less often, the prefix "orang-" is used to denote hybrids.
   - Citrange (sweet orange × trifoliate orange)
   - Tangor (sweet orange × tangerine)

- Orangelo (sweet orange × grapefruit)
- Citrangequat (citrange × kumquat)

4. Tangerine—The prefix "tang-" or suffix "-andarin" is used to denote hybrids.
   - Tangelo (tangerine × grapefruit)
   - Tangor (tangerine × sweet orange)
   - Tangtangelo (tangerine × tangelo)
   - Citrandarin (tangerine × trifoliate orange)
   - Lemandarin (tangerine × lemon)
5. Grapefruit—The suffix "-elo" or "-umelo" *(not to be confused with pummelo)* is used to denote hybrids.
   - Tangelo (grapefruit × tangerine)
   - Citrumelo (grapefruit × trifoliate orange)
6. Lemon—Simply the root "lemon" or the prefix "lem-" is used to denote hybrids.
   - Lemonange (lemon × sweet orange)
   - Lemonime (lemon × lime)
   - Lemandarin (lemon × mandarin)
7. Lime—Simply the root "lime" or the suffix "-ime" is used to denote hybrids.
   - Lemonime (lime × lemon)
   - Limequat (lime × kumquat)

In addition to the aforementioned hybrids are the following: The alemow (termed *C. macrophylla*) is a limelike fruit but is probably a hybrid between pummelo and *C. celebica,* which has become an important rootstock for lemon. The calamondin or calamansi (× *Citrofortunella mitis*) is used for its small, bright orange, acid fruit and makes an attractive ornamental. It is thought to be a hybrid between kumquat and tangerine and may be termed *Citrus mitis* or *Citrus madurensis*. One exception to the prefix and suffix conventions previously noted is the orangequat, which is actually a citrangequat × tangerine hybrid, not a sweet orange × kumquat hybrid, as the name might suggest.

Cultivars

Sweet Oranges

Cultivars are subdivided into four groups based on fruit characteristics:

1. *Common or round oranges.* As the name suggests, this is the most common and widespread group of sweet orange cultivars. Most orange juice comes from these cultivars, with a smaller proportion of the fruits marketed fresh. 'Valencia', the top sweet orange cultivar in the world, is a late-season cultivar having excellent internal quality and juice characteristics. Also, it produces excellent fruits for fresh consumption, and about one-third of California's fresh-market orange crop is made up of sports of 'Valencia', such as 'Frost' and 'Olinda'. 'Hamlin', widely planted in Florida, is an early season, very productive plant, but it produces lower-quality fruits that are used mostly for juice. 'Pineapple' is a high-quality, midseason cultivar grown for juice in Florida. In Brazil, 'Natal' and 'Pera' are major cultivars.
2. *Blood oranges.* These red-juiced variants contain anthocyanin pigments in the peel and juice, giving them a distinctive appearance (Plate 11.3) They achieve good pigmentation only in Mediterranean climates and have been produced in California only recently. 'Torocco' and 'Moro' are the major cultivars.

3. *Navel oranges*. This group is distinguished by a small, secondary ovary embedded within the usual ovary, yielding a small fruitlet at the stylar end of the fruit at maturity—a fruit within a fruit. They are generally seedless and make excellent-quality fresh fruits, but they have poor juicing characteristics and therefore are not processed. 'Washington' is the major cultivar, but there are dozens of others, several of which arose as mutations or chance seedlings of 'Washington'.
4. *Acidless oranges*. This is an odd group of sweet oranges with low acid content in the juice, which has an insipidly sweet flavor. These oranges are rarely grown outside of the Mediterranean region.

Tangerines

Similar to sweet oranges, tangerine cultivars fall into four main groups:

1. *Common*. This group of cultivars produces small fruits with a slightly more adherent peel than is found in other tangerine types. 'Clementine', the most widespread cultivar, is small, seedless (there are seeded sports), and generally flavorful. 'Clementine' tangerines, which have become popular in the United States recently, are often sold in small wooden boxes around the holiday season. They grow best in Mediterranean climates, and Spain is the major producer. 'Dancy' is a popular Florida cultivar that tolerates humid subtropical climates.
2. *Satsuma*. Satsumas are distinguished by being cold hardy, easily peeled, early ripening cultivars. Extreme coldhardiness and fall maturation allow them to be produced in areas where other citrus species cannot grow (such as Japan and the Gulf Coast of the United States). Fruits are often irregular or oblate in shape, and they are larger than common tangerines. 'Owari' is one of the most important cultivars worldwide, although over 100 cultivars exist.
3. *Mediterranean or willowleaf*. This group is named for its leaf characteristics, which resemble the linear, straplike leaves of willow trees. The fruits are small and seedy, and alternate bearing and poor postharvest characteristics have caused a decline in production.
4. *Hybrids*. Crosses of tangerine with orange (tangors) or grapefruit (tangelos) have produced many cultivars of importance in the United States. 'Temple' and 'Murcott' (the latter also known as 'Honey' or 'Honey Murcott') are of unknown origin, but both are probably tangors. 'Orlando' and 'Minneola' are the most common tangelos. 'Minneola' is also called 'Honeybell', due to the pronounced neck on the stem end of the fruit. 'Ugli' fruit, sold as a novelty item in many supermarkets, is probably a tangelo that originated in the Caribbean (Plate 11.4). Several other hybrids of lesser importance have been bred by USDA scientists in Florida; these include 'Robinson', 'Page', 'Nova', 'Ambersweet', and 'Sunburst'. Unlike other citrus species, most hybrid cultivars require cross-pollination for adequate fruit set.

Grapefruit

The major white-fleshed cultivars are 'Duncan' and 'Marsh'; the former is seedy and the latter seedless. 'Marsh' is more popular due to its seedlessness. 'Thompson' (syn. 'Pink Marsh'), a mutation of 'Marsh' with pink flesh, is also seedless. 'Star Ruby' is a deep-red cultivar produced by mutation breeding. Other deep-red cultivars include 'Ruby Red', 'Rio Red', and 'Flame', grown largely in Texas.

Limes

'Key' (syn. 'Mexican', 'West Indian') and 'Tahiti' (syn. 'Persian') are the major cultivars. 'Key' limes are small, round, and seedy and turn yellow under Mediterranean conditions. 'Tahiti' limes are larger, green, and shaped like lemons (Plate 11.5). 'Bearss' is a more cold-hardy variant of 'Tahiti' that turns yellow under Mediterranean conditions.

Lemons

The main cultivars are 'Lisbon' (oval to round, more pronounced stylar end furrow and point) and 'Eureka' (oval, less pronounced stylar end). Both are grown in California, with 'Lisbon' being the major cultivar, due to superior coldhardiness and yield (Plate 11.5). 'Meyer' is a cold-hardy, larger-fruited cultivar used as an ornamental or containerized plant; it is probably a lemon hybrid. 'Femminello' and 'Verna' are the major cultivars in Italy and Spain, respectively.

ORIGIN, HISTORY OF CULTIVATION

The center of diversity for *Citrus* ranges from northeastern India eastward through the Malay archipelago and south to Australia. Sweet oranges probably arose in India, the trifoliate orange and mandarin in China, and acid citrus types in Malaysia. Oranges and pummelos were mentioned in Chinese literature in 2400 BC, and later in Sanskrit writings (800 BC) lemons were mentioned. Theophrastus, the "father of botany", gave a taxonomic description of the citron in 310 BC, classifying it with apple as *Malus medica* or *Malus persicum*. At the time of Christ and shortly thereafter, the term *citrus* arose as a mispronunciation of either the Greek word for cedar cones, *kedros*, or *callistris*, the name for the sandalwood tree.

At this time, citrus fruits were spread throughout Asia, North Africa, and Europe along trade routes. The dissemination was carried out by many cultures, indicating widespread appeal of the fruits at this time. From the first centuries BC to medieval times, orangeries and citrus "groves" were established in Europe, and cultivation became more sophisticated. Christopher Columbus, Juan Ponce de León, and Juan de Grijalva carried various citrus fruits to the new world in the late 1400s and early 1500s. Citrus culture proliferated in Florida in the late 1700s, when the first commercial shipments were made. Right about this time, citrus was introduced to California, although it was much later that commercial production began in the west. With the advent of large-scale irrigation projects in the 1940s, citrus culture increased greatly in western states. Today, citrus fruits are grown commercially in Florida, California, Arizona, and Texas.

FOLKLORE, MEDICINAL PROPERTIES, NONFOOD USAGE

The word *orange* is said to have derived from the Spanish word *naranja;* English-speaking folks applied the indefinite article to the Spanish word to give "a *naranja*" which was corrupted to "an orange." The term *golden apples* (presumably citrus fruits) arose from the myth of Hercules and the golden apples. The Hesperides, Mediterranean islands where the giant Atlas lived, were a haven for the golden apples because the gods were fearful of the golden apples being stolen. Hercules managed to obtain some of the golden apples as one of his 12 tasks, despite Atlas's attempt to trick him into holding up the sky. Later, Perseus visited the Hesperides to obtain some golden apples and succeeded by turning Atlas to stone using the head of Medusa, thus creating the Atlas Mountains in northern Africa.

Another myth involves the Grecian maiden Atlanta, "who was as fleet of foot as she was beautiful." To obtain her hand in marriage, her suitor had to beat her in a footrace; if he lost, he was beheaded. Hippomenes, an apparently slow but sly individual, obtained some golden apples and during the race rolled them at her feet as she passed. She stopped to pick them up and Hippomenes scooted to the finish line the victor.

Because the orange tree bears flowers and fruits at the same time, it was used in fertility rituals and weddings—the white flower symbolized virginity, and the fruits symbolized fertility. The citron is used in religious ceremonies by Hebrews. Orange blossoms are the state flower of Florida.

Citrus has been used for more medicinal purposes than most other fruit crops. An impressive list of folk medicine treatments was compiled by Duke and duCellier (1993) for several citrus species. Remedies range from chew sticks for oral hygiene and toothache relief to contraceptives, laxatives, purgatives, sedatives, and as a treatment for a wide variety of common ailments, such as diarrhea and vomiting. Two citrus relatives are being studied for their potential as cancer therapy drugs: *Acronychia baueri* for acronycine, and *Fagara macrophylla* for 8-methoxy-dihydronitidine. Hesperidin (syn. citrin, vitamin P) is a bioflavonoid, found at doses of up to 8 percent in dried peels. It is a strong vasopressor agent (reduces blood pressure). Hesperidin is also found in rose hips and black currants, which, oddly enough, are used for vitamin C sources, as is citrus. Citrus pectin is reported to reduce cholesterol 30 percent, aortal plaque 85 percent, and narrowing of coronary arteries by 88 percent, as shown in studies using laboratory animals. On the negative side, the limonene oils in citrus peels mentioned previously may cause contact dermatitis. These are also "photosensitizing compounds," meaning that they enhance sunburn on skin exposed to ultraviolet light (the active principle is furocoumarin). Citromellal, another volatile oil, has mutagenic properties.

In addition to various juice products from the pulp, citrus peels are candied, fed to livestock, and used to scent perfumes and soap products. Limonene and petitgrain oils from peels or leaves are used as essential oils and also have an insecticidal property, which was recently discovered. Lemon oil is cold-pressed from lemon peels and used in everything from baked goods and candy to furniture polish and insecticides. Curacao and Cointreau liqueurs are made from sour orange fruits. Bergamot, a sour orange variant, is used to make perfumes and massage oils. Nectar is converted to honey by bees, and "orange blossom honey" is another by-product of citrus production.

The compounds naringin (a flavonoid) and neohesperidin dihydrochalcone from grapefruit and pummelo may have application as artificial sweeteners. The compounds are said to have 1,000 times the sweetness of sugar and produce a long-lasting sweetness, slow to develop, with an aftertaste similar to that of licorice or menthol.

PRODUCTION

Sweet Oranges

World

Total production in 2004, according to FAO statistics, was 63,039,736 MT or 139 billion pounds. Oranges are produced commercially in 112 countries worldwide, on about 9 million acres. Production has increased about 15 percent in response to a 6 percent increase in acreage and a 5 percent increase in yield over the past decade. Worldwide average yields are just over 15,000 pounds/acre, but the highest yields can reach double that amount. The top ten sweet orange–producing countries (percent of world production) follow:

1. Brazil (29)
2. United States (19)
3. Mexico (6)
4. India (5)
5. Spain (4)
6. China (3)
7. Iran (3)
8. Italy (3)
9. Egypt (3)
10. Indonesia (2)

United States

Total production in 2004, according to USDA statistics, was 11,729,900 MT or 25.8 billion pounds. Production has increased 17 percent in the past decade. Several hurricanes during 2004 and 2005 reduced production by 29 percent in Florida. The industry value was $1.6 billion. The leading states in terms of production follow:

1. Florida (82 percent): 565,000 acres
2. California (18 percent): 182,000 acres
3. Texas (<1 percent): 8,800 acres
4. Arizona (<1 percent): 5,800 acres

Total acreage for 2004 was approximately 762,000 and has been declining slowly since the 1960s (Figure 11.2). Yields vary considerably by cultivar, state, year, and intended market but

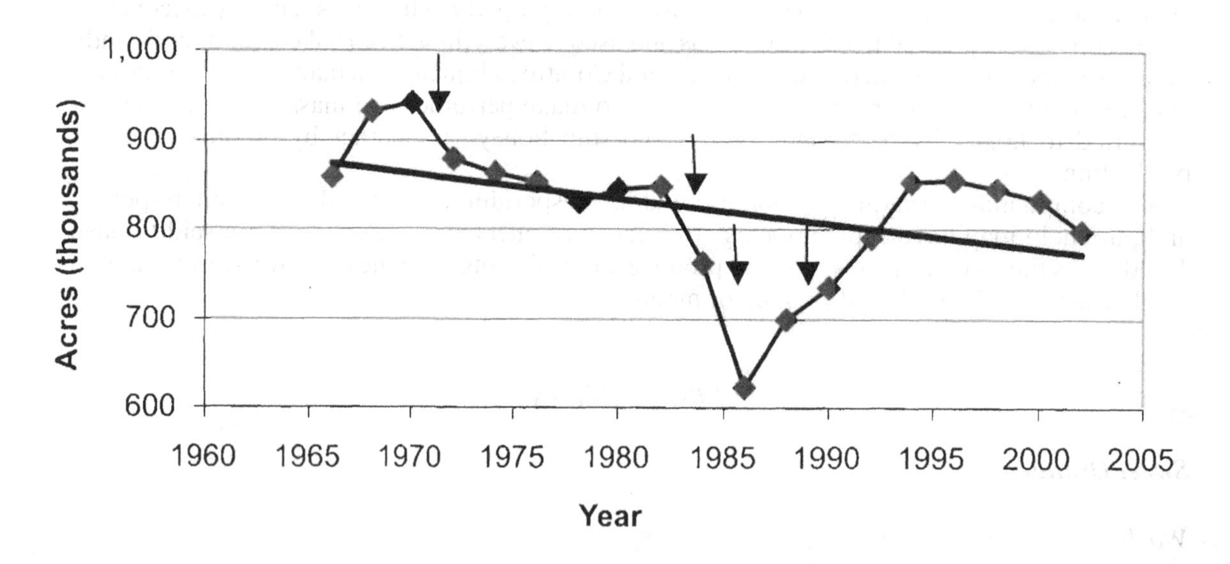

FIGURE 11.2. Florida citrus acreage. Subtropical climates, such as in Florida, make excellent citrus-growing regions, but they are susceptible to periodic freezes. Arrows indicate major freezes, causing the loss of 80,000 or more acres in 1971, 1983, 1985, and 1989. Over 150,000 acres of citrus were killed on a single night in the December 1983 and January 1985 freezes, when temperatures reached the midteens (°F) in the Orlando area. Although acreage recovered through the 1990s, the overall trend *(straight line in graph)* in Florida citrus acreage is downward since the mid-1960s, at a rate of about 2,800 acres/year. *Source:* USDA statistics.

range from about 10,000 pounds/acre in Arizona to over 30,000 pounds/acre in Florida. Prices received by growers are among the lowest for any fruit crop in the United States: $0.05 to $0.15 per pound, with fresh fruit receiving $0.10 to $0.15 per pound, and processed fruit about $0.05 to $0.06 per pound.

The United States exported 61 million gallons of orange juice (single-strength equivalent) in 2004, half as frozen concentrate, and imported 101 million gallons of concentrate. About 3 percent of production was exported fresh, and fresh imports amounted to 0.5 percent of production.

Grapefruit

World

Total production in 2004, according to FAO statistics, was 4,874,910 MT or 10.7 billion pounds. FAO statistics include pummelo with grapefruit, so a small fraction of this amount is actually from *Citrus grandis*. Grapefruit and pummelo are produced commercially in 75 countries worldwide, on about 631,000 acres. Acreage has increased 8 percent while yield has remained roughly constant or declined over the past decade; hence, production has been fairly constant. Yields average 17,000 pounds/acre but range from just a few thousand pounds/acre to over 32,000 pounds/acre in the United States and Israel. The top ten grapefruit-producing countries (percent of world production) follow:

1. United States (40)
2. China (10)
3. South Africa (8)
4. Mexico (5)
5. Israel (5)
6. Cuba (5)
7. Argentina (3)
8. India (3)
9. Turkey (3)
10. Tunisia (1)

United States

Total production in 2004, according to USDA statistics, was 1,952,300 MT or 4.3 billion pounds. The United States dominates world production of grapefruit, although production has decreased 21 percent over the past decade. Several hurricanes during 2004 and 2005 reduced Florida's grapefruit production by 68 percent. The industry value was $297 million. The leading states are as follows:

1. Florida (78 percent): 82,300 acres
2. Texas (11 percent): 18,500 acres
3. California (10 percent): 12,500 acres
4. Arizona (<1 percent): 1,500 acres

Yields vary considerably by cultivar, year, and location but range from about 8,000 pounds/acre in Arizona to 38,000 pounds/acre in Florida. Prices received by growers are extremely low, about $0.05 to $0.10 per pound.

In 2004, the United States exported an amount of fresh fruits equal to 18 percent of domestic production, representing a 10 percent decrease over the past decade. Fresh imports were less than 1 percent of production. Juice exports were 12.6 million gallons (single-strength equivalent juice), which is a sixfold increase over the past decade.

Tangerines and Hybrids

Hybrids include tangelos and tangors. FAO statistics may place some hybrids in an alternate category, "Citrus NES," so these data may be a slight underestimate.

World

Total production in 2004, according to FAO statistics, was 22,198,791 MT or 48.8 billion pounds. Tangerines are produced commercially in 64 countries worldwide, on about 4.3 million acres. Production has increased 51 percent over the past decade due to the combined effects of a 24 percent increase in acreage and a 22 percent increase in yield. Worldwide average yields are 11,350 pounds/acre, ranging from under 3,000 to over 25,000 pounds/acre. The top ten tangerine-producing countries (percent of world production) follow:

1. China (47)
2. Spain (9)
3. Brazil (5)
4. Japan (5)
5. Iran (3)
6. Thailand (3)
7. South Korea (3)
8. Italy (3)
9. Turkey (2)
10. Egypt (2)

United States

Total production in 2004, according to USDA statistics, was 492,600 MT or 1.1 billion pounds. The industry value is $140 million. Total production of tangerines and hybrids has been stable over the past decade, but tangerine production has increased by 40 percent, while tangelos, 'Temple' (a tangor), and other hybrids have seen decreases in production of about 30 to 70 percent. Production of 'K-early', an early maturing, poor-quality hybrid, decreased to virtually zero in 2004. The leading states are as follows:

1. Florida (72 percent): Total acreage is 31,900. Florida grows mainly tangerines but also produces some tangelos and is the only producer of 'Temple' tangors. Acreage has decreased 20 percent in the past decade.
2. California (22 percent): Total acreage has increased 15 percent over the past decade to 10,500 acres in 2004. However, production has remained fairly constant, perhaps reflecting a shift to lower-yielding cultivars. 'Minneola' tangelos have increased in importance over the past decade.
3. Arizona (6 percent): Total acreage is 5,200.

Yields vary considerably by cultivar, year, and location but range from 9,000 pounds/acre in Arizona to 25,000 pounds/acre in Florida. The average price paid to U.S. growers is about $0.13 to $0.14 per pound but varies by state. Arizona and California fruit growers receive $0.21 to $0.24 per pound, whereas Florida fruit growers receive on average only $0.13 per pound. This reflects differences in cultivars grown as well as the better external appearance of western fruits. The United States generally exports about 3 percent of production and now imports an amount equivalent to 15 percent of domestic production, which has recently surged due to imports of 'Clementine'.

Lemons and Limes

World

Total production in 2004, according to FAO statistics, was 12,126,233 MT or 26.7 billion pounds. Lemons and limes, as we know them in the United States, are not distinguished by the FAO. Thus, the following data are totals of lemon- and limelike fruits. In Mediterranean climates (such as Spain, Italy, and California), production of lemons dominates; in tropical and subtropical regions (such as Mexico, Brazil, and Florida) lime production dominates.

Lemons and limes are produced commercially in 95 countries worldwide, on about 2 million acres. Acreage and yield have increased 30 and 9 percent, respectively, and hence production is up almost 42 percent since 1992. Yields average 13,500 pounds/acre but range from 2,500 to 25,000 pounds/acre. The top ten lemon/lime-producing countries (percent of world production) follow:

1. Mexico (15)
2. India (12)
3. Iran (9)
4. Spain (9)
5. Argentina (8)
6. Brazil (8)
7. United States (6)
8. China (5)
9. Italy (4)
10. Turkey (4)

United States

Lemons. Total production in 2004, according to USDA statistics, was 725,000 MT or 1.6 billion pounds. The industry value has declined about 20 percent to $270 million in just 3 years, since prices have dropped and production has decreased 4 percent over the past decade.

1. California (86 percent): 45,000 acres
2. Arizona (14 percent): 14,800 acres

Yields are about 16,000 pounds/acre in Arizona and 32,000 pounds/acre in California. Prices range from $0.12 to $0.17 per pound. The United States imported an amount equal to 2 percent of domestic production, which has more than doubled in the past decade.

Limes. The USDA last reported statistics for limes in 2002, as the industry has practically disappeared. There were 6,360 MT or 14 million pounds at that time, valued at $1.7 million. Florida produced 100 percent of the limes in the United States. Several thousand acres of limes once existed in the Homestead area of extreme south Florida. The area was devastated by hurricane Andrew in 1992, and production dropped from 88 million pounds to less than 20 million the following year. Production slowly rose back to 52 million pounds in 2000 and then dropped to new lows in 2002 as prices fell. Yields were about 16,000 pounds/acre. Estimates showed less than 400 acres in 2003, as trees were being removed due to citrus canker disease. The average price received by growers was $0.12 per pound, but prices ranged from $0.12 to $0.16 per pound. Lime imports have risen 25 percent over the past decade, as domestic production has declined; 310 million pounds were imported in 2004.

BOTANICAL DESCRIPTION

Plant

All citrus plants are small, spreading, evergreen trees or tall shrubs (Figure 11.3). Grapefruit trees are the largest-statured trees of the group, and limes the smallest. Trees may reach 20 to 30 feet in height in nature, but most cultivated trees are less than 15 feet. Stems are often armed with long thorns, particularly in limes, but in all types when young. Current season's shoots are angular and green, with a 3/8 phyllotaxis in all cases except grapefruit, which has 2/5 phyllotaxis. As wood ages, it becomes round and lenticeled and develops thin, gray-brown bark. Leaves are unifoliate (sometimes termed "compound unifoliate" to indicate the loss of lateral leaflets over time), relatively thick, ovate with acute to obtuse tips, having entire or crenulate margins, and stout, winged petioles of various width, depending on species (grapefruit = wide; tangerine = narrow). Leaves contain characteristic citrus oils in glands, termed "pellucid dots," which makes them fragrant when crushed.

Flowers

Fragrant, white flowers are solitary or in short cymes, borne axillary on the current flush of growth (then termed "leafy bloom"), and also without leaves from the previous flush of growth (then termed "bouquet bloom") (Plate 11.6). Flowers of lemon and 'Tahiti' lime contain a purplish hue. Flowers are perfect, with five petals and sepals; petals are linear, sometimes curved lengthwise, waxy, and thick; sepals are fused at the base to form a small cup. Stamens number 20 to 25 and are arranged in a tight, columnar whorl around the gynoecium. A globular, green ovary subtends a thin style that terminates in a pronounced, donut-shaped stigma. The ovary is compound with 10 to 14 locules in most commercial cultivars; its position is superior and subtended by a raised nectary disc.

FIGURE 11.3. Citrus trees have a sympodial growth habit, branching on their own at an early age *(left)*, and then forming dense canopies with fruits borne on the outside *(right)*.

Pollination

Most cultivars are self-pollinated. Some are parthenocarpic, setting and maturing commercial crops of seedless fruit without fertilization and seed set. Examples of parthenocarpic cultivars are 'Marsh' grapefruit, 'Tahiti' lime, and some 'Navel' oranges and tangelos. Fruit size is related to seed number in all cultivars. Cross-pollination is necessary for only some tangerines and tangerine hybrids.

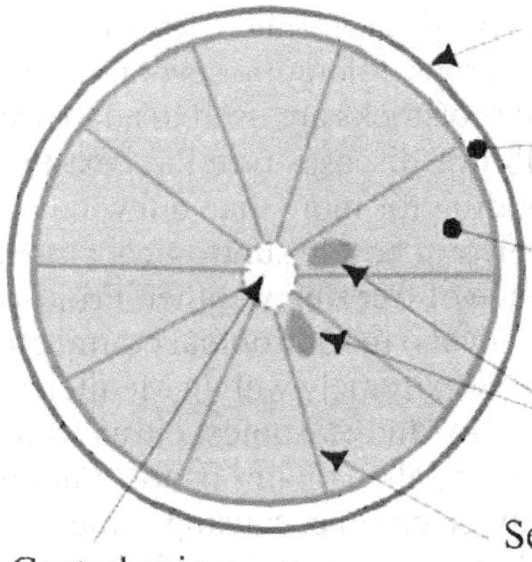

FIGURE 11.4. Schematic cross section of a hesperidium fruit.

Fruit

Citrus fruits are so important that they have received a special name—hesperidium (Figure 11.4; see also Plate 11.3). The name derives from the Hesperides, the Mediterranean islands where citrus fruits were thought to have originated. A hesperidium is basically a leathery-rinded berry. The endocarp is the edible portion, divided into 10 to 14 segments separated by thin septa, each containing up to eight seeds, but usually only one. Placentation is axile. The central axis may be open, as in tangerines. Each segment is composed of juice vesicles ("pulp"), with long stalks attached to the outer wall, containing juice. The mesocarp is the white tissue usually adherent to the outer surface of the endocarp, except in tangerines; it is also called the albedo. The exocarp, or flavedo, is the thin, pigmented outer portion of the rind, with numerous oil glands.

Citrus seed is unusual compared to most fruit crops because it forms nucellar embryos (maternal clones) in addition to the zygotic embryo produced through fertilization. Exceptions include some tangerines and the pummelo, in which only a zygotic embryo forms. Nucellar embryos generally outgrow the zygote as the seed develops, so high percentages of seeds produce only maternal clones when they germinate. The condition is termed "nucellar embryony" or "nucellar polyembryony," and it allows clonal propagation of citrus by seed, which is rare in horticultural crops. This condition is exploited by nurserymen but presents obvious difficulties in citrus breeding. Multiple embryos per seed ensures that germination rates often exceed 100 percent, and production of uniform seedlings in high percentages simplifies clonal rootstock production greatly. In other fruit crops, clonal rootstock production requires some form of layering and an extra year in the nursery, raising costs of tree production. For breeders, variation in seedling progeny is low or nonexistent because most seedlings arise from nucellar embryos and are thus clones of the maternal parent. Lack of nucellar embryony in some tangerines has simplified the production of hybrid cultivars; this is one reason why there are comparatively many tangerine hybrids.

GENERAL CULTURE

Soils and Climate

Citrus is adapted to a wide variety of soil types and conditions. Trees are grown on almost pure sand in central Florida, to organic muck near the Everglades, to loamy, heavy, high-pH soils in the San Joaquin Valley of California. Citrus is more tolerant of high or low pH and salinity than are most other tree crops, tolerating several times the soluble salt levels that rosaceous tree crops tolerate. Specific rootstocks are used to adapt trees to conditions of drought, high pH, salinity, and various soilborne diseases (see the Pest Problems section). Citrus species generally do not tolerate soil flooding for more than a few days without injury.

Most citrus plants perform best in subtropical climates, where there is a slight change of season but little or no chance of freezing weather. Fruits obtain their highest internal quality (juice content, sugar and acid levels) in subtropical humid climates, such as Florida. However, with irrigation, citrus fruits grow extremely well in Mediterranean climates, such as California, achieving the best external quality. In the tropics, citrus fruits accumulate less sugar and acid, and the peel usually remains green; also, bloom is not synchronized, so several stages of maturity are present on the tree at any given time, causing some immature fruits to be harvested (Plate 11.7). The exception is with limes, which perform better in the tropics than in the subtropics because they lack coldhardiness and tolerance of arid, cool conditions.

Coldhardiness is the major limiting factor for citrus production in subtropical areas. Flowers and fruits are killed at about 28°F; larger fruits can withstand a longer freeze duration because they have greater thermal mass than do small fruits. Fruits freeze from the stem end to the tip, and partially frozen fruit can be salvaged for juice after a freeze. Leaves and stems are killed by only a few minutes of exposure to temperatures from 0 to 28°F, depending on stage of acclimation, species, and age of tissue. Hardiness increases in the following order (ranked least to most hardy):

Citron
Limes
Lemons
Grapefruit and pummelo
Sweet oranges
Most tangerines and hybrids
Sour orange
Satsuma tangerine
Kumquats
Trifoliate orange and its hybrids

Cold acclimation occurs in citrus in response to decreasing temperatures in autumn. Experimentally, optimal hardening of sweet orange over a period of 4 weeks changes the killing temperature from 28°F to about 20°F. Historically, freezes that follow relatively warm weather cause more damage to trees than do freezes that are preceded by 1 to 2 weeks of cool weather. Trees acclimate better in California than Florida, since mean temperatures and temperature fluctuations are lower in California. There is also a significant effect of rootstock on the coldhardiness of a given scion cultivar (see the Rootstocks section).

Citrus has no chilling requirement and does not attain a truly dormant state, but it becomes quiescent at temperatures below about 55°F. Flowering is induced following emergence from quiescence, and sometimes by relief from drought.

There is perhaps no better illustration of the influence of climate on fruit quality than in citrus. Internal and external quality differ greatly between humid subtropical and Mediterranean climates. Temperature and humidity are the main environmental factors controlling quality. The following changes are easily seen:

- Peels become thicker and have a more pebbly or rough texture in Mediterranean climates than in subtropical climates.

- Peel color is best in Mediterranean climates due to cool winters enhancing chlorophyll destruction and fewer pests that blemish the peel.
- Juice content is higher in subtropical than in Mediterranean climates.
- Acid content is higher and sugar content is generally lower in Mediterranean than in subtropical climates, due to warmer temperatures during ripening. Acids break down faster with warm nights, and warmer day temperatures allow greater photosynthesis. Hence, the sugar-acid ratio is higher in subtropical than in Mediterranean climates, and fruits are said to be richer in flavor.
- Within arid climates, rate of maturation is faster in hot, desert areas of California and Arizona than in cool, coastal areas.
- On-tree storage is generally better in Mediterranean than in subtropical climates. 'Valencia' oranges maintain color and quality well into the summer if left on trees in California but will regreen if the same is done in Florida.

The previous influences dictate that processed citrus dominates in Florida (fair external quality, good internal quality) and fresh production dominates in California (excellent external quality, fair internal quality). This is particularly true for sweet orange, tangerine, and lemon. However, grapefruit is less dependent on Mediterranean conditions for external quality development and is grown for fresh consumption as well as processing in Florida. Limes do better in Florida because they are marketed green and obtain higher juice content.

Propagation

Although citrus seedlings will produce fruits identical to the parent tree due to nucellar embryony, trees are generally budded onto various rootstocks to avoid the long juvenile period for seedlings. Budding can be performed during most of the year, when pencil-sized, round bud sticks are available and bark slips on rootstocks. Bud unions on citrus are generally higher than in many other tree fruits (8-12 inches above soil line) to avoid any contact of the scion to the ground.

Limes are sometimes air layered, a technique also known as "marcottage" in the citrus trade. There is no rootstock in this case, and trees are said to be "own-rooted." Citrus plants as ornamentals are often propagated from rooted cuttings to avoid potential problems of rootstock suckering. Own-rooted trees are useful when citrus is planted in areas where periodic freezes kill the tops back; regrowth is true to type in this case.

Rootstocks

Citrus was grown from seed until the mid-1800s, due to ease of propagation from seed and trueness of type. Phytophthora foot rot appeared in the Azores in 1842 and later in other parts of the world, which stimulated a search for resistant stocks. Sour orange arose as the predominant stock for sweet orange worldwide but had the drawback of being susceptible to the tristeza virus, problematic in Brazil, Argentina, South Africa, Australia, and California. 'Rangpur' lime and the citranges began replacing sour orange and are growing in popularity worldwide today.

Common rootstocks are described in Table 11.1. Almost all species of citrus fruits can be grafted on any of these, but there are preferences. For example, sour orange makes an excellent stock for grapefruit and sweet orange, particularly in areas where phytophthora foot rot is a problem. 'Cleopatra' mandarin imparts excellent fruit quality characteristics to tangerines, but fruits tend to be smaller, so it is avoided in cultivars where size is important for fresh markets. 'Rangpur' lime is tolerant of salt, high pH, and tristeza virus, so it is used in arid areas that have these problems.

Planting Design, Training, Pruning

Citrus traditionally has been grown in rectangular arrangements that eventually become tall hedgerows. Spacings are typically 20 × 25 feet for grapefruit and vigorous trees, 15 × 20 feet for oranges and tangerines, and 12 to 15 × 18 to 20 feet for limes and smaller cultivars. A trend toward higher tree numbers per acre in both sweet orange and grapefruit has been apparent over the past 30 years. Tree densities are now about 100 to 110 trees/acre for grapefruit and 130 to 140 for sweet orange.

Citrus has a naturally sympodial growth habit, forming a large, bushy canopy that extends to the ground if left unpruned. Therefore, very little training is needed. Young trees may be headed at about 30 inches to induce branching and stripped of trunk sprouts and suckers for the first 2 years. They may be defruited for 1 to 2 years to induce vegetative growth. At maturity, trees are mechanically hedged and topped to form hedges about 12 feet tall and wide. Almost no hand pruning is done. Typically, hedging and topping is done every other year (Figure 11.5).

Pest Problems

Insects

Scale insects and mealybugs. At least 20 species of scale insects infest citrus fruits (e.g., *Pseudococcus, Ceroplastes, Lepidosaphes* spp.). Scales are small (<⅛ inch), round or oval, generally sedentary insects that appear as bumps on twigs, fruits, or leaves; they suck the sap (similar to aphids), debilitating the tree. They may be purple, red, white, yellow, or cottony in appearance. Mealybugs are similar to scales; they are generally white with a cottony or foamy appearance and are often found in leaf axils, between adjacent fruits, and in hard-to-reach places in the canopy. Natural enemies, such as *Aphytis* wasps or lady beetles, often control scales and mealybugs, but oil and insecticide sprays are used for heavy infestations.

TABLE 11.1. Major citrus rootstocks and their characteristics.

				Tolerance to		
Rootstock	**Yield/ tree**	**Fruit size**	**Tree vigor**	**Phytophthora foot rot**	**Drought**	**Freeze damage to tree**
Rough lemon	high	large	high	poor	good	poor
Sour orange	medium	medium	medium	good	fair	good
'Cleopatra' mandarin	low	small	high	good	fair	good
Trifoliate orange	low	small	low	very good	poor	good
'Carrizo' citrange	high	medium	high	good	good	fair
'Troyer' citrange	medium	medium	medium	very good	fair	good
Swingle citrumelo	high	medium	high	good	good	fair
Dwarfing rootstocks	'Flying Dragon', a form of trifoliate orange is the most widely used. Other less-dwarfing rootstocks include 'Rubidioux' trifoliate orange, 'Rangpur' × 'Troyer' hybrid, 'Rusk' citrange, and procimequat (limequat × kumquat).					

FIGURE 11.5. *(left)* Mature citrus orchards in Yuma, Arizona, seen from the air resemble parallel hedgerows. *(right)* A grapefruit orchard in the Indian River district of Florida has been hedged and topped mechanically to control size.

Citrus leaf miner. Phyllocristis citrella is the larva of a small moth recently introduced to the United States. It tunnels through the leaves, feeding as it goes, and leaves erratic "trails" about 1/16 inch wide in leaves. Leaves become severely distorted, and canopy appearance and photosynthetic potential can be greatly reduced. Since the pest has been recently introduced, natural predators are few. It is difficult to kill because its larvae are inside the leaves, and sprays do not reach them after they enter; thus, timing targets new flush growth in early spring and summer when eggs are laid.

Aphids. Several types of aphids (*Aphis, Toxoptera* spp.) may attack young leaves, shoot tips, and fruits. These insects (1/32 to 1/8 inch) come in various colors, from white to black, green, and brown. They congregate in large groups usually and move slowly when disturbed. Their piercing, sucking mouth parts locate the sugar-conducting tissues of the tree, and they rob the young, growing leaves, twigs, and fruits of this food source. In citrus, the feeding itself is less of a problem than is the virus transmission and sooty mold that may occur. The brown citrus aphid *(Toxoptera citricida)* is a citrus tristeza virus vector. Sooty mold *(Capnodium citri)* fungus often grows on leaf and fruit surfaces where aphid honeydew has been secreted, forming a black coating that limits light penetration to leaves and makes fruits unsightly. Ladybugs, lacewings, syrphid fly larvae, and other organisms are natural enemies of aphids that provide control. "Soft" insecticides, such as soaps and light oils, control low populations.

Thrips. Of the few species that attack citrus, western flower thrips *(Frankliniella occidentalis)* and citrus thrips *(Scirtothrips citri)* are the most common. Largely a problem in arid climates, these insects are difficult to control, as they feed inside flowers, between fruits, and on developing shoots. Feeding on developing tissues causes distortion and fruit scarring. Thrips generally will require sprays at petal fall in western states in most years. Biological control agents, such as predaceous mites, often control populations in processed citrus.

Mites. Mites feed on the lower leaf surface, causing a characteristic "mite stipple" or bronzed appearance to the foliage. If severe, leaf browning and drop may occur, ultimately affecting yield. Spider mites (*Tetranychus, Eutetranychus* spp.) may be white, red, or yellow and two-spotted; they generally occur on leaf undersides and are worst in hot, dry weather. The eriophyid species (bud and rust mites), too small to see, are particularly problematic in citrus. Citrus bud mite *(Aceria sheldoni)* is a problem in arid regions, where bud feeding causes fruit losses and damages new foliage. Citrus rust mite *(Phyllocoptruta oleivora)* generally causes cosmetic damage to the fruit peel, although yield reduction may occur from the mites feeding on young fruits. Emulsifiable oil or miticides are applied when visible injury occurs.

Diseases

Greasy spot. Mycosphaerella citri is the main cause of greasy spot, which is largely a disease of foliage; it normally causes decreased productivity but can cause severe defoliation. It occurs in humid, subtropical citrus areas, such as Florida. It is more severe on grapefruit, lemon, and early oranges and less of a problem on tangerines and late oranges, such as 'Valencia'. Leaf symptoms begin as chlorotic areas on the upper surface, with small raised brown areas underneath; later the brown areas become darker and greasy in appearance, and the chlorosis disappears. However, leaves often drop prior to full symptom development. Fruit symptoms, called "greasy spot rind blotch," are less common but occur on grapefruit. Superficial, necrotic specks show up between oil glands on the skin. The disease is controlled by summer oil sprays, which actually only delay symptom development and do not kill the fungus, or by copper and other fungicides, which do kill the organism.

Melanose. Similar to greasy spot, melanose, caused by *Diaporthe citri*, is important in only subtropical citrus areas with high humidity and rainfall during early fruit development. The organism creates raised, brown, corky lesions on leaves, sometimes causing distortion or dieback. Fruit symptoms are superficial, black spots that coalesce if infection occurs early but remain pin-sized and discrete in late infections. Lesions often occur in vertical lines down the fruit surface, since spores overwinter in dead wood and often splash onto and run down the fruits from above. Grapefruit and lemons are most severely affected, but processing oranges generally do not require sprays for this disease. The fungus can be controlled by copper and other fungicides applied after fruit set, or by pruning out dead wood from older or freeze-damaged trees.

Citrus scab or sour orange scab. Caused by *Elsinoe fawcettii*, scab does not occur in Mediterranean climates but can be a problem in subtropical climates, especially on tangerine and its hybrids. Also, seedling rootstocks of rough lemon, sour orange, 'Rangpur' lime, and 'Carrizo' can be seriously stunted by scab. If infection occurs early, shoots are distorted and stunted; if infection occurs later, lesions remain as small, corky pustules, noticeably raised on the surface, with indentations occurring on the opposite side of the leaf. Scab is only a superficial problem, mostly on grapefruit and rarely on sweet oranges; however, scab may seriously reduce yields of 'Temple' and related cultivars. Control is by fungicide application during growth flushes; nursery stock must be sprayed several times per year.

Tristeza virus or quick decline. This is the worst of several virus diseases of citrus and one of the few that can be transmitted by aphids as well as grafting. Trees on sour orange rootstock are most susceptible. Infected trees show no symptoms initially but transmit the virus into the rootstock at the bud union, girdling the rootstock and slowly killing the tree. Resistant rootstocks include rough lemon, 'Rangpur' lime, trifoliate orange and hybrids, and 'Cleopatra' mandarin. Quarantine and bud wood certification programs are generally in place in areas where tristeza is endemic.

Phytophthora foot rot. The organism *Phytophthora parasitica* causes several soil-related diseases of citrus, although the girdling of trees at the soil line, or foot rot, is most common. All cultivars, especially sweet orange, are susceptible. Control is through prevention: trees are budded high (~8 inches) to avoid soil contact with the scion, and resistant rootstocks, such as sour orange, can be used. Irrigation water is kept away from the trunk, and trunks can be painted with copper fungicides. Systemic fungicides are curative agents that can be used only on young, nonbearing trees.

Citrus canker. This bacterial disease, caused by *Xanthomonas campestris* pv. *citri*, leads to mild defoliation and rind blotching of susceptible cultivars, and defoliation and fruit drop only when severe. Susceptible types include grapefruit, 'Key' lime, and trifoliate orange, with sour orange, lemons, and sweet orange being moderately resistant, and tangerines relatively resistant. However, eye appeal of fresh fruits may be impaired. It is difficult to control once established, and government eradication and quarantine measures have been quite severe—infected trees are generally removed or burned, along with healthy adjacent trees for good measure. Control is achieved through preventing leaf lesions from developing as a result of wind scar, sand, and so forth, since the bacterium enters through wounds. A copper spray program can reduce the spread of the disease.

HARVEST, POSTHARVEST HANDLING

Maturity

All citrus varieties are nonclimacteric fruits, meaning that they ripen gradually over weeks or months and are slow to abscise from the tree. Contrast this with climacteric fruits, such as peaches, where the ripening period is several days to two weeks and concludes with fruit drop if fruits are not harvested. External color changes during ripening but is a function of climate more than ripeness, and a poor indicator of maturity. The best indices of maturity for citrus are internal: Brix scale (percent sugar by weight), acid content, and the Brix-acid ratio. This is particularly true for processed fruits. The °Brix (or sugar content) of juice is measured with a hydrometer or hand refractometer. It increases slowly during maturation and falls within the following ranges: oranges—10 to 14 percent; grapefruit—8 to 10 percent; tangerines—12 to 16 percent. Sugar content is not a factor for lemons and limes and is generally under 10 percent. Acid content is measured by titration with sodium hydroxide (NaOH) and decreases with maturation: oranges—0.5 to 1.5 percent; grapefruit—1.0 to 2.0 percent; tangerines—0.5 to 2.0 percent; lemons and limes—6.0 percent. The ratio between the two determines flavor, and minimum standards are about 9:1 for sweet oranges and tangerines and 6:1 for grapefruit. As with sugar content, the Brix-acid ratio is not important for lemons and limes. For lemons and limes, juice content and fruit size must meet minimum standards prior to market (about 40-45 percent juice and a diameter of at least 1⅞ inches). External quality is a function of color and blemishes caused by wind scar as well as disease or insect damage. The highest quality standards require less than 10 percent of the fruit surface to be blemished.

Harvest Method

Citrus fruits are hand harvested, whether processed or marketed fresh. Mechanical harvesters were used on about 3 percent of processed citrus acreage in Florida in 2003 on a trial basis. Mechanical harvesters can reduce harvest costs up to 50 percent but may require new planting and training systems for juice oranges. Tangerines and some fresh oranges must be clipped, not pulled, from the tree to prevent plugging the peel.

Postharvest Handling

Citrus fruits can be handled fairly roughly from tree to packinghouse because they are tough and resilient, with the exception of tangerines. Fruits are dumped from picking bags into large bulk bins, which are moved by forklift onto trucks.

For fresh fruits, standard packing-line operations are used (in order): dumping, culling, washing, brushing, waxing, drying, grading (by eye), sizing, and boxing (Figure 11.6). For processed fruits, growers are paid for "pounds solids" or the quantity of sugar in a load of fruits, based on juice analysis. Harvested fruits are culled for rot, and remaining fruits are washed prior to juicing. Juice is extracted by inserting a cylindrical strainer into the center of the fruit and compress-

FIGURE 11.6. A packinghouse for fresh citrus handling. The fruits are washed, then waxed and brushed *(foreground)*. Fruits are then sorted and packed according to size *(background)*.

ing the fruit hydraulically. Extracted juice contains some pulp and oils, which are separated from the juice by centrifugation and screening. Juice is then evaporated to 48 to 70 °Brix for storage or shipment. Juice is pooled into lots of various colors and sugar levels; some mixing is done to produce a uniform product. Sales of frozen concentrate have been outpaced by single-strength juice products in recent years, due to the superior flavor of the latter.

Storage

Citrus fruits may be stored for periods of up to 2 months at low temperatures (32-40°F). Tangerines have the shortest shelf life of the citrus fruits. Lemons can be harvested green and cured at 55 to 60°F for several weeks. During this time, they turn yellow and increase in juice and acid contents. Chilling injury is common in grapefruit, lemons, and limes when stored below 50°F, but rare in oranges and tangerines. Brown pitting and staining of the rind occur, sometimes with a watery breakdown of peel and pulp. Several pathogenic rots, including the common green mold (*Penicillium* spp.) can occur postharvest. A mild postharvest fungicide is often applied in the packinghouse.

A unique aspect of citrus is the ability to store fruits on the tree. Fruits may reach minimum maturity standards in early winter, but since they are nonclimacteric, they ripen slowly and will not soften or abscise for periods of up to several months. This allows growers to pick over an extended period of time, choosing prices when they are at their highest. Fruit quality steadily improves with time as well. The danger, of course, is potential freeze damage or crop loss due to pest pressure. Tangerines, in general, store poorly on the tree, while oranges and grapefruit hold up extremely well.

CONTRIBUTION TO DIET

Americans derive about 26 percent of total vitamin C from citrus fruits, the highest proportion from any single food group. All other noncitrus fruits contribute another 16 percent, for a total of 42 percent from fruit consumption. Citrus contributes only 0.9 percent of total daily calories and 1.7 percent of daily carbohydrate intake, despite recent efforts to characterize citrus fruits as "high carb" foods.

Per capita consumption of citrus is higher than that for any other fruit crop when juice and fresh consumption are combined. In 2004, U.S. citizens consumed 78.8 pounds of sweet oranges (86 percent as juice), 7.9 pounds of grapefruit (48 percent as juice), 6.7 pounds of lemons (54 percent as juice), 2.6 pounds of limes (28 percent as juice), and 3.9 pounds of tangerines (28 percent as juice) per year. Lime consumption has increased three- to fourfold, while grapefruit consumption has decreased by over 50 percent since 1980; consumption of oranges, lemons, and tangerines has remained relatively constant over time. Total fruit juice consumption was 8.9 gallons per capita in 2004, of which 65 percent is citrus (the remainder being apple, cranberry, grape, and other fruit juices).

Dietary value per 100-gram edible portion:

	Sweet Orange	Grapefruit	Tangerine	Lemon	Lime
Water (%)	87	91	85	89	88
Calories	46	32	53	29	30
Protein (%)	0.7	0.6	0.8	1.1	0.7
Fat (%)	0.2	0.1	0.3	0.3	0.2
Carbohydrates (%)	11.5	8.1	13.3	9.3	10.5
Crude Fiber (%)	2.4	1.1	1.8	2.8	2.8
	% of U.S. RDA (2,000-calorie diet)				
Vitamin A	4	18	14	<1	1
Thiamin, B_1	7	2	4	3	2
Riboflavin, B_2	2	1	2	1	1
Niacin	2	1	2	<1	1
Vitamin C	75	57	44	88	48
Calcium	4	1	4	3	3
Phosphorus	2	1	3	2	3
Iron	<1	<1	<1	3	3
Sodium	0	0	<1	<1	<1
Potassium	5	4	5	4	3

BIBLIOGRAPHY

Castle, W.S. 1987. Citrus rootstocks, pp. 361-399. In: R.C. Rom and R.F. Carlson (eds.). *Rootstocks for fruit crops.* New York: John Wiley and Sons.

Davies, F.S. and L.G. Albrigo. 1994. *Citrus.* Wallingford, UK: CABI.

Dugo, G. and A. Di Giacomo (eds.). 2002. *Citrus: The genus* Citrus. New York: Taylor and Francis.

Duke, J.A. and J.L. duCellier. 1993. *CRC handbook of alternative cash crops.* Boca Raton, FL: CRC Press.

Jackson, L.K. and F.S. Davies. 1999. *Citrus growing in Florida.* Fourth edition. University Press of Florida. Gainesville.

Kimball, D.A. 1999. *Citrus processing: A complete guide.* Second edition. Gaithersburg, MD: Chapman and Hall.

Klotz, L.J. 1973. *Color handbook of citrus diseases.* Berkeley, CA: University of California Division of Agricultural Sciences.

Monselise, S.P. Citrus and related genera, pp. 275-294. In: A.H. Halevy (ed.), *CRC handbook of flowering,* Volume 2. Boca Raton, FL: CRC Press.

Morton, J.F. 1987. *Fruits of warm climates.* Miami, FL: Julia F. Morton.

Mukhopadhyay, S. 2004. *Citrus: Production, postharvest, disease and pest management.* Washington, DC: United States Science Pubs., Inc.

Nagy, S., P.E. Shaw, and M.K. Veldhuis (eds.). 1977. *Citrus science and technology,* Volume I. Westport, CT: AVI Publishing.

Ray, R. and L. Walheim. 1980. *Citrus: How to select, grow and enjoy.* Los Angeles, CA: Price Stern Sloan, Inc.
Reuther, W., H.J. Webber, and L.D. Batchelor (eds.). 1967. *The citrus industry,* Volumes 1-3. Berkeley, CA: University of California Press.
Spiegel-Roy, P. and E.E. Goldschmidt. 1996. *Biology of citrus.* Cambridge, UK: Cambridge University Press.
Statewide Integrated Pest Management Project. 1991. *Integrated pest management for citrus,* Second edition (University of California Division of Agricultural and Natural Resources Publication 3303). Berkeley, CA: University of California Press.
Sturrock, D. 1959. *Fruits for southern Florida.* Stuart, FL: Southeastern Printing Co.
Timmer, L.W., S.M. Garnsey, and J.H. Graham (eds.). 2000. *Compendium of citrus diseases,* Second edition. St. Paul, MN: American Phytopathological Society Press.
Ting, S.V. and R.L. Rouseff. 1986. *Citrus fruits and their products: Analysis and technology.* New York: Marcel Dekker.

Chapter 12

Coconut *(Cocos nucifera)*

TAXONOMY

The Arecaceae (or palm) family is a large, distinct family of monocotyledonous plants, containing up to 4,000 species distributed among over 200 genera. The coconut palm, *Cocos nucifera* L., is undoubtedly the most economically important plant in the family. The genus *Cocos* is monotypic; thus, there are no "close" relatives of coconut. Among the more distant relatives are two of the top 20 fruit crops in the world: *Elaeis guineensis,* African oil palm, and *Phoenix dactylifera,* the date palm. Several other palms are cultivated or wild-harvested for their fruits, including peach palm *(Bactris gasipaes),* ungurahui palm *(Jessenia batua),* jelly palm *(Butia capitata),* betel nut *(Areca catechu),* Chilean wine palm *(Jubaea chilensis),* saw palmetto *(Serenoa repens),* and aguaje palm *(Mauritia flexuosa).* An even greater number of taxa are cultivated as ornamentals, including coconut palm itself, as well as members of the genera *Phoenix, Sabal, Serenoa, Washingtonia,* and *Roystonea,* to name just a few.

Cultivars

Coconut palms have two natural subgroups, simply referred to as "tall" and "dwarf." Most commercial plantings use high-yielding, longer-lived tall cultivars, and each region has its own selections, such as, 'Ceylon Tall', Indian Tall', 'Jamaica Tall' (syn. 'Atlantic Tall'), and 'Panama Tall' (syn. 'Pacific Tall'). The tall cultivar group is sometimes given the name *Cocos nucifera* var. *typical,* and the dwarf cultivar group, *C. nucifera* var. *nana.* Dwarf cultivars are used largely as ornamentals, tend to be self-pollinating, and are more precocious than tall cultivars. Dwarf types were thought to be more tolerant of lethal yellowing disease, which kills many of the tall cultivars. However, extended field observations proved that many dwarf types are also susceptible to lethal yellowing and revealed only one that was resistant—'Fiji Dwarf' (syn. 'Niu Leka'). Hybrids between the tall and dwarf types have intermediate stature and characteristics; in other words, they are slow growing but ultimately grow to about the size of tall cultivars, as in the case of 'Maypan'. Some hybrids combine the high yield of the tall type and the precocity of the dwarf type. Further classification of cultivars is done according to color; for example, 'Dwarf Malayan Gold' is a dwarf Malayan variant having a golden color on immature fruits and petioles. Fruit morphology is also used to distinguish cultivars; selections for cultivation purposes favor the characteristics of large seeds and a thin mesocarp (termed *niu vai*) versus a thick mesocarp and small, triangular-shaped seeds (termed *niu kafa*), which are geared toward ocean dispersal.

ORIGIN, HISTORY OF CULTIVATION

The origin of the coconut palm is obscured by the ability of the fruit to disseminate the species naturally over distances of thousands of miles. Coconuts can float on the ocean for months

doi:10.1300/5547_12

and still germinate when beached, so they may have arisen anywhere between the eastern Indian and western Pacific Oceans. Prior to the age of discovery, coconuts were dispersed from East Africa to the Pacific Coast of Panama. Controversy remains over whether most of this dispersal occurred naturally or mechanically through transmission by early Polynesian voyagers. The Polynesians were master seamen, colonizing such remote areas as Hawaii and the Easter Island thousands of years ago. They were known to have brought coconuts with them on long voyages, planting them on the islands they visited or colonized. Whether they or the equatorial countercurrent were responsible for bringing coconuts to the Americas prior to Columbus remains a mystery. Coconuts provided the only source of food and water on many of the atolls of the equatorial region of the Pacific Ocean, and the natural distribution of coconut may have influenced the initial colonization of this region. It is clear that no coconut palms existed along the east coast of the Americas, in western Africa, or in the Caribbean prior to European exploration in the sixteenth century. Today, coconut is distributed pantropically, extending also into extratropical areas, such as southern Florida and the Bahamas.

Coconut palms have been used since ancient times as a source of food, fiber, fuel, water, and shelter, and many of these uses are still important today. Coconut oil was one of the first, if not the first, plant oil to be used by humans, and it was the leading vegetable oil until 1962, when it was eclipsed by soybean oil. The earliest written references to the coconut date from about 545 AD, by the Egyptian Monk Cosmos Indicopleustes, who described them while traveling in India; followed by the writings of Marco Polo, who encountered and wrote of the coconut while exploring India and Sumatra around 1280. The use of coconut oil for soap manufacture was patented as early as 1841; this was followed by patents for several other industrial and food uses of the oil (e.g., margarine) that continue today. Unlike many tropical fruits, coconuts are still grown largely by small landholders instead of on large plantations, although plantations have become more popular recently. In the Philippines, the leading producer of coconut, an estimated 18 million people are involved in coconut production in some way. Coconut production carries a "poor man's" connotation in many areas because the production methods are crude and often confined to manual labor, such as the planting of trees and harvesting fruits from the ground.

FOLKLORE, MEDICINAL PROPERTIES, NONFOOD USAGE

The coconut palm is referred to as the "tree of life," "tree of heaven," or "tree of abundance," reflective of its essentiality to everyday life in the tropics. It is considered one of the ten most useful trees in the world and has far more uses than any other fruit crop, for example:

> The immature nut provides a pleasant beverage (coconut milk or liquid endosperm—also used in tissue culture), the raw kernel is an important article of food; pared, shredded and dried it provides the desiccated coconut of commerce. The oil is used for cooking, for illumination and lubrication, and in the manufacture of margarine, bakery fats, soaps, detergents and toiletries. Coconut cake, the residue after extracting the oil from copra, is valuable as cattle and poultry feed. Tapping the inflorescence produces sap which can be used to provide sugar, vinegar, sweet or fermented toddy (like beer or wine) and, when distilled, arrack (liquor). The timber can be used for building and furniture construction; the plaited leaves for roofing; fiber from the husk for the manufacture of ropes and matting; the shell for charcoal or the manufacture of artifacts (pots, buttons, bowls, etc.). Even the roots are used in dyes and traditional medicines. (cited in Greene, 1991, p. 6)

This quote gives some idea, but not a full inventory, of the uses of coconut palms, which also have several medicinal uses: The resin of the inner husk is used in Panama for toothache relief.

Coconut milk (liquid from endosperm) is used as an anthelmintic (eliminates intestinal worms). The milk or water obtained from immature coconuts contains sugars and is sterile; it is said that field surgeons in World War II used coconut water when supplies of intravenous glucose solution ran out. Duke and duCellier (1993) list over 60 different folk remedies that utilize coconut fruits or plant parts, including treatment of: abscesses and tumors, dysentery, cold and flu, constipation, scurvy, and venereal diseases. Lauric and capric acids found in coconut may be converted to monolaurin and monocaprin in the body; these compounds have shown antibacterial, antiviral, and antiprotozoal activity, useful in defending against various bacteria, HIV and herpes viruses, and the protozoan *Giardia*. Coconut oil may protect the liver from the damaging effects of alcohol and improve the anti-inflammatory response of the immune system.

Coconut oil is 92 percent saturated fat, and the remainder is largely monounsaturated rather than polyunsaturated fat. Major fatty acids include lauric, myristic, palmitic, caprylic, and oleic acids (in descending order of percent composition). Although coconut oil contains no cholesterol, it does contain saturated fats, which are believed to contribute to the elevation of serum cholesterol in the body and are associated with obesity. However, it appears that coconut oil has been unfairly criticized in this regard, as studies have shown that short- and medium-chain fatty acids in coconut oil do not increase cholesterol, nor are they incorporated into body fat when consumed normally. In fact, unlike common hydrogenated oils, coconut oil does not form harmful trans fatty acids when used in cooking, so it does in fact contain fewer compounds documented to have detrimental health effects.

Such a useful plant finds its way into folklore and religion frequently. In Samoa, it was believed that coconut trees grew at the entrance to the spirit world, Pulotu. Unripe coconut represented both heaven and the underworld, and in the New Hebrides, coconut is eaten at funerals to allow communication with the dead. So sacred were coconut trees to the Wanika of East Africa that cutting one down was tantamount to murder. Coconut shells were used to house the souls of newborn babies for protection during their first year of life in Borneo, and they were used to bury the afterbirth in the Philippines. Palms, including coconut, are symbolic of "victory" in the "language of flowers."

PRODUCTION

World

Total production in 2004, according to FAO statistics, was 53,473.584 MT or 118 billion pounds. Coconuts are produced in 93 countries worldwide on about 26 million acres, a greater land area than is covered by virtually any other fruit crop. Production has increased about 13 percent over the past decade, largely due to yield increases, as area cultivated has been relatively constant. Average yield is 4,450 pounds/acre, ranging from less than 1,000 to 14,700 pounds/acre. The top ten coconut-producing countries (percent of world production) follow:

1. Indonesia (29)
2. Philippines (27)
3. India (18)
4. Brazil (6)
5. Sri Lanka (4)
6. Thailand (3)
7. Mexico (2)
8. Vietnam (2)
9. Malaysia (1)
10. Papua New Guinea (1)

Coconuts are used to derive a number of products, each of which is often produced in quantities greater than the total amount of production for minor fruit crops. Copra is the dried endosperm or "meat" of the coconut, commonly used in cakes and candies and thus sold mainly for the confectionary trade, from which coconut oil is extracted. Although data have been discontinued, world copra production generally amounted to about 10 percent of total coconut production. Coconut oil production in 2002 was 3.2 million MT, slightly more than olive oil production. Coconut cake, the residue left after pressing oil from copra, is often used as livestock feed; 1.8 million MT of coconut cake was produced in 2002. Coir, another product derived from coconut, is the fiber from the husk that is used as packing material, rope, matting, and fuel and also in potting mixes. About 642,000 MT of coir was produced in 2004, with 90 percent coming from India and Sri Lanka. Water in immature coconuts provides a refreshing, nutritious drink. No data are available for water coconuts, but they are frequently sold and/or used in the tropics.

United States

No U.S. production records exist for coconut. Coconut palms are grown in Hawaii and extreme southern Florida, but largely for their use as ornamentals.

BOTANICAL DESCRIPTION

Plant

Trees are typical, single-trunked palms, reaching up to 100 feet in height, but generally they grow 20 to 50 feet under cultivation (Plate 12.1). Trunks may reach a diameter of 16 inches and often have swollen bases. Leaves are among the largest of any plant (up to 20 feet long), pinnately compound with 200 or more leaflets, and borne in a spiral arrangement at the apex of the trunk. Leaf life span may be 3 years, and mature, healthy palms have about 30 leaves, forming a new one and dropping the oldest one each month. Age of the palm is therefore approximated by the total number of leaves and leaf scars divided by 12. Tall cultivars can live 60 to 80 years, while dwarf cultivars generally live only about 30 years. The petioles are grooved to channel water toward the crown. Roots form adventitiously at the base of the trunk, become thickened, and spread for 40 to 60 feet to anchor the palms in even loose beach sand.

Flowers

Separate male and female flowers are borne in the same inflorescence, which is a compound spadix arising in the leaf axil. Flowers are off-white to gray and inconspicuous. Coconuts are polygamomonoecious, having hundreds or thousands of male flowers and a few female flowers in the same inflorescence (Plate 12.2). They are generally protandrous, meaning that male flowers release pollen before females become receptive. Flowering occurs continuously, since each leaf axil produces one inflorescence, and new leaves are produced approximately monthly. Female flowers have superior, trilocular ovaries, and potentially three ovules, although only one develops into a seed.

Pollination

Since coconuts are protandrous, they are believed to be largely cross-pollinated. Dwarf cultivars, particularly the popular ornamentals, are largely self-pollinating, as opposed to the tall cultivars of commerce, which rarely pollinate themselves. Since trees are grown from seed and each is genetically distinct, cross-pollination does not present a problem from a management standpoint. Flowers are nectariferous and attract bees, and wind is likely important in pollination as well. Fruit set is about 10 to 50 percent, with half of the female flowers dropping due to lack of pollination or fertilization. About three to six fruits per inflorescence develop into mature coconuts (Plate 12.3).

Fruit

Coconuts are large, dry drupes, ovoid in shape, up to 15 inches long and 12 inches wide. The exocarp or skin is green, yellow, or bronze-gold, turning to brown, depending on cultivar and maturity. The mesocarp is fibrous and dry at maturity; the product coir is derived from this layer. The endocarp is the hard shell enclosing the seed. Seeds are the largest of any plant and have a thin, brown seedcoat. Seeds are filled with endosperm, which is solid and adherent to the seedcoat, and also in liquid form, called "milk." Copra is derived from the solid endosperm. The seed has three germ pores in the proximal end, such that the seed resembles a miniature bowling ball. Two pores are generally plugged, and the embryo lies beneath the third functional pore (Figure 12.1). A "blind coconut" has all three pores plugged. The embryonic root and shoot emerge through the functional pore, while the cotyledon forms a spongy mass of tissue within the seed cavity, absorbing the endosperm, which fuels initial growth. Seeds germinate within intact coconuts, and for up to a few months, young palms can remain rooted within the mesocarp while leaves emerge, giving the appearance of a "self-potted plant" (Figure 12.2). Dwarf cultivars are precocious and begin fruiting in 3 years, whereas tall cultivars require 6 to 10 years. Fruits require about 1 year to mature.

GENERAL CULTURE

Soils and Climate

Coconuts are considered "beach plants," and it is sometimes assumed that they need beach habitat and saltwater to survive. In fact, coconuts are adaptable to many soil types and can be grown inland when provided with adequate drainage and soil pH between 5 and 8. Although more salt tolerant than most fruit crops, coconuts do not require salt for growth, and saltwater will kill trees that are inundated for long periods. Coconuts on beaches actually tap the freshwater lenses that persist above the salty groundwater; these thin reservoirs of freshwater move up and down with the tide, successively watering and then aerating the root zone. Coconuts are naturally confined to beach habitats because they are poor competitors, intolerant of shade and competition, and seeds generally do not travel far inland.

Coconut palms are found in the hot, rainy tropics, generally below 1,000 feet elevation. They can be planted in frost-free subtropical areas and in climates with extended dry seasons when provided with irrigation or soils with sufficient water-holding capacity. Coconut palms may tolerate short durations of light frost with only minor injury, but they grow best in areas that have mean temperatures of 70 to 80°F and are free of frost and cool weather. They require 40 to 60 inches of water per year and perform better in high-humidity climates, particularly in areas with rainfall levels at the lower end of this range. Supremely tolerant of hurricane-force winds and driving rain, they are rarely uprooted. Also, they tolerate salt spray better than most crops and can tolerate brief flooding associated with tropical storms.

FIGURE 12.1. Coconuts contain the largest seeds of flowering plants. The top two photos show the intact nut (endocarp plus seed) with three germ pores; the seedling emerges through one of the pores. *(bottom left)* The seedcoat is thin, brown, and wrinkled, appressed to the inner wall of the endocarp. *(bottom right)*. The embryo is tiny (¼ to ½ inch) and sits beneath one of the three germ pores.

Propagation

As with most palms, coconuts are seed propagated. Seeds undergo a dormant period of 4 to 5 months from the time of abscission, probably to prevent premature germination while fruits are floating at sea for long periods. The entire fruit is generally planted, not just the seed, as the husk provides a porous medium for initial root growth. Fruits are planted close together in seedbeds and then transplanted to plantations or growing sites (with fruits still intact) when eight to ten leaves have formed. Fruits may be selected from trees with high yields and good quality, or special seed orchards may be used to produce relatively uniform seed.

FIGURE 12.2. Coconuts have many uses. This one was painted white and used as a tee marker on a Caribbean golf course. It has sprouted, illustrating the viability of coconut seed after many months lying on the ground, and its ability to grow as a "self-potted plant."

Rootstocks

No coconut rootstocks exist.

Planting Design, Training, Pruning

In plantations, palms are set about 25 feet apart in all directions. Since coconut palms take several years to bear fruit, they are often intercropped with staple crops, such as maize, or even with other tree crops (Figure 12.3). Enough light penetrates the palm canopy to allow grasses to grow beneath, and thus coconut is often a component of tropical silvopastoral systems for raising livestock.

Palms require no pruning or training; they form monopodial trees naturally. They tend to lean toward the light or into the prevailing wind and may need to be guyed or propped if leaning is excessive. In plantations, they tend to grow more or less straight. In landscape settings, older fronds or those injured by frost or disease are sometimes pruned off. Injury to the apical bud is fatal.

Pest Problems

Insects

Rhinoceros beetle. Rhinoceros beetles (*Orcytes rhinoceros*) are easily identified; they are large (up to 2 inches long), brown beetles with prominent horns on their heads, hence the name. They are found throughout Southeast Asia and the Pacific Islands. Adults tunnel into the bases of developing leaves and can cause severe defoliation. Beetles may also tunnel into the meristem, severely deforming the palm or sometimes killing a young palm, although this is not as common as defoliation. They breed in rotting logs and composting vegetation, where larvae complete their development. Some control measures are targeted at the removal of breeding sites. Another cultural control involves intercropping with tall-growing crops; this is effective because beetles are attracted to young palms by sight. Biological control using the fungus *Metharhizum anisopilae* or *Baculovirus orcytes* has also been effective.

FIGURE 12.3. Coconut palms intercropped with sugarcane on an alluvial soil of volcanic origin, well inland from the south coast of Guatemala.

Other beetles and weevils. Several species of *Rhynchophorus* weevils may attack coconuts, particularly *R. ferrugineus*. They lay eggs in wounds on trunks or crowns, and the resulting larvae tunnel through the tissue, causing damage. They grow up to 2 inches in length, similar to rhinoceros beetles. *Rhinostomus* species are similar to those of *Rhynchophorus,* but they do not require wounds for entry, and the weakening caused by the larvae's tunneling can actually topple the palm. Baits and pheromone traps have been used for control.

Foliage feeders. Species from at least four different orders of insects are known to feed on young and mature leaves. Locusts (*Locusta, Schistocerca* spp.) have been known to severely defoliate palms. Several butterflies and moths in at least 15 families within the order Lepidoptera produce caterpillars (larvae) that feed on coconut palm leaves. Chemical insecticides are applied when threshold populations are reached, or, in some cases, beneficial insects can keep populations in check. Leaf miners, particularly *Promecotheca* spp., can cause considerable leaf damage in coconut.

Diseases

Bud rot and nut fall. The fungus *Phytophthora palmivora,* widespread throughout the tropics, causes a fatal disease of the apical meristem of coconut. Generally only a small percentage of trees are affected each year, but mortality may reach 50 percent and fruit loss 25 percent in severely affected areas. *Phytophthora hevae* and another unnamed *Phytophthora* species cause the same disease in Cote d'Ivoire and Hawaii, respectively, but *P. palmivora* is the primary causal agent. The disease starts on leaves, particularly the youngest or "heart" leaf, producing sunken lesions, followed by chlorosis, browning, and collapse. The disease progresses to other leaves from the middle outward; thus, infected trees have green leaves at the periphery and chlorotic and dead leaves in the center of the canopy. Young coconuts fall to the ground, but older fruits lower on the trunk often remain. The disease is worse in areas of high rainfall and may gain entry through wounds created during tropical storms. There are no resistant cultivars, and spraying with traditional fungicides is ineffective. Systemic fungicides, such as fosetyl Al and metalaxyl, can be applied to individual trees as a preventative measure before disease symptoms occur, but this is not economical in commercial production. Infected trees are removed once the disease symptoms occur because the spores are rain splash disseminated to other trees.

Ganoderma butt rot or basal stem rot. Three species of *Ganoderma* fungi can cause a fatal, systemic rot of coconut and other palms, such as African oil palm and areca palm. *Ganoderma boninense* is the primary pathogen. It kills the roots and grows within the trunk, affecting water supply and creating symptoms resembling drought stress. Older leaves die and collapse, younger leaves are chlorotic and slow growing, and fruits are smaller and sometimes distorted. It may take 3 to 4 years for trees to die. The fungus is a weak parasite or saprophyte, entering through wounds and thus affecting older plantations or individuals. Control centers around sanitation of infected or dead trees, with prompt removal, burning, or the digging of trenches around infected sites being used to reduce inoculum levels. Fungicide injections or drenches are sometimes successful in slowing the disease but may not be economically feasible.

Cadang-cadang. This lethal disease is currently confined to the Philippines, where it has killed tens of millions of coconut palms. A viroid composed of RNA is the causal agent, and a similar viroid causes a similar disease, tinangaja, in Guam. Both are related to the potato spindle tuber viroids. The viroid is found in coconut and other palms in many regions, suggesting that the disease may become a problem elsewhere and on other crops in the future. The disease takes several years to kill a tree, but early on, fruits are smaller and rounded, with scarring in an equatorial band. Fruiting stops completely in 2 to 4 years. The disease spreads slowly, with boundaries moving about 500 yards per year. It is thought that the viroid is spread by workers on machetes, but it may also spread through pollen. There is no way to control the disease, and no resistant cultivars exist at present, but a rapid diagnostic test for the presence of the viroid has been developed.

Lethal yellowing. Unlike cadang-cadang and *Ganoderma*-related rot, which take years to kill coconut palms, lethal yellowing disease generally kills within 3 to 5 months of the onset of symptoms. The incubation period may be several months to over a year, however. The disease was first noticed in Jamaica and is currently confined to tropical America. A mycoplasma organism causes the disease and is apparently spread by a grasshopper *(Myndus crudus)*. The disease can spread very quickly, covering over 75 miles in southern Florida in 3 years time. The disease is far worse on tall cultivars, where it first causes premature fruit drop, followed by yellowing of leaves, starting with the oldest and progressing inward. The entire crown then dies and falls off, leaving a bare, dead trunk. Dwarf cultivars show leaf browning instead of yellowing and may take longer to die than tall ones. 'Malayan Dwarf' and 'Maypan' were reported to be resistant to lethal yellowing, but long-term trials in Florida have shown their resistance to about as poor as that seen in tall cultivars. 'Fiji Dwarf' has shown considerable resistance. Trunk injection with antibiotics at intervals of 4 months can effectively control the disease, if caught early, but this is justified for only landscape specimens.

HARVEST, POSTHARVEST HANDLING

Maturity

Fully mature fruits require about 1 year to ripen and are brown or black, depending on cultivar. The endosperm, from which the copra and oil are derived, is mature at 10 months after bloom. For coir production, fruits must be harvested about 1 month before full maturity because the mesocarp fiber turns brittle and dark at maturity. Water coconuts are harvested when about 7 months old, just after fruits reach their full size and prior to mesocarp drying. Coconuts are classified as nonclimacteric fruits with respect to ripening.

Harvest Method

Coconuts fall from trees when fully mature and are easily collected from the ground if the intended use is the copra or oil. Normally, one or two fruits fall per week per tree. Alternatively, fruits are often harvested by hand prior to abscission, particularly those not used for copra or oil. Fruits can be cut from the ground using knives on long poles to avoid climbing tall trees. It is said that, in Southeast Asia, monkeys are trained to harvest ripe coconuts.

Postharvest Handling

Coconuts intended for copra or oil production are split open with a machete, discarding the milk and exposing the endosperm to the sun to dry. Drying takes 2 to 5 days in the sun or can be done more quickly in kilns. Fruits do not have to be dehusked prior to splitting; the seed is easily removed from the split husk. Dehusking can be done in the field, rather than at the processing site, so that the husks can be used for mulch in the field. Alternatively, husks may be used as fuel for kilns, so dehusking at the processor may be indicated. Once dry, the copra is removed from the seeds with metal tools and further dried to reach a water content of 5 to 6 percent.

Coconut oil for is extracted mechanically by pressing or chemically by the use of solvents; mechanical pressing is most common. Pressing is sometimes followed by solvent extraction to obtain higher oil yields. The most common solvent used is hexane due to low cost and low toxicity. Hexane is separated from oil by heating and trapping the hexane in mineral oil; hexane is then recycled for another extraction. Solids are removed from the extracted oil by settling and screening, followed by filtration. The remaining solids, or coconut cake, are dried, bagged, and sold for livestock feed.

As an alternative to the aforementioned process using dried copra, several "wet processing" techniques have been used to obtain oil and other food products from fresh endosperm of coconuts. Although too complex to describe here, the various methods end up producing these products: "virgin" coconut oil, which, similar to olive oil, has low fatty acid content; coconut cream, a white cream with 20 to 30 percent fat; coconut skim milk, a white coconut drink product with 1 to 3 percent protein and 0.5 to 3.0 percent fat; and coconut water, a clear, sometimes sweetened drink product.

Storage

Whole coconuts can be stored for several weeks prior to dehusking and extracting copra. Fruits are kept dry but stored at ambient temperatures. Some nuts will germinate when stored in this way, which reduces the copra content, since this is the reserve tissue that supports embryo growth. Short-term storage may be advantageous in that moisture content decreases, facilitating drying, the copra gets thicker, and oil content increases. Fruits are also easier to dehusk when stored. Coconuts are not sensitive to chilling injury and can be stored at 32°F for 1 to 2 months.

CONTRIBUTION TO DIET

Several food uses or products exist for coconut. The primary product is copra, the white "meat" found adhering to the inner wall of the shell. It is dried to 2.5 percent moisture content, shredded, and used in cakes, candies, and other confections. Alternatively, coconut oil is expressed from copra and used in a wide variety of cooked foods and margarine. About 1,000 coconuts will yield 500 pounds of copra, which in turn yields about 25 gallons of oil (copra contains about 70 percent oil). Since the oil is saturated fat, it will be an off-white semisolid fat at room temperature or below (75°F) but melt into a brown-yellow oil at higher temperatures. The oil has as many industrial uses (e.g., lubricants, lamp oils, body and hair oils) as it does food uses. The raw copra can be grated and squeezed to obtain coconut "milk." Coconut water is obtained from immature coconuts, providing a welcome source of fresh, sterile water in hot, tropical environments. In overripe fruits, the seed may begin to germinate, producing a mass of soft, pleasant-tasting, edible tissue inside the endocarp, known as coconut "bread" or the coconut "apple." The terminal growing point (sometimes called "cabbage") can be used as a vegetable resembling heart o' palm (*Bactris gasipaes*). The sap from the cut end of an inflorescence produces up to 1 gallon per day of brown liquid, rich in sugars and vitamin C, that can be boiled down into a brown sugar called "jaggery," which is used as a sugar substitute in many areas. Left to ferment, the sap makes an alcoholic toddy, and later vinegar; "arrack," made by distilling the toddy, is a common, potent alcoholic spirit.

Per capita consumption of coconut is 0.6 pound/year in the United States. Coconut oil is probably consumed in greater quantities than are confectionary coconut products, but coconut oil would be only a small percentage of the 47 pounds of vegetable oils consumed annually. Coconut oil is used in a wide variety of processed foods, so indirect consumption may be higher than direct.

Dietary value per 100-gram edible portion:

	Dried Coconut (copra)	Coconut Water	Coconut Oil
Water (%)	3.3	95	0
Calories	556	19	862
Protein (%)	3.6	0.7	0
Fat (%)	39.1	0.2	100
Carbohydrates (%)	53.2	3.7	0
Crude Fiber (%)	4.1	1.1	0
	% of U.S. RDA (2,000-calorie diet)		
Vitamin A	0	0	0
Thiamin, B_1	4	0	0
Riboflavin, B_2	1	0	0
Niacin	3	0	0
Vitamin C	6	5	0
Calcium	1	3	0
Phosphorus	16	2	0
Iron	14	3	<1
Sodium	<1	2	0
Potassium	10	5	0

BIBLIOGRAPHY

Duke, J.A. 2001. *Handbook of nuts*. Boca Raton, FL: CRC Press.

Duke, J.A. and J.L. duCellier. 1993. *CRC handbook of alternative cash crops*. Boca Raton, FL: CRC Press.

Fisher, J.B. 1985. Palmae (Arecaceae), pp. 337-354. In: A.H. Halevy (ed.), *CRC handbook of flowering*. Volume 1. Boca Raton, FL: CRC Press.

Green, A.H. (ed.). 1991. *Coconut production: Present status and priorities for research* (World Bank Technical Paper No. 136). Washington, DC: The World Bank.

Harries, H.C. 2001. The coconut palm (*Cocos nucifera*), pp. 321-338. In: F.T. Last (ed.), *Tree crop ecosystems*. New York: Elsevier.

Ohler, J.G. (ed.). 1999. *Modern coconut management: Palm cultivation and products*. London, UK: Intermediate Technology Publ.

Persley, G.J. 1992. *Replanting the tree of life*. Wallingford, UK: CAB International.

Ploetz, R.C., G.A. Zentmyer, W.T. Nishijima, K.G. Rohrbach, and H.D. Ohr (eds.). 1994. *Compendium of tropical fruit diseases.* St. Paul, MN: American Phytopathological Society Press.

Popenoe, J. 1969. Coconut and cashew, pp. 315-320. In: R.A. Jaynes (ed.), *Handbook of North American nut trees.* Knoxville, TN: North American Nut Growers Association.

Rosengarten, F. 1984. *The book of edible nuts.* New York: Walker and Co.

Sauer, J. 1983. *Cocos nucifera* (Coco, coconut), pp. 216-219. In: D.H. Janzen (ed.), *Costa Rican natural history.* Chicago, IL: University of Chicago Press.

Vandermeer, J. 1983. Coconut (Coco), pp. 85-85. In: D.H. Janzen (ed.), *Costa Rican natural history.* Chicago, IL: University of Chicago Press.

Woodroof, J.G. 1979. *Coconuts: Production, processing, products.* Westport, CT: AVI Publishing.

Chapter 13

Coffee *(Coffea arabica, Coffea canephora)*

TAXONOMY

Coffee is a member of the Rubiaceae (or madder) family, which is closely allied with the Caprifoliaceae (or honeysuckle) family. It is a large, diverse family of mostly woody plants, with 500 genera and about 6,000 species. Coffee is by far the most economically important species, but two genera contain plants of medicinal value: *Cinchona,* used to derive the malarial medication quinine, and *Cephaelis,* from which we get ipecac. A number of ornamentals derive from the family as well, notably gardenia *(Gardenia),* madder *(Rubia),* patridgeberry *(Mitchella),* and ixora *(Ixora).* Genipap *(Genipa americana)* is another fruit crop in the Rubiaceae that is used largely for beverages, on a limited scale, in Latin America.

Of the 25 to 100 species in the genus *Coffea* (the number is still debated), two main species are used in production: *C. arabica* L., generally called "arabica" coffee, and *C. canephora* Pierre ex Froehner, called "robusta" coffee. Two other species, *C. liberica* and *C. excelsa,* are used on a small scale in West Africa but are unimportant in the global coffee market. All four species are found in one of three sections of the genus *Erythrocoffea.* About 70 percent of the world's coffee is derived from *C. arabica,* which is considered to produce higher-quality fruits than *C. robusta.* Main characteristics of each species are shown in Table 13.1. Hybrids between *C. arabica* and *C. robusta* are rare because they differ in ploidy level, and most hybrids are sterile triploids. One natural hybrid cultivar, 'Hibrido de Timor', may have been the product of an unreduced pollen grain of a robusta plant fertilizing an arabica plant. It resembles arabica coffee and has resistance to coffee rust disease, so it is gaining favor. Hybrids have been produced artificially using colchicine to double the chromosome number of *C. robusta* species. These hybrids, termed *arabusta,* are said to be closer to robusta in characteristics. Backcrossing the interspecific hybrids with 'Mundo Novo' or 'Caturra' has produced the icatu hybrids, closer to *C. arabica* in characteristics but with disease resistance. "Congusta" hybrids are those derived from *C. canephora* × *C. congensis,* making a robusta-type coffee plant somewhat better adapted to higher altitude and rainfall.

Cultivars

Cultivars of *C. arabica* were derived from two botanical varieties, *C. arabica* var. *bourbon* and *C. arabica* var. *typica* (syn. *C. arabica*). The *bourbon* variety is more slender in habit than the *typica* variety, has smaller beans but higher yield, resists coffee berry disease better, and lacks the bronze cast of the new foliage. *Coffea arabica* var. *bourbon* derives its name from the former name of the French colony now called Reunion, an island in the Indian Ocean where it was selected. Today, a plethora of modern cultivars, derived from *C. arabica* var. *typica* and *C. arabica* var. *bourbon,* are now grown in greater quantities than the originals, and the botanical variety designation lacks utility outside of breeding. Different coffee regions have selected cultivars (or produced hybrids) that suit the local conditions, and no single cultivar predomi-

Introduction to Fruit Crops

doi:10.1300/5547_13

TABLE 13.1. Selected characteristics of the two main species of coffee.

Characteristic	*C. arabica*	*C. canephora*
Growth habit	shrubby, moderate vigor	small tree, vigorous
Ploidy level	tetraploid	diploid
Flowering time	beginning of rainy season	sporadic
Pollination	self	cross
Fruit maturation time	6-9 months	9-11 months
Fruit abscission at maturity	yes	no
Yield potential	lower	higher
Optimal temperature	60-75°F	75-85°F
Optimal rainfall	48-60 inches/year	60-120 inches/year
Altitude grown	3,000-6,500 feet	0-2,500 feet
Disease and pest resistance	low	high
Coffee flavor	rich, acid	bland, lacking acidity
Caffeine content	1.2%	2.0%
Price on global market	higher	lower

nates worldwide. For example, 'Blue Mountain', a famous cultivar selected in Jamaica, is still grown there and in Kenya. 'Mundo Novo' is a cross between the varieties of *typica* and *bourbon* made in Brazil. Other Brazilian cultivars include 'Maragopipe', 'Cera', 'Laurina', and 'Catuai'. Several dwarf cultivars are used, including 'Caturra', 'Mokka', 'Sao Bernardo', and 'San Ramon'. Today, it is popular to label coffees according to origin instead of cultivar, analogous to wines, yielding "varietals," such as Kenyan AA, Ethiopian Yirgacheffe, Guatemalan Antigua, Kona, and Sumatran. Such varietals may be mixtures of cultivars grown in a particular area.

As with *C. arabica*, *C. canephora* has two botanical varieties: *C. canephora* var. *robusta*, an upright-growing form, and *C. canephora* var. *nganda* (also called *kouillou*), a spreading form. Once again, the utility of retaining these botanical form designations is questionable, as 90 percent of cultivars grown today are erect forms stemming from the *robusta* variety. *Coffea canephora* cultivars differ across growing regions. A number of cultivars with alphanumeric names have been bred and selected (e.g., B11, IF200, J21). Traditional cultivars include 'Conilon', 'Niaouli', 'Robusta Ebobo', and 'Gamè'.

ORIGIN, HISTORY OF CULTIVATION

Coffee is an understory shrub or small tree native to tropical Africa. *Coffea arabica* is native to the highlands of southwestern Ethiopia, whereas *C. canephora* is native to the lowland forests, from Liberia east and south to Kenya and the Congo basin. The word *coffee* may be a corruption of Kaffa, the province of Ethiopia where *C. arabica* originated and may have been domesticated. Coffee fruits may have been eaten in the native area, but the beverage is a much more recent invention. Coffee seeds were transported to southern Arabia (modern-day Yemen) as slaves were taken from the Sudan region to Arabia around 600 AD. The Arabs are credited with the discovery of the beverage coffee from roasted seeds as we know it today, although it is unclear how far in advance of the fifteenth century that this occurred.

The Arabs were protective of coffee and, for many years, did not allow germinable seed to be exported. However, coffee cultivation finally grew beyond Arabia in the early 1600s, when seeds were taken by the Dutch back to Holland and grown in greenhouses. Later, coffee seed was smuggled to Chikmagalgur, India, where it was successfully cultivated; this is the origin of the popular Indian cultivar 'Old Chick'. The Dutch had interests in Indian coffee cultivation at this time and established it on other Dutch colonies in Southeast Asia, most notably in Java (modern-day Indonesia), where coffee obtained one of its nicknames. French efforts to cultivate coffee initially failed due to frost, but King Louis XIV was honored with a live tree by the Dutch that was grown in a greenhouse and came to be known as the "noble tree." A seedling from the noble tree was taken to Martinique in 1720, where millions of trees were eventually grown and further distributed throughout the Caribbean and Mexico. This is the origin of the *typica* variety. At approximately the same time, two seedlings were brought from Mocha in modern-day Yemen to the island of Bourbon (now called Reunion) in the Indian Ocean, from which the botanical variety *bourbon* was derived. Many modern cultivars have been derived from the original *typica* and *bourbon* varieties, such that coffee underwent one of the most narrow genetic bottlenecks of any crop species known. The Dutch brought coffee to Surinam in 1718, and it quickly spread to neighboring French Guyana. A Brazilian nobleman, Francisco de Mello Palheta, obtained seeds by charming the wife of the governor of French Guyana, thereby circumventing the governor himself, who was reluctant to allow Brazilians access to coffee. The governor's wife placed coffee seeds in a bouquet of flowers presented to Francisco, and from these few seeds, many plantations in Para, Brazil, were established. Brazil now leads the world in coffee production. Coffee was introduced to Hawaii in 1825, and Hawaii remains the only coffee-producing state in the United States.

Today, coffee is grown by an estimated 25 million small farming families in many tropical countries worldwide. In some countries, coffee receipts account for 80 percent of foreign trade earnings. Most use traditional culture methods, and individual fruits are still picked by hand. However, research conducted in Hawaii and Brazil may cause coffee growing to change dramatically in the near future. Coffee has been genetically engineered to produce such things as caffeine-free coffee (no decaffeination required): nematode-, disease-, and herbicide-resistant trees; as well as uniform ripening that may open the door to once-over mechanical harvest. The Brazilians have mapped the coffee genome, identifying about 35,000 genes; the map will facilitate even more genetic engineering and breeding research. Fear of and backlash against genetically modified crops have placed the application of this research on hold, and no genetically engineered coffee was being grown commercially as of 2005. Questions have arisen about the fate of small coffee growers, and whether increased productivity is warranted, given the world coffee glut and the recent downturn of prices. Major coffee retailers, such as Starbucks, have vowed not to use genetically modified coffee in their products, and thus the fate of such coffee is uncertain at present.

FOLKLORE, MEDICINAL PROPERTIES, NONFOOD USAGE

Coffee is one of the great social drinks of human culture. However, coffee began as a medicine and an object of religion and evolved into a social drink later, as coffeehouses sprang up in Mecca and other Arabian cities. It is not clear what the specific medicinal uses of coffee were at the time, although the stimulant effects of the beverage were well known and appreciated. Coffee was introduced to Europe from Arabia in 1615 and quickly assimilated into the culture. Tea and cocoa beverages were also introduced to Europe around this time. By the late 1600s, coffeehouses were found in many major European cities, as well as in American cities, such as New York, Boston, and Philadelphia. It is said that the New York Stock Exchange began in a coffeehouse, and the Boston Tea Party was planned in one, a testimonial to the social nature of the drink. Coffee consumption has always been associated with work and activities requiring thought and attention. In Turkey, coffeehouses were known as "schools of the wise," and in England as "penny universities." The latter arose from the cost of admission to a coffeehouse in seventeenth-century England, where one could read newspapers and books and even attend seminars by the great thinkers of the day.

The human health impact of coffee revolves largely around caffeine. Caffeine (1,3,7-trimethylxanthine) is an alkaloid thought to be important in plant pest defense and possibly allelopathy. It is also found in tea, cacao, and other plants. Caffeine is the most active stimulant of the xanthine alkaloids and also acts as a diuretic, vasodilator, laxative, and appetite suppressant. A lethal dose is about 10,000 mg, or about 100 times that found in the average cup of coffee. Doses of 1,000 mg can cause headache, nausea, vomiting, insomnia, elevated blood sugar, increased intraocular (eye) pressure, and tachycardia (rapid heart rate). Caffeine is sold commercially as a component in pain medications, diet pills, over-the-counter stimulants, and medicines administered to hyperactive children, for whom, ironically, it has a calming effect. While the short-term effects of coffee consumption are clear cut, the long-term effects on health are debated. Some studies have shown a greater risk of birth defects, osteoporosis, various cancers, and heart attack in coffee drinkers, but for just about every one of these studies, another study shows no effect. At rates of under 300 mg per day (3 cups of coffee), there are few documented long-term health effects of any kind. On the positive side, moderate coffee consumption has been reported to reduce the risk of colon cancer, cirrhosis, gallstones, Parkinson's disease, asthma, and suicide.

Aside from caffeine, a number of compounds that are potentially toxic are produced during roasting, but in small quantities that are likely to have no effect. These include acetaldehyde, furfural alcohol and mercaptan, hydrogen sulfide, methylpyrrole, pyridine, and vinyl guaiacol. Workers that inhale the smoke from roasting coffee or coffee bean dust can develop chronic lung problems.

Other than the beverage, coffee has few uses. The wood can be made into furniture, and cattle feed can be made from the pulp left over from processing. The dried leaves are chewed or made into teas in some countries. Coffee is said to have antiaphrodisiacal effects.

PRODUCTION

World

Total production in 2004, according to FAO statistics, was 7,719,600 MT or 17.0 billion pounds of dried, green (not roasted) coffee beans. Coffee is produced commercially in 82 countries worldwide on over 25 million acres. Coffee is therefore one of the most important fruit crops in terms of area cultivated. Production has increased 34 percent over the past decade, largely due to an increase in yield, while area in cultivation has remained roughly constant. Average yields are 684 pounds/acre, ranging from 100 to over 5,000 pounds/acre. The top ten coffee-producing countries (percent of world production) follow:

1. Brazil (24)
2. Vietnam (10)
3. Indonesia (9)
4. Colombia (9)

5. Mexico (4)
6. India (4)
7. Guatemala (3)
8. Ethiopia (3)
9. Honduras (2)
10. Peru (2)

Vietnamese and Indonesian production has increased greatly in the past decade, and this has partly led to overproduction and, consequently, lower prices (Figure 13.1). The World Bank encouraged Vietnam to increase its production to relieve debt, and the country rose to second place in the world, from outside the top ten, in the past 10 years. Recently, wholesale prices hit their lowest levels in over 30 years (Figure 13.2), and for many countries, they have dipped below the costs of production. This has caused an economic crisis for the estimated 25 million families who make their living growing coffee in developing countries. For 2003, prices paid to growers averaged about $0.48 per pound in five of the top six producing countries. About 1 percent of coffee is now sold in the organic and fair-trade markets, where prices of about $1.26 per pound were received in fiscal year 2003-2004.

The coffee crisis was precipitated by a simple imbalance in supply and demand. Although demand has increased over time at a rate of about 1.5 percent per year, production has increased about 3.6 percent per year. In 1989, the International Coffee Agreement, which had kept prices high and somewhat stable, was dissolved. In 1993, the Association of Coffee Producing Counties (ACPC) was formed to act as OPEC does for oil, and prices surged for a few years. However, few member countries, other than Brazil and Colombia, held to an agreement to reduce production by 20 percent, and countries that previously had not been significant in world markets, such as Vietnam, greatly increased production. In 2001, prices dipped below $0.50 per pound, as Brazil, the leading coffee producer, also abandoned the agreement and began selling its coffee reserves. Preliminary data for 2005 suggest that the coffee crisis may have ended, as composite prices moved above $1.00 per pound for the first time since 1998.

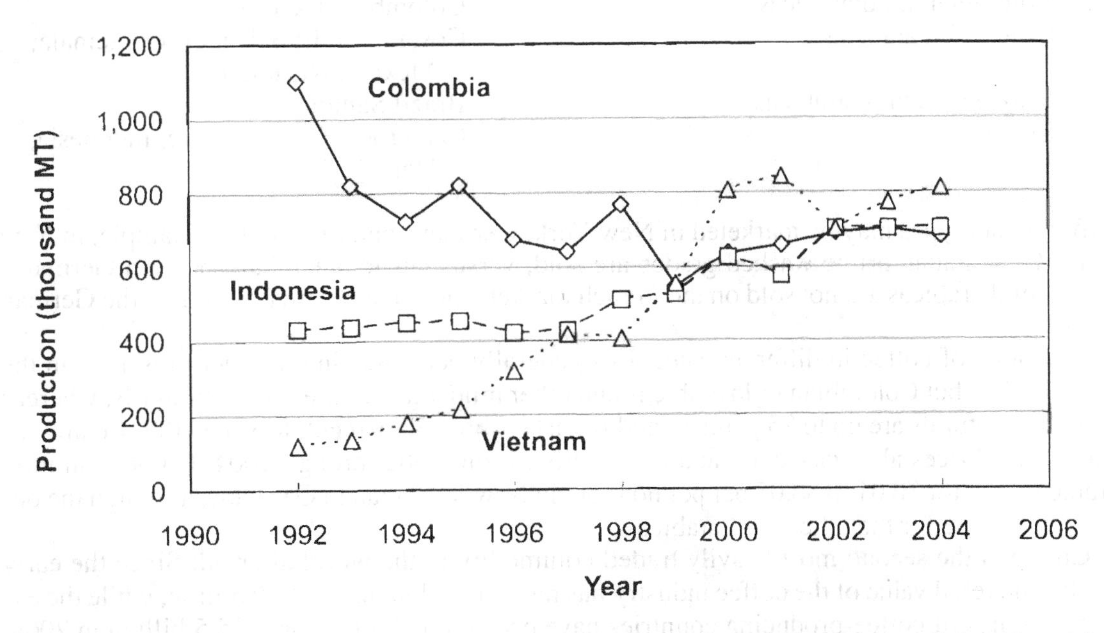

FIGURE 13.1. Coffee production is declining in some traditional countries, such as Colombia, while Southeast Asian production has surged. *Source:* FAO statistics.

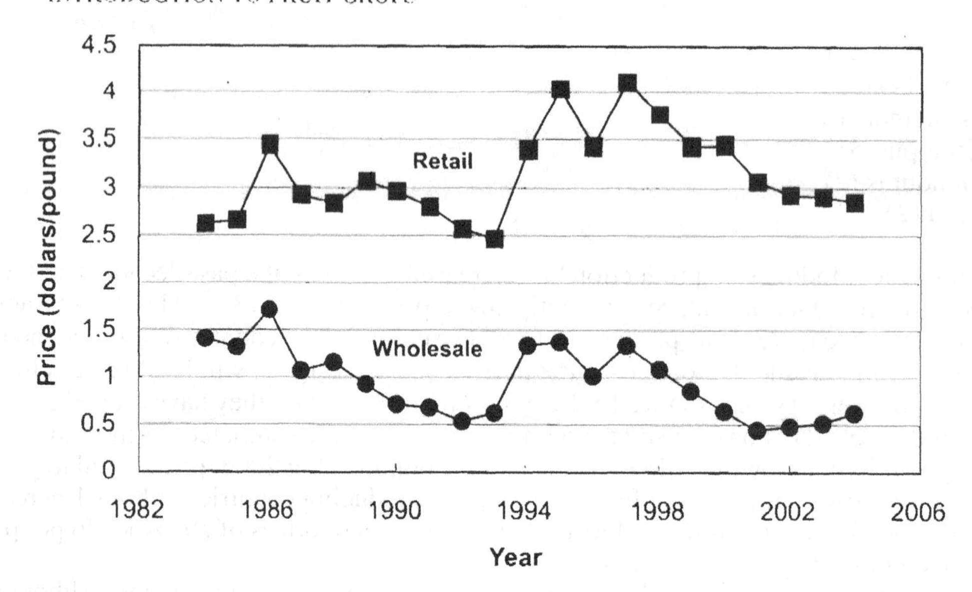

FIGURE 13.2. Coffee prices hit their lowest levels in over two decades in recent years due to overproduction. Note that the correspondence between wholesale and retail prices is good, but the margin has increased from $1.50 to $2.00 per pound in the 1980s to $2.00 to $2.50 per pound more recently. *Source:* International Coffee Organization; retail prices are for the United States.

Currently, coffee marketing is controlled by the International Coffee Organization (ICO), headquartered in London. Markets are located in the United States (New York), Germany (Bremen/Hamburg), and France (Le Havre/Marsielles). Coffee is divided into four categories, with the following types within each:

ICO Marketing Group	Type of Coffee or Origin
Colombian mild arabicas	Colombian Excelso
Other mild arabicas	Costa Rica, El Salvador, Guatemala, Mexico, Nicaragua
Brazilian natural arabicas	Brazil Santos
Robustas	Cameroon, Cote d'Ivoire, Indonesia, Uganda, Vietnam

Different subtypes may be marketed in New York, Germany, and France. For example, in New York, Guatemalan prime washed grades are sold, versus Guatemalan hard bean in Germany. Other mild arabicas are not sold on the French market, and robustas are not sold on the German market.

The price of coffee in different categories generally decreases in the order presented in the previous list, but Colombian mild arabicas and other mild arabicas are priced similarly, whereas Brazilian naturals are up to 25 percent, and robustas about 50 percent, less than the two arabica categories. Prices also vary daily and by location. From 2002 through 2004, Colombian mild arabicas sold for $0.01 to $0.05 per pound more in New York than in Germany, although the opposite was true for the other mild arabicas.

Coffee is the second most heavily traded commodity in the world after oil. Since the early 1990s, the retail value of the coffee industry has more than doubled, to $70 billion, while the export earnings of coffee-producing countries have been cut in half, to only $5.5 billion in 2004. Four countries—the United States, Germany, France, and Japan—consume over half of the world's coffee.

United States

Total production in 2004, according to USDA statistics, was 3,954 MT or 8.7 million pounds. All coffee is grown in Hawaii on about 5,900 acres. Both acreage and yield have increased over the past decade, raising production twofold. The industry value is $23.5 million, as Hawaiian growers enjoy prices of $2.00 to $3.00 per pound, two- to threefold the world average. Yields are double the world average at 1,500 pounds/acre. Much of the coffee is grown in the Kona region of the big island.

Foreign Trade

The United States is the largest importer of coffee in the world, importing 2.6 billion pounds of coffee in 2002, 79 percent from Latin America. The United States imports an additional 180 million pounds as instant coffee or roasted beans. In all, imports are valued at over $2 billion per year.

BOTANICAL DESCRIPTION

Plant

The coffee plant is a small, evergreen tree, reaching 30 feet in the wild, but generally maintained at 6 to 8 feet in cultivation. Leaves are ovate, 3 to 8 inches long, dark green, prominently veined, with interveinal areas raised, a waxy and smooth texture, and entire margins (Plate 13.1). Leaves are arranged oppositely on stems and decussate, with successive pairs of leaves arising at 90-degree angles from each other around the stem. Leaves live about 1 to 2 years. Fruiting lateral shoots arise at nearly 90-degree angles from the main stem in the *typica* variety but are more acute (55-degree angle of attachment) in the *bourbon* variety.

Coffee is an unusual plant with respect to axillary bud formation. As many as six tiny, indistinct buds form serially in each leaf axil, one above the other, yielding up to 12 at a node. Most other plants produce only one bud per axil. The two to four buds furthest from the leaf axil are the oldest and generally produce inflorescences, but the topmost bud can produce a branch.

Flowers

Flowers are white, fragrant, about ½ to 1 inch across, with five linear petals fused into a slender tube at their base (Plate 13.2). The ovary is inferior and bicarpellate and contains two ovules and is thus capable of producing two seeds. There are five stamens and a two-lobed stigma extending beyond the floral tube. "Star flowers" are male-sterile flowers produced in response to drought stress or high light intensity. Flowers are borne in highly compressed, headlike racemes of two to four flowers each. Up to 20 individual flowers form per axil, yielding a profuse, white cluster of flowers at each node. *Coffea arabica* flowers on only the previous season's growth, whereas *C. canephora* flowers on the current season's growth.

Flowering normally occurs at the beginning of the rainy season in tropical, wet/dry climates. Flowers are initiated during the dry season but do not develop due to water stress. Flowering is sudden and synchronized; it occurs over a period of a few days in *C. arabica*, and within almost the same day in *C. robusta*. Coffee is a short-day plant with a critical photoperiod of 13 hours. In most coffee regions, this is of little consequence because the daylength never exceeds 13 hours; a notable exception is southeastern Brazil, where coffee is grown at 20 to 25°S latitude, and flowering is regulated by both daylength and rainfall.

Pollination

Coffea arabica is self-pollinating, often cleistogamous (pollinates itself when flower is closed), whereas *C. canephora* is self-incompatible and must be cross-pollinated. Flowers of both species are wind and insect pollinated.

Fruit

Coffee fruits are epigynous berries, often called "cherries," that are erroneously considered to be drupes. A true drupe has a single, hard, central endocarp or pit enclosing the seed(s) and is the product of a superior ovary. Coffee fruits are not drupes, but epigynous berries, because they derive from inferior ovaries and lack a true pyrene or pit.

Fruits are ½ to ¾ of an inch long, ovoid, and borne axillary in clusters on growth from the previous season (arabicas) or current season (robustas) (Plate 13.3). Two ovoid seeds are found in each fruit, with their flat sides facing each other; seeds occupy the bulk of the fruit's volume. Single-seeded fruits occur and are called peaberries. Seeds are pale green but surrounded by a thin, silver seedcoat called the "silverskin." The endocarp is the "parchment" or hull, a papery covering around each seed; the mesocarp is the mucilaginous pulp; and the exocarp is the relatively thick, red skin (the skin is yellow in some cultivars). Fruits mature in 6 to 11 months, depending on species, cultivar, and climate. Conveniently, ripening occurs during the dry season, which facilitates drying and processing. Fruits turn from green to yellow and then red, which signals peak maturity.

Fruiting begins 2 to 4 years after plant establishment and may continue for decades. Plants are pruned periodically to rejuvenate the fruiting wood and keep it close to the ground for harvest. About 40 to 60 percent of flowers set fruit in arabicas, and a lesser percentage, down to only 10 to 15 percent, sets in robustas. Low fruit set in robustas may occur if weather at bloom is not conducive to cross-pollination. Coffee is known to exhibit alternate bearing if too large of a crop is set.

GENERAL CULTURE

Soils and Climate

Coffee is grown on a wide range of soil types worldwide, from coarse volcanic sands to loamy alluvial soils. Light-textured, well-drained volcanic soils with slightly acid pH (5-6) are best because coffee cannot tolerate flooding and develops nutrient deficiencies at extremes of pH. Deep rooting is considered important because most growing areas have distinct dry seasons that coincide with fruit maturation and harvest. Irrigation is provided in extremely dry areas, such as Yemen, and can be used to increase yields, induce earlier flowering, and offset the tendency for alternate bearing by stimulating shoot growth. Most coffee is nonirrigated.

Optimal temperatures, rainfall levels, and altitudes for the two main species of coffee are listed in Table 13.1. As one would anticipate, *C. robusta* is less tolerant of cold than *C. arabica*, but both are killed or severely injured by light frost. Neither require any chilling.

One of the unique aspects of coffee is its cultivation with some degree of shade, stemming from its origins as an understory plant (Figure 13.3). Its leaves undergo photoinhibition or high light stress at full sun intensities, particularly at warm temperatures. Almost all other fruit crops are grown in full sunlight (cacao is also shade grown). However, coffee can be grown in full sunlight and often is in large plantations where the focus is on high yield. Sun-grown coffee will outyield shade-grown coffee if fertilized more heavily. Small coffee growers who lack access to

FIGURE 13.3. *(top, left and right)* Sun- and shade-grown coffee in Costa Rica. *(top left)* A planting near Turrialba that had *Erythrina* shade trees but was converted to a full-sun plantation by severe pruning. *(top right)* Eucalyptus trees used for shade in a planting near Buenos Aires. *(bottom left)* Drought-stressed coffee trees in an exposed area of an otherwise shaded planting are defoliated in the middle of the dry season in Guatemala, whereas the shaded plants in the background are foliated. *(bottom right)* Shade trees provide mulch, compost, fenceposts, and firewood when pruned during the rainy season in Antigua, Guatemala.

such inputs achieve moderate yields with minimal chemical fertilizers, but they must maintain the shade crop as well as the coffee. The type and degree of shade varies among coffee regions, often in proportion to the degree of natural cloudiness. Regions that routinely get afternoon cloud cover can cultivate coffee with little or no shade and still avoid the negative effects of sun-grown cultivation. Shade trees alter the microclimate considerably in coffee plantings by reducing temperature and wind speed and increasing humidity. Measurements have shown that shade trees reduce evapotranspiration (hence water use) of coffee by 25 to 50 percent, while intercepting only 3 to 13 percent of incident rainfall. Shade coffee is regarded as more sustainable by some, as it promotes greater on-farm biodiversity. Shade coffee plantings play host to a number of migratory birds and a wide range of insects not found in sun-grown coffee plantings.

Propagation

Unlike most fruit crops, *C. arabica* coffee is fairly true when breeding from seed and is therefore seed propagated. Seeds can be planted directly in the field but are generally grown in nurseries for several months to a year and then transplanted (Figure 13.4). Direct seeding requires

FIGURE 13.4. Coffee propagation. Coffee is generally grown from seed on its own roots. *(above left)* Seeds are germinated in nursery bags and *(above right)* planted after about 1 year. *(left)* Arabica coffee is sometimes grafted onto robusta rootstock to gain nematode resistance; a simple wedge graft is performed just after the stock and scion germinate.

more attention to weed control and results in lower establishment rates. In Brazil, several seeds or transplants are planted in each site to account for losses, and all seedlings that grow are retained to produce a multistemmed plant from multiple seedlings. Seed propagation produces highly variable plants of heterozygous hybrid *C. arabica* and *C. robusta* cultivars; these are propagated from cuttings or by grafting.

Cuttings of coffee will root in well-aerated media, often without growth regulator application. Cuttings are used for propagation of *C. robusta*, and to a lesser extent for *C. arabica*. Cuttings are taken near the tip of upright stems and reduced to a single node, with the leaves cut in half. About 60 to 80 percent successful rooting is obtained in 2 to 3 months, during which time the cuttings are grown under mist or in a plastic enclosure. Grafting is infrequently used for arabicas when a nematode-resistant rootstock is needed. Scions of *C. arabica* or *C. robusta* can be wedge grafted onto robusta seedling rootstocks when both are just recently germinated, prior to cotyledon expansion (Figure 13.4).

Rootstocks

When used, rootstocks are generally a *C. robusta* cultivar, since this species is more resistant to nematodes and disease. The vast majority of the time, coffee is grown from seed or cuttings on its own roots.

Planting Design, Training, Pruning

Arabica coffee plants are traditionally planted at distances of 4 to 8 feet in rows 8 to 10 feet apart, yielding plant densities of 500 to 1,300 trees/acre. Dwarf cultivars can be planted at densities of up to 2,000 trees/acre, particularly in high-input systems. Robusta coffee plants are more vigorous and therefore are spaced 6 to 8 feet × 12 to 13 feet (about 350 trees/acre). Terrain is often hilly in highland areas suited to arabicas, and rows are thus laid out on contours. Raised beds are used in areas with poorly drained soils and/or high rainfall. In some cases, large drainage ditches must be provided periodically throughout the field to remove the excess rainfall.

Shade trees may be planted in regular rows, periodically within every other or every third coffee row, or scattered throughout the planting. Leguminous trees are most frequently used because they fix nitrogen. Species of *Acacia, Albizzia, Cassia, Erythrina, Gliricidia, Inga,* and *Leucaena* are frequently used. Nonlegumes may also be used, particularly if they bear an economic product. Citrus, avocado, mango, guava, sapote, banana, coconut, pejibaye, and rubber are examples of other crops used to shade coffee plants. If used, shade plants are often pruned heavily in the rainy season because clouds provide natural shading. Regrowth is vigorous and in place by the start of the sunnier dry season. The small prunings provide mulch and nutrients for the coffee plants, and large pruned limbs can be used as fence posts or fuel wood.

Coffee has a unique, predictable growth and bearing habit that is managed by pruning to maintain good production. A new plant will produce one or more upright stems that do not fruit but produce horizontally growing lateral stems that eventually will fruit. The laterals grow several inches in their first year, cease growth and initiate flowers in the dry season, and then flower and fruit in their second year. As the 1-year-old section of the lateral is fruiting, the apical bud breaks and extends the lateral vegetatively. Since only the portion of stem that grew the previous year can fruit, in successive years, the bearing wood moves progressively farther and farther from the upright stem. Lower laterals eventually shade out, leaving the lower and central portions of the canopy devoid of leaves and fruits.

The type of pruning practiced is called renewal or rejuvenation pruning, which basically constitutes cutting the entire plant back to about 18 inches and allowing it to reestablish a canopy of fruiting laterals (Figure 13.5). This is done every 4 to 7 years. Depending on grower preference, cultivar, and spacing, each tree is thinned to one, two, or several upright stems after pruning, and these generally grow unchecked for another 4 to 7 years. If a single stem is retained, it may be tipped periodically to keep it from becoming too tall before the next pruning cycle. In multiple-stemmed trees, one or two stems may be kept for the year after pruning to provide photosynthate for the new growth and the bearing of a small crop. The stems left behind are called "lungs" or "breathers." Robusta plants are commonly trained to a multiple-stem system because they tend to overshade themselves and become unproductive when trained to a single stem. Entire fields may be pruned at once, or every fifth row can be pruned each year if, for example, a 5-year cycle is being employed. High-density plantings are generally "stumped" or rejuvenated in solid blocks; shading of stumps by unpruned trees produces weak, etiolated shoots on pruned trees.

FIGURE 13.5. A coffee plant 1 year after rejuvenation pruning. Two upright shoots were left, one on each pruning stub.

Pest Problems

Insects

Borers. The larvae of several species of beetles can infest the fruits or vegetative parts of the tree, causing severe crop losses or tree death. The coffee berry borer *(Hypothenemus hampei)* is regarded as the most important insect pest of coffee. This tiny black beetle ($<\frac{1}{10}$ inch long) bores a hole in the tip of a fruit and lays eggs in the tunnel; the resulting larvae riddle the seeds and fruit. Since larvae are inside the fruits most of their lives, insecticide sprays have little effect. Several beneficial insects and birds may help to keep populations under control, but by themselves they are insufficient. Cultural controls include pruning and removal of shade, since beetles tend to proliferate in damp, shady areas. Removal of fallen fruits and frequent harvest of ripe fruits reduce potential breeding sites and are important control measures. Stem and branch borers can be serious pests, causing high rates of tree death in certain countries. The West African coffee borer *(Bixadus sierricola),* white stem borer *(Anthores leuconotus),* and coffee white stem borer *(Xylotrechus quadripes)* all bore beneath the bark in the main stem and roots. Trunks are painted with a persistent insecticide in areas that allow this practice; otherwise, infested trees are removed and burned. The lesser coffee bean borer *(Araecerus fasciculatus)* is a postharvest pest of coffee and other stored foods. Fortunately, it can be controlled by sun drying beans and keeping stored beans clean and fumigated.

Leaf miners. The larvae of three species of *Leucoptera* and *Perileucoptera coffeella* moths tunnel through leaves, creating large lesions and defoliating trees. The pests are found in the Americas and Africa, but not in Southeast Asia or India. Several generations may occur per year, and outbreaks are worse in plantings with high light and low humidity, which the insects prefer for breeding (sun coffee). Plantings that are mulched and sprayed with copper fungicides experience more severe outbreaks. Insecticide sprays are timed to adult moth flights, or granular systemic insecticides are applied to the soil.

Scale and mealybugs. At least 50 species of these sap-sucking insects can infest coffee. The more common pests are green coffee scale *(Coccus viridis),* helmut scale *(Saissetia coffea),* Star scale *(Asterolecanium coffea),* verruga *(Cerococcus catenarius),* and root mealybugs (e.g., *Geococcus, Rhizoecus* spp.). They can be controlled with insecticides, but such treatments are toxic to the wide variety of natural predators and may increase scale infestations. Oil sprays are used to suffocate scales and are less toxic to beneficial organisms. Ants tend scale in many cases, killing off parasites and predators and receiving honeydew in return. Grease or oil rings at the base of the trunk can prevent ants from climbing trees and thus reduce scale populations.

Nematodes

Nematodes that affect coffee are some of the more common soilborne pests of fruit crops in general: root-knot *(Meloidogyne),* root lesion *(Pratylenchus),* and burrowing *(Radopholus)* nematodes. Nematodes debilitate but generally do not kill coffee. Control is accomplished by limiting spread through sanitation of old infested plants, replanting with clean nursery stock, or, in some cases, use of fumigants and nematocides. *Coffea robusta* cultivars are often tolerant of nematodes and thus used as rootstocks in heavily infested areas.

Diseases

Leaf rust. Hemileia vastatrix causes leaf rust, which is one of the most notorious fungal diseases of fruit crops in the world. First noted in 1869 in Ceylon (Sri Lanka), rust has spread to almost all coffee-growing regions, defoliating coffee trees and virtually eliminating coffee as a crop from some regions. It is confined to the leaves, where orangish, circular spots develop on the undersides, giving way to necrotic areas on the upper sides, which coalesce and eventually cause leaf drop. Copper fungicides, particularly Bordeaux mixture, were used extensively but had to be applied to the leaf undersides before infection, as the fungus lives within the tissue. Breeding coffee for resistance started decades ago but has been hampered by the existence of many races of the fungus. Hybrids are often resistant to some, but not all, races. *Coffea robusta* is resistant, and 'Hibrido de Timor', a natural hybrid between *C. arabica* and *C. robusta*, has shown resistance to many races and is used in breeding programs. 'Catimor' and 'Ruiri 11' are also resistant to most races, and 'Ruiri 11' has resistance to coffee berry disease as well.

Coffee berry disease. Colletotrichum coffeanum is a form of anthracnose that affects the flowers and fruits of arabica coffee plants (robusta plants are resistant). Currently, it is confined to Africa and thus does not have the impact of coffee leaf rust. The spores produced on the bark and diseased fruits provide the inoculum for berry infection, which is generally worst when fruits are expanding rapidly a few months after flowering. Early attack produces lesions that spread over the fruits and inhibit seed production, although in late attack, one may salvage the beans. Copper fungicides as well as synthetics are applied frequently in the rainy season for control. 'Blue Mountain', 'Geisha 10', 'Hibrido de Timor', 'Rume Sudan', and 'Riuri 11' show some resistance to the disease.

Brown eye spot and berry blotch. Caused by *Cercospora coffeicola*, this widespread fungal disease affects most species of *Coffea*, yet it is generally a problem of economic significance in only young trees. The disease gets its name from the appearance of the resulting round lesions, which have grayish white centers and brown margins. On fruits, lesions are sunken and dry. Fungicide sprays for coffee leaf rust will control this disease, and nursery stock is generally sprayed as a prophylactic measure.

HARVEST, POSTHARVEST HANDLING

Maturity

Individual fruits turn from green to completely bright red and glossy over a period of 1 to 2 weeks when mature. The seeds are easily squeezed from mature fruits by moderate finger pressure. Both immature and overripe fruits are difficult to process, and seeds obtained may produce off-flavors or off-colors to the coffee. Coffee is classified as a climacteric fruit with respect to ripening.

Harvest Method

Coffee is hand harvested, which accounts for most of the labor input and costs associated with coffee production (Figure 13.6). On individual trees, fruits ripen over a period of weeks and must be picked by hand several times, at 7- to 10-day intervals, in order to harvest only the ripe fruits. Mechanical

FIGURE 13.6. Coffee being hand harvested in Antigua, Guatemala.

harvesters resembling those used for brambles have been used on large plantations, such as those in Brazil, but they harvest many green, unripe fruits, producing a lower-quality product. Pickers average 50 to 100 pounds of coffee beans per day in selective handpicking, but a single mechanical harvester can cover 7 to 8 acres in a day, harvesting thousands of pounds of coffee beans. Another approach, still in the experimental stage, involves using coffee plants that have been genetically engineered to ripen uniformly when sprayed with a ripening chemical. If such cultivars were made available to the industry, once-over harvesting could be used to produce high-quality coffee at greatly reduced expense, although many laborers would be left jobless. Once-over harvest by hand is used in Brazil to produce low-quality coffee. All of the fruits are stripped from the tree, and ripe and overripe fruits are separated from the green berries before processing. If "dry" processing is used (see the following Dry Processing section), overripe fruits can be used for high-quality coffee, and in those cases, fruits are left on the trees to become overripe and picked only once.

Postharvest Handling

Dried coffee beans are obtained from the fruits by either dry or wet processing. The dry method is the oldest and simplest but produces lower-quality coffee; it is used for robustas and for arabicas in Brazil, where trees undergo once-over harvesting. One exception is the "mocha"-type coffees from Brazil, which are high-quality coffees obtained from dry processing. Wet processing began in the 1700s in Indonesia and the Caribbean, where ripe fruits were stomped like wine grapes to remove the seeds, and the mucilaginous pulp was allowed to ferment for a period of time before drying. This was refined in the nineteenth century, and today it is the most popular method for processing high-quality arabica coffee. One problem with wet processing is the large volume of wastewater produced, which can contaminate rivers and streams. Wastewater is sometimes placed in settling ponds before discharging.

Dry Processing

The major difference between dry and wet processing is when the seeds are dried. In the dry method, seeds are dried while still in the fruits, which is why under- and overripe fruits can be accommodated. Pulp removal in the wet method requires ripe fruits. Harvested fruits are sun dried on benches or on concrete coffee yards or "barbeques" (Figure 13.7). Fruits are spread in a layer 1 to 2 inches deep and frequently turned to prevent fermentation of the covered fruits. The process takes 10 to 30 days, depending on weather, during which time mold may grow on the fruits, producing off-flavors, and seeds may become discolored from contact with the deteriorating mesocarp. Machine drying can speed the process to about 3 days or is sometimes used after partial sun drying. The dried fruits may be sieved to remove small fruits and debris, and they are then hulled in a machine and polished.

FIGURE 13.7. Coffee beans are dried in cement yards by spreading into layers a few inches thick and turning periodically with rakes.

Wet Processing

Ripe fruits are dumped into a tank of water, where debris is removed and poor-quality fruits are floated off and discarded. The good fruits sink and are moved by a stream of water to a pulping machine. Pulping machines rub off the skin and some of the mesocarp by forcing fruits between a rubber strip and a perforated rotating drum. The expelled fruits still have the mucilaginous mesocarp adhering to the endocarp around the seeds. The pulp is dried and used as mulch or animal feed. The next step, called "fermentation," is actually just enzymatic degradation of the pectinaceous mucilage. Beans are carried in a stream of water to a tank, where they undergo this process for 12 to 48 hours, depending on temperature. Beans are washed either during fermentation or afterward to prevent staining. Beans, still inside the endocarp or "in parchment," are then dumped from wash tanks into the coffee yard for sun drying. Drying takes 10 to 15 days in dry weather and may be accompanied by mechanical drying at the end to bring the water content down to 12 percent, which allows safe storage. Parchment coffee is generally bagged and taken to a plant that dehulls and polishes the beans. Since the process uses a large quantity of water, coffee-processing facilities are generally sited near a stream and designed to use gravity to carry the fruits through the process. Water flumes may also be used to carry the fruits from the fields to the site, since ripe coffee beans must be processed immediately.

Grading is often done by hand to remove poor-quality beans, but machines have been developed to grade by size and color. In many producing regions, coffee is subjected to tasting before shipping. Experienced tasters remove samples, roast and brew coffee meticulously, and smell and taste the coffee. If off-flavors or off-aromas are detected, the coffee cannot be sold under the top-grade or "varietal" designation.

Storage

Properly dried, coffee beans can be stored for years under optimal conditions. If stored in parchment, storage should not exceed 6 months, or the coffee develops a woody flavor. Storage is best (several years) at 40°F and 55 percent humidity, but coffee beans will keep 3 to 4 years even at room temperature. Green coffee beans may actually improve in flavor when stored ("aged" coffee), but they will develop off-flavors if storage rooms are contaminated or contain odoriferous materials.

CONTRIBUTION TO DIET

Virtually all coffee fruits are used to make the familiar beverage, but small amounts are used to flavor ice cream, confectionary products, and liqueurs, such as Kahlúa. Per capita consumption of coffee is about 10 pounds of beans per year, roughly what it was in 1980.

Correct roasting can make or break even the finest-quality coffee beans. Roasting is done in machines consisting of a rotating drum with a heat source beneath. The coffee is tumbled and stirred constantly to promote even roasting. Temperatures of 450 to 500°F, for anywhere from a few minutes up to a half hour, are used to produce the main types of roasts. A few minutes or a 10 to 20°F variance can ruin the quality. Too little roasting produces a nutty, breadlike flavor, and roasting too long produces burned, baked, thin-tasting coffee. During roasting, beans lose virtually all remaining water and 3 to 15 percent of their dry weight. Some of the oils volatilize, proteins and organic acids are destroyed, sugars are caramelized, and some 200 different chemicals are produced, giving the characteristic taste and aroma of the roast. Light roasts are produced in a matter of minutes; beans are light brown, without oil on the surface, and the coffee is acidic. Medium roasts are darker brown but not oily, have more caramelized sugars and lower acid contents, and are thus richer and sweeter flavored. Dark roasts have low acidity and oily beans, and espresso roast, which takes the most time to roast, produces very oily beans with no acids and highly caramelized sugars. Quenching with cool air or water sprays must be done at a precise time to achieve a particular roast type.

Decaffeination of coffee was accomplished as early as 1820. These early efforts produced decaf from roasted coffee and had the drawback of adversely affecting flavor. About 100 years ago, it was discovered that caffeine could be extracted from green beans if they were first treated with steam or soaked in hot water. Heat causes the beans to swell and increases efficiency of extraction, and since it is done before roasting, most of the flavor compounds have not yet been synthesized. Three basic methods are used: solvent extraction, supercritical gas extraction, and water extraction. In solvent extraction, methylene chloride or ethyl acetate is applied to the steamed beans under pressure at temperatures close to the solvent's boiling point. After extraction, the coffee is steam treated again to remove the solvents. Trace amounts of solvents are left in the coffee, but both have been stringently reviewed by the FDA and have no harmful effects on humans at the levels found in coffee. In fact, ethyl acetate is a natural component of many fruits, where it occurs in levels higher than those found in decaf coffee. Supercritical gasses are gasses subjected to high pressure that behave like liquids. Supercritical CO_2 can extract caffeine very efficiently, leaving almost all constituents behind, except caffeine. Water extraction involves steeping beans in hot water to leach caffeine and other water-soluble substances. The water extract is passed over activated carbon, which adsorbs the caffeine, and the caffeine-free leachate is added back to the beans before drying. Water extraction causes the loss of some sugars and acids, however. All of the methods are at least 97 percent efficient, reducing caffeine levels from 100 milligrams per cup in normal coffee down to 1 to 5 milligrams per cup in decaf.

Dietary value per 100-gram edible portion (*note:* an 8-ounce cup of coffee weighs about 230 grams, so multiply the following amounts by 2.3 to yield amounts per cup):

	Black Coffee
Water (%)	99
Calories	1
Protein (%)	0.1
Fat (%)	0.1
Carbohydrates (%)	0
Crude Fiber (%)	0
	% of U.S. RDA (2,000-calorie diet)
Vitamin A	0
Thiamin, B_1	1
Riboflavin, B_2	4
Niacin	1
Vitamin C	0
Calcium	<1
Phosphorus	<1
Iron	<1
Sodium	<1
Potassium	1

BIBLIOGRAPHY

Alvim, P.T. 1985. Coffea, pp. 308-316. In: A.H. Halevy (ed.), *CRC handbook of flowering*. Volume 2. Boca Raton, FL: CRC Press.

Boucher, D.H. 1983. Coffee (Cafe). pp. 86-88. In: D.H. Janzen (ed.). *Costa Rican natural history*. Chicago, IL: University of Chicago Press.

Clarke, R.J. and R. Macrae. 1985. *Coffee*, Volume 4: *Agronomy*. New York: Elsevier Applied Science.

Clarke, R.J. and O.G. Vitzhum. 2001. *Coffee: Recent developments*. Oxford, UK: Blackwell Publ.

Clifford, M.N and K.C. Willson (eds.). 1985. *Coffee: Botany, biochemistry and production of beans and beverage*. Westport, CT: AVI Publishing.

Debry, G. 1994. *Coffee and health*. Paris: John Libbey Eurotext.

Duke, J.A. and J.L. duCellier. 1993. *CRC handbook of alternative cash crops*. Boca Raton, FL: CRC Press.

Ferwerda, F.P. 1976. Coffee, pp. 257-260. In: N.W. Simmonds (ed.), *Evolution of crop plants*. London: Longman Scientific and Technical.

Gilman, E.F. 1999. *Coffea arabica* (Florida Cooperative Extension Services Fact Sheet FPS-135). Gainsville, FL: Institute of Food and Agricultural Sciences.

Haarer, A.E. 1963. *Coffee growing*. London: Oxford University Press.

Maestri, M., R.S. Barros, and A.B. Rena. 2001. Coffee, pp. 339-360. In: F.T. Last (ed.). *Tree crop ecosystems*. New York: Elsevier Press.

Willson, K. 1999. *Coffee, cocoa, and tea*. Wallingford, UK: CAB International.

Wrigley, G. 1988. *Coffee*. London: Longman Scientific and Technical.

Chapter 14

Cranberry *(Vaccinium macrocarpon)*

TAXONOMY

The cultivated cranberry, *Vaccinium macrocarpon* Ait., is a member of the Ericaceae (or heath) family. It is placed in the *Oxycoccus* section of the genus *Vaccinium*, along with its close relative the mossberry or small cranberry *(V. oxycoccus)*. The lignonberry *(V. vitis-idaea)* is closely related to cranberry and mossberry, and, as with the latter, it is wild-harvested and used for similar purposes as the cultivated cranberry. All are evergreen, creeping shrubs native to cool temperate or boreal areas. The so-called highbush cranberry, *Viburnum trilobum*, is similar to cranberry only in fruit size, color, and food usage. However, it is a member of the Caprifoliaceae, not the Ericaceae, and thus not a close relative of cranberry.

The Ericaceae family contains several important ornamentals: rhododendrons and azaleas *(Rhododendron)*, mountain laurel *(Kalmia latifolia)*, heather *(Calluna)*, heath *(Erica)*, and leatherleaf *(Leucothoe)*. Blueberry, huckleberry, and bilberry are other economically important fruit crops within *Vaccinium*. The genus is very diverse, containing 150 to 450 species, mostly in the tropics at high elevations.

Cultivars

Comparatively few cultivars of cranberry are in production, relative to other fruit crops, and most are native selections or first-generation crosses. Estimates of the total number of cultivars are as high as 130, but only about a dozen constitute 99 percent of the industry. 'Ben Lear', 'Early Black', 'Howes', 'McFarlin', and 'Searles' are the most popular cultivars derived from native selections. 'Early Black', the leading Massachusetts cultivar, has high anthocyanin content and, consequently, good color, which is needed for making high-quality juices and processed products. In Wisconsin, 'Stevens' ('McFarlin' × 'Potter') is a major cultivar giving reliable yields of large berries.

ORIGIN, HISTORY OF CULTIVATION

The cranberry is native to acidic bogs and peat wetlands of the northeastern United States and southern Canada. It was long appreciated by Native Americans, and its fruits have been collected since the English colonized the Massachusetts area in the 1600s. Today's industry has evolved slowly from the manipulation of the plant in its ecological niche, the bogs, to increase yields, provide winter protection, and facilitate harvest. Similar to lowbush blueberry cultivation, cranberry production in Massachusetts first involved the management of native stands. This evolved into the building of ditches and dikes for water control, and eventually to the sophisticated artificial bogs used today. Cranberry production moved to Wisconsin in the 1860s,

Introduction to Fruit Crops

doi:10.1300/5547_14

and later to New Jersey and Quebec, Canada. More recently, the Pacific Northwest has added a few thousand acres, but over 80 percent of production still remains in Wisconsin and Massachusetts.

FOLKLORE, MEDICINAL PROPERTIES, NONFOOD USAGE

The term *cranberry* is derived from "crane berry," which was coined by early settlers who saw a resemblance between the cranberry flower and the neck, head, and beak of a crane. The cranberry fruit is symbolic of a cure for heartache.

Cranberries have long been touted as being useful for urinary tract infections. They are high in organic acids (quinic, benzoic), which are converted to hippuric acid in the human body. Hippuric acid is believed to have antibacterial activity. Cranberry (and the related *V. oxycoccus*) may have hypoglycemic activity (lowering of blood sugar) as well.

PRODUCTION

World

Total production in 2004, according to FAO statistics, was 344,000 MT or 757 million pounds. Production of cranberry is largely limited to the northern United States and Canada, the native range of cranberry. Other areas of the world are only beginning to produce the crop; Belarus and the Ukraine had their first crops in 1997 and 2000, respectively. World area under cultivation is about 73,000 acres, increasing by about 24 percent in the past decade. Yields average 10,400 pounds/acre, with the highest yields in the United States and Canada, and yields as low as 1,000 pounds/acre in other countries where production is just beginning. The top cranberry-producing countries (percent of world production) follow:

1. United States (78)
2. Canada (16)
3. Belarus (3)
4. Latvia (2)
5. Azerbaijan (<1)
6. Ukraine (<1)
7. Tunisia (<1)
8. Turkey (<1)

United States

Total production in 2004, according to USDA statistics, was 289,136 MT or 636 million pounds. Production has increased 36 percent in the past decade. Five states produce the entire crop:

1. Wisconsin (55 percent): 17,400 acres
2. Massachusetts (28 percent): 14,100 acres
3. Oregon (8 percent): 2,900 acres
4. New Jersey (6 percent): 3,100 acres
5. Washington (3 percent): 1,700 acres

The total U.S. acreage stands at 39,200, increasing about 26 percent in the past decade. Massachusetts had been the leading state until the recent downturn in prices. Acreage has declined steadily since 1900, especially in the Cape Cod area, due to urbanization, and also in New Jersey, where poor yields and conversion of land to blueberry cultivation caused reductions in area. However, yield per acre has increased dramatically during this time, from 1,500 pounds/acre in 1900 to over 16,000 pounds/acre today. Unfortunately, increased production has outpaced demand, and the overall industry value has decreased greatly due to record low prices. The 2004 value was $222 million, down from the record of $350 million in 1997. Prices received by growers in 1999 and 2000 were less than one-third the prices in the mid-1990s and well below the costs of production of $0.35 to $0.45 per pound (Figure 14.1). A marketing order that restricted the amount of fruit a grower could sell to 65 percent of historical averages was lifted in 2001; revocation prompted some to increase production even further. However, prices have been recovering steadily since 2001 despited increased production.

BOTANICAL DESCRIPTION

Plant

Cranberries are creeping, evergreen shrubs that spread by rhizomes, similar to their close relatives the lowbush blueberries (*V. angustifolium*). Upright shoots that produce the flowers and fruits form from rhizomes after about 2 years (Plate 14.1). Uprights may grow 2 to 4 inches annually, with bases of the stems sagging down as the uprights elongate; hence, only the terminal 5 to 8 inches remains in the vertical position. Leaves are tiny (¼-½ inch long), evergreen, thick, and oval/oblong in shape, with entire margins. Leaves persist for two seasons, being shed in late summer of the year after development. Some may remain for a third season. Premature leaf drop indicates poor health. Roots are very fine and shallow, extending no more than 1 to 3 inches into the soil; hence, plants are prone to drought stress during dry summer periods.

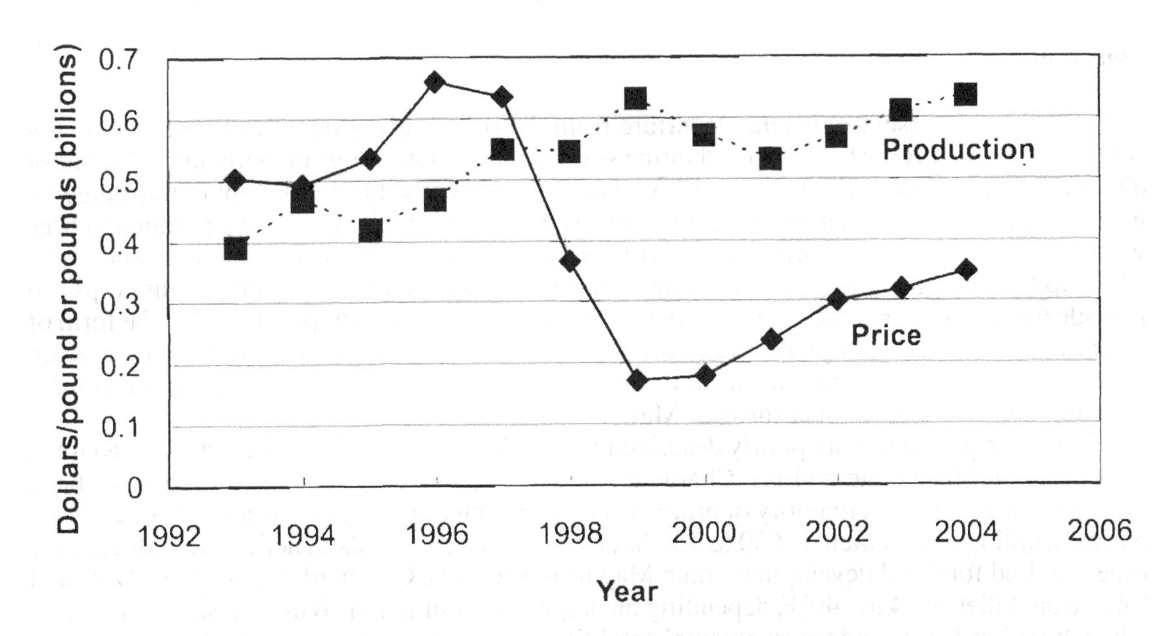

FIGURE 14.1. Overproduction through the 1990s eventually led to more than a threefold drop in the price of cranberries in the United States. *Source:* USDA statistics.

Flowers

Flowers are borne singly in leaf axils on the basal portion of a newly expanded, terminal mixed bud. Flowers are initiated the previous year. Flowering occurs over a 2- to 4-week period in late June to early July, following budbreak (Wisconsin), which is the latest blooming period of any temperate fruit crop. As in blueberry, flowers are inverted and point downward. However, cranberry flowers are distinct from blueberry by being borne on relatively long pedicels and by having a four-merous (or tetramerous) corolla that is not fused but strongly reflexed (Plate 14.1), as opposed to a five-merous (or pentamerous) corolla in blueberry. The four petals are whitish pink in color and strongly reflexed. The inferior ovary contains four locules and potentially 50 ovules, although few of these develop into seeds.

Pollination

Cranberries are self-fruitful. Insect or wind disturbance is necessary for pollen release. Seedless fruits develop occasionally, but most fruits have some seeds and fruit size is related to seed number.

Fruit

The fruit is an epigynous or "false" berry, as in blueberry. Fruits are bright red with waxy bloom at maturity, giving a dark red to black appearance; color changes from green to white, and then red during development (Plate 14.2). Fruits do not abscise naturally, and some may remain attached into the winter; hence, a portion of the pedicel may remain attached to fresh fruits. Fruits mature in 60 to 120 days after fertilization, depending on cultivar and weather.

GENERAL CULTURE

Soils and Climate

Cranberries represent a striking departure from the rule of using deep, well-drained, loamy soils with neutral pH for fruit crops. Planting sites are constructed bogs or wetlands, which have either peat moss or sand floors (Plate 14.3). The pH is typically 4 to 5 and kept in this range by using fertilizers with an acid reaction. Sites have a high water table. Cranberry production uses water for winter protection, irrigation, and harvest, so plentiful water must be available to the site—up to 6 acre-feet per year, more than double that used by other temperate fruit crops. To provide the water, every acre of cranberry bog has 3 to 10 acres of "support land," in the form of reservoirs, wetlands, and uplands, for storing or transporting water to the bog. Cranberries are extremely flood tolerant when dormant; this is exploited for the purposes of winter protection by flooding bogs from November through March.

Climatic requirements are poorly described for cranberries, but they are clearly well adapted to cool or cold-temperate regions. Characterization of the climatic requirements has not been necessary due to the vast majority of production being centered in regions where cranberries are native. Chilling requirement is 600 to 700 hours, with longer periods beneficial in reducing the time required for floral development (late May in Wisconsin). Coldhardiness is good; buds and foliage are killed at –4 to –40°F, depending on degree of acclimation. Winter desiccation occurs when the soil is frozen and environmental conditions favor transpirational water loss, such as on cold, clear, windy days. Water loss from the evergreen vines is not replaced by root uptake and thus desiccation occurs. This may cause more damage than the outright killing of tissues by low temperatures and is the primary reason for winter flooding. The winter flood is applied when water will freeze within a few days of application. It is applied in successive layers, allowing complete freezing between layers. Areas with mild winters, such as the Pacific Northwest, do not need winter flooding for protection, and in Massachusetts, flooding duration is shorter because winters are milder.

Bogs have unusual microclimates; they are very frost prone by virtue of topography, and yet they can get very hot on calm, sunny, summer days. Overhead irrigation has replaced flooding for frost protection and evaporative cooling, due to lower water and labor requirements. Unusually cold frost nights (10-20°F) require flooding of bogs, however. Evaporative cooling on hot days improves pollination and fruit set by increasing bee activity. Irrigation is run for a few minutes intermittently during the hottest part of the day.

Propagation

Shoot cuttings are used as propagation material for new bogs. Upright shoots from an existing site are mowed, bundled, and sold by the ton for propagation material. The cuttings are simply spread over the surface and pushed into the sand with a blunt disc. After discing, the bog is flooded to settle the soil around the cuttings, and intermittent sprinkler irrigation is applied thereafter for 2 to 4 weeks, until cuttings are rooted. Under good management, runners cover the surface by the end of the first season and uprights are formed the second year. Full production occurs in about 4 to 7 years after propagation of a new bog.

Rootstocks

Cranberry has no rootstocks.

Planting Design, Training, Pruning

Cranberry bogs are solid surfaces of plants, as vines spread naturally by rhizome and upright extension (Plate 14.3). Mature bogs have densities of over 500 upright shoots per square foot, but optimal densities are about 200 to 300 uprights per square foot. No training of plants is necessary, but occasionally uprights and runners must be thinned. The mechanical harvest operation causes some pruning and thinning of vines as a by-product, and bogs harvested this way may not require additional pruning. Judicious control of water and fertilizer substitutes for pruning in some cases, since these processes control vine vigor.

Since flooding is a major cultural practice, land must be carefully graded and smoothed prior to planting. Drainage ditches and dikes are constructed, and the bog is "scalped" to a depth of 6 to 18 inches to remove the roots of woody species, potentially troublesome weeds. Then 2 to 4 inches of sand is applied to the bog floor, which stabilizes the surface, facilitates leveling and equipment movement, and provides a well-aerated propagation media. Sand is reapplied every few years to rejuvenate the vines and help control weeds and pests. Conveniently, dump trucks can be driven on the bogs in winter, when they are fully iced over, spreading sand in an even layer that settles to the bog floor when the ice melts.

Cranberry planting sites are generally located in wetlands, which raises some important environmental issues. Wetlands are protected natural resources that are often vulnerable to agricultural activities in and around them. Cranberry bog construction can cause siltation of waters downstream, and the use of water for cranberry production can impact stream flows and water tables. Agricultural chemicals may run off from cranberry operations, impacting fish and other species. Dike and dam construction may interfere with fish spawning, and increased water temperature caused by discharging water from cranberry bogs also negatively affects fish and aquatic life. Although the local impacts of cranberry growing may be high, the small acreage involved in cranberry production translates to one of the smallest impacts on wetlands. University of Wisconsin researchers showed that, historically, their state has lost over 5 million acres of wetlands to urbanization and agriculture, but only about 0.2 percent of this was attributable to cranberry production (Wisconsin is the leading producer of cranberry). Furthermore, the support lands that provide water to cranberry bogs are host to a wide variety of animal and plant species, even some endangered species. River otters, sandhill cranes, ducks, geese, bald eagles, fox, mink, and deer are just a few of the hundreds of species known to inhabit cranberry support lands.

Pest Problems

Insects

Leaf and shoot feeders. The black-headed fireworm *(Rhopobota naevana)* is the most serious insect pest of cranberry. The lepidopterous larvae feed on old leaves first and then migrate to new shoots, webbing leaves together and eating upper leaf surfaces. Injured leaves turn brown, giving a "fire-swept" appearance to heavily damaged areas of fields. Other species of fireworms may also be present occasionally. Cranberry tipworm *(Dasyneura vaccinii)* larvae feed on developing shoot tips, which precludes fruit production on infested terminals during that season. Pheromone traps are used to capture adults, and sprays are timed just after peak flights are observed.

Cranberry fruitworm. Acrobasis vaccinii is the most serious pest attacking cranberry fruits. The adult female lays eggs under the calyx lobes of young fruits; the emerging larvae tunnel completely through several fruits as they reach pupation stage. Pheromone traps and frequent scouting of bogs is used to assess populations, and a number of insecticides can be applied if populations warrant such a measure.

Root feeders. Several species attack roots, but no one species is widespread across all production regions. Cranberry root grub *(Lichnauthe vulpina)*, cranberry rootworm *(Rhabdopterus picipes)*, and common white grub larvae *(Phyllophaga anxia)* are found in eastern areas, and the black root weevil *(Brachyrynus sulcatus)* is found in the Pacific Northwest. Their occurrence is sporadic and often does not require treatment.

Diseases

Fruit rots. Several fruit-rotting fungi can attack cranberry, but most are limited in distribution or importance across producing regions. Early rot *(Phyllosticta vaccinii)* and end rot *(Godronia cassandrae* f. *vaccinii)* cause serious losses in bogs and storage, respectively, if left unchecked. Early rot is a problem largely in Massachusetts and New Jersey. End rot is generally not a problem if fruits are frozen immediately after harvest. Bitter rot *(Colletotrichum gleosporioides)* can be an occasional problem on fruits in Massachusetts and New Jersey. Cottonball disease *(Monolinia oxycocci)* is the most serious disease in Wisconsin, but it occurs only there and in British Columbia. The seed cavities of infected fruits fill with white, cottony mycelia, and fruits turn yellow with brown stripes instead of turning red in autumn. The cottonball fungus can also cause a shoot tip blight, which itself is not as damaging but serves as a cue for spray timing. Fungicides are used to control fruit rots unless postharvest practices, such as freezing, minimize their impact.

HARVEST, POSTHARVEST HANDLING

Maturity

Red color is the primary determinant of harvest maturity and fruit quality. Color increases over time, and harvest is delayed as long as possible to allow color development. Cranberry is classified as a climacteric fruit with respect to ripening but exhibits very low rates of respiration.

Harvest Method

Berries are harvested either dry or wet ("on-the-flood"), with wet harvesting the predominant method (Plate 14.4). Dry harvesting consists of mechanical rakes that remove berries by a "hair combing"-like action. Dry harvest is less damaging to berries than wet harvest and is used for the small percentage of fruits to be sold fresh. Wet harvesting involves flooding bogs with several inches of water and beating the berries from the vines with water reels. Cranberries float on the surface and are corralled into a corner, where they are lifted by a conveyor belt into bins destined for processing plants.

Postharvest Handling

Wet-harvested berries must be dried before storage. Berries are cleaned and sorted by passing them over sizing screens. Culling is performed using "bouncing boards" initially (good-quality berries are resilient and bounce), and by hand afterward. Berries for processing into juice are frozen in 100-pound barrels, and those for fresh-market sales are packaged in polyethylene bags and sold around the holidays.

Storage

Frozen berries can be stored for several months until made into juice or other products, with no problems. Fresh berries can be stored without refrigeration for a few weeks if marketed early. Fresh cranberries have a storage life of several months in refrigerated storage (40°F), unusually long for a small fruit crop. They are sensitive to chilling injury if stored below 40°F and will develop a rubbery texture as a result.

CONTRIBUTION TO DIET

About 90 to 95 percent of the cranberry crop is processed into juices and sauces. Juice blends have become more popular in recent years. A small amount of the cranberry fruit crop is marketed fresh around the Thanksgiving and Christmas holidays. A small portion of the crop is dried and used as a raisin substitute, "cran-raisins," used in muffins, scones, cookies, and breads.

Per capita consumption of cranberries is 2.0 pounds per year, nearly all as juice. Cranberry consumption has risen about 50 percent since 1980.

Dietary value per 100-gram edible portion:

Water (%)	87
Calories	46
Protein (%)	1.4
Fat (%)	0.1
Carbohydrates (%)	12.2
Crude Fiber (%)	4.6

	% of U.S. RDA (2,000-calorie diet)
Vitamin A	1
Thiamin, B_1	<1
Riboflavin, B_2	1
Niacin	<1
Vitamin C	22
Calcium	<1
Phosphorus	2
Iron	1
Sodium	<1
Potassium	2

BIBLIOGRAPHY

Caruso, F.L. and D.C. Ramsdell. 1995. *Compendium of blueberry and cranberry diseases.* St. Paul, MN. American Phytopathological Society Press.

Chandler, F.B. and I.E., Demoranville. 1959. *Cranberry varieties of North America* (Massachusetts Agricultural Experiment Station Bulletin 513). Amherst, MA: University of Massachusetts.

Dana, M.N. 1983. Cranberry cultivar list. *Fruit Var. J.* 37:88-95.

Dana, M.N. 1990. Cranberry management, pp. 334-362. In: G.J. Galleta and D.G. Himelrick (eds.), *Small fruit crop management.* Englewood Cliffs, NJ: Prentice-Hall, Inc.

Dana, M.N. and G.C. Klingbiel. 1966. *Cranberry growing in Wisconsin* (University of Wisconsin Cooperative Extension Series Circular 654). Madison, WI: University of Wisconsin.

Dittl, T.G. and L.D. Kummer. 1997. *Major cranberry insect pests of Wisconsin* (University of Wisconsin Publication 8-97). Madison, WI: University of Wisconsin.

Eck, P. 1990. *The American cranberry.* New Brunswick, NJ: Rutgers University Press.

Galletta, G.J. 1975. Blueberries and cranberries, pp. 154-196. In: J. Janick and J.N. Moore (eds.), *Advances in fruit breeding.* West Lafayette, IN: Purdue University Press.

Luby, J.J., J.R. Ballington, A.D. Draper, K. Pliszka, and M.E. Austin. 1990. Blueberries and cranberries *(Vaccinium),* pp. 391-456. In: J.N. Moore and J.R. Ballington (eds.), *Genetic resources of temperate fruit and nut crops* (Acta Horticulturae 290). Belgium: International Society for Horticultural Science.

Mainland, C.M. 1985. *Vaccinium,* pp. 451-455. In: A.H. Halevy (ed.), *CRC handbook of flowering.* Volume 4. Boca Raton, FL: CRC Press.

Stang, E.J. and M.N. Dana. 1984. Wisconsin cranberry production. *HortScience* 19:478.

Turner, D. and K. Muir. 1985. *The handbook of soft fruit growing.* London: Croom Helm.

Vander Kloet, S.P. 1983. The taxonomy of *Vaccinium* section *Oxycoccus. Rhodora* 85:1-43.

Vander Kloet, S.P. 1988. *The genus* Vaccinium *in North America* (Agriculture Canada Publication No. 1828). Ottawa: Agriculture and Agri-Food Canada.

Chapter 15

Date *(Phoenix dactylifera)*

TAXONOMY

The Arecaceae (or palm) family is a large, distinct family of woody monocotyledonous plants, containing up to 4,000 species distributed over 200 genera. The date palm, *Phoenix dactylifera* L., is one of three economically important fruit crops in the palm family. Coconut *(Cocos nucifera)* and African oil palm *(Elaeis guineensis)* are the other two crops, both of which are true tropical species, whereas date palm is considered subtropical. The genus *Phoenix* contains 12 species, five of which bear edible fruit in addition to *P. dactylifera (P. atlantica, P. reclinata, P. farinifera, P. humilis,* and *P. acaulis),* and all of which are grown as ornamentals. A sixth species, *P. sylvestris,* is used in India to obtain sugar. Several interspecific hybrids have been produced. The date palm is unique among other species in *Phoenix* in producing offshoots, or suckers, from the base of the trunk, which allows for exceptionally easy vegetative propagation.

Date, coconut, and African oil palm are the major fruit crops in the Arecaceae, but several minor palms are also cultivated for their fruits or wild-harvested, including peach palm *(Bactris gasipaes),* ungurahui palm *(Jessenia batua),* jelly palm *(Butia capitata),* betel nut *(Areca catechu),* Chilean wine palm *(Jubaea chilensis),* saw palmetto *(Serenoa repens),* and aguaje palm *(Mauritia flexuosa).* An even greater number of taxa are cultivated as ornamentals, including coconut, date and its close relatives, and species within the genera *Sabal, Serenoa, Washingtonia,* and *Roystonea.*

Cultivars

There are an estimated 3,000 cultivars of date palm worldwide; thus it is difficult to generalize and impossible to discuss the breadth of germplasm available here (Figure 15.1). As for many fruit crops, just a few are important in the global market. 'Medjool' is probably the best known and one of the highest-quality dates available. Fruits are large, soft, and ship well, making it one of the most preferred for exportation. 'Medjool' was introduced to California from Morocco in the 1920s and is still a major cultivar there. 'Deglet Noor', a high-yielding, semidry cultivar, is popular in northern Africa and the major cultivar in California. 'Barhi' (also 'Barhee'), which originated in Basra, Iraq, is another high-yielding cultivar. It is unusual in that it lacks the normal astringency of other dates at the Khalal (middle) stage of development and can be eaten when crisp and immature; other cultivars are eaten at later stages when dried and brown. 'Zahidi' is an important Middle Eastern cultivar that can be eaten either firm or soft.

ORIGIN, HISTORY OF CULTIVATION

Dates are native to the Persian Gulf area of the Middle East, where they have been cultivated for at least 6,000 years, longer than many other fruit crops. The date does not exist in a truly wild

Introduction to Fruit Crops

doi:10.1300/5547_15

state and may have originated as a hybrid between *P. reclinata* from Africa and *P. sylvestris* from India. Date palms were one of the few crop plants that could survive desert conditions and thus became a reliable source of food in an otherwise inhospitable climate (Plate 15.1). Sumerian, Babylonian, Egyptian, and other ancient peoples used the palm for house construction and thatching as well as for food. It was spread by nomadic people across northern Africa, along the coast, and at oases, where it became a staple crop. From there, it was introduced to Southern Europe, where it is grown chiefly as an ornamental. Iraq, which has led the world in date production from antiquity until recently, is the origin of many major cultivars. Dates were introduced to northern Mexico and California by Spanish missionaries in the late 1700s. However, it was not until the turn of the twentieth century, when superior cultivars were introduced, that the California industry was born. A research station was established in Indio, California, in 1904, to study date and citrus cultivation. Cultivars collected from northern Africa and the Middle East were brought to the station, studied by USDA scientists, and later released to growers. A small but healthy date industry is located around Indio today.

FIGURE 15.1. A number of different date cultivars on display at a market in Fez, Morocco.

FOLKLORE, MEDICINAL PROPERTIES, NONFOOD USAGE

Date palms have many uses other than those centered on their fruits. The seeds are often fed to livestock, along with culled fruits, but an oil can be derived from them and used for soap manufacture. As with many palms, dates exude copious sap, when cut, that is fermented into a palm wine. The wood is often used in construction. Leaves are used for thatch and woven into mats, baskets, and other objects. The petioles and leaf sheaths are made into crude tools, fuel, and fibers for making rope and cloth.

The relationship between dates and humans goes back to antiquity, so it is not surprising that the fruit has been used in several dozen folk remedies. Ailments treated include everything from asthma and bronchitis to cancer and venereal disease. Saharan people who consume large quantities of dates have low cancer rates, and this is attributed to the high magnesium content of dates. Although of debatable efficacy, several other compounds in dates are known to have medicinal value: tannins are used as astringents for coughs or a sore throat; fiber is used as a digestive aid; serotonin affects mood, emotion, sleep, and appetite; anthocyanins act against cancer and as antioxidants; and estrones act as estrogenic hormones in mammals. The estrone content of seeds and pollen may add validity to their use for delaying child bearing in Egypt.

Date palms, particularly the leaves, are used in religious ceremonies in Christian, Jewish, and Islamic faiths. For example, date palm leaves are used on Palm Sunday to commemorate Jesus' entrance into Jerusalem. The botanical name *Phoenix* may relate to either Phoenicia, where dates originated, or the phoenix bird of Egyptian mythology, which arose from the ashes of a fire alive and renewed. It is said that date palms can withstand fire.

PRODUCTION

World

Total production in 2004, according to FAO statistics, was 6,772,068 MT or 15 billion pounds. Dates are produced in 35 countries worldwide on about 2.9 million acres. Average yields are just over 5,000 pounds/acre but range from less than 1,000 pounds up to over 10,000 pounds/acre. World production has doubled since the late 1980s. The top ten date-producing countries (percent of world production) follow:

1. Egypt (16)
2. Iran (13)
3. Saudi Arabia (12)
4. United Arab Emirates (11)
5. Iraq (10)
6. Pakistan (10)
7. Algeria (7)
8. Sudan (5)
9. Oman (4)
10. Libya (2)

United States

Total production in 2004, according to USDA statistics, was 18,818 MT or 41.4 million pounds. All dates are produced in extreme southern California, near Indio, on about 4,500 acres. Acreage is down from 5,500 in 1992, but the industry value is up 38 percent, to $31.5 million, in the same period, due to a doubling of the price paid to growers ($0.76 per pound in 2004) (Figure 15.2). Average yields in California are 9,000 pounds/acre, almost double the worldwide average. The United States imported 5,253 MT in 2002, or an amount equivalent to 28 percent of domestic production.

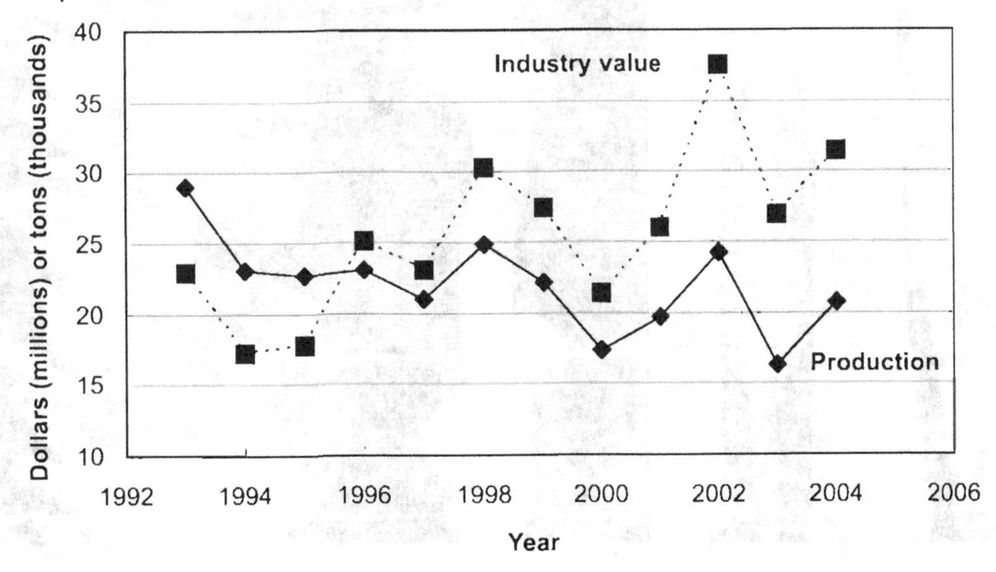

FIGURE 15.2. The value of the California date industry relative to production. A doubling of price paid per ton allowed industry value to reach an all-time high in 2002 despite a decrease in production over the past decade. *Source:* USDA statistics.

BOTANICAL DESCRIPTION

Plant

Dates are tall, straight-trunked palms, reaching 60 feet in the wild, but up to 30 feet when used in cultivation. Trunks reach 3 feet in diameter, retain persistent leaf bases for years, and produce offshoots or suckers, either at the base or higher up (Plate 15.2). Unlike many palms, dates can branch and have two or three crowns, although this is uncommon. Leaves are 10 to 15 feet long, pinnately compound palm "fronds," with spines occupying the lower third of the rachis and leaflets the distal two-thirds. Leaves are long-lived, 3 to 7 years, and persistent when they die; dead leaves are pruned off to facilitate harvest and pollination. Mature palms have over 100 leaves, producing one to two new leaves per month. About 100 to 200 leaflets are found on each leaf; each leaflet is up to 3 feet long, ½ to 2½ inches wide, folded longitudinally, with entire margins.

Flowers

Dates are dioecious, meaning that they have separate male and female plants. Whether staminate or pistillate, flowers are borne on a compound spadix having 50 to 150 lateral branches in leaf axils (Figure 15.3). A dozen or more inflorescences are produced annually, more in males than in females. Inflorescences are sheathed in a bract or spathe until just prior to anthesis. Male inflorescences are shorter and wider than female ones. Each sex produces thousands of tiny flowers per inflorescence. Male flowers are white, fragrant, about ¼ to ½ inch wide, with six stamens each, and female flowers are more yellowish or cream colored, smaller (¼ inch), with trilocular superior ovaries and a three-lobed stigma.

FIGURE 15.3. *(left)* The date inflorescence is a compound spadix borne in leaf axils for both male *(on left)* and female *(on right)* trees. *(right)* Close-ups of female *(on left)* and male *(on right)* flowers. *Source:* Photos courtesy of USDA.

Pollination

While some cultivars are polygamous or even have hermaphroditic flowers, dates are largely dioecious. They are wind or insect pollinated naturally, and natural pollination can be practiced in seedling orchards with 1:1 ratios of males to females. In most commercial orchards, however, only 1 male is grown for every 50 females, and pollination is accomplished artificially. In this way, most of the orchard space is occupied by females, increasing yield per acre tremendously. Traditionally, a few strands of male inflorescences are laid among the female inflorescences by hand (Figure 15.4). Hand pollination is laborious, requiring up to 700 man-days of work for an average-sized planting. More commonly, pollen is collected from male inflorescences, dried, and stored up to a year in refrigeration, if females are not immediately ready for pollination. Pollen is generally blown onto receptive female inflorescences by spray machine or hand-held puffers, or it is transferred first onto cotton balls that are then placed into the inflorescences. The period of female receptivity is as few as 3 and as many as 30 days after the female spathe opens. Pollen is often mixed with fillers, such as talc or walnut shell dust, and sometimes sprayed onto the leaves above the female inflorescences, if they are not yet opened. Fruit set is often lower with artificial pollination, but it is adequate, and it decreases the need for thinning and reduces expense.

FIGURE 15.4. Artificial pollination of dates. *(left)* Pollen being applied with a hand-held puffer. *(right)* A few strands of a male inflorescence placed in the middle of a female inflorescence by hand. The inflorescence to the left of the one being handled has already been pollinated and tied. *Source:* Photos courtesy of USDA.

Fruit

Fruits are drupes, 1 to 3 inches long, dark brown, and thick skinned, with thick, sweet flesh and a large seed in the center (Plate 15.3). Unlike other drupes, the endocarp is thin and membranous, instead of thick and woody, prompting some to classify the fruit as a single-seeded berry. Immature fruits are green, yellow, or red. Several hundred to just over 1,000 fruits are borne in each bunch, which can weigh up to 80 pounds. Bunches are often tied to leaves to prevent breakage of the peduncle under the weight of the fruits. As they mature over 6 to 8 months, fruits go through five distinct stages, given the following Arabic names: Hababouk, Kimri, Khalal, Rutab, and Tamar. Fruits grow rapidly in the first two stages; then turn their characteristic color, lose water, and accumulate sugar in the Khalal stage; and finally ripen completely in the last two stages (Plate 15.4). At the Tamar stage, fruits have only 10 to 25 percent water content and thus are resistant to spoilage and fermentation and shrivel like raisins. Craft paper covers or shade cloths are often applied to fruit bunches in the Kimri stage to protect them from rain, insects, and birds.

Fruit thinning is practiced at pollination time, or sometimes 6 to 8 weeks after pollination. Thinning is accomplished by bunch thinning, whereby some strands or portions thereof are removed, or by complete bunch removal or by a combination of both methods. Thinning individual strands helps improve fruit size and allows better air circulation within the bunches. Complete bunch removal is needed if too many clusters are produced because overcropping causes alternate bearing. About eight to ten leaves are needed to mature one bunch of fruits, so on a mature tree with 100+ leaves, 10 to 15 bunches of fruits would be left, at maximum. No fruits are left on young trees (<3 years), so they can grow vegetatively, and the number of bunches allowed is increased each year from year 4 to maturity at 10 to 15 years.

GENERAL CULTURE

Soils and Climate

Deep, well-drained loamy soils are best for dates, but they can be grown on a range of soil types. They are among the most tolerant of crop plants when exposed to high pH and salinity. Although the tree itself is highly drought tolerant, commercial fruit production requires considerable water, and little, if any, rainfall occurs in the most desirable production areas. Flood irrigation is applied several times per year.

Dates require a hot, dry climate for proper pollination and fruit maturation. Pollination and seed set are disfavored by temperatures below the mid-70s°F. Fruits are often marketed dry (25 percent water content or less), and hot, arid conditions are needed from August through November for most cultivars. Rainfall is generally detrimental during the growing season. At bloom, rain either washes away pollen or creates conditions unfavorable for fruit set by reducing temperature. At harvest, rain causes fruit rot. Most areas where dates are cultivated experience rain in the winter months only. Dates tolerate higher temperatures than most other fruit crops: midsummer temperatures of 100 to 110°F, which would injure many temperate crops, are typical in date production. High humidity is detrimental to fruit maturation, making fruits sticky and soft and predisposed to fungal problems. On the other hand, perpetually arid conditions are conducive to outbreaks of date mite. Dates do not require chilling to break dormancy but can tolerate light freezing (to 20-23°F) in winter. A cool winter helps to synchronize flowering and ripening of fruits. Dates, as with coconut palms, are extremely resistant to wind damage. However, windborne sand can stick to fruits and create problems at harvest.

Propagation

Dates are propagated by offshoots, which are suckers growing from the trunks. A typical date palm will produce 10 to 20 offshoots during its lifetime, mostly when it is young. Offshoots arising at the base of the trunk are best, as those arising higher up are difficult to root. There are differences among cultivars in ability to form offshoots; for example, 'Zahidi' produces numerous offshoots, while 'Barhi' produces few. Large offshoots, 3 to 5 years of age, 20 to 50 pounds, and having diameters of 8 to 12 inches are best. Offshoots are removed in the early spring and placed in nurseries to grow for 1 to 2 years. Care is taken not to damage any roots present, and planting sites in nurseries are often amended to promote rooting. Leaves are cut back and bundled at removal to facilitate handling and reduce water loss.

Seed propagation is not used because 50 percent of the plants will be males, and trees are slow to come into production. Tissue culture techniques have been developed in the past decade or so, and several laboratories around the world offer tissue-cultured plants. The advantages over offshoot production are less disease in propagules, more consistent/reliable supply, and higher survival rates, but the costs are upward of $20.00 to $25.00 for each plant, and plants have to be acclimated or hardened before transplanting in the field.

Rootstocks

There are no date rootstocks.

Planting Design, Training, Pruning

Dates are planted about 25 to 35 feet apart, which yields about 50 plants per acre on average. A tendency toward slightly higher densities of up to 60 or so per acre has occurred more recently. Dwarf cultivars, such as 'Khadrawy', can be planted at much higher densities.

Very little pruning and training is involved in the production of any palm fruits. Dates have persistent leaves, which must be pruned off to allow access to fruit and crowns; old fruit stalks are pruned off as well. As discussed previously under Propagation, offshoots are removed annually, generally during the first 10 to 15 years of the palm's life, and used as planting stock. Undesirable offshoots are simply pruned off.

Pest Problems

Insects

Beetles and weevils. Some of the same pests of coconut and oil palm also attack date. The rhinoceros or black palm beetle (*Orcytes rhinoceros*) attacks all three species, causing injury to young leaves and occasionally the meristem. Date fruit clusters may also be attacked. This pest is a particular threat on young trees, which may be killed by the beetles' tunneling in the meristem. These beetles are easily identified by being large (up to 2 inches long), brown beetles with prominent horns on their heads, hence the name. They are found throughout the production regions of Asia and Africa. Some control measures are targeted at removal of rotting logs, compost, and debris, environments where they commonly breed. Rhinoceros beetles are nocturnal and attracted to light, so light traps are used as another nonchemical control measure. Here, a mercury vapor light is erected above a pot of diesel or kerosene, into which the insects fall. Biological control using the fungus *Metharhizum anisopilae* or *Baculovirus orcytes* has also been effective.

Several species of *Rhynchophorus* weevils attack palms, with the red or Indian palm weevil, *R. ferrugineus*, being a serious pest of date. Once regarded as a pest of coconut, until a few decades ago, the weevil spread to date-producing areas in the Middle East throughout the 1980s and 1990s. The weevil is similar to the rhinoceros beetle in size, but the larvae, not the adults, cause damage by tunneling through young leaf and meristem tissue. Damage is more often fatal with the weevil than with the rhinoceros beetle. *Rhynchophorus phoenicis*, the African palm weevil, is a similar pest that was first discovered feeding on date in 1999. Quarantines on international shipments of offshoots from weevil-infested areas are in place. Prompt removal of dead or infested palms and avoidance of wounding trunks (eggs are laid in wounds) help to reduce infestation. Holes where larvae have entered are sometimes treated with insecticides, and baits and pheromone traps have been used for control as well.

Old World date mite. Unlike many mites that infest leaves, this mite *(Oligonychus afrasiaticus, O. pratensis)* is primarily a fruit pest. The larvae feed on fruit surfaces and cover them with webbing, which in turn retains sand particles, rendering fruits unmarketable. Fruit drop may also occur. As many as six generations are produced per year, with populations peaking in the hottest, driest months. Sulphur is applied when infestations occur. 'Sayer' is said to be resistant to date mite.

Scale insects. White scale, *Parlatoria blanchardii*, is a widespread pest of date and many palms. It is most serious in northern Africa, exclusive of Egypt, where it is considered a minor pest. As with all scale insects, they suck sap from leaves, damaging or killing individual leaves but not entire palms. They cause greater problems for young palms. Several beneficial insects (e.g., *Aphytis mytilaspidis* wasps) help control populations, and this is augmented by pruning out infested leaves and occasional chemical sprays. Red scale, *Phoenicococcus marlatii*, primarily affects leaf bases but is not as serious a pest as white scale.

Carob moth. The larvae of this moth *(Ectomyelois ceratoniae)* infest fruits in the field and can cause serious postharvest losses as well. Pheromone traps are used to monitor populations in the field; *Bacillus thuringiensis*, the biological control known as Bt, can be applied in the field; stored dates can be fumigated.

Diseases

Bayoud disease. This disease, caused by the fungus *Fusarium oxysporum* f. sp. *albedinis*, is widespread in Morocco and Algeria, where it was first noted in 1870, and continues to move eastward through northern Africa. The fungus is present in soil, roots, trunks, and offshoots of palms and is transported by either humans or irrigation water. The disease caused by this fungus is fatal, leading to death in a period of weeks to months once symptoms surface. Foliar symptoms are a whitening or graying of foliage in a particular pattern—acropetally on leaflets on one side of the leaf rachis, and then basipetally down the other side. Adjacent leaves are affected and the palm is killed when the meristem dies. Control is achieved by quarantine of international plant shipments from infested countries and the rapid destruction of affected palms. The disease is harbored in dead roots and asymptomatic orchard floor vegetation, such as alfalfa and henna. Thus, borders of infected areas can be trenched and irrigated separately from uninfected areas. Soil fungicides are futile, but fumigation of isolated spots where the fungus has occurred is marginally successful. The major cultivars 'Deglet Noor' and 'Medjool' are both susceptible, but resistance has been found among other genotypes that have survived epidemics in Morocco, and resistant cultivars may soon be available.

Black scorch or Medjnoon. Aptly named, the fungus *Ceratocystis paradoxa* causes the blackening of leaves but also attacks inflorescences, the trunk, and meristems. If the meristem is infected, the palm usually dies, but some will continue growing from a lateral bud and develop a bend in the stem. Sanitation of infected leaves followed by fungicide application is done on an individual basis; severely affected palms are removed and burned. Most major Middle Eastern cultivars are reported to be susceptible.

Graphiola leaf spot. The fungus *Graphiola phoenicis* infects leaves, causing spotting and premature death. Severe infections reduce yield. It is more common in humid growing areas, such as the Egyptian delta region, coastal Algeria, California, and the southern Sahel region of Africa. Control is achieved by pruning infected leaves and applying fungicides. Several resistant cultivars exist, including 'Abdad', 'Barhi', 'Gizaz', 'Khastawy', 'Rahman', and 'Tadala'.

Khamedj disease or inflorescence rot. This disease, caused by the fungus *Mauginiella scattae,* is widespread in date culture, but worst in more humid regions, particularly those with rainfall prior to flowering. It has caused losses of 50 to 80 percent in unusually wet years in such arid regions as Iraq and Saudi Arabia. The fungus affects the spathe surrounding the inflorescence and is indicated by brown/rust-colored areas on the outside prior to spathe splitting. The fungus is also on the inside, infecting flowers before the spathe opens; severely infected spathes do not ever open. The fungus spreads during pollination, as this is often practiced by hand. Control is accomplished with fungicides, and resistant cultivars, such as 'Hallawi', 'Hamrain', and 'Zahdi', do exist. The major cultivars 'Medjool', 'Khadrawy', and 'Sayer' are highly susceptible.

HARVEST, POSTHARVEST HANDLING

Maturity

At full maturity, dates are dark brown and have very low water content. However, they may be harvested prior to full maturity for special markets. In the Khalal stage, or the third of five developmental stages, dates are firm, crisp, and moist (50-85 percent water content) and have their characteristic color, usually red or yellow. This "fresh" market is minor and limited largely to 'Bahri', which has low tannin content at this stage and is therefore not bitter tasting. In the Rutab or fourth stage, dates begin to turn brown and soft and have lost an appreciable amount of water (30-45 percent water content) and tannin. Many prefer the fruits at this stage of development over the Tamar or fifth stage, when fruits are completely browned and dry (10-25 percent water content). However, Rutab stage dates are more perishable and difficult to handle. 'Deglet Noor' and 'Medjool' are harvested at Rutab, but most other dates (e.g., 'Halawy', 'Khadrawy', 'Sayer', and 'Zahidi') are harvested at the Tamar stage, or full maturity. Dates are classified as climacteric fruits with respect to ripening but have comparatively small peak rates of ethylene production and respiration.

Harvest Method

Dates are hand harvested, generally by cutting entire bunches at one time (Figure 15.5). In some cases, individual fruits may be picked, or bunches are shaken to remove only the ripe fruits. Pickers climb trees or use ladders, picking platforms, or hydraulic picking aids to harvest fruits. Not all bunches or dates within a bunch mature at the same time, and each palm may be picked up to eight times over a period of 2 to 3 months to obtain consistent maturity.

Postharvest Handling

Fresh dates (Khalal stage), still on the stems, are washed and packed in 10-pound cardboard boxes. Immature and overmature fruits are removed, as are damaged or small fruits at the proper maturity. They are refrigerated for transport and storage. Dates at a more advanced maturity level can be packed as bunches, but variation in maturity may dictate the handling of individual fruits. Fruits are washed and sorted and have their moisture content adjusted by hydration or dehydration. In regions that can expect rain at harvest, fruits are harvested a bit early and dehydrated; areas with hot, dry winds at harvest may rehydrate fruits in water to improve texture and taste. Dates may be pitted prior to packing, depending on intended market (particularly for the U.S. market).

FIGURE 15.5. Hand harvest of dates in California. Paper covers are placed over bunches in fall to prevent damage from rainfall. *Source:* Photo courtesy of USDA.

Storage

Dates at full maturity store remarkably well compared to other fruits, owing to their low water content. Ripe dates do not experience chilling injury and can be stored at 32 to 34°F for a year or more, if humidity is kept below 20 percent. Fresh dates (Khalal stage) can be stored for up to 8 weeks under the same conditions. Fumigation of fruits with methyl bromide is sometimes practiced if stored long periods, especially with fruits for export.

CONTRIBUTION TO DIET

Dates are generally eaten out of hand. Pitted dates are sold for out-of-hand consumption and often used in baking. Date paste is used for cake fillings and other processed products. Date syrup is used as a sweetener; fruits are pressed, and the juice is concentrated and filtered. Alcoholic beverages can be made from fermenting date syrup or juice. Dates are low in fat and protein, but higher in vitamins and minerals than are many other fruits and vegetables. The date provides more than just its fruit, however; the meristem can be eaten like heart o' palm, and flour can be derived from the pithy trunk.

Per capita consumption of dates is extremely low, only 0.12 pound/year. This is about one-half to one-third the consumption rate of the 1970s and 1980s.

Dietary value per 100-gram edible portion:

	Moist Dates (Khalal and Rutab stages)	Dried Dates (Tamar stage)
Water (%)	30-85	20.5
Calories	142	282
Protein (%)	1.8	2.4
Fat (%)	1.0	0.4
Carbohydrates (%)	37	75
Crude Fiber (%)	3.5	8.0

	% of U.S. RDA (2,000-calorie diet)	
Vitamin A	18	<1
Thiamin, B_1	2	4
Riboflavin, B_2	2	4
Niacin	31	6
Vitamin C	50	<1
Calcium	3	4
Phosphorus	50	9
Iron	33	6
Sodium	<1	<1
Potassium	7	19

BIBLIOGRAPHY

Barreveld, W.H. 1993. *Date palm products* (FAO Agricultural Services Bulletin 101). Rome: FAO.

Carpenter, J.B. and H.S. Elmer. 1978. *Pests and diseases of the date palm* (USDA Agricultural Handbook No. 527). Washington, DC: U.S. Department of Agriculture.

Duke, J.A. 1983. *Phoenix dactylifera* L., In: *Handbook of energy crops*. Available at www.hort.purdue.edu /newcrop/duke_energy/Phoenix_dactylifera.html; accessed April 2004.

Horticultural Crops Group. 1982. *Date production and protection: With special reference to North Africa and the Near East* (FAO Plant Production and Protection Paper 35). Rome: FAO.

Magness, J.R., G.M. Markle, and C.C. Compton. 1971. *Food and feed crops of the United States* (Interregional Research Project IR-4, IR Bulletin 1; Bulletin 828 New Jersey Agricultural Experiment Station). East Rutherford, NJ: Rutgers University.

Morton, J.F. 1987. *Fruits of warm climates*. Miami, FL: Julia F. Morton.

Reuveni, O. 1985. *Phoenix dactylifera*, pp. 343-354. In: A.H. Halevy (ed.), *CRC handbook of flowering*, Volume I. Boca Raton, FL: CRC Press.

Sauer, J. 1993. *Phoenix*—Date palms, pp. 182-186. In: *Historical geography of crop plants*. Boca Raton, FL: CRC Press.

Wrigley, G. 1995. Date palm, *Phoenix dactylifera*, pp. 399-403. In: J. Smartt and N.W. Simmonds (eds.), *Evolution of crop plants*, Second edition. London: Longman.

Zaid, A. (ed.). 2002. *Date palm cultivation* (FAO Plant Production and Protection Paper 156). Rome: FAO.

Zohary, D. and M. Hopf. 1993. Date palm, *Phoenix dactylifera*, pp. 157-162. In: *Domestication of plants in the Old World*, Second edition. Oxford: Clarendon.

Chapter 16

Grape (*Vitis* spp.)

TAXONOMY

Grapes are members of the Vitaceae family. The genus *Vitis* is broadly distributed, largely between 25 and 50°N latitude in eastern Asia, Europe, the Middle East, and North America. *Vitis* is by far the most economically important genus in the family, which also contains Virginia creeper *(Parthenocissus)* and pepper vine *(Ampelopsis)*. A few other species of *Vitis* are found in the tropics—in Mexico, Guatemala, the Caribbean, and northern South America. There are over 100 species documented in the literature, 65 of which are thought to be genuine and another 44 that are questionable and probably interspecific hybrids. *Vitis* is split into two subgenera:

1. *Euvitis:* Known as "true grapes," members of this subgenus are characterized by elongated clusters of fruits, berries that adhere to stems at maturity, forked tendrils, loose bark that detaches in long strips, and diaphragms in the pith at nodes. Also called "bunch grapes," members of *Euvitis* include *Vitis vinifera*, the European grape, and *V. labrusca*, the concord grape, and all but two to three species of the genus.
2. *Muscadinia:* Known as "muscadine grapes," members of this subgenus are characterized by small fruit clusters, thick-skinned fruits, berries that detach one by one as they mature, simple tendrils, smooth bark with lenticels, and the lack of diaphragms in the pith at nodes. There are only two to three species in this section, and *V. rotundifolia* is the main species of commerce. Related species include *V. munsoniana* (bird grape), which grows wild in the understory of forests of the southeastern United States, and *V. popenoeii*, native to Florida. The validity of the latter is questioned by some, as it may be a subspecies of *V. munsoniana*.

Vitis *Species and Cultivars*

Three important species and one hybrid group account for most of grape production worldwide:

1. *Vitis rotundifolia* Michx.: Known as the muscadine grape, this species is used primarily for its fresh fruits, but the grapes also make good juice and sweet wines of local importance in the southern United States (Plate 16.1). The species is extremely vigorous and disease tolerant compared to *V. vinifera* grapes and is well adapted to the southeastern United States. It lacks cold-hardiness, so is not grown in the mid-Atlantic, Northeast, or Midwest states. It has a different chromosome number (40 instead of the 38 in *Euvitis* species), which makes interbreeding it with *V. vinifera* or concord grapes difficult. Muscadines are not graft compatible with *Euvitis* either.

A few dozen cultivars of muscadines are cultivated in the southeastern United States. 'Scuppernong', one of the oldest, has bronze skin. Many bronze-skinned cultivars are called "scuppernong," and the name is often used as a synonym for all muscadines. Other skin colors exist, the most common being black (dark purple) and red. In nature, muscadine grapes are often dioecious, meaning that male and female flowers are found on separate vines. Two classes of

Introduction to Fruit Crops

doi:10.1300/5547_16

cultivars occur: (1) pistillate or female and (2) perfect flowered or hermaphroditic. Pistillate types are still grown with cross-pollination from perfect-flowered cultivars since many are of high quality. 'Cowart', 'Hunt', 'Noble', 'Jumbo', 'Nesbitt', and 'Southland' are popular black cultivars, and 'Carlos', 'Higgins', 'Fry', 'Dixieland', and 'Summit' are popular bronze-skinned cultivars other than 'Scuppernong'. There are no seedless cultivars of muscadine grapes.

2. *Vitis labrusca* L. (syn. *V. labruscana* Bailey): Known as the concord, American bunch, or fox grape, this species is used primarily for sweet grape juice and associated products, such as jellies, jams, and preserves (Plate 16.1). The most popular name brand in the United States is Welch's. The wines that are made from these grapes, particularly in New York, are generally sweet and fruity. "Cold Duck" is a sparkling wine made from concord grapes. In the United States, concord grapes are more important economically than are muscadine grapes, but far less important than those of *V. vinifera*.

'Concord' is the major cultivar, responsible for about 80 percent of production, hence the common name "concord" to represent the entire species. Other important cultivars include 'Niagara', 'Isabella', 'Delaware', and 'Catawba'. There are dark blue, red, and white skin colors available, with 'Concord' being dark blue. Several seedless cultivars are available for all skin colors. Some cultivars are interspecific hybrids with other native American grapes. Several cultivars, termed "eastern seedless table grapes," are derived from 'Thompson Seedless' (male parent) × *V. labrusca* or a French-American hybrid cultivar.

3. *Vitis vinifera* L.: Commonly known as the European or wine grape, *V. vinifera* is also called "Old World grape" because most production occurs in Europe and this name provides a contrast for New World natives *V. labrusca* and *V. rotundifolia* (Plate 16.1). *Vitis vinifera* and its hybrids account for over 90 percent of world grape production. Most of the production is used to make wine, but it is also the primary species used for table (fresh eating) and raisin grape production. White grape juice concentrate from *V. vinifera* grapes (mostly 'Thompson Seedless') finds it way into several juice blends and jellies.

There are at least 5,000 cultivars of *V. vinifera* grapes grown worldwide, and some estimates put the number of known cultivars as high as 14,000. However, less than 100 make up the vast majority of production. The most popular white wine cultivar is 'Chardonnay', and major reds include 'Cabernet Sauvignon', 'Merlot', and 'Pinot Noir'. As their names suggest, many important wine grape cultivars were developed in France, and major wines take their names from the principal cultivar used to make them. 'Thompson Seedless' is the most common cultivar used for table grapes and raisins because it is seedless and productive. A discussion of *V. vinifera* cultivars is well beyond the scope of this text, but a listing of major cultivars used for wine is given in Table 16.1, and detailed cultivar descriptions can be found in several of the references listed for this chapter.

4. *French-American hybrids:* The introduction of the phylloxera or grape root louse to Europe in 1860 devastated vineyards composed of the susceptible *V. vinifera* species and created a need for resistant rootstocks. *Vitis labrusca* and other species native to the host range of the phylloxera (north-central United States) were hybridized with *V. vinifera* to produce a range of rootstocks with resistance. In addition to their use as rootstocks, some of the hybrids showed both phylloxera resistance and good wine quality attributes. Cultivars such as 'Marechal Foch', 'Vidal Blanc', 'Chambourcin', and 'Seyval' make good wine and allow wine grape growing in areas where pure *V. vinifera* grapes do not perform well, such as in the eastern United States. French-American hybrids are more fruitful than *V. vinifera* grapes, often requiring cluster thinning to obtain proper quality. Also, they have a propensity to produce higher yields than *V. vinifera* grapes from secondary shoots if the primary shoots are killed by frost; thus, they are regarded as being more frost tolerant.

TABLE 16.1. Important wine cultivars from major wine grape countries of the world.

Country	Reds	Whites
France	Cabernet Sauvignon, Merlot (Bordeaux is a mix of these two), Pinot Noir, Syrah, Cabernet Franc, Gamay, Grenache	Chardonnay (white burgundy), Semillon, Sauvignon Blanc, Chenin Blanc, Aligote, Viognier
Italy	Sangiovese, Nebbiolo, Canaiolo, Vernatsch (= Schiava) Barbera, Lagrein, Pinot Nero (= Pinot Noir), Aglianico	Trebbiano (= Ugni Blanc), Malvesia, Chardonnay, Vernaccia
Germany	Spatburgunder (= Pinot Noir), Portugieser	Riesling, Silvaner, Muller-Thurgau, Gewurztraminer
United States (California)	Zinfandel, Cabernet Sauvignon, Merlot, Petite Sirah (= Duriff), Pinot Noir	Chardonnay, Sauvignon Blanc, Riesling, Gewurztraminer, Chenin Blanc, Colombard
Spain	Airen, Garnacha Tinta (= Grenache), Tempranillo, Bobal, Monastrell	Macabeo (= Viura), Garnacha Blanca
Australia	Shiraz (= Syrah), Cabernet Sauvignon, Merlot, Pinot Noir, Malbec	Chardonnay, Sauvignon Blanc, Semillon, Rhine Riesling (= Riesling)

Grape species and cultivars are also classified by food usage:

1. *Table grapes:* These are consumed as fresh fruits. Cultivars have an attractive appearance and are generally seedless. Taste is said to be secondary, and good flavor may not be as important as production output, shipping tolerance, and shelf life. 'Thompson Seedless', also called 'Sultanina' or 'Sultana' in Australia, is a major cultivar of table grapes. Other seedless cultivars include 'Flame Seedless' and 'Ruby Seedless' (both red), as well as 'Perlette', another white cultivar similar to 'Thompson Seedless'. Major seeded cultivars in the United States include 'Emperor', 'Ribier', and 'Calmeria'. In Italy, 'Italia' is a major white table grape, and in Spain, 'Almeria'. 'Tokay' is a seeded, red, Algerian cultivar that was once important in California but was later supplanted by 'Thompson Seedless'.

Table grapes can include any of the three major grape species or hybrids, but *V. vinifera* is by far the most important species worldwide. Several cultivars of *V. labrusca* and *V. rotundifolia* are also consumed as table grapes.

2. *Raisin grapes:* These are seedless cultivars that obtain a soft texture and pleasing flavor upon drying (Plate 16.2). 'Thompson Seedless', the major cultivar worldwide, makes up 90 percent of raisin production in the United States. 'Black Corinth' (syn. 'Zante Currant' or just 'Currant'—not to be confused with *Ribes* spp.) and 'Muscat of Alexandria' are important in Europe.

3. *Sweet juice grapes:* Traditionally, this classification was primarily dominated by concord grapes. In addition to juice, jellies, jams, preserves, and some wines are produced from sweet juice grapes. Recently, white grape juice concentrate, from 'Thompson Seedless' and other *V. vinifera* cultivars, has been used increasingly in blends with many other fruit juices and beverages.

4. *Wine grapes:* Some wine is produced from all grape species (and many other fruits), but the bulk of commercial wine production is dominated by *V. vinifera* cultivars. Several French-American hybrids also produce good-quality wine.

Wine cultivars vary by country and region within a given country (Table 16.1). Adaptation and climatic requirements dictate which cultivars can be grown, and winemaking consortia or governments may place further restrictions on what cultivars can be used for particular wines; for example, red Burgundy from France must be 100 percent 'Pinot Noir'. Although other cultivars may grow well in the Burgundy region of France, higher prices received for wines labeled "Burgundy" favor cultivation of 'Pinot Noir'. The same is true in the Chianti region of Italy—by decree, the main cultivar used for Chianti Classico must be 'Sangiovese'.

ORIGIN, HISTORY OF CULTIVATION

Vitis vinifera is thought to be native to the area near the Caspian Sea, in southwestern Asia, the same region to which apple, cherry, pear, and many other fruit crops are native. Grape was one of the first fruit crops to be domesticated for at least three reasons: First, it is native to the region where agriculture had its origin—the Fertile Crescent. Early farmers would have gathered grapes from the wild and had thousands of years over which to select and improve them. Second, grapes lent themselves to a variety of uses; they stored reasonably well, so they could be eaten fresh over fairly long periods, with the remainder being dried (raisins) or made into wine, which would keep until the following season at least. Third, grapes are easily propagated by cuttings, allowing superior selections to be easily cloned. Seed propagation would at least yield edible grapes , as they are self-pollinating and fairly uniform, unlike other fruit species that are cross-pollinating and too variable from seed. Although support is required for the vines, this was probably not an impediment to cultivation, since vines would grow on other trees, fences, or dwellings without much training.

Seeds of grapes were found in excavated dwellings of the BronzeAge in south-central Europe (3500-1000 BC), indicating early movement beyond its native range. Egyptian hieroglyphics detail the culture of grapes and winemaking in 2440 BC. The Phoenicians carried wine cultivars to Greece, Rome, and southern France before 600 BC, and Romans spread the grape throughout Europe. Grapes moved to the Far East via traders from Persia and India. Grapes came to the New World with the early settlers along the East Coast, but they quickly died out or performed poorly, mainly due to the poor coldhardiness and weaker insect and disease resistance of *V. vinifera* types. Spanish missionaries brought *V. vinifera* grapes to California in the 1700s and found that they grew very well there. Today, U.S. wine production is dominated by California, although Washington, Oregon, New York, Pennsylvania, and Michigan also have significant commercial wine industries based on *V. vinifera* grapes or French-American hybrids.

Vitis labrusca is found growing wild from Maine to the South Carolina Piedmont, and as far west as the Tennessee mountains. Close relatives of *V. labrusca* were first seen by Viking explorers before Columbus's voyages. The Vikings named the maritime provinces of Canada "Vinland," meaning "grape land," due to the abundance of wild grapes growing in the forests. Early settlers to the Northeast domesticated this species after *V. vinifera* grapes failed to grow, probably due to cold injury. The species became useful as a rootstock and in breeding for phylloxera resistance in the middle of the nineteenth century, when it was carried to Europe. Today, most concord grapes are grown in New York and the surrounding states.

Vitis rotundifolia is native in regions from Virginia, south through central Florida, and west to eastern Texas. This species has been enjoyed by southerners since antebellum times, and has received little attention outside of the southeastern United States. Several thousand acres are cultivated in the southeastern states, mostly in Georgia, Alabama, and Mississippi.

FOLKLORE, MEDICINAL PROPERTIES, NONFOOD USAGE

Grapes have been used to make wine since prehistoric times. It is not clear how or when wine was discovered, but it was likely the result of an accident in grape storage. Yeasts occur naturally on the skin of grapes, and placing grapes into a vessel that excluded air but allowed CO_2 to escape (a clay pot with a lid) created all conditions necessary for their fermentation into wine. Wine can be made from virtually any fruit juice, but since the *V. vinifera* grape was among the earliest fruit crops to be domesticated, wine became most closely associated with grapes.

Wine became the beverage of the aristocrats and upper class in ancient Europe. The Greco-Roman wine god was Dionysus-Bacchus; the "Oschophoria" (which means "carrying of the grapes") was his festival. Some believed the drunken state resulted from possession by the gods, allowing one to see the future. Others believed wine revealed the truth, as exemplified by the Roman slogan *in vino veritas* (in wine, the truth). Wine is symbolic of the blood of life in several religions. Grapes themselves are symbolic of charity.

Historically, "grape therapy" was used to treat cancer. About 15 pounds of grapes eaten *per day* for 6 weeks was thought to cure cancer. However, this could have caused more cancer than it cured, since grape seeds and skins contain tannins, which are carcinogenic in high quantities. The ashes of burned grape branches were once used as an abrasive dentifrice in England. The juice of *V. compressa*, a tropical grape relative, is used for general healing in tropical Asia.

In 1996, scientists discovered the chemical in grapes that shows anticarcinogenic activity—resveratrol. Originally identified in a tropical legume, this compound was found later in grapes (particularly red) and wine. It can inhibit tumor formation in three ways: stopping DNA damage, slowing/halting cell transformation from normal to cancerous, and slowing tumor growth. Resveratrol has anti-inflammatory properties and may be very useful for colon cancer prevention, and in the prevention of a wide variety of other tumors. Also, resveratrol may be important in reducing heart disease. Red wine consumption has been associated with lower LDL (low-density lipoprotein = "bad") and higher HDL (high-density lipoprotein = "good") forms of cholesterol, and resveratrol may be the active principle involved. Its use is still experimental, but it may become a useful drug in the future.

The ellagic acid that occurs in grapes may have a number of health effects on humans. It has anticancer properties and may act as a free radical scavenger. Strawberries also contain ellagic acid (see Chapter 29 for more on strawberries and ellagic acid).

PRODUCTION

World

Total production in 2004, according to FAO statistics, was 65,486,235 MT or 144 billion pounds. Grapes are grown commercially in 90 countries worldwide. Production has ranged from 55 to 65 million MT over the past decade, down from a peak of over 72 million MT per year in the early 1980s. In leading countries, there has been a trend to reduce or stabilize production, although other areas of the world continue to grow. Italy typically produced 11 to 12 million MT, and France 9 to 10 million, in the 1980s. In 2004, Italy produced 8.4 and France 7.8 million MT. With the emergence of the European Economic Community (EEC), pressure has been placed on growers to downsize by reducing area planted and setting lower yield maxima.

The top four producers have been constant for many years, but China has risen to fifth from being well outside the top ten just 20 years ago. As with many other crops, Chinese production of grapes has increased almost tenfold since 1980. Chile and Australia surpassed the perennial grape producer Germany, knocking that country out of the top ten only recently, with rapidly expanding wine and table grape industries in the Southern Hemisphere.

In contrast to Europe, production in the United States increased until the 1980s but then leveled off to current amounts. The United States enjoys one of the highest production efficiencies in the world—yields of 6 to 8 tons/acre, twice the world average. This reflects a greater proportion of acreage used for table grapes and raisins instead of wine grapes. In France and Italy, average yields are 4.4 and 3.8 tons/acre, respectively, due to emphasis on wine quality (low yields are associated with better-quality wine). Average yields worldwide are 3.8 tons/acre, increasing slightly with time, due to greater emphasis on intensive cultivation.

Grape is the second most extensively cultivated temperate fruit crop in the world after olive. Worldwide acreage equaled 18,955,000 in 2004, down from 1980 levels of 23 million acres. Over 60 percent of grape production area is in Europe. Acreage in Spain has fallen by 1 million acres since 1980, but Spain still holds the highest ranking in cultivated grape acreage in the world (3 million acres). The top ten grape-producing countries (percent of world production) follow:

1. Italy (13)
2. France (12)
3. Spain (10)
4. United States (8)
5. China (8)
6. Turkey (6)
7. Iran (4)
8. Argentina (4)
9. Australia (3)
10. Chile (3)

United States

Total production in 2004, according to USDA statistics, was 5,429,545 MT or 11.9 billion pounds. The grape industry is valued at $2.9 billion, making it the highest-value fruit crop in the United States. It is also cultivated on more land area than any other fruit crop, except pecan (*note:* pecan acreage data are inflated somewhat by the high proportion of wild-harvesting of native stands). There are 800,000 acres of grapes in production in California, and anywhere from 50,000 to 150,000 acres of nonbearing vines ready to come into production. The major grape-producing states in terms of production include the following:

1. California (90 percent)
2. Washington (4 percent)
3. New York (2 percent)
4. Michigan (1 percent)
5. Pennsylvania (1 percent)

In California, grapes are the second-place agricultural commodity in terms of value, surpassed only by milk production. In terms of tonnage produced, the utilization of California grapes is as follows: wines = 43 to 50 percent; raisins = 24 to 44 percent; table grapes = 10 to 12 percent.

The value of the industry has increased greatly over the past decade, most sharply through the 1990s, when prices received for wine grapes increased (Figure 16.1). Acreage also increased during this period. The value of raisins has declined substantially, largely due to lower prices, not lower production. Prices paid to grape growers average $0.16 to $0.24 per pound, but this varies greatly with intended market, demand, and production level in any given year. Table

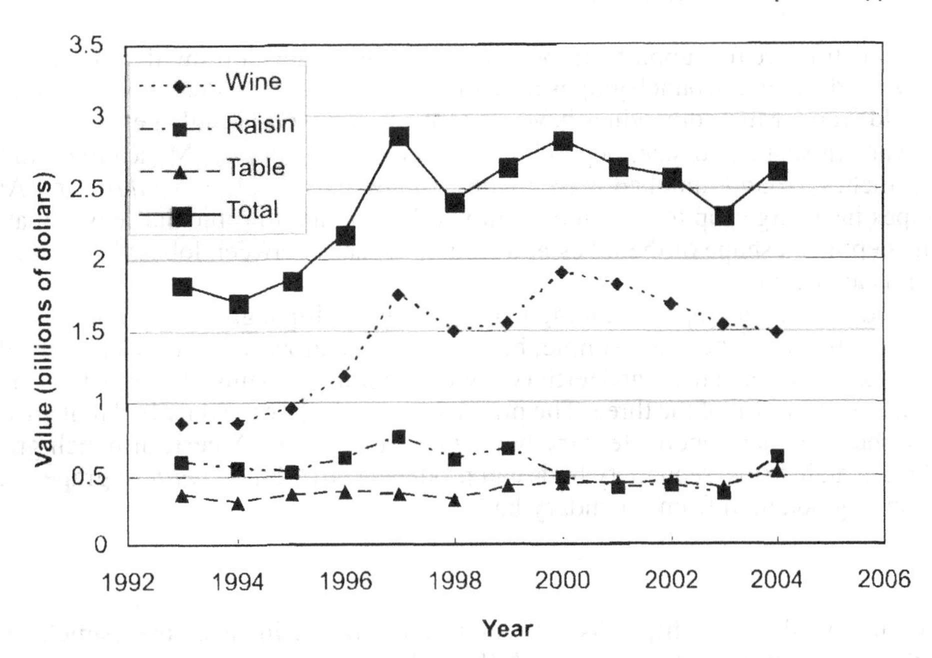

FIGURE 16.1. The value of the California grape industry over the period 1993-2004. Wine grape value largely dictates the total industry value, but it is generally less than 50 percent of California grape production.

grapes receive the highest prices of $0.25 to $0.36 per pound, wine grapes enjoy intermediate prices at $0.18 to $0.30 per pound, and raisins are priced the lowest at $0.08 to $0.30 per pound.

In 2002, the United States exported amounts of fresh grapes and raisins equal to 6 percent and 2 percent of domestic production, respectively. Imports were equivalent to 10 percent of fresh grape production and less than 1 percent of raisin production. Wine trade has increased greatly in the past decade. The United States imported 608,245 MT (~167 million gallons) and exported 329,330 MT (~91 million gallons) of wine in 2004. Imports have increased over twofold, and exports almost threefold, since 1994. The dollar value of wine imports in 2002 was $3.4 billion, up from just $1.1 billion in 1994. Export value was $610 million, fourfold greater than a decade ago. The top countries importing wine to the United States are France, Italy, and Australia (in order of value).

The United States produced 375,000 MT (825 million pounds) of *V. labrusca* in 2004, 85 percent of which was the 'Concord' variety. The industry value is about $80 million. Muscadine production is less certain but is estimated to be 6,000 to 7,000 MT based on data from Georgia, the largest producer of muscadine grapes in the country. Muscadine value is estimated at about $5 million to $10 million.

BOTANICAL DESCRIPTION

Plant

All *Vitis* species are "lianas," or woody, climbing vines. Unlike trees, they do not expend energy to make large, self-supporting trunks but instead use tendrils to attach themselves to other tall-growing plants. Their shoots can extend several feet per year because most of the plant's energy goes into growth in length, not girth. Tendrils occur opposite leaves at nodes and automatically begin to coil when they contact another object. Grapes are generally cultivated on a trellis, fence, or other structure for support, although it is possible to develop small, freestanding vines. *Vitis vinifera* and American bunch grapes have loose, flaky bark on older wood, but smooth bark on 1-year-old wood. Muscadine vines have smooth bark on wood of all ages.

Leaves vary in shape and size, depending on species and cultivar. Muscadine grapes have small (2-3 inches), round, unlobed leaves with dentate margins. *Vitis vinifera* and American bunch grapes have large (up to 8-10 inches in width), cordate to orbicular leaves that may be lobed. The depth and shape of the lobes and sinuses (spaces between lobes) varies by cultivar. Leaf margins are dentate.

Buds are compound in grapes, meaning that they have multiple growing points or meristems. In most other fruit crops, buds are simple, having only one growing point. Generally, there are three buds—primary, secondary, and tertiary—with the primary being the largest, most well developed, and most fruitful of the three. The primary bud is usually the only bud that grows, but if it is killed, the secondary and/or tertiary buds will grow out. In American bunch grapes and French-American hybrids, secondary buds can produce a crop, but *V. vinifera* grapes have very limited cropping potential from secondary buds.

Flower

Flowers are small (~⅛ inch), indiscrete, and green, borne in racemose panicles opposite leaves at the base of current season's growth (Plate 16.3). There are five each of sepals, petals, and stamens. Ovaries are superior and contain two locules, each with two ovules. The calyptra, or cap, is the corolla, in which the petals are fused at the apex; it abscises at the base of the flower and pops off at anthesis. Species in *Euvitis* may have 100+ flowers per cluster, whereas muscadine grapes have only 10 to 30. *Vitis vinifera* and concord grapes are perfect-flowered and self-fruitful, whereas some muscadine cultivars have only pistillate flowers. Grapes will flower in their second or third year after planting but will not produce a full crop for several years.

Pollination

Most grapes are self-fruitful and do not require pollinizers; however, pistillate muscadines (e.g., 'Fry', 'Higgins', 'Jumbo') must be interplanted with perfect-flowered cultivars for pollination. Since parthenocarpy does not exist, all grapes require pollination for fruit set. Even seedless cultivars, such as 'Thompson Seedless', are not parthenocarpic; rather, the embryos abort shortly after fertilization and fruit set. This condition is called "stenospermocarpy," which is biologically different from seedless fruit production in such crops as citrus, banana, or pineapple. Pollination is accomplished by wind and, to a lesser extent, insects.

Fruit

Grapes are true berries that are small (<1 inch)and round to oblong, with up to four seeds (Plate 16.1). Berries are often glaucous, having a fine layer of wax on the surface. Skin is generally thin and is the source of the anthocyanin compounds that give rise to the red, blue, purple, and black (dark purple) colors of grapes. Thus, dark-colored grapes, such as 'Zinfandel', can be made into a white or blush wine by limiting contact of the clear fruit juice with the colored skins. Green- and yellow-skinned cultivars are often termed "white" grapes. Muscadines differ from other types by having a thick skin that is sometimes bitter and tough. The fruits of muscadine grapes ripen one by one and detach from the plant at maturity. The berries detach from the vine, leaving a dry stem scar, unlike bunch grapes, which remain attached to the cluster at maturity. In bunch grapes, the small stem that holds the berry "plugs" the fruit when the berry is detached, yielding a wet stem scar. Fruits are borne in clusters, with two clusters per shoot in most cultivars, but up to five clusters per shoot in French-American hybrids.

Thinning is not practiced for most types; crop load is controlled through meticulous pruning (see Planting Design, Training, and Pruning). However, French-American hybrids may require cluster thinning for development of quality and proper vine vigor.

Fruit size and cluster length are increased through gibberellic acid (GA) application on 'Thompson Seedless' and other table grape cultivars. GA is applied at 10 to 15 ppm at 50 percent bloom, and again at a higher concentration 1 to 2 weeks later. This opens the cluster, prevents crushing of berries, and reduces disease.

GENERAL CULTURE

Soils and Climate

Grapes and their rootstocks are adapted to a wide variety of soil conditions, from high pH and salt to acidic and clayey. Deep, well-drained, light textured soils are best for wine grapes. Highly fertile soils are unsuited to high-quality wine production, since vigor and yield must be controlled. Irrigation can be detrimental to the internal quality of grapes if applied in excess and is sometimes *illegal* for wine grapes, though beneficial for table and raisin grapes, for which high yields are desirable. It is debatable whether soils per se affect wine quality; climate, cultivar, training and pruning, as well as irrigation and fertility may also be more important.

Vitis vinifera can be characterized as requiring Mediterranean climates, which are typical in their native range. This means warm, rainless summers, low humidity, and mild winter temperatures. Concord and muscadine grapes are better adapted to humid, temperate climates, whereas muscadines require longer growing seasons and milder winters than concord types.

Coldhardiness is a major limiting factor for *V. vinifera* grapes. Damage to primary buds occurs at 0 to −10°F, and trunks may be injured or killed below −10°F. Concord grapes are more cold hardy than those of *V. vinifera* or French-American hybrids but will experience some damage at −10 to −20°F. Muscadine grapes are the least cold hardy, being killed at temperatures below 0°F and injured in single-digit temperatures.

Winter chilling requirement is highly variable among grape species. Concord grapes generally have high chilling requirements of 1,000 to 1,400 hours. *Vitis vinifera* grapes have low chilling requirements of 100 to 500 hours and are frost prone in many regions. Muscadines have intermediate chilling requirements, but they do require several weeks of warm weather following chilling in order to initiate budbreak; spring frost is rarely a problem with muscadines.

The number of days from bloom to maturity required during the growing season ranges from under 150 to over 200, depending on species, cultivar, and climate. Maturation time increases as follows:

V. labrusca (least)
French-American hybrids
V. vinifera
V. rotundifolia (greatest)

Grape internal quality and hence wine quality are affected by summer temperature. Cool climates favor lower sugar and higher acid contents, whereas higher temperatures favor the opposite. Generally, longer growing seasons and warmer temperatures are required for red wine cultivars than for white wine cultivars. This is why red wines are produced in warm summer regions, such as Italy, southern France, and Spain, while white wines are produced in cooler regions, such as northern France and Germany. As always, there are exceptions, such as the red cultivar 'Pinot Noir', which prefers cooler microclimates than do most other red wine cultivars.

Concord grapes do not tolerate high temperatures as well as *V. vinifera* or muscadine grapes. Production is lower, and grapes ripen unevenly within a cluster in warm regions.

Humidity is another limiting factor for *V. vinifera* grape culture, due to disease susceptibility. *Vitis vinifera* grapes cannot tolerate high humidity or rain during harvest. Internal quality is reduced, fruits may crack when swollen with water, and fungal diseases are far more severe. Concord grapes are not as ill affected by high humidity and rainfall during the growing season as are *V. vinifera* grapes. The thick-skinned, disease-tolerant muscadines are well adapted to humid, rainy climates.

Propagation

The most common method of grape propagation is bench grafting, although rooted cuttings, T-budding, layering (for difficult-to-root types, such as muscadines), and, to a limited extent, tissue culture are used in various situations. The most common method of muscadine propagation is trench layering, which is done by specialized nurseries. Thus, muscadine vines are "own rooted" and have the advantage of coming back true from the roots if they are killed back during winter.

Bench grafting is unusual in that the rootstock is either bare rooted or unrooted at the time of grafting. In other fruit and nut crops, grafts are made on fully rooted rootstocks in the nursery. The ease with which grape rootstocks root from dormant hardwood cuttings allows the propagator to handle the plants in this manner. The basic steps in bench grafting are as follows:

1. Dormant scion and rootstock canes are collected in late winter/early spring and grafted immediately, or they are collected in late fall and stored in refrigeration for 1 to 2 months. Canes are cut to 12- to 14-inch lengths and sorted by diameter. The diameters of rootstock and scion cuttings should match. The rootstock cuttings are disbudded to prevent suckers from forming.
2. Grafts are usually made by machines that make accurate, tight-fitting, complementary cuts in stocks and scions. If done by hand, whip-and-tongue grafts are used. The scion is waxed by dipping in molten paraffin (and cooling in water immediately) down to the union to prevent dehydration.
3. Vines are allowed to callus and form roots for 3 to 4 weeks at 80°F in special rooms. Moist peat moss is packed around the rootstock portion of the graft.
4. Vines are then planted in nursery rows, or first into containers of soil-less media, grown in greenhouses for a short time, and then planted in the nursery.

Rootstocks

Vitis vinifera was propagated on its own roots from the time of its domestication until about the 1870s. The grape phylloxera (*Dactylosphaera vitifolii,* Homoptera), also called the grape root louse (actually an aphid), was introduced to Europe from eastern North America in the 1860s, where it caused the most significant pest-related disaster in all of fruit culture. The phylloxera nymphs feed on leaves after hatching, and then drop to the ground to burrow into the soil, where they destroy the root system. A winged form of the insect emerges at the end of the season and flies to other vines; thus, the spread of phylloxera is rapid and difficult to control. In 1868, phylloxera was implicated as the cause of the devastation of at least one-third of the vineyards in France and countless thousands of vineyards elsewhere, including California. The search for resistant rootstocks led horticulturists to the native range of the phylloxera, where various species of American grapes had coexisted with the pest for millennia and thus were resistant to it.

Initial selections from *V. riparia* ('Riparia Gloire') and *V. rupestris* ('Rupestris St. George') were exported to Europe for use as rootstocks with resistance to phylloxera, but, unfortunately, they had low tolerance for the high-pH soils of France. *Vitis berlandieri*, native to Texas, is tolerant of both high pH and phylloxera, but difficult to root from cuttings; hence, hybrids of *V. berlandieri* and other easy-to-root grapes were made, many of which are used today. Most grape rootstocks used today are numbered clonal selections of hybrids of *V. riparia*, *V. rupestris*, and *V. berlandieri* and other American grape species (Table 16.2). *Vitis champini* selections ('Dog Ridge', 'Salt Creek') are highly resistant to certain nematodes but lack sufficient resistance to phylloxera to be widely used.

TABLE 16.2. Characteristics of common grape rootstocks.

Rootstock	Vigor	Wet soil	Drought	Phylloxera	Nematodes	Lime/high-pH tolerance
Riparia Gloire	2	3	1	5	2	poor
Rupestris St. George	4	1	2	4	2	medium
420 A	2	2	2	4	2	good
5 BB	2	3	1	4	3	good
SO 4	2	3	3	4	4	medium
5 C	3	—	1	4	4	medium
110 R	3	3	4	2	4	medium
99 R	4	1	3	4	3	medium
140 Ru	4	2	4	4	3	good
1103 P	3	3	3	4	2	medium
3309 C	2	3	2	4	1	poor
3306 C	3	3	1	4	1	poor
101-14	2	3	1	4	2	poor
44-53 M	3	3	4	4	4	poor
1616 C	3	2	1	3	1	poor
1202 C	3	—	2	2	1	medium
AXR #1	3	—	2	2	1	medium
41 B	2	1	3	4	1	very good
333 EM	1	1	2	2	1	very good
1613 C	3	2	2	2	4	poor
Dog Ridge	4	2	2	2	4	good
Salt Creek	4	2	2	2	4	good
Harmony	3	—	2	2	4	—
Freedom	3	—	2	2	4	—

Source: Adapted from Howell, 1987, and Ahmedullah and Himelrick, 1990.

Note: 1 = low resistance, poor vigor, or very susceptible to a given problem; 5 = high resistance, high vigor, or very tolerant of a given problem.

Planting Design, Training, Pruning

Most grapes are trellised and grown in long, narrow rows, spaced about 9 to 15 feet apart, depending on training system. Typically, there is 3 to 8 feet between vines in a row. Muscadines use similar trellises and row spacings but are planted 20 feet apart in a row because they are extremely vigorous (Figure 16.2).

Training Systems

Training refers to the permanent parts of the vine, not the 1-year-old wood. Two basic forms are used: head and cordon. In head training, the permanent part of the vine consists of the trunk and some fattened stubs or a bulbous "head" at the top (Figure 16.3). Spurs or canes develop directly from the head. In cordon training, the permanent part of the vine consists of the trunk and

FIGURE 16.2. Examples of the many different vineyard arrangements for grapes. *(left)* 'Thompson Seedless' vines are closely spaced, head trained, and grown on a T-bar trellis. *(right)* The commonly used bilateral cordon system employing three wires, the lower one for the cordons and the upper two as training wires for new growth.

FIGURE 16.3. Two forms of head training in grapes. *(left)* Trunks are trained to a stake, and a single cane of 1-year-old wood is trained along a wire for fruiting. Note the renewal spurs left on the head to produce next year's cane. *(above)* The trunk is staked for support, but no trellis is used to train shoots.

one to four long, straight arms or cordons trained along a wire. Spurs or canes are spaced at regular intervals along cordons. There are two basic cordon systems:

1. *Bilateral (single-curtain) cordon.* The most popular and simplest cordon system is the bilateral or "single curtain" cordon. This trellis consists of two or three wires, placed one above the other, with the top wire at 5 to 7 feet (Figure 16.2, right). Shoots grow vertically along the wires, and a single "curtain" of foliage is produced along the trellis. The vine has two cordons running in opposite directions from the trunk along the wire. This is a popular system for wine grape production.
2. *Geneva double-curtain cordon.* Four arms are developed from a single trunk and trained along two parallel wires, placed 5 to 6 feet above the ground. This system produces high yields per acre due to the large bearing surface, but it requires special harvesting equipment and shoot positioning and is more prone to disease development than are the single-wire systems. It is used for certain wines and juice/table grape production, due to the high yields obtained.

Grape Pruning Basics

Grapes are pruned more severely and methodically than any other fruit crop. Pruning not only controls vine growth but also sets crop load. Without severe, annual pruning, grape vines become tangled masses of unproductive shoots that decline in yield and quality very quickly. A considerable amount of research and time-honored practice has been devoted to vine pruning, and the following is a condensation of a considerable amount of literature. A few important points follow:

- Pruning is done in the late winter when vines are dormant.
- Pruning involves *only* last year's growth, or 1-year-old shoots. Vine form is accomplished by training.
- *Balanced pruning* is a method of determining the severity of pruning or amount of 1-year-old wood to remove. One "balances" next year's crop load with last year's vigor. This method eliminates the need for fruit thinning in most cases.
- Placement of cuts involves the choice of *spur* versus *cane* lengths of pruned shoots, with the former having a shorter stub than the latter.

Another important concept is that each bud on 1-year-old wood is basically a compressed, dormant shoot waiting to grow out and produce fruits. Each shoot produces about two fruit clusters. Thus, one can predict fruit load based on a bud count made after pruning. Leaving fewer buds means that the vine will have fewer grape clusters containing higher-quality grapes; many buds means more clusters of low-quality grapes. The best wines are made from high-quality grapes, whereas internal quality is less important than yield for table and raisin grapes.

Balanced Pruning

Early research showed that the amount of buds to be left after pruning could be determined by a simple formula that employs the weight of shoots pruned. To start, the pruner removes about 90 percent of the previous season's growth and then bundles and weighs it (or estimates its weight). For a given grape type, the following formulas are applied (*note:* muscadine grapes are not balance pruned):

	Vitis labrusca	French-American Hybrids	*Vitis vinifera*
Number of buds retained for first pound of prunings	20-40	20	20
Number of buds retained for each additional pound of prunings	10	10	20
Maximum number of buds regardless of pruning weight	50-70	40-50	40

Spur versus Cane Pruning

Once the number of buds per vine is calculated from the formula, one must decide whether to leave the requisite number on several short stubs or on just a few longer shoots. Short stubs with few buds are referred to as spurs, whereas longer pieces of 1-year-old wood are referred to as canes (Figure 16.4). In general, spur pruning is for use on

- cultivars with fruitful basal buds (i.e., the buds at the base of 1-year-old wood will produce fruit clusters instead of only leaves);
- cultivars that are excessively vigorous; and
- wine cultivars for which quality is more important than quantity.

Cane pruning is for use on

- cultivars with unfruitful basal buds (e.g., 'Thompson Seedless' will not produce clusters on shoots that arise at the base of last year's wood) and
- cultivars that lack vigor or are low yielding.

FIGURE 16.4. Cane and spur pruning. *(left)* A small head-trained vine has been cane pruned, leaving one shoot with several buds. The cane will be bent downward at the tip into the "Tuscan bow" system. Note the renewal spur on the left side of the head for next season's growth. *(above)* A close-up of an arm that has been spur pruned, leaving stubs with only two to three buds.

For example, a mature 'Sangiovese' vine, a *V. vinifera* grape used for wine, is to be spur pruned (pictured in Figure 16.4, right). All of the shoots that grew last year are pruned to a length of about 12 inches, and the prunings are judged to weigh about 2 pounds. Since *V. vinifera* vines are pruned on a "20 + 20" basis, this dictates leaving about 40 buds on the vine. If 2 buds per spur are left, then about 20 spurs would be spaced evenly along the permanent wood of the vine, or if 4 buds, then 10 spurs, and so on.

Grape growers make their own modifications of these rules of thumb; most common are these:

- Small buds within ½ inch of the base of the cane are not counted.
- Spurs are spaced at least 9 to 12 inches apart on the arms of a vine.
- More buds are left on thicker (½-inch) canes or spurs, less on thin (¼-inch) ones.
- For cane pruning, two to six shoots are retained per vine, each with several buds.
- A "renewal spur" is retained for each cane when cane pruning. The renewal spur will produce a shoot to be used as next year's cane.

Additional summer pruning is needed only with overly vigorous cultivars that tend to produce many shoots which shade the fruits and other canes of the vine. Undesirable shoots arising from the trunk are also removed in summer. Leaves from around the fruit clusters are often removed to allow color and flavor enhancement (direct sun on the fruits often increases wine quality), and to reduce bunch rot in humid climates.

Muscadine Training and Pruning

All muscadines are spur pruned because they are highly vigorous and have fruitful basal buds. Spurs are 4 to 6 inches long, containing one to four buds. Muscadines are trained to one of the two cordon systems listed earlier. Spurs are spaced 6 inches apart on cordons, with complete removal of the weakest, thinnest shoots if overcrowding occurs (Figure 16.5). Spurs become spur clusters in older vines because, in muscadines, new canes do not grow easily from the older wood of the cordons. Some spur clusters will need to be thinned completely in crowded parts of the vine, allowing adjacent spurs to grow into the bare space.

FIGURE 16.5. Muscadine grapes are most commonly grown as cordon-trained, spur-pruned vines. *(left)* A vine has developed a trunk and two permanent arms in its first 2 years; it has been spur pruned for its first crop in the coming year. *(right)* A mature muscadine vine.

Pest Problems

Insects

Grape phylloxera or root louse. Dactylosphaera vitifolii is by far the most destructive pest of grapes where resistant rootstocks are unavailable. Ironically, introduction of American grapes to Europe in the mid-1800s as a source of powdery mildew resistance led to the introduction of phylloxera to France. It destroyed at least one-third of all vineyards in France before it was described and control measures were developed. Signs of the pest are most visible as leaf galls, containing eggs of the insect, although this is the least debilitating to vines. The nymphs feed on leaves during the summer, then drop to the ground, enter the soil, and feed on roots, causing eventual vine collapse. Root galls or "nodosities" and "tuberosities" form on affected plants. A winged form of the insect that is produced near the end of the season flies to other vines and starts the cycle over. Phylloxera requires soils with high clay content that crack upon drying, which allows easy entry to the root zone. The grape root borer *(Vitacea polistiformis)* causes similar debilitation or death of vines through larval feeding on large roots and trunks; this has been a particular problem in the southeastern United States. Control is achieved through the use of resistant rootstocks, derived from American species. The most widely used phylloxera-resistant rootstocks are 'Rupestris St. George', 99R, and 1202. Sandy soils produce fewer and less severe phylloxera infestations.

Grape leafhopper. This insect *(Erythroneura elegantula)* is the most damaging to grapes in California, and often the only insect that requires regular control measures. These leafhoppers are about ⅛ inch long and yellow, with red and brown markings. The nymphs are semitransparent, have red eyes, and run sideways when disturbed. Nymphs feed on leaves, causing defoliation, sunburn injury to fruits, and excrement deposition on fruits. They fly when disturbed, sometimes into the eyes, nose, or mouth of the person picking the fruits. Insecticides are needed for control if more than about ten nymphs per leaf are found when scouting. This insect is not a problem in the eastern United States.

Mites Several mites (*Tetranychus, Calepitremeris, Eriophyes* spp.) can attack grape, feeding on mature or expanding leaves, causing a characteristic "mite stipple" or bronzed appearance. In colored grapes, leaves may turn red as well. Grape bud mites damage overwintering or emerging buds, causing stunting and distortion. A dormant oil spray before budbreak (combined with sulfur or lime sulfur) reduces overwintering populations. In hot, dry summers, miticides may be required.

Grape root borer. Vitacea polistiformis, the larva of a clear-winged moth, is about 1 inch in length and dark brown/yellowish orange in color. Eggs are laid on leaves in summer, and then the larvae hatch, drop to the soil, and bore into the crown or large roots, where they may remain for up to 3 years. Adults emerge from the soil in June or July to repeat the cycle. All grapes, even muscadines, are susceptible. This pest is difficult to control with insecticides because it is within the plant or beneath the soil much of the time. Effective control is achieved by mounding soil around the trunk in mid-June to prevent emergence of adults and/or larvae from finding the crown or roots. The mounds are removed in late November.

Grape berry moth. This insect *(Polychrosis viteana)* is found only east of the Rocky Mountains. Larvae feed on blossoms and developing berries, killing the berries and webbing the cluster where they feed. The larvae are ⅜ inch long, dark green/purplish, with a light brown head. Insecticides can be applied once webbing or larvae are noticed. Since the cocoons are laid on the soil surface, one can till the soil in spring before bloom, leaving a 2- to 3-inch layer of tilled soil over the soil surface; this prevents the emergence of the adults from the ground. Pheromone mating disruption is also available as a control measure for this pest.

Grape leaf roller or leaf folder. These moths *(Desmia funeralis)* lay eggs in April to May that hatch into larvae which web leaves together and then feed between them. They may also affect fruits. As they grow, they spread out and roll the edge of a leaf under to feed on it, causing defoliation. The larvae are bright green with a brown head, about 1 inch long, and they wiggle violently when disturbed. The omnivorous leaf roller *(Platynoda stultana)* is a similar pest. Pesticides can be applied before the larvae form the leaf rolls, in about mid-May in California. The insecticide *Bacillus thuringiensis,* or Bt, is a natural spray material used to control leaf rollers. These moths are not a problem in muscadine grapes.

Orange tortrix. Argyrotaenia citrana is a ½-inch-long caterpillar that feeds on leaves and fruits, predisposing fruits to fungal decay or causing parts of clusters to shrivel. They produce webbing on leaves and in fruit clusters. Caterpillars are green to yellow with a gold head, and they wiggle when disturbed. Beneficial insects provide some control, but insecticides are needed in many cases.

Diseases

Powdery mildew or oidium. Uncinula necator is responsible for this serious disease of grapes, found in all countries where grapes are grown, but particularly in drier climates. The organism is unusually susceptible to rainfall, which removes conidia, prevents sporulation, and disturbs hyphae. The disease affects all green tissues of the plant, including fruits, until they begin to accumulate sugars in late summer. The fungus overwinters inside buds and covers emerging tissues, upper and lower leaf surfaces, with masses of powdery-looking hyphae; yield and vine vigor are seriously reduced if no control measures are used. Infected tissues have a whitish, "powdery" appearance, and interior shoots and clusters are usually the first infected. The disease causes off-flavors in wine and is worst during the cooler weather of spring. The disease was introduced to Europe via England in the mid-1800s, where it caused losses up to 80 percent throughout France. Mr. Tucker, an Englishman who first noted mildew in England (*Oidium tuckeri* was the first scientific name), also discovered that a mixture of lime and sulfur controlled the disease quite well. To this day, sulfur is still the major control measure, with dust applications preferred in drier climates; other fungicides are used as well. Good canopy ventilation and light exposure help reduce the incidence of the disease. Leaves near fruit clusters are removed, and summer pruning and/or shoot positioning is done, so that light and air travel through the canopy and grape clusters. Fungicides are often needed, starting just after budbreak and applied at 10- to 14-day intervals.

Downy mildew. Plasmopara viticola is the organism responsible for downy mildew. This type of mildew is less of a problem in grape cultivation than is powdery mildew because it occurs in warm, humid regions, and the American grape species grown in these climates are less susceptible. The fungus attacks all green parts of the vine, as does powdery mildew, but signs of the disease appear only on lower leaf surfaces; lesions are yellowish and oily, and spore masses appear as delicate, white, cottony patches. Leaves may drop if vines are severely infected, thus reducing yield and quality. A "shepherd's crook" appearance of shoot tips is characteristic of the disease. Portions of the fruit cluster may drop due to rachis infection, and berries are susceptible until they approach maturity. Control is mostly by fungicides applied during rapid shoot growth and repeated at 10- to 14-day intervals. Pruning out infected shoots and selecting soils with good winter drainage (since fungus overwinters in upper layers of moist soil) can be beneficial. Good canopy ventilation and light exposure help reduce the incidence of the disease.

Botrytis bunch rot/"noble rot." Botrytis cineria is a problem in damp, cool climates, near harvest, but it can infest developing shoots and inflorescences. It also causes postharvest losses in table grapes. Infected berries turn off-colors and sometimes are covered with brown or white mycelia. Rotted berries occur in groups within the clusters. Cultivars with tight clusters and thin-skinned berries are most susceptible. As many as four sprays may be required for successful control: the first at fruit set, the second at berry touch, the third at veraison (color change), and the fourth 3 weeks prior to harvest. Canopy ventilation will improve air flow and light in the leaf canopy, reducing disease incidence. This involves leaf removal from around the clusters and some shoot pruning, if canopies are unusually thick.

This disease has been exploited in the making of very sweet white wines, such as French Sauternes, Hungarian Tokays, and the sweet German wines Auslese, Beerenauslese, and Trockenbeerenauslese. The disease is allowed to develop on clusters, which causes partial berry dehydration and greatly increases sugar concentration, while sacrificing total yield. When fermentation ceases, much residual sugar is left over, making the wines sweet. The wines are typically very expensive, since only small quantities are made each year.

Phomopsis cane and leaf spot. The fungus *Phomopsis viticola* generally infects canes near their bases. Lesions are brown/black and elongated, and growth of the cane is stunted. On leaves, small yellow spots with dark centers may occur, with a "shot hole" appearance once the dead central tissue drops out. The fungus may cause some berry rot in late summer or early fall when temperatures decline. The lesions that harbor the fungus can be seen when pruning in winter; wood with dark, sunken lesions on the lower cane is pruned out. Fungicides can be applied, beginning when shoots are just emerging and ending when temperatures are warm.

Eutypa dieback. Eutypa lata is the fungus that causes this disease of older wood, which is generally a problem only in vines 8 years or older. The fungus enters through pruning cuts that have been made in 2-year-old or older wood. The infected shoot will have small, cupped, and often spotted leaves; it may be surrounded by healthy shoots and thus is most easily seen in spring before a complete canopy forms. Pruning out infected shoots and surrounding older wood reduces diseases progress. Pruning late in winter, just before budbreak, allows cuts to heal faster, reducing the possibility of infection. Using two trunks or multiple arms on a vine in areas where this disease occurs frequently reduces the impact of severe pruning on yield.

Black rot. Mostly a disease of developing fruits, black rot is caused by *Guignardia bidwellii* and is very severe in the eastern United States. Infected fruits shrivel, turn black, and remain attached to the clusters. Small, brown spots may form on leaves as well. The raisinlike, mummified fruits should be removed and destroyed because the pathogen overwinters in them. Good canopy ventilation and light exposure help reduce the incidence of the disease. Leaves near fruit clusters are removed, and summer pruning and/or shoot positioning is done, so that light and air can travel through the canopy and grape clusters. Fungicides are almost always needed, starting just after budbreak and applied at 10- to 14-day intervals. The most important sprays are those applied between budbreak and bloom. Sprays for this disease will control virtually all other fungal diseases as well.

Late-season fruit rot. At least three fungi cause rot of fruits in the final 6 to 8 weeks before harvest: bitter rot *(Greeneria uvicola)*, macrophoma rot *(Botryosphaeria dothidea)*, and ripe rot *(Colletotrichum gleosporoides)* are occasional problems. Fungicides are applied at 10- to 14-day intervals, beginning in midsummer, if rot occurs.

Pierce's disease. Xylella fastidiosa is behind this disease, a problem of only regional importance in the Gulf states of America and isolated "hot spots" in California. It is a major factor limiting production of *V. vinifera* and *V. labrusca* in the southeastern United States, and it is a potential threat to the California grape industry. It is caused by a xylem-limited, fastidious mycoplasma (small bacterium with several layers of extracellular fibrous strands). Clusters of organisms, along with plant gums, gradually plug the xylem and cause drought stress symptoms on foliage (marginal scorch) and eventual decline of vines. Vines may die within the first 12 months of infection, or they could live 5+ years, depending on the age of the vine when infected and climate. In warmer areas, vines die sooner. The organism is vectored by sharpshooter leafhoppers, and it has a very broad host range. Hence, control is almost impossible where populations are endemic in native vegetation. The organism is confined to mild winter areas and apparently lacks tolerance to freezing temperatures below about 15°F. Muscadines are resistant, but *V. vinifera* and concord grapes are susceptible. There are no effective controls, other than avoiding areas where the disease is present and removing infected vines immediately. Antibiotic treatments are ineffective, as are insecticides targeted at sharpshooter leafhoppers.

HARVEST, POSTHARVEST HANDLING

Maturity

Table Grapes

Time of harvest is determined mostly by appearance, including the color and size of berries. The stems of the cluster also turn a wood or straw color when berries are mature. Grapes are classified as nonclimacteric fruits with respect to ripening.

Raisins

Since drying ratio should be minimized for maximum yields and quality, it is best to allow grapes to ripen fully on the vines before picking. Water content decreases and sugar content (Brix) increases during maturation, so waiting as long as possible maximizes yield. 'Thompson Seedless', the major raisin cultivar, reaches 22 to 23 °Brix at full maturity, and this is easily determined with a refractometer. In the central San Joaquin Valley of California, damaging rains can be expected in late September, so time of harvest is a compromise between maximizing Brix and avoiding damaging rains.

Wine Grapes

Harvest criteria, and hence date, depend on the type of wine to be made. For any given type, sugar content (Brix) is perhaps the most important parameter, although pH is also very important. The pH is more difficult to adjust than sugar in the winery, and wines with a pH above 3.6 are potentially unstable. Sugar-acid ratio and total acidity are also important.

Generally (California standards), white wine grapes are harvested at 19 to 23 °Brix, ≥ 0.7 percent acidity, and pH ≤ 3.3. Criteria for red wine grapes are 20.5 to 23.5 °Brix, ≥ 0.65 percent acidity, and pH ≤ 3.4. A random sample of a few hundred berries from different vines is often taken for lab quality evaluations, over intervals of a few days, as harvest approaches. Government agencies in Europe often set rigid guidelines for assessing maturity and actually perform independent quality assessments.

Sweet Juice Grapes

The most important parameter is sugar content; grapes must reach a minimum of 15 °Brix for processing.

Harvest Method

Table Grapes

Individual clusters are judged for maturity by pickers, clipped from vines with minimal handling, and placed in plastic lugs. Vines are harvested two to three times over a period of several weeks, as grapes ripen rather slowly and hold their condition on the vine for weeks following ripeness.

Raisin, Wine, and Sweet Juice Grapes

Mechanical harvest is the norm for all processed grapes, except for high-quality wine grapes. Wine grape clusters are clipped by hand, and special wines are made from handpicked berries from particular areas of a cluster. Mechanical harvesters generally have horizontally pivoting arms that smack the canopy or vibrating fingers that dislodge clusters and berries from clusters. Machines harvest approximately 5 tons/hour, the equivalent of the output of about 100 handpickers.

Raisin grapes are often harvested 4 to 8 days after pruning canes, which causes berries to abscise more easily from the clusters. Canes are pruned but left in place and then harvested mechanically when berries are loose. This allows a neat, single layer of destemmed berries for harvest onto continuous paper trays.

Muscadines

Time of harvest is determined mostly by appearance, including the color and size of berries. Muscadines ripen unevenly compared to bunch grapes, often over a period of weeks, so individual berries are harvested instead of clusters. Picking the largest, best-colored fruits twice a week is commonly practiced.

Postharvest Handling

Table Grapes

Packing and shipping occurs immediately after harvest when possible, but some storage may be necessary when production volume exceeds demand. Grapes are precooled in forced-air rooms and fumigated with sulfur dioxide (SO_2) to reduce fungal decay.

Raisins

Prior to picking or mechanical harvest, the soil between rows is smoothed to accommodate trays or papers used for field drying. Grapes are picked and placed onto drying papers in single layers or in small clusters. The top layer browns and shrivels in 7 to 10 days, and the berries are then turned. When they are dry enough (13-15 percent moisture) that juice cannot be squeezed out when they are pressed, they are ready for curing. Machines collect raisins from continuous paper trays, and fruits are placed in storage. During the first few weeks of storage, water is transferred from wetter to drier berries and moisture content equalizes within the lot; this is referred to as "sweating." Raisins are then boxed for sale.

Wine

The science of winemaking is called enology. Entire courses and curricula are available at some universities for specializing in this area. A detailed explanation of winemaking is beyond the scope of this text, but the basic processes are these (Plate 16.4):

1. Harvested grapes are destemmed and crushed, and the "must" is extracted ("must" = juice + skins + seeds ± stems for red; juice only for white). Initial quality evaluations are made

and the vinification process is decided upon. Adjustments of sugar level and acid content can be done at this point.

2. SO_2 is added to prevent oxidation and color deterioration, selectively activate certain yeasts for fermentation, and kill bacteria and other undesirable microbes.
3. Must is transferred to fermentation vats for sugar-alcohol conversion. Red wine is produced by partial fermentation with skins where red pigments reside. Rosè is produced by limited contact of must with skins. White wine is produced through no contact with skins. Fermentation temperature is higher for reds (74-82°F) than for whites (58-68°F). Fermentation is vigorous initially but slows as sugars are used up and alcohol reaches levels toxic to the yeast. Fermentation is stopped by racking or dispensing wine into containers; sometimes SO_2 is used to stop the process and to act as a preservative.
4. Wines are filtered and then may be aged in oak barrels for some time prior to bottling. Time limits are imposed by wine appellations or individual enologists. White wine is generally aged for only short periods of time (<1 year), whereas some reds may be aged for up to 10 years.
5. Bottling takes place at the appropriate age for the wine, after ultrafiltration for clarity and antisepsis.

Storage

Fresh *V. vinifera* grapes may be stored for relatively long periods of time at around 31°F and 95 percent relative humidity (3-6 months). This is unusually long for a small-fruit crop, most of which are generally perishable within only days or weeks of harvest. *Vitis labrusca* grapes are more perishable, lasting only 2 to 8 weeks under similar conditions. Muscadine grapes can be stored for up to 3 weeks at 32°F and 90 percent relative humidity. Raisins may be stored for several months at room temperature due to low moisture content. Wine quality generally improves with storage up to several years, with red wines capable of longer storage periods than white wines (in general).

CONTRIBUTION TO DIET

The major food products made from grapes are reflected in the utilization data (USDA 2003):

Product	Precentage of Crop
Wine	50-55
Raisins	25-30
Table	10-15
Juice, jelly, etc.	6-9
Canned	<1

A powerful alcoholic drink, grappa, is distilled from the fermented skins, seeds, and stems that are left over from pressing the juice in winemaking. Grappa is often used as an after-dinner drink in Italy. Many types of flavorings are added (e.g., orange or lemon peel) to improve flavor. In addition to the fruit or its pulp, young grape shoots and leaves are also edible. Grape seed oil is used as an edible oil, and also for making soaps.

In 2004, per capita consumption of grapes was 19.1 pounds/year, 41 percent as fresh fruit, 36 percent as raisins, 22 percent as juice, and 1 percent as canned fruit. Consumption has increased 16 percent since 1980. Per capita consumption of wine is 2.3 gallons/year and has remained roughly the same for the past 25 years.

Dietary value per 100-gram edible portion:

	Table Grapes	Raisins	Wine (100 g = 4 oz)
Water (%)	80	15	87
Calories	69	299	84
Protein (%)	0.7	3.1	0.1
Fat (%)	0.2	0.5	0
Carbohydrates (%)	18.1	79.2	1-3
Crude Fiber (%)	0.9	3.7	0
	% of U.S. RDA (2,000-calorie diet)		
Vitamin A	1	0	0
Thiamin, B_1	5	7	<1
Riboflavin, B_2	4	7	1
Niacin	1	4	1
Vitamin C	18	4	0
Calcium	1	5	<1
Phosphorus	3	14	3
Iron	2	10	2
Sodium	<1	<1	<1
Potassium	6	21	3

BIBLIOGRAPHY

Ahmedullah, M., and D.G. Himelrick. 1990. Grape management, pp. 383-471. In: G.J. Galleta and D.G. Himelrick (eds.), *Small fruit crop management*. Englewood Cliffs, NJ: Prentice-Hall, Inc.

Alleweldt, G., P. Spiegel-Roy, and B. Reisch. 1990. Grapes (*Vitis*), pp. 289-328. In: J.N. Moore and J.R. Ballington (eds.), *Genetic resources of temperate fruit and nut crops* (Acta Horticulturae 290). Belgium: International Society for Horticultural Science.

Bowling, B.L. 2000. *The berry grower's companion*. Portland, OR: Timber Press.

Clark, O. 1987. *The wine book*. New York: Portland House.

Cooke, G.M. and J.T. Lapsley. 1988. *Making table wine at home* (University of California Division of Agricultural and Natural Resources Publication 21434). Davis, CA: University of California, Davis.

Cox, J. 1999. *From vines to wines*. Pownal, VT: Storey Books.

Duke, J.A. and J.L. duCellier. 1993. *CRC handbook of alternative cash crops*. Boca Raton, FL: CRC Press.

Hedrick, U.P. 1945. *Grapes and wines from home vineyards*. London: Oxford University Press.

Howell, G.S. 1987. *Vitis* rootstocks, pp. 451-472. In: R.C. Rom and R.F. Carlson (eds.), *Rootstocks for fruit crops*. New York: John Wiley and Sons.

Lavee, S. 1985. *Vitis vinifera*, pp. 456-471. In: A.H. Halevy (ed.), *CRC handbook of flowering*. Volume 4. Boca Raton, FL: CRC Press.

Morton, L.T. 1985. *Winegrowing in eastern America*. Ithaca, NY: Cornell University Press.

Mullins, M.G., A. Bouquet, and L.E. Williams. 1997. *Biology of the grapevine*. Cambridge: Cambridge University Press.

Nelson, K.E. 1979. *Harvesting and handling California table grapes for market* (University of California Division of Agricultural and Natural Resources Publication 4095). Davis, CA: University of California, Davis.

Pearson, R.C. and A.C. Goheen. 1988. *Compendium of grape diseases*. St. Paul, MN: American Phytopathological Society Press.

Peynaud, E. 1984. *Knowing and making wine*. New York: John Wiley and Sons.

Ponracz, D.D. 1985. *Rootstocks for grapevines*. Capetown, South Africa: David Philip Publ.

Robinson, J. 1986. *Vines, grapes, and wines*. New York: Alfred A. Knopf.

Rombough, L. 2002. *The grape grower*. White River Junction, VT: Chelsea Green Publ.

Shoemaker, J.S. 1978. *Small fruit culture*, Fifth edition. Westport, CT: AVI Publishing.

Weaver, R.J. 1976. *Grape growing*. New York: John Wiley and Sons.

Winkler, A.J., J.A. Cook, W.M. Kliewer, and L.A. Lider. 1974. *General viticulture*. Berkeley, CA: University of California Press.

Wolf, T.K. and E. B. Poling. 1995. *The mid-Atlantic winegrape grower's guide*. Raleigh, NC: North Carolina Cooperative Extension Service.

Chapter 17

Hazelnut or Filbert *(Corylus avellana)*

TAXONOMY

The hazelnut, *Corylus avellana* L., belongs to the Betulaceae (or birch) family. This family contains other important forest tree species and ornamentals: *Betula* (birch), *Alnus* (alder), *Carpinus* (hornbeam, a landscape tree), and *Ostrya* (hop hornbeam or ironwood, a landscape tree). *Corylus avellana* is 1 of 9 to 20 recognized species in the genus, all of which produce edible nuts that are collected from the wild by humans and are important sources of food for wildlife. Nuts are subtended by a leafy involucre that gives rise to the names for the crop. The root "hazel" comes from the Anglo-Saxon word for bonnet, *haesel*. The genus name probably derives from the Greek *korys* (as in *Corylus*), meaning "helmet" or "hood." The name "filbert" may have derived from "full beard." Other *Corylus* species of importance include the following:

- *C. americana*—American filbert, a shrub found in the eastern United States
- *C. cornuta*—Beaked filbert, a stoloniferous shrub, found in central United States to the West Coast
- *C. ferox*—Tibetan hazel, a small tree that is a high-altitude species
- *C. colurna*—Turkish tree hazel, a large tree that is both an ornamental rootstock and a wild nut species
- *C. jacquemontii*—Indian tree hazel, a smaller version of *C. colurna*
- *C. chinensis*—Chinese tree hazel, a large subtropical tree that is both an ornamental and a timber species

Cultivars

All major cultivars of hazelnut have been selected from the wild. The primary cultivars differ for each production region. In Turkey, the leading hazelnut producer, 'Tombul' is the primary cultivar. The nuts are small but have a relatively high percent kernel and blanch easily. Blanching is the removal of the seedcoat around the kernel by exposure to heat and friction. Seedcoats are undesirable due to bitterness and poor appearance. The major cultivars of Italy and Spain also have these traits and are sold as shelled nuts ('Tonda Gentile della Langhe', 'Tonda di Giffoni', 'Negret'). Worldwide, these types of cultivars are most important because about 90 percent of the crop is shelled before sale. In the United States, the in-shell market predominates, and cultivars with large nuts and attractive shells are grown. About 70 to 80 percent of production in the United States is from 'Barcelona', although 'Ennis' is gradually replacing 'Barcelona' because it has larger nuts and produces fewer "blanks" (nuts with no kernels). However, 'Ennis' is more susceptible to eastern filbert blight and big bud mite than is 'Barcelona', which may limit its use. About 10 percent of the Oregon industry involves other cultivars that are used as pollinizers, such as 'Daviana' and 'Butler'. 'Casina', which produces small nuts with a high percent kernel can pollinate both 'Barcelona' and 'Ennis' and is more suitable for the shelled-

Introduction to Fruit Crops

doi:10.1300/5547_17

kernel market (Plate 17.1). Several new cultivars with resistance to eastern filbert blight are being released, and the mix of cultivars in Oregon will likely show a trend toward those with greater disease resistance in the future.

ORIGIN, HISTORY OF CULTIVATION

The cultivated hazelnut is native to Europe and Asia Minor, preferring regions with mild, moist winters and cool summers. For this reason, most production is located near large bodies of water at middle latitudes in the Northern Hemisphere (along the Black Sea in Turkey, the coastlines of Spain and Italy, and the Willamette Valley in Oregon). Hazelnut production failed initially in the eastern United States due to eastern filbert blight, a fungal disease that attacks young shoots first but later invades older wood, girdling and killing the tree. In the 1970s, it was found in the Pacific Northwest, where it now threatens production.

Today, hazelnuts are grown exclusively in the Pacific Northwest, virtually all in Oregon's Willamette Valley. The industry began in the 1920s and has steadily increased over time. The United States produces only about 2 percent of the world's crop, with most being produced in Turkey, along the southern coast of the Black Sea. The Oregon industry is valued at $15 million to $40 million dollars per year, and increasing demand over the past two decades has caused both imports and exports to increase greatly.

FOLKLORE, MEDICINAL PROPERTIES, NONFOOD USAGE

Hazelnut wood was used for divining rods and witching rods used to locate buried treasure or valuable soil minerals. Nuts were associated with the occult and said to possess mystic powers. Nuts were burned by priests to enhance clairvoyance, used by herbalists for various remedies, and used in marriage ceremonies as a symbol of fertility. The hazelnut's powers were believed to be strongest on Halloween, which was referred to as "Nutcrack Night" in England. Hazelnut trees and/or nuts were thought to impart knowledge, wisdom, or inspiration. Hazelnut means "reconciliation" in the "language of flowers." Hazelnuts are high in vitamins E and B_6.

PRODUCTION

World

Total production in 2004, according to FAO statistics, was 699,939 MT of shelled nuts or 1.5 billion pounds. Hazelnut is produced in 25 countries worldwide on about 1.2 million acres. World hazelnut production has been relatively constant over the past decade, as have acreage and yield. However, yield fluctuates between about 1,150 and 1,450 pounds/acre in consecutive years because hazelnut is an alternate-bearing species. Yield ranges from less than 800 pounds/acre up to about 1,600 pounds/acre in Turkey, the leading producer. The top ten hazelnut-producing countries (percent of world production) follow:

1. Turkey (70)
2. Italy (12)
3. United States (6)
4. Azerbaijan (3)
5. Spain (2)
6. Iran (2)

7. China (2)
8. Georgia (1)
9. France (<1)
10. Russia (<1)

United States

Total production in 2004, according to USDA statistics, was 33,600 MT or 74 million pounds of in-shell nuts. Statistics pertain only to Oregon production, which accounts for at least 98 percent of the U.S. crop. Washington produced the remaining 2 percent, as of 2002, the last year data were reported for that state. Acreage is 28,600 and has remained steady over the past decade. The industry value was at an all-time high, $50.7 million, in 2004 but fluctuates with the bearing cycle, generally from $18 million to $35 million. Yields overall are 2,300 pounds/acre, but heavy alternate bearing causes great variation (Figure 17.1). Growers received all-time-high prices of $0.68 per pound in 2004 because worldwide production was unusually low. Prices generally range between $0.32 and $0.50 per pound, lower than those for most other nut crops.

Hazelnut foreign trade has been active in recent years. Imports were 6,441 MT or 19 percent of domestic production in 2002, doubling in the past decade. Exports range from 13 to 33 percent of the crop, increasing five- to tenfold since 1982.

BOTANICAL DESCRIPTION

Plant

Hazelnut plants are large shrubs (12-18 feet) in Europe but trained to a single trunk in the United States to facilitate mechanical harvest (trunk shakers). Leaves are 2 to 3 inches long, broadly ovate, acuminate, and slightly lobed, with doubly serrate margins (Figure 17.2).

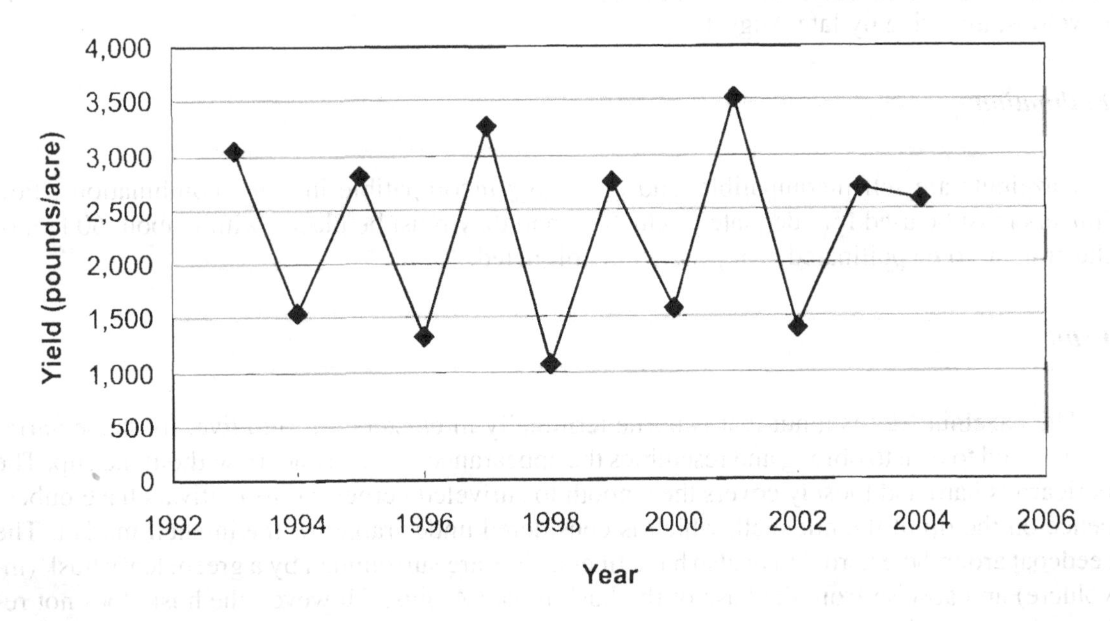

FIGURE 17.1. Hazelnut yield data from Oregon clearly show classic alternate bearing, typical of many nut tree species. Unusually good conditions for pollination, fruit set, and growth prevented an anticipated downturn in yield in 2004. *Source:* USDA statistics.

Flowers

Hazelnuts are monoecious; female flowers are borne terminally in headlike inflorescences on short shoots developing on 1-year-old wood (borne on current season's growth, although it appears to be 1-year-old wood, as extension growth at flowering is severely limited). Male catkins are borne from unmixed lateral buds on 1-year-old wood (Plate 17.2). Most cultivars are self-unfruitful. Hazelnuts are strongly dichogamous, and about 90 percent of all cultivars are protandrous (male flowers are mature before females).

Flowering habit is unusual. Both male and female flowers are initiated the summer prior to harvest. Male catkins are long, pendulous spikes with highly reduced male flowers consisting basically of anthers. Female flowers lack perianth and ovaries at the time of pollination in mid-January to mid-February; they are merely stigmatic surfaces connected to short styles. The stigma may be receptive to pollen for up to 3 months, which is highly unusual for a fruit crop. The pollen tube grows to the base of the style and becomes quiescent until 5 months later (June), when the ovary and ovule develop. Fertilization then takes place in July, and the nut rapidly develops, maturing by late August.

FIGURE 17.2. A mature hazelnut tree trained to a single trunk in Oregon's Willamette Valley.

Pollination

Hazelnuts are self-incompatible, and also cross-incompatible in some combinations. Pollinizers must be used for adequate production, and they must be placed within about 60 feet of the cultivar to be pollinated. They are wind pollinated.

Fruit

The hazelnut fruit is a nut that is borne terminally in clusters of up to five. Its shape varies from round to oval to oblong and resembles the appearance of an oak acorn without the cup. The pericarp is hard and loosely covers the smooth to shriveled kernel. Some cultivars have pubescence on the tip of the nut shell, which is considered undesirable for the in-shell market. The seedcoat around the kernel may also have fibers. Nuts are surrounded by a green, leafy husk (involucre) and abscise from the base of the husk in late August. However, the husk does not release the nut until 6 weeks later, when it dries and opens (Plate 17.3). Trees begin bearing when 3 to 4 years old and can bear for up to 40 to 50 years. Trees commonly produce 20 to 25 pounds of dried nuts each.

GENERAL CULTURE

Soils and Climate

Hazelnuts are grown in the United States on deep, well-drained, alluvial silt/loam soils with a pH of 6 to 7.5. In Turkey and Europe, trees are generally nonirrigated, being planted in areas with cool, maritime climates. In the Pacific Northwest, the deep, fertile soils of the Willamette Valley generally do not require irrigation for mature trees. In arid regions, irrigation is provided, and supplemental water is helpful for young tree establishment in most climates.

Hazelnuts have peculiar climatic preferences. They require mild, humid winters but need long periods of chilling at around 45°F for proper budbreak. The high chilling requirement prevents their cultivation in subtropical areas. Otherwise, they are cold hardy enough to be grown in northern Europe and in the upper Midwest of the United States. Strangely, the flowering dates are very early, often in February, so temperatures below freezing from midwinter through early spring can cause frost damage. Summer temperatures are moderated by large bodies of water in most production regions of the world, and too much heat causes stress and sunscald of the trunk. Summers should be dry as well, since several diseases can attack and kill trees in rainy, humid summer climates. In the United States, the most favorable climates for hazelnuts are in Oregon and Washington States. Trees can be grown as a winter-interest specimen in gardens in many locations of the eastern United States, although nut production is low or nonexistent and diseases often kill off trees every few years.

Propagation

The most common propagation method is simple layerage. In Oregon, rooting is induced on layers by sliding a metal ring over the dormant shoot, which girdles the shoot and induces root growth at the constriction site in the following season. Traditionally, hazelnuts were propagated by removing suckers from existing trees. However, suckers do not produce as efficiently as layers. Grafting has been largely unsuccessful, yet possible, due to poor callusing, poor scion wood, lack of graft survival, and suckering of rootstocks. A nonsuckering rootstock (*C. colurna*) was selected, but it tends to overgrow the scion, and its yield efficiency is low. Cutting propagation has also been unsuccessful, due mostly to poor survival, despite 60 to 80 percent rooting.

Rootstocks

Hazelnut has no rootstocks.

Planting Design, Training, Pruning

In cultivated orchards, tree spacing is generally triangular or square, with 10 to 20 feet between trees, yielding densities of 150 to 250 trees/acre (Figure 17.3). Pollinizer placement was 8:1 (every third tree in every third row) but has been increased to 17:1 in many cases (every sixth tree in every third row), since pollen can travel about 60 feet. The level of management in Turkey is low, with bushes planted in clumps, often on steep hillsides, and harvested by hand.

Vigor gradually declines over about a 5-year period, so every fifth row is severely pruned each 5 years to stimulate new productive wood; only thinning cuts are made. Yields are drastically reduced the year of pruning, since flower buds are removed, but trees yield high for 2 to 3 years after pruning. As an alternative, trees can be lightly pruned every other year, preferably in the "on" year, to maintain more consistent yields.

FIGURE 17.3. A hazelnut orchard in Oregon. Note that the orchard floor has been flail mowed and dragged in preparation for harvest, when nuts are windrowed and swept from the ground.

Pest Problems

Insects

Big bud mite. Bud damage caused by mites *(Phytoptus avellanae, Cecidophyopis vermiformis)* is a serious problem worldwide. Mites enter and feed upon buds before they emerge. Feeding kills the buds or severely distorts their growth. Spray programs are futile because the mites are inside the buds. 'Barcelona', 'Halls Giant', and, to a lesser extent, 'Casina' are naturally resistant to big bud mites. 'Ennis', 'Daviana', and 'Butler' are susceptible, and their replacement with resistant cultivars is recommended as a control measure.

Filbertworm. Mellissopus latifferanus is a serious pest of hazelnuts, walnuts, chestnuts, oaks, and even stone fruits. It is the major insect pest of hazelnut in Oregon. Larvae of this moth feed on developing kernels in midsummer. All cultivars are susceptible, particularly the thin-shelled 'Daviana'. Sprays are timed to the appearance of adult moths in the trees. Traps that attract adults can be placed in trees; when five moths are caught, sprays are applied thoroughly. If traps are not used, the timing of insecticide application is about July 10 in Oregon; second sprays are applied about 3 weeks later.

Leaf rollers/leaf tiers. At least three types of leaf-rolling or leaf-tying insects can cause damage in some years. The filbert leaf roller *(Archips rosanus)* and omnivorous leaf tier *(Cnephasia longana)* larvae feed on developing buds in spring, resulting in malformed shoots and reduced growth. Heavy infestations of leaf rollers can reduce growth and yield; they damage the trees in April and May. The oblique-banded leaf roller *(Choristoneura rosaceana)* is the only one to cause nut damage, and its feeding may only stain nuts in less severe cases. Sprays are based on levels of infestation; in some years, none are required. If shoot tips are malformed by leaf tiers in spring, sprays applied in early and late May often control the pest. Leaf rollers may also be controlled by early May sprays. Adult moths of the oblique-banded leaf roller fly and mate in mid-

June, and sprays at this time are needed if a high degree of nuts are stained or have aborted in past years.

Other pests. Filbert weevil (*Balaninus nucum*) is the most serious pest in Europe; its larvae tunnel through nuts. Its presence has not yet been reported in the United States. Blue jays, crows, wood ducks, and squirrels can remove considerable quantities of mature nuts and are difficult to control. Shooting and trapping are the only methods of control, and both are fairly impractical. Lagerstedt (1979) stated that a single squirrel nest had been found to contain 129 pounds of hazelnuts.

Diseases

Eastern filbert blight. This disease is caused by the fungus *Anisogramma anomola,* which attacks the wood, causing large cankers that may enlarge and eventually girdle the limb, killing it. Trees are killed if the cankers spread from limbs to the trunk. Signs of the disease include small, football-shaped pustules aligned vertically on branches. Spores are produced and are spread by wind and rain. The initial infection occurs on leaves in spring, but the wood damage is the greatest problem. Eastern filbert blight kills trees in 7 to 10 years, although the roots shortly produce suckers after canopy death. Pollinizers and 'Ennis' are more susceptible than are 'Barcelona' and 'Halls Giant'. 'Gasaway' and recent releases from Oregon State University are highly resistant to blight. Control is achieved by removing infected trees or wood, and through fungicide sprays, starting at budbreak and again every 2 to 3 weeks thereafter.

Filbert bacterial blight. This disease, caused by the bacterium *Xanthomonas corylina,* is a serious problem on young trees (<6 years old). The damage usually occurs on small, herbaceous shoots, whereas eastern filbert blight is a problem on older wood. However, young trees can be girdled and killed. Older trees lose some shoot growth but are rarely killed. The incidence of bacterial blight is strongly related to rainfall and can be reduced by sanitation of pruning equipment and fungicide sprays (fungicides have bacteriostatic activity). 'Ennis', 'Daviana', and 'Halls Giant' are somewhat resistant to bacterial blight. Sprays of copper compounds can be applied to trees in late summer to prevent infection. Blighted, dead, or dying shoots are pruned out several inches below where the dead tissue stops and then removed and buried or burned.

HARVEST, POSTHARVEST HANDLING

Harvest is by hand in most of Europe but mechanized in the United States. Nuts drop naturally during a 6-week period beginning in September. Winds or helicopters can remove the last remaining nuts. The orchard floor is flail mowed and floated (leveled) to keep it free of debris, to facilitate windrowing and sweeping of nuts.

Debris is blown from the nuts, which are washed in mild chlorine to reduce surface contamination, and then bleached by exposure to sulfur dioxide (SO_2) to enhance appearance. Filberts are dried to 8 to 10 percent moisture, permitting storage for up to 1 year at 36 to 40 F.

CONTRIBUTION TO DIET

Most of the hazelnut crop is utilized immediately for the in-shell holiday market (Halloween, Thanksgiving, and Christmas). In 2003, 42 percent of the U.S. crop was sold as shelled kernels, and the proportion of the crop sold as shelled kernels has been increasing over time. A large portion of the crop is cracked and the kernels are used for cereals, as a confectionery (mostly in baked goods), and in canned mixed nuts. Hazelnut is one of the most popular coffee flavorings.

Per capita consumption is the lowest of all tree nuts, only 0.07 pound/year in 2004; the consumption rate has remained unchanged for decades.

Dietary value per 100-gram edible portion:

Water (%)	5
Calories	628
Protein (%)	15
Fat (%)	61
Carbohydrates (%)	17
Crude Fiber (%)	9.7

	% of U.S. RDA (2,000-calorie diet)
Vitamin A	<1
Thiamin, B_1	43
Riboflavin, B_2	7
Niacin	9
Vitamin C	10
Calcium	11
Phosphorus	41
Iron	26
Sodium	0
Potassium	19

BIBLIOGRAPHY

Duke, J.A. 2001. *Handbook of nuts.* Boca Raton, FL: CRC Press.

Lagerstedt, H.B. 1979. Filberts, pp. 128-147. In: R.A. Jaynes (ed.), *Nut tree culture in North America.* Hamden, CT: Northern Nut Growers Association.

Lagerstedt, H.B. 1984. Filbert production. *Fruit Var. J.* 38:95-100.

Mehlenbacher, S.A. 1990. Hazelnuts *(Corylus)*, pp. 789-836. In: J.N. Moore and J.R. Ballington (eds.), *Genetic resources of temperate fruit and nut crops* (Acta Horticulturae 290). Belgium: International Society for Horticultural Science.

Mehlenbacher, S.A. (ed.). 2001. *Proceedings of the fifth international congress on hazelnut* (Acta Horticulturae 556). Belgium: International Society for Horticultural Science.

Mehlenbacher, S.A. 2003. Hazelnuts, pp. 183-216. In: D.W. Fulbright (ed.), *Nut tree culture in North America.* Hamden, CT: Northern Nut Growers Association.

Painter, J.H. 1969. Filberts in the northwest, pp. 294-298. In: R.A. Jaynes (ed.), *Handbook of North American nut trees.* Knoxville, TN: North American Nut Growers Association.

Rosengarten, F. 1984. *The book of edible nuts.* New York: Walker and Co.

Slate, G.L. 1969. Filberts including varieties grown in the east, pp. 287-293. In: R.A. Jaynes (ed.), *Handbook of North American nut trees.* Knoxville, TN: North American Nut Growers Association.

Teviotdale, B.L., T.J. Michailides, and J.W. Pscheidt. 2002. *Compendium of nut crop diseases in temperate zones.* St. Paul, MN: American Phytopathological Society.

Woodroof, J.G. 1967. *Tree nuts: Production, processing, products,* Volume 1. Westport, CT: AVI Publishing.

Chapter 18

Macadamia *(Macadamia integrifolia, Macadamia tetraphylla)*

TAXONOMY

Macadamia nuts belong to the relatively obscure Proteaceae or Protea family. This family contains 55 genera and 1,200 species divided into two subfamilies: Proteoidae and Grevilleoidae; nuts of commerce are grouped in the latter. The entire family is of Southern Hemisphere origin, and over 50 percent of the family members are found in Australia, with 35 to 40 percent located in South Africa. Dozens of species are cultivated as ornamentals, due to their often strikingly beautiful flowers. Examples include members of the genera *Leucospermum* (pincushions), *Leucodendron* (conebushes), *Diastella* (day stars), and *Mimetes* (bottlebrushes).

The genus *Macadamia* is divided into four subgenera and contains nine species, only two of which produce edible nuts. *Macadamia integrifolia* Maiden & Betche, called "smooth-shell macadamia," is the main species of commerce, but there is some production of rough-shell macadamia, *M. tetraphylla* L. Johnson. Both are native to the east coast of Australia and from rain forest–like climates. *Macadamia integrifolia* is more tropical in its requirements than is *M. tetraphylla*. The Hawaiian industry is based largely on *M. integrifolia*, whereas a small cottage industry in southern California is based on *M. tetraphylla*. Hybrids of the two species exist naturally, and some cultivars, such as 'Beaumont' and 'Vista', are man-made interspecific hybrids.

Cultivars

Few cultivars of macadamia exist because only limited breeding efforts have been employed with this species. Several cultivars that are grown in Hawaii include 'Kakea', 'Kau', 'Keaau', 'Makai', 'Mauka', 'Pahala', and 'Purvis'. Cultivars are distinguished by crown shape, and upright types, such as 'Kau' and 'Keaau', are being favored over more spreading types recently. 'Keaau' produces medium nuts with high percent kernel (44 percent) and is high yielding. 'Cate' is one of the most widely grown *M. tetraphylla* cultivars in California; it has medium-large nuts with a crisp texture and good flavor.

ORIGIN, HISTORY OF CULTIVATION

Macadamia nut is the only agricultural crop native to Australia. It is a riparian species found in the forests along the east coast of Queensland and New South Wales. Although considered a tropical tree and grown in the tropics commercially, it is technically subtropical in origin, found between 23.5 and 29°S latitude. Aboriginal people appreciated the nuts as a food source, calling it *Kindal Kindal*. The first European to collect a specimen of the plant was German explorer Frederich Leichart, although the plants were not named by him. English botanist Ferdinand Von

Introduction to Fruit Crops

doi:10.1300/5547_18

Mueller, working with Walter Hill of Australia in 1858, named the genus after Dr. John Macadam, said to be the first European to recognize the edible quality of the nut.

Trees were brought to Hawaii and California in the late 1800s, but only the Hawaiians developed nut culture, whereas the Californians used it initially as an ornamental. In 1910, the Hawaiian Agricultural Experiment Station encouraged planting of macadamia on Hawaii's Kona coast, as a crop to supplement coffee production in the region. Breeders released several improved cultivars that formed the basis of a new industry. As land and labor prices have increased greatly in Hawaii, the major companies that produce nuts have diversified their holdings to include Latin America and other regions where costs of production are low. Australia, South Africa, and a few other countries outside of Latin America have developed production in the past 20 to 30 years, and Australia now leads the world in macadamia nut production and exportation. In Hawaii, most macadamia nuts are grown on the big island (Hawai'i). In San Diego County, California, a small industry of about 200 growers is based on the rough-shell macadamia, *M. tetraphylla*. The area is too dry for production of *M. integrifolia*. One small co-op collects the nuts from growers and then dries, shells, and culls bad nuts before selling them to specialty produce firms.

FOLKLORE, MEDICINAL PROPERTIES, NONFOOD USAGE

Macadamia nuts are high in fats but contain 84 percent monounsaturated fats ("good" or "healthy" fats), more than any other plant source, and no cholesterol. A University of Hawaii study found that consumption of macadamia nuts in a controlled diet (same total calories from fat) actually reduced cholesterol by 10 milligrams/deciliter compared to a typical American diet with higher levels of saturated fats.

Shells and husks from nuts are used as mulch, compost, potting medium, and fuel for processing plants. Shells can be used in plastics and sandblasting. Macadamia nut oil is used in cosmetics, soaps, sunscreen, and shampoo. The wood is fine grained and reddish and used in cabinetmaking.

PRODUCTION

World

Total production in 2003/2004, according to USDA Foreign Agricultural Service statistics, was 92,923 MT or 204 million pounds. Production has increased 42 percent since 1996, largely due to expansion in Australia, South Africa, and Guatemala. Worldwide acreage is unknown but likely to be well under 100,000 acres. The top macadamia-producing countries (percent of world production) follow:

1. Australia (36)
2. United States (23)
3. South Africa (16)
4. Guatemala (11)
5. Kenya (6)

Australian production surpassed the United States in 1997, the first time in history that the United States did not lead the world in production. Australian production continues to increase slightly, while that of the United States is steady or slightly declining. Production in other countries will continue to rise as trees mature and nonbearing acreage comes into production. However, Hawaiian plantings are almost all of bearing age, and growers show little interest in expansion, due to low prices in recent years.

Two-thirds of the total world production is exported from the countries of production. Australia also leads the world in exports. The United States is the single largest export market for all other countries that produce macadamia, and it consumes up to 80 percent of Hawaiian production.

United States

Total production in 2004, according to USDA Foreign Agricultural Service statistics, was 21,133 MT or 46 million pounds. (*Note:* All statistics are for in-shell nuts.)

Virtually all production is in Hawaii on about 18,000 acres, mostly on the big island. The small volume of rough-shell macadamia produced in California is not included in USDA statistics. The value of the industry in 2004 was $33.1 million, down from $44 million in the mid-1990s. Most of the drop in value is reflected by a drop in the price paid to growers, which decreased from $0.78 per pound in 1996 to $0.65 per pound in 2004. Yields average about 3,000 pounds/acre. The United States consumes the majority of the world's macadamia nuts, importing an amount almost equal to domestic production. Exports are 20 to 35 percent of production, and they go chiefly to Japan, Hong Kong, and other Asian markets.

BOTANICAL DESCRIPTION

Plant

The macadamia is a medium-sized, tropical evergreen tree, with a spreading, full canopy that reaches widths of 30 feet and heights of 30 to 40 feet. Trees begin bearing in 4 to 5 years, bear full crops at 12 to 15 years, and have productive lives of 75 to 100 years. Leaves are up to 12 inches long and 1 to 2 inches wide, sparsely toothed, with sharp teeth on otherwise entire margins, and thick. The overall impression is that of a long, narrow holly leaf (Plate 18.1). Leaves of smooth-shelled cultivars are shorter and less spiny than those of rough-shelled types, and they lack the pinkish-bronze coloration when young. Leaves are whorled at the nodes, with two to three leaves per node in smooth-shelled and four in rough-shelled types.

Flowers

Small, perfect flowers are borne on long, fragrant racemes (4-8 inches) of dozens of individuals. Racemes are borne laterally on wood produced two to three flushes previous to the current leafy growth (Plate 18.2). Racemes are longer in rough-shelled types and are off-white or pink, instead of white, as in smooth-shelled types. Flowering occurs in midwinter, and nuts are harvested 7 to 8 months later in July through November (Northern Hemisphere), although some flowering and nut maturation occurs more or less year-round, depending on cultivar and location.

Pollination

Insects (bees) are the pollinators. Cross-pollination is necessary for full production, but the degree of self-incompatibility varies widely among cultivars.

Fruit

The macadamia fruit is a drupe; the shiny green hull (exocarp + mesocarp) is 1/8 to 1/4 inch thick, and the nut shell or endocarp is brown, thick, and about the diameter of a nickel (Plate 18.3). Fruits number from 10 to 30 on crowded stalks. Hulls are adherent to the shell but split along one suture at maturity. Nuts have shelling percentages of about 40 percent (40 percent of the weight of the whole nut is edible). Nuts of smooth-shelled types are generally higher in oil content and quality than those of rough-shelled types.

GENERAL CULTURE

Soils and Climate

Deep, well-drained soils with a pH of 5.5 to 6.5 are best, but trees are grown on a wide variety of soils. In Hawaii, they grow well on lava rock soils. Trees are intolerant of poor drainage and salinity.

Macadamias can tolerate mild freezing (28-32°F) and some drought but do not tolerate excessive heat (>90°F). In Hawaii, cool ocean breezes allow cultivation of certain cultivars near sea level, but inland in the tropics, trees perform best at 1,500 to 3,500 feet, where temperatures are cooler. Elevations suitable for coffee are often suitable for macadamia. The rough-shelled cultivars are slightly more tolerant of frost and heat than are the smooth-shelled types. They have no chilling requirement, but a seasonal change in temperature may help synchronize flowering. Trees are susceptible to wind damage and grown with windbreaks or in areas protected from prevailing winds.

Propagation

Macadamias can be grown from seed but will take several years to bear nuts and may not produce good-quality nuts. Grafting is used to enhance bearing and retain the desired nut characteristics. Whip grafts or side wedge grafts are made on young seedling rootstocks (Figure 18.1).

Rootstocks

Seedlings of rough-shelled (*M. tetraphylla*) cultivars are the best rootstocks because they are more vigorous and hardy than smooth-shelled seedlings.

Planting Design, Training, Pruning

Tree spacing depends on growth habit of the cultivar and the use of filler trees. Final spacings of 30 to 35 feet apart are typical (Figure 18.2). Closely spaced hedgerow orchards are sometimes used in Australia, with trees planted at 10 × 20 feet and later thinned to prevent overcrowding.

Macadamias are trained to a central-leader framework over a period of several years, as they grow slowly. The objective is to have only one central bole, an extension of the trunk, with fruiting limbs radiating off of it, spaced about 12 to 18 inches apart on the leader. Pruning the central stem of a young tree will generally cause at least three buds to break (buds are whorled at nodes in groups of three), and all will compete for the role of the central leader. The strongest, straightest one is allowed to grow, and others are pruned back to stubs, so they branch and bear fruits. If the leader grows about 18 inches per year, then this would be done each year.

FIGURE 18.1. A side wedge graft used for propagation of macadamia. *(left)* A successful graft several weeks after grafting. The union is waxed to prevent desiccation. *(right)* An unsuccessful graft reveals the technique.

FIGURE 18.2. A macadamia nut orchard in Hawaii with columnar-shaped trees allows closer in-row spacings than one using cultivars with rounded canopies.

Pest Problems

Insects

Stinkbugs/plant bugs. These large, shield-shaped insects pierce the hull and shell of very young fruits, causing staining, distortion, and possibly decay of nuts. The southern green stinkbug *(Nezara viridula)* can be particularly problematic. These insects are large and easily seen, but damage occurs just after bloom. Postbloom insecticide sprays are required, if there is a history of damage and large populations.

Thrips. These insects *(Thrips hawaiiensis, Selenothrips rubrocinctus)* are occasional problems that are difficult to control because they feed inside flowers and emerging shoots. Feeding on developing tissues causes leaf distortion and fruit scarring. A thorough spray, targeted to developing leaves and buds especially, may be needed, if damage is severe.

Diseases

Phytophthora foot rot. The organism *Phytophthora cinnamoni* causes several soil-related diseases of many tree crops, although the girdling of trees at the soil line, or foot rot, is most common. Macadamias may get trunk cankers from the disease but are fairly tolerant, particularly when mature. Root rot is generally not a problem. Control is achieved through providing good soil drainage. Irrigation is kept away from the trunk. Fungicides are not required in most cases.

Anthracnose. Colletotrichum spp. cause this disease and can be serious pests of many tropical fruit crops, including macadamia, particularly in Florida, where hot, humid conditions prevail and the disease is widespread. Fungal infections can occur on leaves and young fruits, and diseased fruits do not drop but instead rot on the tree. Control is through a fungicide program, requiring weekly sprays from bloom to half-size fruit stage. A few cultivars, such as 'Keauhou', are resistant.

Macadamia quick decline (MQD). This syndrome causes tree death, usually in areas where trees are stressed by a number of abiotic factors (e.g., poor drainage, pH, or nutritional imbalance). Ambrosia beetles *(Xyleborus affinis)* and wood fungi may be involved as well.

HARVEST, POSTHARVEST HANDLING

Maturity

Macadamia nuts will fall to the ground when mature, and they may fall over a period of several months. Shaking or knocking the trees may dislodge immature nuts and is not recommended.

Harvest Method

Hand harvest is used in small plantings or on hillsides, but mechanical harvesters are used in large operations (terrain permitting) to handle the tons of fruits produced. Nuts are picked up every few weeks.

Postharvest Handling

Harvested nuts are dehulled mechanically and dried to low water contents (<2 percent) for processing. Hulling is done immediately to prevent spoilage once nuts are collected. Nuts are cracked and sorted to remove off-color kernels and pieces of shell. Kernels are graded into two classes by flotation: grade I kernels (> 72 percent oil) float and grade II kernels (< 72 percent oil) sink. Nuts are roasted in hot oil (275°F for 10-15 minutes) or can be dry roasted in ovens in about 40 minutes (Figure 18.3).

FIGURE 18.3. *(left)* Macadamia nuts are roasted in hot oil. *(top)* Candied nuts are graded and singulated, and then dipped in chocolate. *(bottom)* Nuts are often glazed with various flavorings and sold in vacuum-packed cans.

Storage

Nuts are hulled and dried immediately after they fall. Unshelled nuts can be stored 2 to 4 weeks at room temperature once water content is reduced to 7.5 to 10 percent. Storage life can be extended to 1 year by drying nuts to less than 3.5 percent water and/or by storing them at 32 to 40°F.

CONTRIBUTION TO DIET

Most of the macadamia crop is used for confectionery purposes, but whole kernels are roasted and salted and sold in jars, usually in the "gourmet" sections of markets. Macadamias are considered to be among the finest table nuts in the world. They contain high quantities of oil (67-72 percent) and are therefore very fattening.

In 2004, per capita consumption of macadamia was 0.11 pound/year, the second lowest of all tree nuts (hazelnut is lowest). Consumption has increased threefold since 1980.

Dietary value per 100 grams of roasted kernels:

Water (%)	2
Calories	718
Protein (%)	7.9
Fat (%)	76
Carbohydrates (%)	13.8
Crude Fiber (%)	8.6
	% of U.S. RDA (2,000-calorie diet)
Vitamin A	0
Thiamin, B_1	80
Riboflavin, B_2	10
Niacin	12
Vitamin C	2
Calcium	8
Phosphorus	27
Iron	20
Sodium	<1
Potassium	10

BIBLIOGRAPHY

Bittenbender, H.C. and H.H. Hirae. 1990. *Common problems of macadamia nut in Hawaii* (University of Hawaii, College of Tropical Agriculture and Human Resources, Research Extension Series 112). Hilo, HI: University of Hawaii, Hilo.

Buyers, R. 1982. *The marvelous macadamia nut.* New York: Irena Chalmers Cookbooks.

Duke, J.A. 2001. *Handbook of nuts.* Boca Raton, FL: CRC Press.

Duke, J.A. 1983. *Macadamia integrifolia* Maiden & Betche, *Macadamia tetraphylla* L. Johnson. In: *Handbook of energy crops.* Available at www.hort.purdue.edu/newcrop/duke_energy/Macadamia.html; accessed April 2004.

Hawaii Agricultural Statistics Service. 2004. *Hawaii macadamia nuts.* Honolulu, HI: Hawaii Agricultural Statistics Service.

Ito, P.J. 1984. Macadamia nut production in Hawaii. *Fruit Var. J.* 38:101-102.

Malo, S.E. and C.W. Campbell. 1975. *The macadamia* (Florida Cooperative Extension Service Fruit Crops Fact Sheet FC-9). Gainesville, FL: University of Florida, Gainesville.

Rosengarten, F. 1984. *The book of edible nuts.* New York: Walker and Co.

Storey, W.B. 1985. Macadamia, pp. 347-355. In: A.H. Halevy (ed.), *CRC handbook of flowering,* Volume 3. Boca Raton, FL: CRC Press.

Thomson, P.H. 1979. Macadamia, pp. 188-202. In: R.A. Jaynes (ed.), *Nut tree culture in North America.* Hamden, CT: Northern Nut Growers Association.

Wagner-Wright, S. 1995. *History of the macadamia nut industry in Hawai'i, 1881-1981.* Lewiston, NY: Mellen Press.

Woodroof, J.G. 1967. *Tree nuts: Production, processing, products.* Volume 1. Westport, CT: AVI Publishing.

Chapter 19

Mango *(Mangifera indica)*

TAXONOMY

The mango, *Mangifera indica* L., is the most economically important fruit crop in the Anacardiaceae (cashew or poison ivy family). Other important members of this family include cashew, pistachio, and the mombins (*Spondias* spp.). The family contains 73 genera and about 600 to 700 species, distinguished by their resinous bark and caustic oils in leaves, bark, and fruits. Several species, including mango, can cause some form of dermatitis in humans. It is therefore ironic that two of the most delectable nuts and one of the world's major fruit crops come from this family.

Mangifera contains about 30 species, although some authors put the number as high as 69. Up to 15 other species produce edible fruit, including the water mango, *M. laurina*, and *M. sylvatica*, the wild, forest mango from which *M. indica* is thought to have descended.

Cultivars

Hundreds of mango cultivars exist throughout the world, most of which were derived from chance seedlings. However, a few cultivars derived from a breeding program in Florida are among the most popular for international trade. In home gardens throughout the tropics, seedling trees are grown as a backyard food source.

There are two classes of cultivars: Indo-Chinese and Indian (syn. West Indian) (Plate 19.1). The Indo-Chinese group is characterized by flattened, kidney-shaped, somewhat elongated fruits, with light green or yellow skin and little or no red blush color. Indo-Chinese mangos generally have polyembryonic seeds, and many are resistant to anthracnose, the major fungal disease of mango. In contrast, the Indian group contains cultivars that are more rounded and plump, with generally a bright red blush to the skin. Indian mangos are usually susceptible to anthracnose and have monoembryonic seeds, which facilitates breeding efforts. Many of the so-called "Florida cultivars" are Indian types selected or bred in Florida. Hybrids between the two groups have been produced using the Indian type (monoembryonic) as the female parent.

Each producing country has its own suite of cultivars. However, the export market is dominated by Indian cultivars, many of Florida origin. 'Haden' was a chance seedling of 'Mulgoba', one of the original mango cultivars brought to the United States from India by the USDA in the late 1800s. 'Haden' is round and relatively small, with a bright red blush to the skin; this became the standard of high-quality mangos in many areas of the world. Unfortunately, 'Haden' proved to be fairly susceptible to anthracnose and has been surpassed by other cultivars in many regions. 'Tommy Atkins', 'Keitt', 'Kent', 'Palmer', and 'Irwin', all similar to 'Haden' in appearance, are grown for export in many areas of the world. Southern Hemisphere countries produce and export mangos from October through March, and Northern Hemisphere countries produce

Introduction to Fruit Crops

doi:10.1300/5547_19

from April through October, making these cultivars available year-round. Excellent cultivar descriptions are found in the text by Campbell (1992).

ORIGIN, HISTORY OF CULTIVATION

The cultivated mango is probably a natural hybrid between *M. indica* and *M. sylvatica* that occurred in southeastern Asia to India. Selection of wild types has occurred for 4,000 to 6,000 years, and vegetative propagation has been practiced for at least 400 years in India. Mangos were brought to England and Europe after the English occupied India in the 1800s. They were brought to Brazil and the West Indies in the 1700s, with exploration of the area, and to Florida in the late 1800s. Portuguese and Spanish traders took mangos from India to East Africa, the Philippines, and western Mexico. From western Mexico, mangos were taken to Hawaii in the early 1800s. A cottage industry of a few hundred acres exists there today. Florida production grew through the early 1900s in southern Florida, but urbanization, freezes, and hurricanes reduced production from 7,000 acres to 2,500 acres. In 1992, hurricane Andrew reduced the Florida acreage to 1,000 to 1,500, where it remains today. Despite the small scale of the Florida industry, it has been extremely valuable to mango cultivation worldwide, due to the research and cultivar development carried out through the 1900s.

FOLKLORE, MEDICINAL PROPERTIES, NONFOOD USAGE

The mango is known as the "apple of the tropics" and is as important there as are apples in the temperate zone. It is common to see mango trees in kitchen gardens, in pastures, or as street trees in the tropics because they are delicious and a good source of vitamins.

Mango sap is toxic, causing a rash similar to poison ivy on the skin. The active principles causing sap toxicity are named for mango: mangiferin, mangiferic acid, and mangiferol. The allergenic principle is identified as 3-pentadecyl catechol, which occurs throughout the Anacardiaceae. Livestock may be killed by excessive ingestion of mango leaves. Breathing the smoke from burning leaves or the wood of mango can cause severe irritation. Despite the known toxicity, mango is used in a variety of folk remedies for various ailments. Chew sticks are made from twigs and leaves in India and Panama. Astringents,and remedies for bronchitis, internal hemorrhage, bladder ailments, diarrhea, syphilis, ringworm, and warts are made from twigs and leaves. The wood is used in a number of ways, including for building dugout canoes and as construction lumber and a substitute for charcoal. The bark is high in tannins and used for tanning animal hides.

PRODUCTION

World

Total production in 2004, according to FAO statistics, was 27,043,155 MT or 60 billion pounds. Mangos are produced in 90 countries worldwide on about 9.4 million acres. Production has increased 27 percent in the past decade, while acreage has increased 46 percent. Worldwide, yields have declined slightly, to about 6,400 pounds/acre, ranging from a few thousand pounds per acre to over 23,000 pounds/acre in intensive plantations. The top ten mango-producing countries (percent of world production) follow:

1. India (40)
2. China (13)
3. Thailand (6)
4. Mexico (6)
5. Indonesia (5)
6. Pakistan (4)
7. Philippines (4)
8. Brazil (3)
9. Nigeria (3)
10. Egypt (1)

Note: Chinese production has increased about 15-fold since 1980; China displaced Mexico as the second-place country in 1993.

United States

Florida Agricultural Statistices Service data (2003) show that mango production has declined significantly in the United States, largely due to a decline in Florida, the major producer. FAO data (2004) show total U.S. production of 2,800 MT or 6.1 million pounds in 2004. Florida produced greater than 90 percent of the U.S. crop on 80 percent of the acreage, with Hawaii producing the remainer. The total industry value was about $1.8 million. Acreage in Florida and Hawaii totaled about 1,600 acres in 2004, roughly the same since hurricane Andrew devastated Florida's production region in 1992. Production in 1993 was one-tenth that of prehurricane levels and has never recovered. Prices received by growers differ greatly: Florida—$0.26 per pound; Hawaii—$0.92 per pound. Yields are poor relative to other areas of the world, only 1,400 pounds/acre in Hawaii and just under 4,000 pounds/acre in Florida. (*Note:* Data from Hawaii are from 2002.)

FAO statistics show that the United States imported over 276.800 MT (609 million pounds) of mango in 2004, valued at $193 million, a 150 percent increase since 1994. Two-thirds of mango imports comes from Mexico, with small amounts from Brazil, Guatemala, and Ecuador.

BOTANICAL DESCRIPTION

Plant

The mango is a large, long-lived tree with a broad, rounded canopy that generally grows 20 to 100 feet tall. Trees in cultivated orchards are kept at about 20 to 30 feet. Leaves are lanceolate to linear (4-16 inches long × 1-2 inches wide), dark green, with prominent light-colored veins and entire margins. Emerging leaves on new growth flushes are bronze-red initially and appear wilted. One or two growth flushes occur per year, with flushes placed sporadically across the canopy of a given tree (Figure 19.1). Leaves may persist several years on mango trees.

Flowers

Tiny (⅛ to ¼ inch), red-yellow flowers are borne in large, terminal panicles of up to 4,000 individuals (Plate 19.2). About 25 to 98 percent of the flowers are male, depending on cultivar, and the remaining are hermaphroditic. Panicles that arise later in the bloom season or in shaded parts of the canopy tend to have more hermaphroditic flowers. Panicles are initiated in terminal buds 1 to 3 months prior to flowering; this is triggered by low temperatures or seasonally dry conditions. Mangos are distinct from most fruit crops in that chemical application is used to promote flowering and fruiting. Ethephon, potassium nitrate (KNO_3), and naphthalene acetic acid (NAA) are used either to induce flowering or to enhance fruit set or the proportion of hermaphroditic flowers.

FIGURE 19.1. A large mango tree on a farm in southern Guatemala. Individual trees may reach 100 feet tall and live well over 100 years.

Pollination

Mangos are considered self-fertile and do not require pollinizers, but research indicates that some cultivars are self-unfruitful or at least benefit from cross-pollination. Fruit set is generally just a few percent, with an average of only one mango borne per panicle (Plate 19.3). The pollen is incorrectly said to cause eye irritation and dermatitis; there is almost no airborne pollen because the pollen is heavy and adherent. The irritation probably results from volatile, irritating oils. Pollination is achieved by wild insects and, to a lesser extent, honeybees.

Fruit

Mangos are large drupes (Plate 19.4). The large, flattened, kidney-shaped, central stone contains a seed with one or more large, starchy embryos and can constitute up to 20 percent of the fruit's weight. The skin has a yellow or green background color, with a red blush in many cultivars, and is thicker than is usual for drupaceous fruits. The skin contains irritating oils, particularly in unripe fruits. The flesh is yellow to orange in color, sometimes astringent (turpentine-like) and can have fibers extending from the endocarp (stone). The endocarp can be lignified and fairly thick (e.g., as in 'Tommy Atkins'), or almost paper thin (e.g., as in 'Nam Doc Mai'), depending on cultivar.

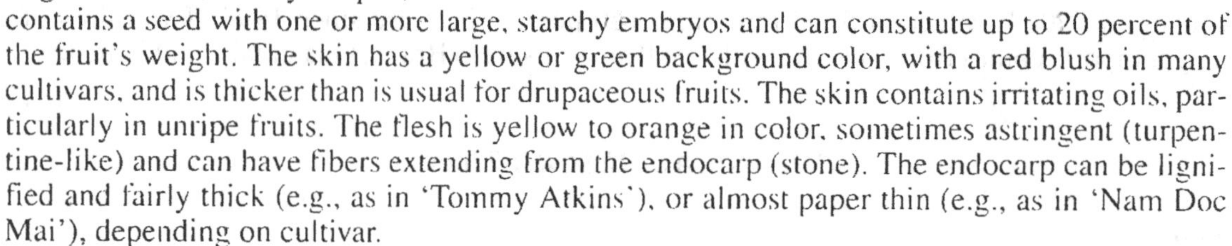

Fruit ripens in 80 to 180 days after bloom, depending on cultivar, generally late May to September in southern Florida. Fruiting begins in 6 to 10 years for seedlings, and 3 to 5 years for grafted trees. Alternate bearing may occur in trees over 10 years old and may be asynchronous across the canopy.

GENERAL CULTURE

Soils and Climate

Mangos are adapted to many soil types, provided they are adequately drained and mildly acidic (pH 6-7).

Mangos grow best in seasonally wet/dry climate zones of the lowland tropics, or in frost-free subtropical areas, such as extreme southern Florida or South Africa. A dry and/or cool season causes uniform floral initiation and tends to synchronize bloom and harvest. Mango does not attain a truly dormant state but ceases growth at temperatures below 55 to 60°F. Temperatures below 60 or above 100°F at flowering can cause flower abortion, loss of pollen viability, and occasionally seedless (and small) fruit development. Leaves and fruits are injured by mild frost (28-32°F), but wood is not killed unless temperatures drop to the mid- to low 20s°F. Trees have a high water requirement during fruit maturation but tolerate "winter" drought well. High winds can knock fruits off trees or cause scarring, since the fruits hang on long, pendulous floral branches at the periphery of the canopy.

Propagation

As mentioned earlier, Indo-Chinese cultivars are often polyembryonic and will produce true-to-type trees from seed. In less-developed countries, seedling trees are grown, and these come into bearing later than do grafted trees. Indian mangos tend to be monoembryonic and are mostly grafted on seedling rootstocks. Techniques vary from inarching and approach grafting in India, to veneer grafting and chip budding in Florida and the American tropics (Figure 19.2).

Rootstocks

Vigorous mango seedlings of various cultivars are used as rootstocks. Polyembryonic cultivars are often preferred because the rootstocks are genetically identical when grown from seed. In Florida, 'Turpentine' and 'Number 11' are used because they perform well on the limestone-based, higher-pH soils. Research in Puerto Rico has shown that 'Eldon' rootstock reduced tree height and canopy volume at least twofold for the scion cultivar 'Irwin'. However, results with other cultivars were inconsistent, and yields were also reduced. Thus, the search for a dwarfing rootstock continues.

Planting Design, Training, Pruning

Mangos are planted in square or rectangular designs and spaced 30 to 50 feet apart, yielding 18 to 35 trees/acre (Figure 19.3). In Florida, where roots are restricted by rocky soils, trees are planted in limestone at the intersections of trenches at 20 to 30 feet apart (up to 100 trees/acre). Trees in alternate rows can be removed from higher-density plantings once crowding occurs.

Very little pruning or training is necessary for mangos. In formative years, trees may be pruned to have one main trunk clear of branching, up to about 3 feet. After that, they assume a desirable rounded canopy shape naturally. Later, trees may be hedged and topped to control size. This is done in the summer after harvest and, if light, does not impact next year's crop because fruits are borne terminally on growth flushes that occur after pruning. Severe pruning will decrease fruiting the following year, however.

FIGURE 19.2. A side graft being employed for mango propagation. Rootstocks are 1 year old and grown from seed.

Pest Problems

Insects

FIGURE 19.3. A mature orchard of 'Manila' mango, with trees about 30 feet tall, spaced 30 to 40 feet apart.

Fruit flies. At least eight species of fruit fly (*Anastrepha, Dacus, Ceratitis* spp.) can attack mango, although most regions have only one or two species. The Mediterranean fruit fly *(Ceratitis capitata)* is the most widespread and is cause for some of the most stringent quarantine and phytosanitary standards in the global fruit industry. Fruit fly adults are attracted to ripening fruits, and females lay eggs on or under the fruit skin. The larvae then tunnel and feed on the fruits. However, damage can be avoided by timely harvest and postharvest heat treatment, in many cases. Fruits are dipped in hot water (115°F) for 65 minutes, which kills the eggs or larvae of the fly but does not injure the fruits. This process is overseen by government officials just prior to export and is vital to the global mango trade.

Mango seed weevil. Sternochetus mangiferae is a major problem for mango exporters in some areas, such as Hawaii, India, and Southeast Asia. The female lays eggs on the developing fruits near the stem end; eggs hatch and the tiny larvae tunnel into the fruit flesh to the seed, where they feed and develop into adults. Normally, the insects do not emerge until the fruits decompose, so the fruits remain edible and show no blemishes. Quarantines are used to limit the pest from spreading worldwide. All cultivars appear to be susceptible.

Others. Mango hopper (*Idioscopus* spp.) is a widespread pest and a particular threat in India, the world's leading mango producer. As with aphids, these insects suck sap, mostly from panicles at flowering time, causing flower abortion and poor fruit set. Scale insects (*Chrysomphalus, Pseudaulacaspis, Protopulvinaria* spp.) and mealybugs *(Pseudococcus)* infest wood and leaves, debilitating trees. Red-banded thrips *(Selenothrips rubrocinctus)* feed on developing leaves, causing leaf distortion and defoliation. Other thrips (*Frankliniella* spp.) can cause blemishes on the fruit skin through early infestation, lowering cosmetic quality. Mites (*Oligonychus, Aceria* spp.) can be a problem in dry seasons, as with many fruit crops. Natural enemies often control many of these insects, but insecticides are used when populations are high.

Diseases

Anthracnose. Caused by *Colletotrichum gleosporioides,* this is the most serious fungal disease of mango, requiring weekly sprays from bloom to half-size fruit stage in some regions. Infections can occur on all plant parts but are most serious on panicles, causing young fruits to be killed, and on ripening fruits, which will rot on trees or decay postharvest. The fungus invades young fruits but remains latent until ripening begins, when rot occurs. Tiny lesions often coalesce into spots or "tear stains" on mature fruits, covering almost the entire surface, but limited mostly to the skin. The disease is worst in conditions of frequent rain and dew. Control is through a diligent fungicide program or the use of Indo-Chinese cultivars, which are relatively resistant. 'Tommy Atkins', 'Keitt', 'Van Dyke', and 'Edward' are Indian types that are relatively resistant to anthracnose.

Powdery mildew. The fungus *Oidium mangiferae* can infect panicles in cool, dry conditions, causing low fruit set. Thus, it is more often seen in subtropical growing regions, such as Florida. It is only a sporadic problem, however, and is easily controlled by fungicide applications at bloom. Less-susceptible cultivars include 'Tommy Atkins', 'Sensation', and 'Kensington'.

Verticillium wilt. The fungus *Verticillium albo-altrum* invades the root system of trees, eventually causing collapse of the scion through water stress. It is worst when orchards are planted on sites previously planted to solanaceaous vegetables, especially tomatoes. The only practical control is to avoid these sites, since spores remain viable in soils for up to 15 years.

Mango scab. The fungus *Elsinoe mangiferae* can cause damage to flowers, leaves, twigs, and fruits. Gray-brown lesions may form on expanding tissues. Usually it is not a problem in mature trees, but nursery stock must be sprayed several times per year. Control is achieved by fungicide application during growth flushes; fungicide application for anthracnose controls this disease as well.

Alternaria rot or black spot. Alternaria alternata causes a fungal disease of the blossoms or ripe fruits, particularly on fruits in storage. Black lesions form at the stem end and may enlarge and coalesce. Symptoms can be confused with those for anthracnose, but lesions are generally firmer, darker, and smaller in size than are anthracnose lesions. A fungicide spray program, starting 2 weeks after fruit set, controls the disease. Sometimes this diseases is not a problem until after harvest.

HARVEST, POSTHARVEST HANDLING

Maturity

Color change from green to yellow and the development of "shoulders" on the stem end of the fruits are the best indicators of maturity. Also, the fruit flesh turns from white to yellow, starting at the endocarp and progressing outward to the skin during maturation. When the flesh is yellow to half that distance, the fruits can be harvested and will ripen normally off the tree. Fruits are not ripe at this stage but are picked firm to withstand shipment. If picked before this stage, flavor never develops and fruits are more susceptible to chilling injury and hot water damage during postharvest fruit fly control. If fruits are left on trees until ripe, they often develop a physiological breakdown called "soft nose," "jelly seed," or "spongy tissue" (*note:* these disorders are also linked to calcium deficiency). Fruit removal force may also be assessed by individual pickers to determine ripeness. Mangos are classified as climacteric fruits with respect to ripening.

Harvest Method

Mangos are hand harvested, simply by snapping off fruits from peduncles in less-developed plantings, or by clipping peduncles about 4 inches above the fruits when intended for export. This allows the milky, toxic latex to ooze from the stem without touching the fruit surface. Pickers use poles with cloth bags at the end, or ladders and hydraulic lifts in developed countries, to reach fruits high in the trees.

Postharvest Handling

Stems are trimmed to ¼ inch prior to packing in boxes containing 8 to 20 fruits, depending on size. Fruits are culled by hand to remove diseased and off-grade fruits. In countries where fruit flies are endemic, fruits are dipped in hot water for fruit fly and anthracnose control. Hot-air treatments can be used to meet export requirements as well. Resins left on fruits cause black lesions that may lead to rot.

Curing is carried out on immature fruits after harvest to allow quality development. Fruits are stored for 15 days at 70°F and 85 to 90 percent relative humidity. Ethylene is often supplied in more sophisticated operations to accelerate color development (by 3-8 days) and to allow more uniform ripening.

Storage

Mangos are subject to chilling injury and must not be stored at lower than 55°F. Storage life is only 2 to 3 weeks under optimal conditions.

CONTRIBUTION TO DIET

Mangos are one of the finest fresh fruits in the world, but they can be dried, pickled, or cooked as well. Mangos are higher in vitamin C than citrus fruits. Green mangos are the tropical equivalent of green apples—tart, crisp, and somewhat dry, often eaten with salt. They are cooked or used in salads in the tropics. About 25 percent of mangos are processed into juices, chutneys, sauces, or dried fruits. The large seeds can be processed into a flour, and the fat it contains can be extracted and substituted for cocoa butter.

In 2004, per capita consumption of mango was 2.1 pounds per year, increasing about tenfold since 1980.

Dietary value per 100-gram edible portion:

Water (%)	82
Calories	65
Protein (%)	0.5
Fat (%)	0.3
Carbohydrates (%)	17
Crude Fiber (%)	1.8
	% of U.S. RDA (2,000-calorie diet)
Vitamin A	15
Thiamin, B_1	4
Riboflavin, B_2	3
Niacin	3
Vitamin C	46
Calcium	1
Phosphorus	2
Iron	<1
Sodium	<1
Potassium	4

BIBLIOGRAPHY

Campbell, R.J. (ed.). 1992. *Mangos: A guide to mangos in Florida.* Miami, FL: Fairchild Tropical Garden.

Chandra, K.L. and R.N. Pal. 1985. *Mangifera indica*, pp. 211-230. In: A.H. Halevy (ed.). *CRC handbook of flowering*, Volume 5. Boca Raton, FL: CRC Press.

Chandra, K.L. and R.N. Pal (eds.). 1989. *Second international symposium on mango* (Acta Horticulturae 231). Belgium: International Society for Horticultural Science.

Crane, J.C. and C.W. Campbell. 1991. *The mango* (Florida Cooperative Extension Service FC-2). Gainesville, FL: University of Florida, Gainesville.
Duke, J.A. and J.L. duCellier. 1993. *CRC handbook of alternative cash crops*. Boca Raton, FL: CRC Press.
Gangolly, S.R., R. Singh, S.L. Katyal, and D. Singh. 1957. *The mango*. New Delhi: Indian Council of Agricultural Research.
Indian Council of Agricultural Research. 1967. *The mango, a handbook*. New Delhi: Indian Council of Agricultural Research.
Kostermans, A.J.G.H. and J.-M. Bompard. 1993. *The mangoes: Their botany, nomenclature, horticulture, and utilization*. New York: Academic Press.
Litz, R.E. 1997. *The mango: Botany, production, and uses*. Wallingford, UK: CAB International.
Morton, J.F. 1987. *Fruits of warm climates*. Miami, FL: Julia F. Morton.
Nakasone, H.Y. and R.E. Paull. 1998. *Tropical fruits*. Wallingford, UK: CAB International.
Ploetz, R.C., G.A. Zentmyer, W.T. Nishijima, K.G. Rohrbach, and H.D. Ohr (eds.). 1994. *Compendium of tropical fruit diseases*. St. Paul, MN: American Phytopathological Society Press.
Schaffer, B. (ed.). 1993. *Fourth international mango symposium* (Acta Horticulturae 341). Belgium: International Society for Horticultural Science.

Chapter 20

Oil Palm *(Elaeis guineensis)*

TAXONOMY

The Arecaceae or palm family is a large, distinct family of monocotyledonous plants, containing up to 4,000 species, distributed among over 200 genera. The African oil palm, *Elaeis guineensis* Jacq., is placed in the same subfamily as the coconut palm, the Cocosoideae. The American oil palm, *E. oleifera*, is similar to African oil palm and is used locally in tropical America for cooking oil and soaps. It can hybridize with African oil palm and may be important in breeding for reduced tree size, slow growth, different oil composition, and resistance to lethal yellowing disease. African oil palm is by far the most important fruit crop in the world in terms of production—approximately double that of banana, which is often considered the world's most important fruit. Another important relative is *Phoenix dactylifera*, the date palm. Taken together, fruiting palms account for almost 200 million MT of fruit production annually, used for a variety of products, such as fresh fruits, fiber and textiles, edible and industrial oils, and construction materials.

Several other palms are cultivated for their fruits or wild-harvested, including peach palm *(Bactris gasipaes)*, ungurahui palm *(Jessenia batua)*, jelly palm *(Butia capitata)*, betel nut *(Areca catechu)*, Chilean wine palm *(Jubaea chilensis)*, saw palmetto *(Serenoa repens)*, and aguaje palm *(Mauritia flexuosa)*. An even greater number of taxa are cultivated as ornamentals, including coconut and date, as well as members of *Sabal, Serenoa, Washingtonia*, and *Roystonea*. Perhaps only the Rosaceae family (apples, pears, peaches, plums, cherries, almonds) is more important to the study of fruit culture.

Cultivars

The three naturally occurring forms of the oil palm fruit are called *dura, tenera*, and *pisifera*. The *dura* form has a thick endocarp (up to ¼ inch), such that the mesocarp occupies only 35 to 65 percent of the fruit. The *tenera* form has a thin endocarp (<⅛ inch) and is 55 to 96 percent mesocarp. Yield and fruit weight are highest in *dura* and lower in *tenera*, although *tenera* produces fruits with higher oil content (Plate 20.1). The endocarp is absent in the *pisifera* form, and it may be female-sterile, lacking seeds entirely. Pure *pisifera* are undesirable from a commercial standpoint because they have low yield and are too vigorous. However, they have been extremely important in breeding. Oil palm fruits also fall into two categories based on fruit color, with a *virescens* form that lacks anthocyanins and turns orange at maturity and a *nigrescens* form that turns brown or black on the light-exposed portion of the fruits.

Since palm oil comes from the mesocarp of the fruits, early breeding work focused on *tenera* types or on *dura* types with a thick mesocarp, such as 'Deli'. It was revealed that *tenera* types were in fact first-generation (F_1) hybrids of *dura* × *pisifera* crosses. *Tenera* crossed with itself gave 25 percent *pisifera* types, which would be expected from the segregation of a single gene from the hybrid. Today, oil palm culture is based almost exclusively on *dura* × *pisifera* crosses

Introduction to Fruit Crops

doi:10.1300/5547_20

(i.e., *tenera*), and F_1 hybrid seed for orchards is produced by controlled breeding, similar to hybrid corn seed production. 'Deli' is often the female parent, and a variety of male *pisifera* lines ('AVROS', 'Ghana', 'Ekona', 'Nigeria') are used as pollinizers. Breeding focuses on maximizing oil yield and reducing vigor in terms of height growth and leaf size.

FIGURE 20.1. Wild oil palm growing in southeastern Nigeria, the heart of the native range of the species.

ORIGIN, HISTORY OF CULTIVATION

The African oil palm is native to tropical Africa, from Sierra Leone in the west through the Democratic Republic of Congo in the east (Figure 20.1). It is thought to be a riparian species, as it cannot tolerate shade but does tolerate periodic flooding and high water tables. It was domesticated in its native range, probably in Nigeria, and moved throughout tropical Africa by humans, who practiced shifting agriculture at least 5,000 years ago. European explorers discovered the palm in the late 1400s and distributed it throughout the world during the slave trade period. In the early 1800s, the slave trade ended but the British began trading with West Africans in ivory, lumber, and palm oil. The oil was prized for making soaps, candles, margarine, and industrial oils. Trade between Britain and West Africa continued until World War II, then declined. The oil palm was introduced to the Americas hundreds of years ago, where it became naturalized and associated with slave plantations, but it did not become an industry of its own until the 1960s. Four oil palm seeds were brought to a Javanese botanical garden in 1848, and the resulting progeny formed the basis of the world's largest production region. The 'Deli' cultivar, which was discovered near Deli, on the island of Sumatra, had a much larger oil yield due to its relatively thick mesocarp. The first plantations were established on Sumatra in 1911 and in Malaysia in 1917. The industry grew quickly at first, declined during World War II, and then began to increase exponentially in the 1950s. Oil palm plantations were established in tropical America and West Africa at about this time, and by 2003, palm oil production equaled that of soybean, which had been the number one oil crop for many years.

FOLKLORE, MEDICINAL PROPERTIES, NONFOOD USAGE

Two types of oil are derived from the oil palm fruit: palm oil from the mesocarp and palm kernel oil from the seed. Traditionally, palm oil was used for nonedible products and palm kernel oil was used more often as edible oil. Today, many products can be made from both, and both types of oil have food and nonfood uses. About 10 percent of palm oil is used for nonfood products, primarily in the manufacture of soaps, detergents, and candles; in rubber processing and the tinplating of metals; and as an ingredient in cosmetics, plastics, lubricants, and glycerol. Methyl esters from palm oil can be used as a diesel fuel substitute that has lower particulates in smoke and runs engines smoothly, without the need for modification. A high-protein livestock feed is derived from press cakes, the residue left over from palm kernel oil processing.

Other parts of the oil palm can be used for various purposes. The male inflorescences or plant meristems can be tapped to obtain sap, which is fermented into palm wine. As with many palms, the meristem can be eaten like a heart o' palm. The fronds are used as thatch and other crude tools. Investigations have shown that palm trunks can be used for biomass production, although they are generally felled and returned to the soil in commercial culture. The trunks can be made into particle board for construction, but this has yet to go commercial. The waste from processing fruits to oil—empty fruit bunches, particulates and sludge from liquid waste, fruit shells and fibers—is commonly applied to the ground as mulch and fertilizer.

Medicinal uses are few compared to other crops with long histories of human use. Folk remedies/uses include treatments for cancer, headache, and rheumatism and as an aphrodisiac, diuretic, and liniment.

The impact of palm oil on human health is controversial. Palm oil, similar to coconut or other "tropical" oils, carries a negative connotation due to its saturated-fat content (about 50 percent). Despite the common perception of palm oil being unhealthy, it contains no cholesterol and has less than 1.5 percent of the short-chain fatty acids that are harmful to health. Because it does not require hydrogenation, it does not contain trans fatty acids, which are associated with elevated cholesterol levels and increased risk of heart disease. Palm oil contains over 500 ppm (parts per million) of carotenoids, which have cancer-fighting properties. Specifically, it contains 15 times more carotenoids than are found in carrot, and 300 times more than in tomato. Red palm oil, a new food product, retains about 80 percent of the total carotenoids in crude palm oil, which impart the characteristic color. Palm oil contains more tocopherols and tocotrienols than any other vegetable oil. These compounds are related to vitamin E and have antioxidant and anticancer effects. One study showed that gamma-tocotrienol had a threefold greater suppressive effect on breast cancer cell growth than the popular breast cancer drug tamoxifen.

PRODUCTION

World

Total production in 2004, according to FAO statistics, was 153,578,600 MT or 338 billion pounds. Oil palm is produced in 42 countries worldwide on about 30 million acres. Production has nearly doubled in the past decade (Figure 20.2), and oil palm has been the world's top fruit crop in terms of production for over 10 years. Average yields are 11,400 pounds/acre, ranging from 2,700 to 24,000 pounds/acre. Per acre yield of oil from African oil palm is more than fourfold that of any other oil crop, and this has contributed to the vast expansion of the industry over the past few decades (Figure 20.3). The top ten oil palm–producing countries (percent of world production) follow:

1. Malaysia (44)
2. Indonesia (36)
3. Nigeria (6)
4. Thailand (3)
5. Colombia (2)
6. Ecuador (1)
7. Cote d'Ivoire (1)
8. Cameroon (<1)
9. Papua New Guinea (<1)
10. Congo (<1)

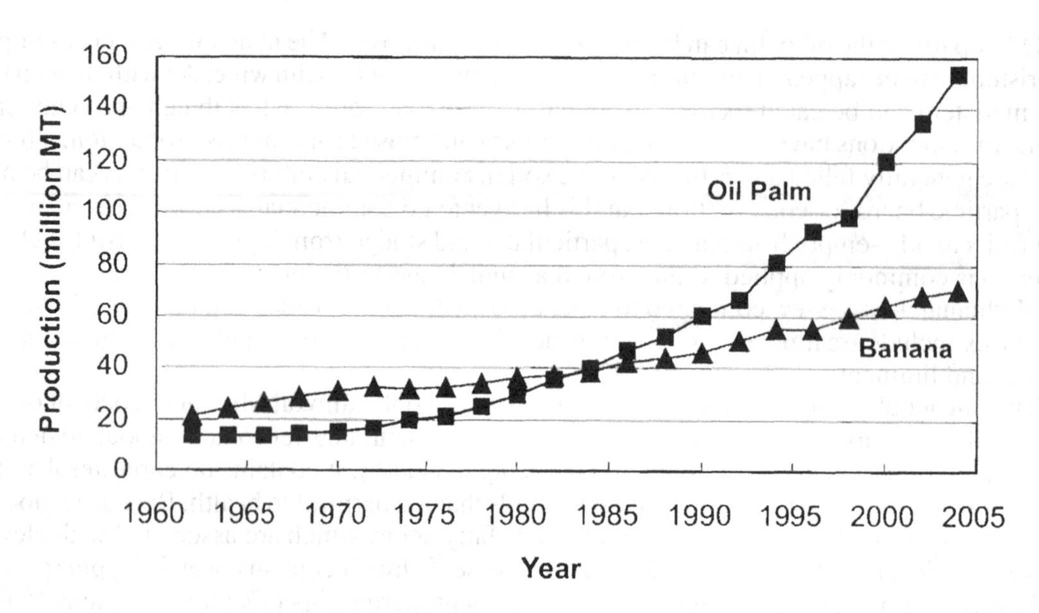

FIGURE 20.2. Historical production of the world's top two fruit crops, African oil palm and banana. Banana is often considered the world's leading fruit crop. Data show that oil palm production has increased exponentially since the 1960s, whereas the tripling of banana production has followed a linear trajectory. *Source:* FAO statistics.

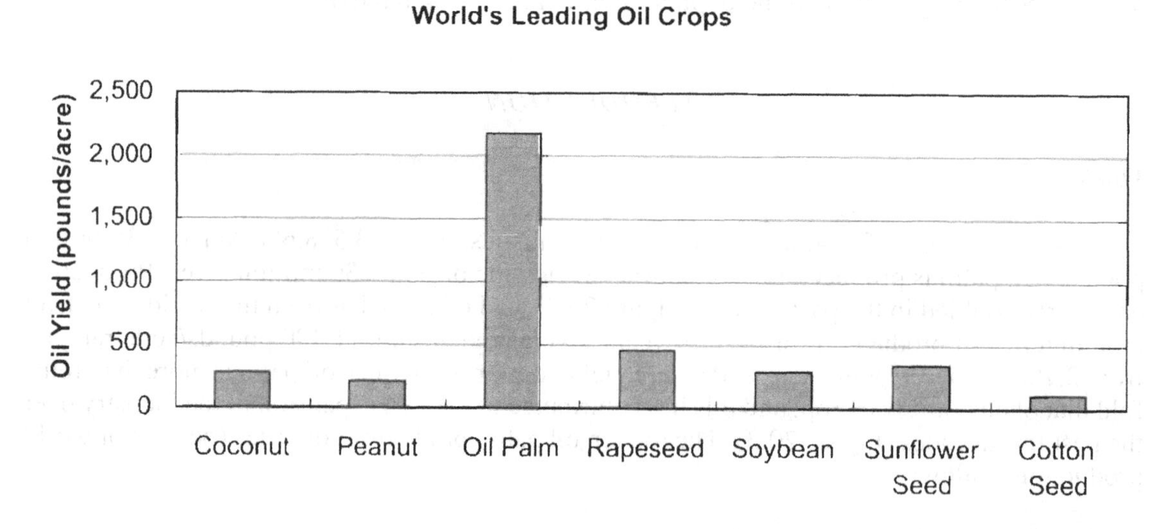

FIGURE 20.3. A comparison of per-acre yields of the world's top seven oil crops shows the huge yield advantage of oil palm. Oil palm and soybean had similar total oil production at 28 million MT in 2003, despite an approximately eightfold greater acreage for soybeans. *Source:* FAO statistics.

United States

No oil palm production takes place in the United States. The United States imported 211,000 MT of palm oil and 220,000 MT of palm kernel oil in 2003, collectively valued at about $200 million.

BOTANICAL DESCRIPTION

Plant

Oil palm trees can reach 60 to 80 feet in height in nature but are rarely more than 30 feet in cultivation. Trunks are stout and straight, about 1 to 2 feet in diameter, and grow about 1.5 to 3 feet per year (Figure 20.4). Leaf bases are persistent for years, and prominent leaf scars are arranged spirally on the trunk of mature palms where bases have fallen. The crown contains 30 to 50 leaves, producing about 20 to 40 leaves per year. Old or dead leaves snap and remain attached on wild palms, but they are pruned off to facilitate harvest in plantations. Wild palms are said to reach 200 years of age, but cultivated palms are removed after about 25 years, when they reach about 30 feet in height, since taller trees present difficulties in harvesting operations. Leaves are up to 25 feet in length, with leaflets numbering 200 to 300 per leaf, each about 3 to 4 feet long, and 1.5 to 2.0 inches wide, with entire margins. Leaflets cover the distal two-thirds of the leaf, and the lower one-third is spined, with spines increasing in length acropetally. The terminal pair of leaflets are ovate and short.

Flowers

Oil palms are monoecious, producing male and female inflorescences in leaf axils (Plate 20.2). The inflorescence of both sexes is a compound spadix, with 100 to 200 branches, that is initially enclosed in a spathe or bract which splits 2 weeks prior to anthesis. Each branch of the male inflorescence contains hundreds of tiny flowers, yielding about 100,000 flowers in total. Individual flowers are less than ⅛ inch long and contain six anthers that protrude beyond the petals, yielding a fuzzy, bronze or gray appearance. The female inflorescence contains hundreds of flowers, larger than those of the male, that are borne in triads, with two abortive males flanking one female, all enclosed in a spiny bract. Females have off-white, trilobed stigmas visible among the bracts. As vigor increases in some cultivars, the ratio of male to female inflorescences drops.

Pollination

Oil palms were originally thought to be wind pollinated, but more recent evidence suggests that they are primarily insect pollinated. In Africa, weevils (*Elaeidobius* spp.) are the pollinators, and they had to be introduced to Southeast Asia because *Thrips hawaiiensis,* a native species, was not an efficient pollinator. In Latin America, a native beetle, *Mystrops costaricensis,* and introduced *Elaeidobius* spp. are the pollinators (Plate 20.2). The insects are attracted to the male inflorescences, where they forage for pollen and lay eggs; they move on to female flowers almost by accident, being attracted by the same scent as is found with male flowers.

FIGURE 20.4. These 8-year-old oil palms are already becoming crowded and tall enough to present difficulties in harvest.

Fruit

As in many palms, oil palm fruits are drupes. The mesocarp and endocarp vary in thickness, with *dura* types having thick endocarps and thinner mesocarps, and *tenera* types the opposite. The exocarp color is green, changing to orange at maturity in the *virescens* form, and orange with brown or black cheek colors in the *nigrescens* form. Fruits range in size from less than 1 inch to 2 inches and are obovoid in shape. The mesocarp, from which palm oil is derived, is fibrous and oily, and the seed is opaque white, encased in a brown endocarp; palm kernel oil is derived from the seeds. The female infructescence contains 200 to 300 fruits, and fruit set is 50 to 70 percent. Fruits ripen about 5 to 6 months after pollination (Plate 20.3).

GENERAL CULTURE

Soils and Climate

Oil palms are grown on a wide range of soil types, when provided with good drainage and a pH between 4 and 7. They do poorly on heavily leached sands or heavy clays that do not drain well and have low amounts of exchangeable cations. The species is riparian in nature and can tolerate periodic flooding or a high water table, as most of the roots form in the upper few feet of soil. Since flat land is preferred for cultivation, many oil palms are planted in soils that are alluvial in nature. Irrigation is generally not practiced, as selection of sites focuses on those which lack an extended dry season.

Oil palm thrives in hot, wet tropical lowlands. The major production regions receive at least 6 feet of rain per year, evenly distributed, with at least 4 inches per month, if a short dry season exists. Areas with a strong dry season and less than 6 feet of rain have yields of 25 to 75 percent of their potential. Optimal temperatures are in the 80s to 90s°F, with temperatures below 75°F slowing growth. Oil palm is generally grown within a few hundred feet of sea level, although some cultivars have been bred to tolerate lower temperatures at higher elevations (e.g., 'Tanzania' × 'Ekona'). High humidity and cloudiness prevail in most regions, but 5 to 7 hours of direct sunlight per day is beneficial.

Propagation

Oil palm is propagated by seed, using F_1 hybrid seeds from controlled crosses that produce *tenera* types (*dura* × *pisifera*). Seed production is undertaken by companies specializing in oil palm breeding. Germination may be poor without pretreatment, and pregerminated seeds are often sold to improve establishment. Pregerminated hybrid seeds are as much as $0.50 each.

Pretreatment of seeds involves drying them to 17 to 18 percent water content, bagging them in plastic, and storing them at about 100°F for several weeks. After heat treatment, seeds are soaked in water for a few days and then germinated in bags at 75to 85°F. When germination reaches the desired level, in 15 to 30 days, seeds are sorted based on radicle length, packed, and shipped. Seeds are sown in black plastic bags and grown in a nursery for about 1 year before transplanting to the field (Figure 20.5).

Tissue culture of oil palm is carried out on a commercial scale to produce clones of superior types. Thus far, clone performance has not been stable across all environments. Efficient tissue culture regeneration of plants opens the door to genetic modification, which is currently being applied to other oil crops to produce higher-quality oils.

FIGURE 20.5. *(left)* An oil palm nursery *(foreground)* with a mature orchard *(background)*. *(right)* Pregerminated oil palm seeds are sold by companies that specialize in hybrid seed production.

Rootstocks

Oil palm has no rootstocks.

Planting Design, Training, Pruning

FIGURE 20.6. Dead trunks of a former oil palm planting stand among young trees planted on the same site.

Optimal plant density is about 58 trees/acre, and trees are planted in triangular patterns, about 30 feet apart. During the first 3 years, few or no fruits are obtained and plantations are often intercropped with staples, such as maize or yams (*Dioscorea* spp.). In later years, palm plantations are too densely shaded to allow successful intercropping of staples, grasses for grazing, or other crops. Ground covers are sometimes sown at planting to prevent soil erosion and to reduce competition from problem weeds. The leguminous, creeping ground cover *Pueraria,* itself a noxious weed in the southeastern United States, is often used as a cover crop for oil palm. Virtually no pruning or training is required of oil palm. Trees grow straight up and do not sucker or branch. Old leaves are pruned off to facilitate access to the bunch at harvest. When palms reach heights of 20 to 30 feet, they become difficult to harvest and are often injected with an herbicide, to kill them, or bulldozed down. New trees are planted among the dead and rotting trunks (Figure 20.6).

Pest Problems

Insects

Rhinoceros or black palm beetle. These large (up to 2 inches long), brown or black beetles (*Orcytes* spp.) are pests of many species of palm, including coconut and date. *Orcytes rhinoceros* is distributed throughout Southeast Asia, and *O. boas* is the main species of Africa, but at least five species can attack oil palm. They are easily recognized by the curved horn on their heads. Adults tunnel into the bases of developing leaves, causing malformed leaves and sometimes killing young palms. They breed in rotting logs and composting vegetation, where larvae complete their development. Some control measures are targeted at removal of breeding sites or handpicking of beetles and larvae from the field. An effective cultural control involves ground covers because beetles are attracted to young palms by sight and the ground cover blurs the silhouette. Biological control by the fungus *Metharhizum anisopilae* or *Baculovirus orcytes* has also been effective.

Foliage-feeding caterpillars. Over 20 species of slug and nettle caterpillars (*Limacodidae* spp.) can cause defoliation of oil palm, and one to several of these pests are found in nearly all oil palm–growing regions. In addition, three species of bagworms *(Mahasena corbetti, Pteroma pendula, Metisa plana)* cause defoliation. Some species reach outbreak levels after insecticides are applied for other pests, suggesting that there are biological controls in place that could be exploited. Synthetic insecticides and Bt have been effective when populations require control.

Leaf miners. Two species of *Coelaenomenodera* beetles can cause severe defoliation of oil palm, largely in Africa. *Coelaenomenodera lameensis* can lay several hundred eggs on the underside of leaves, and these eggs will produce the leaf-tunneling larvae. *Coelaenomenodera elaeidis* produces fewer eggs and is not as damaging. Feeding galleries from a single larva can be 6 inches long and 0.5 inch wide. In severe infestations, the leaflets of older leaves turn brown, die, and then shatter, to leave only the midrib of the leaf. Sprays or trunk injections of insecticides are applied when populations reach thresholds. Several parasites of eggs or larvae exist but have yet to be found effective in controlling infestations.

Rats. Several species of rats feed on oil palm fruits, causing losses of about 10 percent per year. *Rattus* is the main genus, but *Dasymys, Lemniscomys, Lophuromys*, and *Uranomys* are also found. Populations of up to 200 rats per acre exist in many plantations. Anticoagulant baits are used for control by hanging one bait in each tree. Since 1951, the number of barn owls *(Tyto alba)* in Malaysia has increased greatly, and this has been attributed to the availability of rats in oil palm plantations. One pair of owls are capable of killing 1,200 to 1,500 rats per year. Currently, it is unclear whether owls alone can keep rat populations below the economic threshold level of 5 percent fruit damage.

Diseases

Ganoderma butt rot or basal stem rot. Three species of *Ganoderma* fungi can cause a fatal, systemic rot of African oil palm and other palms, particularly coconut. *Ganoderma boninense* is the primary pathogen, although *G. miniatocinctum* and *G. zonatum* are also involved. The disease is widespread in Southeast Asia, the major oil palm region, and is worse in plantations previously planted with coconuts. It kills the roots and grows within the trunk, affecting water supply and creating symptoms resembling drought stress. Older leaves die and collapse at their base and hang vertically along the trunk. The younger leaves fail to unfurl or turn chlorotic and die back. It may take 3 to 4 years for trees to die. The fungus is a weak parasite or saprophyte, entering through wounds, and thus affects older plantations or individuals. Control centers around sanitation of infected or dead trees, with prompt removal, burning, or mounding of soil around trunks, which stimulates new root growth. Fungicide drenches combined with soil mounding may delay death and extend fruit production even more than mounding alone.

Fusarium wilt or vascular wilt. Fusarium oxysporium f. sp. *elaeidis* is considered to be responsible for one of the most devastating diseases of oil palm in Africa, and it has also been found in parts of South America. Presently, quarantine of plant material has prevented spread to Southeast Asia. The fungus enters through roots and grows throughout the vascular tissue of the trunk. The water-conducting tissues eventually become clogged with gums, causing death of leaves and, frequently, the entire tree. Older leaves die first, becoming desiccated and breaking along the midrib; dead leaves on *Ganoderma*-affected palms generally break at the petiole base.

In acute cases, trees die within months of infection, although chronic disease leading to death over a period of years is more common. Fungicide application is futile; diseased palms are generally removed and/or burned, and the site is left unplanted for several years. Disease incidence is higher with *Pueraria* ground covers, and in plantations where empty fruit bunches are used as mulch. Long-term control has focused on breeding for resistance, and germplasm with tolerance to fusarium wilt has been identified.

HARVEST, POSTHARVEST HANDLING

Maturity

As fruits ripen, they change from black (or green in *virescens* types) to orange but have varying degrees of black cheek color, depending on light exposure and cultivar. Thus, color is not specific enough to determine time of harvest. Fruits begin to abscise from bunches at maturity, and the number of loose fruits per bunch is used as a field criterion for harvest. When just a few fruits are detached, the entire bunch is harvested, since more oil is potentially lost by fruit abscission than by harvesting some of the fruits not fully ripe.

Harvest Method

Fruit bunches are harvested using chisels or hooked knives attached to long poles. The leaf or leaves below the bunch are first removed to gain access to the peduncle. Loose fruits must be collected by hand beneath the tree and when bunches fall, so the area around trees is kept free of vegetation and debris. In Africa, wild palms are still hand harvested by pickers climbing the trees. Each tree must be visited every 10 to 15 days, as bunches ripen throughout the year.

Postharvest Handling

Harvested bunches are heavy, and several implements have been developed to transport them to the oil mills. In less-developed areas, animals are used to haul fruits on their backs or by carts, but in most plantations, hydraulic cranes lift fruit bunches into trucks or railcars.

Oil extraction is a complex process that is carried out either by large mills, which may process up to 60 tons of fruit per hour, or small-scale mills in rural villages, which produce only about 1 ton of oil in an 8-hour shift (Figure 20.7). Oil extraction from the fruits follows the same basic steps in either case:

1. Steam sterilization of bunches (inactivates lipase enzymes and kills microorganisms that produce free fatty acids, reducing oil quality)
2. Stripping fruits from bunches
3. Crushing, digestion, and heating of the fruits
4. Oil extraction from macerated fruits (hydraulic pressing)
5. Palm oil clarification
6. Separating fiber from the endocarp
7. Drying, grading, and cracking of the endocarp
8. Separating the endocarp from the kernel
9. Kernel drying and packing

The product of step 5 is called "crude palm oil," which must be refined to remove pigments, free fatty acids, and phospholipids, as well as to deodorize it. Through a combination of chemi-

FIGURE 20.7. Palm oil processing. *(above left)* Bunches are transported to the mill and weighed. *(above right)* Bunches are first steam sterilized in large tanks to inactivate enzymes before oil is extracted. *(right)* Palm oil extraction by hand in a village in Nigeria. Fruits are smashed in a hollowed-out tree stump, the resulting pulp is boiled to remove excess water, and the oil is strained from the fibrous pulp.

cal treatments, bleaching, and steam stripping, the final product, called "refined, bleached, deodorized palm oil," is produced. Palm kernel oil may be pressed hydraulically in the same mill but more often is done elsewhere, in facilities designed for this process. Palm kernel oil extraction is generally done by solvents (hexane), which extract all but 2 percent of the oil from kernel cake. Simple pressing leaves up to 13 percent oil in the cake but is cheaper and easier than solvent extraction.

The fuel used to generate steam and extract oil is the spent fiber and shells (endocarps); thus, the process is self-contained. Palm oil plantations may even generate surplus power that can be sold to municipalities or used in houses, a system analogous to burning bagasse to generate power in sugar mills. Wastewater and solids not used as fuel are utilized on site as fertilizers or mulches; the water is held in settling ponds before being discharged into rivers.

Storage

Palm oil is stored in large steel tanks at 88 to 105°F, to keep it in liquid form during bulk transport. The tank headspace is often flushed with carbon dioxide (CO_2) to prevent oxidation. Higher temperatures are used during filling and draining tanks. Maximum storage time is about 6 months at 88°F.

CONTRIBUTION TO DIET

About 90 percent of the palm oil produced finds its way into food products, with industrial uses accounting for the remaining 10 percent. Palm oils are used in a wide variety of foods, primarily margarine, shortening, and vegetable cooking oil. Palm oil is used as a replacement for cocoa butter and butter fat, and in ice cream and mayonnaise. It is stable at the temperatures used in deep frying and is used quite often for fried foods. Red palm oil is increasing in popularity, as it contains large quantities of carotenoids.

Palm wine is made by tapping the male inflorescence of the oil palm and fermenting the resulting sap. An alternative method that involves felling entire trees and tapping the meristem is often used on old plantations that are being replanted. Palm wine, which has been an important part of West African culture, is still made today in large quantities and is fetching good prices.

Per capita consumption of palm oil specifically is unknown, but Americans consumed 84.7 pounds of fats and oils in 2004. About 45 percent of this is margarine and shortening, two major products containing palm oil. A crude approximation of worldwide per capita consumption is 8.7 pounds/year, obtained by dividing the 61 billion pounds of annual palm oil production by the world population of about 6.3 billion people, and assuming 10 percent is used for nonfood purposes. Using similar math, U.S. consumption of palm oil for 2004 is estimated at 1.5 pounds/year based on import data and population.

Dietary value per 100-gram edible portion:

	Oil Palm Fruit	Palm Oil	Palm Kernel Oil
Water (%)	26	0	0
Calories	540	884	884
Protein (%)	1.9	0	0
Fat (%)	58.4	100	100
Carbohydrates (%)	12.5	0	0
Crude Fiber (%)	3.2	0	0
	% of U.S. RDA (2,000-calorie diet)		
Vitamin A	3.5	0	0
Thiamin, B_1	13.3	0	0
Riboflavin, B_2	5.6	0	0
Niacin	7.0	0	0
Vitamin C	26.7	0	0
Calcium	10.3	0	<1
Phosphorus	5.9	0	0
Iron	45	<1	2
Sodium	2.4	0	<1
Potassium	5.3	0	<1

BIBLIOGRAPHY

Chow, C.K. (ed.). 1992. *Fatty acids in foods and their health implications.* New York: Marcel Dekker.

Corley, R.H.V. 2001. Oil palm, pp. 299-320. In: F.T. Last (ed.), *Tree crop ecosystems.* New York: Elsevier Press.

Corley, R.H.V., J.J. Hardon, and B.J. Wood (eds.). 1976. *Oil palm research.* Amsterdam: Elsevier Scientific Publ.

Corley, R.H.V. and P.B. Tinker. 2003. *The oil palm.* Fourth edition. Oxford, UK: Blackwell Science Ltd.

Duke, J.A. 2001. *CRC handbook of nuts.* Boca Raton, FL: CRC Press.

Gunstone, F.D. (ed.). 1987. *Palm oil* (Critical Reports on Applied Chemistry, Volume 15). New York: Wiley.

Hartley, C.W.S. 1988. *The oil palm* (Elaeis guineensis *Jacq.*). Third edition. Essex, UK: Longman Scientific and Technical.

Poku, K. 2002. *Small-scale palm oil processing in Africa* (FAO Agricultural Service Bulletin 148). Rome: FAO.

Turner, P.D. 1981. *Oil palm diseases and disorders.* Oxford: Oxford University Press.

Vandermeer, J. 1983. African oil palm (Palma de aciete), pp. 73-75. In: D.H. Janzen (ed.), *Costa Rican natural history.* Chicago, IL: University of Chicago Press.

Chapter 21

Olive *(Olea europaea)*

TAXONOMY

The olive, *Olea europaea* L., is placed in the title genus of the Oleaceae family. This family contains about 22 genera and 500 species, most of which are placed in the Oleoideae subfamily with olive. Olive is by far the most economically important member of the family, but several others are valued as ornamentals: *Fraxinus* (ash), *Syringa* (lilac), *Ligustrum* (privet), *Jasminum* (jasmine), *Forsythia* (forsythia), *Osmanthus* (fragrant olive), and *Chionanthus* (fringe tree). The genus *Olea* contains about 20 species, but only the olive produces edible fruits.

Cultivars

Each Mediterranean country has its own unique cultivars of olive, and many seedling trees are cultivated. The recently published world catalog of olive varieties lists over 130 cultivars, with more than 30 cultivated in both Spain and Italy. Some orchards (groves) are hundreds of years old, making it difficult to tell what cultivars were originally planted. Different cultivars are generally used for oil (e.g., 'Picual', 'Leccino', 'Frantoio') and for table olives ('Manzanillo', 'Sevillano', 'Ascolano', 'Calamata').

In California, 'Manzanillo' and 'Sevillano' constitute about 90 percent of olive production, with small amounts of 'Mission', 'Ascolano', and others. 'Manzanillo' is by far the major cultivar, having a small fruit that lends itself to the "black ripe" olive market, but with high enough oil content (>20 percent) that culls can be used for making olive oil. The 'Sevillano' fruit is two to three times the size of the 'Manzanillo' fruit but has low oil content and is used only as a table olive. Both were introduced from Spain in the late 1800s. 'Mission' was formerly the most popular cultivar in California, but small fruit size, a relatively large pit, and susceptibility to diseases and late frost led to its decline in popularity.

ORIGIN, HISTORY OF CULTIVATION

The olive originated in the eastern Mediterranean area and has been cultivated by humans since ancient times. Trees are extremely long-lived (up to 1,000 years) and tolerant of drought, salinity, and almost total neglect; thus, they have been reliable producers of food and oil for thousands of years. The earliest references of olive oil use and international trade date to 2000 to 3000 BC. Oil was used for cooking as well as for burning in lamps. Several references are made to olive oil lamps in the Bible and other ancient writings from Greece and Rome. The olive was spread throughout Mediterranean Europe and North Africa very early, due to its ease of vegetative propagation and cultivation in dry climates. The Romans, building on earlier work on olive culture by Greeks, Arabs, and Egyptians, refined olive oil extraction and improved cultivars used for oil. Today, the industry remains largely confined to the Mediterranean countries of Europe, the Middle East, and North Africa, where it began thousands of years ago.

Introduction to Fruit Crops

doi:10.1300/5547_21

Olives were brought to California shortly after 1769, when the first mission was founded by Franciscan padres in San Diego. The California industry began in the late 1800s, as settlers planted orchards from cuttings taken from the original mission trees. By 1900, about a half million trees were being grown in California, largely for olive oil production. Around this time, pickling and canning procedures were developed for producing black olives, which are the primary olive product from California today. Although there is some interest in producing high-quality virgin oils in California, olive oil is largely a secondary outlet for table olives unsuitable for market.

In Mediterranean or desert climates, olives are frequently used as yard trees. They have attractive, silver-green foliage and full canopies, are evergreen, and require very little water and maintenance. However, the fruits will stain sidewalks and cars (black-purple, oily stains), and the pollen is highly allergenic. Substitute cultivars that do not fruit (e.g., 'Swan Hill', 'Majestic Beauty', 'Little Ollie') have been developed for landscape use.

FOLKLORE, MEDICINAL PROPERTIES, NONFOOD USAGE

Olives were cultivated in ancient times for lamp fuel, lubrication, and dietary fat, as there were few substitutes at the time. The olive is often mentioned in mythology and the Bible. Athene, the goddess for which Athens was named, is said to have won this honor by placing the world's first olive tree on the Acropolis in Athens. The olive is associated with peace and security, and this probably originates from the passage in Genesis (8:11) describing how the dove (another symbol of peace) returns to Noah's ark with an olive branch in its beak, after God had made peace with man and had stopped the flood. In contrast, olives were sometimes items of aggression; Odysseus jabbed an olive branch into the cyclops's eye to blind him and thus escape being eaten. Olive crowns were awarded to brave Roman soldiers, and olive oil was awarded to the winner in ancient athletic competitions.

Olive oil is an important component of the Mediterranean diet and in fact is included in the European food pyramid. Those eating the Mediterranean diet (rich in olive oil, fruits, vegetables, and fish) are known to have lower rates of colon, breast, and skin cancer and coronary heart disease. The active principles in olive oil are thought to be monounsaturated fats (primarily oleic acid), squalene, and phenolic compounds that function as antioxidants in the body. Oleuropein, responsible for the bitterness of raw olives, is one of the phenolic compounds. Other simple phenols (e.g., tyrosol) and lignans (pinoresinol) also function as antioxidants. Extra-virgin oils are higher in these protective compounds than are processed oils. Olive oil may act by reducing the LDL (low-density lipoprotein = "bad") and raising the HDL (high-density lipoprotein = "good") forms of cholesterol in the blood. Olive extracts have been shown to have hypoglycemic activity, and oil reduces gallstone formation by activating the secretion of bile from the pancreas. Olive oil may act as a mild laxative.

PRODUCTION

World

Total production in 2004, according to FAO statistics, was 15,340,488 MT or 34 billion pounds. Olives are produced in 39 countries worldwide on an area of over 21 million acres. Olive is the most extensively cultivated temperate fruit crop in the world, since its acreage surpassed that of grape several years ago. Production has increased 34 percent in the past decade. Average yields are 1,610 pounds/acre but range widely from 300 to over 8,500 pounds/acre.

Olive oil is produced in 29 countries worldwide, and the leading producers of oil are the same as those for overall olive production. Olive oil production was 2.5 million MT in 2002. Over 75 percent of the world's olive oil is produced in just three countries—Spain, Italy, and Greece. The vast majority of the olive crop is used for oil. The top ten olive-producing countries (percent of world production) follow:

1. Spain (30)
2. Italy (20)
3. Greece (15)
4. Turkey (12)
5. Syria (6)
6. Morocco (3)
7. Tunisia (2)
8. Egypt (2)
9. Portugal (2)
10. Lebanon (1)

United States

Total production in 2004, according to USDA statistics, was 94,500 MT or 208 million pounds, produced on around 32,000 acres, or less than 1 percent of world production (Figure 21.1). All production is in California, largely in the Central Valley. The industry value was $60.6 million in 2004 and has varied between $35 million and $102 million over the past decade. Prices were about $0.29 per pound in 2004, higher than average for the past decade. Average yields are 6,500 pounds/acre, about fourfold greater than the worldwide average yield. Imports of table olives in brine were 108,734 MT in 2003, up from about 61,000 MT in the past decade. Thus, the United States imports more table olives than it produces annually. Olive oil imports were 220,000 MT (69 million gallons), almost double that of a decade ago. Dried olives are also imported, but in tiny quantities by comparison (only 523 MT).

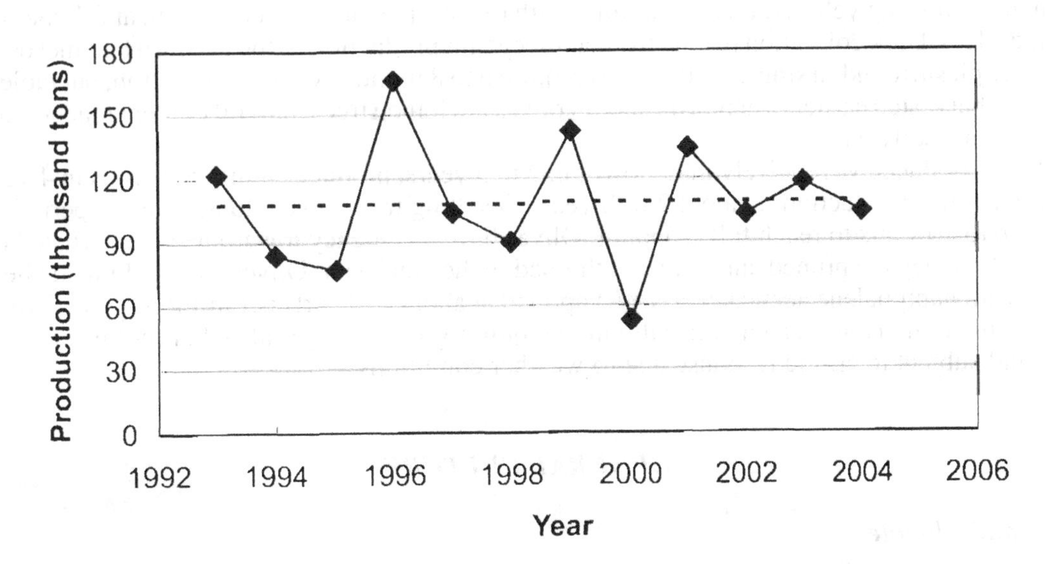

FIGURE 21.1. California olive production shows a pattern characteristic of alternate or irregular bearing. The dashed line shows the mean production of 108,000 tons/year. *Source:* USDA statistics.

BOTANICAL DESCRIPTION

Plant

Olives are large, evergreen shrubs in their native state but are trained as stout trees on massive trunks, especially in older plantings. Most trees have round, spreading crowns, but tall, cylindrical trees are grown in some parts of Europe (Plate 21.1). Trees in neglected groves grow almost imperceptibly slowly, whereas irrigated trees in California may reach 12 feet in 4 years. Olives have the longest-lived trees of any fruit crop; some trees in Europe are believed to be 1,000 years old.

Leaves are small (1½ inches long, ¼ to ½ inch wide), linear, with entire margins and acute tips, silver-green in color, and fairly thick. Leaf arrangement is opposite, as for all members of the Oleaceae. Leaves live about 2 years.

Flower

Small, off-white flowers are borne in racemose panicles of 15 to 30 flowers, in axils of 1-year-old wood (Plate 21.2). Most flowers are staminate by pistil abortion, leaving only a few perfect flowers per inflorescence that may set fruit. The ovary is superior, and there are four sepals and petals, and two stamens. Flowering occurs rather late relative to other tree crops—May, in southern Europe and California.

Pollination

Most olives are self-fruitful, but some cultivars bear heavier crops when cross-pollinated. Wind is the pollinator.

Fruit

The fruit type is a drupe. Fruits are oblong with smooth, waxy surfaces. Color is green when immature, turning yellow-green in autumn, with red, purple, or black coloration at full maturity (Plate 21.3). Dark coloration results from anthocyanin production in the exocarp and mesocarp. A stony pit surrounds a single seed. Olives require 6 to 8 months for full maturation, but table olives are harvested earlier, when firm, and oil olives are left on trees until oil content reaches 20 to 30 percent (early winter).

Trees produced vegetatively may flower in 2 to 3 years, produce significant crops in 4 years, and reach full production in about 7 to 8 years. Seedling trees have a long juvenile period and take up to 10 years to reach full cropping. Olives have a tendency toward heavy alternate bearing, unless they are pruned annually and thinned in the "on" year. Olives can be thinned chemically with naphthalene acetic acid (NAA) applied at about 150 parts per million (ppm) 2 weeks after full bloom. However, chemical thinning requires great managerial skill, is cultivar dependent, and subject to erratic response due to weather conditions.

GENERAL CULTURE

Soils and Climate

Olives are grown on a wide variety of soils, many too poor to support cultivation of other crops. They are tolerant of high pH, salinity, excess boron, and drought but sensitive to flooding.

Olives are supremely adapted to Mediterranean climates and cannot tolerate high humidity, due to disease and physiological disorders. Fluctuations in humidity and temperature 1 to 3 months following fruit set cause a condition known as aseptic apical decay on fruits. Part of the fruit surface turns black, and the fruit usually abscises; even in Mediterranean climates, losses may reach 30 percent.

Chilling requirement varies among cultivars; those grown in California have relatively high requirements (~1,000 hours). Cultivars grown in northern Africa fruit well with only a few hundred hours of chilling. Some texts state that olive flowering occurs in response to vernalization, since flowers develop during or following (as opposed to before) the cold period. However, floral induction occurs in November, prior to winter, so technically olives are chilled, not vernalized. Coldhardiness varies with cultivar and may approach 5 to 15°F in midwinter, when fully acclimated, but foliage and fruits are damaged by frost during active growth.

Propagation

Cuttings

The ease of vegetative propagation of the olive undoubtedly contributed to its early domestication and use by humans. Worldwide, rooting of cuttings is the most popular method of olive propagation. Unusually large cuttings will root: in Spain, cuttings with diameters up to 6 to 12 inches are used to establish new plantings. They are pruned heavily and mounded with soil throughout winter, and rooting takes place prior to summer heat. More commonly, hardwood cuttings are made from 3- to 4-year-old wood taken in midwinter. Leaves are stripped off completely, and cuttings are rooted over the course of several months. Propagators may heat the root zone and use growth regulators to improve rooting.

Softwood cuttings, or leafy stem cuttings, root better and are more common than hardwood cuttings. Leafy shoots are taken from 1- to 2-year-old wood in summer, treated with growth regulators, and stuck in well-aerated media under mist for 2 to 3 months. A 4- to 5-inch cutting will attain a height of about 2 feet by the following summer and be ready to plant. Planting is often delayed until the following spring due to the hot, arid conditions of summer and the consequent poorer field survival.

Grafting and Budding

T-budding and wedge grafting are used for cultivars that root poorly or when rootstock use is warranted. T-budding is done in spring, when bark is slipping on rootstocks propagated the previous year from seed or cuttings. Simple wedge grafts can be made in winter or spring. Scions 2 to 3 inches long, taken from the central portion of 1-year-old shoots, are grafted onto stocks.

Suckers and Ovuli

Suckers are simply shoots that arise from the trunk or roots and thus are similar to a naturally rooted cutting. They can be removed and planted directly, if well rooted, or treated as a softwood cutting, if the root system is poor. Ovuli are masses of callus tissue that often form at the base of trunks of older trees. They produce shoots and, if mounded with soil, adventitious roots. Both are primitive methods of propagation, practiced mostly in areas of less-intensive cultivation.

Rootstocks

Due to the ease of rooting, most olives are grown on their own roots, from cuttings, suckers, or ovuli. There are no particular rootstocks of worldwide importance. Cultivars that germinate well from seed or are easily rooted are generally used as rootstocks. The cultivar 'Oblonga' is resistant to verticillium wilt and is used as a landscape plant but rarely used commercially. Rootstocks are known to affect olive yield, tree size, and fruit quality, but the effects are highly cultivar dependent. The potential for dwarfing olive by using rootstocks appears to be possible but has remained unexploited.

Planting Design, Training, Pruning

Traditional olive groves are usually composed of large trees, scattered at irregular distances from one another, as a result of tree mortality or grafting/rooting in situ to fill in empty spots. There may be as few as 10 trees/acre, and usually no more than 40 trees/acre. Many are situated on steep hillsides or terraces that are unsuitable for other crops. As little as 10 to 15 percent of the land area is covered by tree canopy. Since most groves are not irrigated, wide spacing is necessary for production because trees must survive on stored soil water throughout the summer. When possible, the soil surface is cultivated to eliminate weeds. Intensive orchards are typically planted at densities of 30 to 40 feet apart in all directions, for a yield of about 40 to 60 trees/acre. Filler trees, which initially double the tree density, may be used but are removed after 8 to 10 years. Super-high-density orchards, with hundreds of trees per acre, are used to a limited extent (Figure 21.2).

FIGURE 21.2. Traditional olive groves are widely spaced and generally clean cultivated in arid climates. *(above left)* An olive orchard in La Mancha, Spain. *(above right)* Young trees planted at high density and staked to encourage fast growth in western Sicily. *(right)* A widely spaced table olive orchard with trees trained to an open-center system.

Olives naturally form large shrubs with spreading canopies, similar to citrus trees. Traditionally, trees are trained to single trunks devoid of scaffolds over the lower 6 feet. This allows grazing of animals in groves, intercropping, and ease of movement around trees by cultivation equipment and harvesting labor. Initial productivity is sacrificed by heavy pruning of young trees to form single trunks. In more intensive plantings, trees are pruned very little during the first 3 years. The objective is to develop an open canopy composed of several scaffold limbs. Young trees may be headed to promote initial branching, and shoots below 2.5 to 3 feet are removed to allow shaker attachment for mechanical harvest. Afterward, undesirable limbs are thinned and suckers are removed from the trunk, and trees branch naturally on their own. Mature trees that are mechanically harvested are trained to fewer than five scaffolds to speed the harvest operation because mature trees require individual limbs, not trunks, to be shaken.

Pruning is necessary to stimulate new fruiting wood for the following year's crop, decrease the tendency for alternate bearing, and help control insect and disease problems. In old, traditional groves, olive trees are sometimes burned down or severely cut back and allowed to regrow on an infrequent basis. This eliminates production for some time but accomplishes the task of rejuvenating the trees. It is far more desirable to prune annually, but labor availability and tradition may preclude annual pruning.

Olives tend to form thick canopies of tightly bunched shoots, which limits productivity and encourages pests. Pruning cuts are designed to open the canopies to light, increasing the depth of canopy involved in fruiting and discouraging black scale infestation, as these insects congregate in shaded areas. Spray penetration is also enhanced on trees with open canopies. Thinning cuts are made in dense areas, removing shoots growing downward and long, willowy shoots that do not lend themselves to shake harvesting. Time of pruning is critical; the bacterium causing olive knot disease is spread by rain and can only enter trees through wounds. Thus, unlike most fruit trees, olives are pruned in spring or summer when rain does not occur to minimize this disease. Also, pruning harder in "on" years and lighter in "off" years helps to decrease the degree of alternate bearing.

Pest Problems

Insects

Black scale. Saissetia oleae is a widespread pest of olive, and a particular problem for olive cultivation in California. The scale adults are black, $\frac{1}{10}$ inch to $\frac{1}{4}$ inch in diameter, and easily distinguished by a ridge in the shape of the letter "H" on the back. As with all scale insects, they use piercing sucking mouthparts to remove plant sap, and the honeydew they excrete invites sooty mold fungus to grow on leaves and stems. Sooty mold can block light from leaves and reduce photosynthesis in the canopy. These insects congregate on young shoots and leaves, particularly in shaded, dense areas of the canopy. Their sensitivity to sunlight and heat is exploited as a cultural control—trees are pruned to have open canopies. Pruning is sufficient to prevent severe infestations in most cases, but high populations are controlled by insecticides and/or oil sprays targeted at the crawler stage in early summer. Several other scale insects may attack olive, including olive scale *(Parlatoria oleae)*, mussel scale *(Lepidosphea ulmis)*, ivy scale *(Aspidiotus hederae)*, greedy scale *(Hemiberlesia rapax)*, and latania scale *(Hemiberlesia lataniae)*.

Olive kernel borer. The larvae of this moth *(Prays oleae)* damage leaves, flowers, and fruits, causing fruits to drop. The olive kernel borer has been regarded as one of the most serious pests of olive in the Mediterranean area. The larvae enter the young fruits and bore to the developing seeds, where they feed during the summer; infested fruits drop in September. Limb prunings left on the ground in late winter provide egg-laying sites for the first generation of adult moths and thus should be removed from orchards. Sprays are timed to kill the adults as they lay eggs on flowers.

Olive fruit fly. *Bactrocera oleae* is the worst pest of olive in the Mediterranean region. It is widely distributed across the Mediterranean countries, and, unfortunately, it was introduced into California in 1998 and is now widespread there. It is a large fruit fly, resembling a housefly, with a whitish spot on the back and black spots on the wing tips. Adults lay eggs on fruits, and the maggots feed on the fruit flesh; there are generally three to four generations per year. Infested fruits remain on the trees but will have reduced oil quality and often are infected by secondary fungal pathogens. Fruits destined for table olive markets can be rejected by processors even if only a few maggots are found. Proteinaceous baits laced with insecticide attract adults, which are killed as they try to lay eggs. These are applied to small areas of the canopy as sprays, or for heavier infestations, the entire canopy is sprayed with insecticide. Bait sprays containing the insecticide spinosad are used in California from summer to harvest for control.

Diseases

Olive knot or canker. This bacterial disease, caused by *Pseudomonas syringae* pv. *savastanoi*, occurs worldwide, affecting wood, reducing productivity, and causing tree decline over many years. It also causes incidental defoliation, fruit drop, and off-flavors in fruit, but it is the tumors or knots that form on wood which debilitate trees the most. The disease is carried from tree to tree by pruning tools, and it is recommended that tools be disinfected after pruning infected trees. It is also moved by rain splash within trees. The bacterium can enter the plant only through wounds, including leaf scars, branch gaps, pruning scars, and freeze-induced cracks in tissues. It is often worse after cold winters or in areas where leaf drop coincides with rainfall. Pruning during the dry season reduces the incidence of the disease, but sprays are often needed. In California, where rain occurs primarily in winter, sprays of copper-containing fungicides are applied in fall, prior to rain, and in spring, during leaf shed. There are no truly resistant cultivars, although 'Ascolano' is said to be more resistant than 'Frantoio' and 'Leccino'.

Olive leaf spot or peacock spot. This foliar fungal disease, caused by *Spilocea oleaginea*, leads to defoliation and indirect loss of yield in most regions every year, but it is not important as a fruit pathogen. Several small black lesions form on leaves, some having a yellow halo, giving the appearance of a bird's eye (hence, the name). Copper sprays are used to prevent infection, which is more prevalent on trees with dense, shrubby habits in more humid climates. Sprays are applied in autumn before winter rains. 'Frantoio', 'Picholine Marocaine', 'Arbequina', and 'Mission' are highly susceptible, and 'Koroneiki', 'Farga', 'Sevillano', and 'Ascolano' are resistant.

Fruit pathogens. Olive shield, caused by *Macrophoma dalmatica*, and anthracnose, a result of infestation by *Gleosporium olivarum*, can cause injury to fruit surfaces, complete fruit rot, and premature fruit drop, leading to dehydration and increased acidity of fruits. Anthracnose can also affect wood and leaves but requires higher humidity and rainfall for infection. Copper and synthetic fungicides are applied in late summer for control.

HARVEST, POSTHARVEST HANDLING

Maturity

Most table olives are harvested when they change from green to yellowish green in color and are firm; this is usually midautumn. The mesocarp exudes a white juice when squeezed at this stage. In California, most fruits are processed into "black ripe" olives, so some red coloration is allowable at harvest. Greek-style table olives are harvested at a more mature stage, when dark red to black in color. Oil olives are harvested in late autumn or winter, when they have turned black and have reached their maximum oil content (20-30 percent). Delaying harvest results in poor-quality oil, due to the loss of essential oils and aromas and increased acidity. Delaying harvest also results in increased alternate bearing, and trees used for table olives often fluctuate in yield less than trees used for oil. Olives are classified as nonclimacteric fruits with respect to ripening.

Harvest Method

Olives are traditionally hand harvested, a process that is not only tedious and laborious but also represents the major proportion of the costs of production. Hand harvest is accomplished by three techniques:

1. Collection of fallen fruits from the ground
2. "Milking," or the stripping of fruits from limbs using half-open hands, so that fruits fall into picking bags or onto nets below the tree
3. Beating limbs with large sticks to dislodge fruits, which are also collected on nets

Some table olives and high-quality oil olives are picked individually into baskets around the picker's neck to avoid damage and subsequent quality loss. Collecting fruits that fall naturally to the ground is inexpensive but seriously compromises oil quality. Thus, milking or beating are the most commonly used techniques.

Mechanical harvest of olives has been studied and attempted in various forms for years. It is used to a limited extent in more intensive orchards. Compared to other tree fruits that are mechanically harvested, olives are problematic. Olives require about five times more shaking energy than other fruits, such as prunes and almonds, due to the willowy nature of the trees and the resistance to detachment of fruits. Using mechanical shakers designed for almond harvest in California, at best, 65 to 80 percent of the fruits can be removed from the trees. The remaining fruits are either lost or must be hand harvested. Cullage can be four times higher with mechanical shakers. Initial costs of equipment are also high, precluding this approach for small growers.

Postharvest Handling

Table olives are cleaned and transported to processing plants, where growers are paid based on fruit size, color, and total weight. The fruits are washed and may be stored temporarily before processing (see Storage section). Raw olives are bitter due to the glucoside oleuropein in the fruits; this is neutralized by soaking in a solution of 1 to 2 percent lye (sodium hydroxide), followed by thorough leaching with water to remove the lye. Black table olives are those exposed to air during this process, which allows the oxidation of phenolic compounds, yielding a black color. Green table olives are exposed to lye in the absence of aeration and remain green. Spanish green olives are fermented in brine to which lactic acid bacteria have been added after lye is leached. A hot brine solution is added to fruits in cans or jars and then heated to 240°F for 1 hour for pasteurization. Many canned olives are pitted prior to canning, and green olives often have pimentos (small slices of peppers) placed into pit cavities.

Oil olives are brought to mills, where they are crushed whole, usually by two large, round stones rotating in opposite directions (Figure 21.3). The resulting paste is spread onto round mats of coconut fiber or nylon mesh with holes in the center. The mats are stacked onto a dowel under a hydraulic press, and the liquid (water and oil) is pressed out in a process known as "cold-pressing." Alternatively, olive paste can be centrifuged to extract oil. Percolation is a process

FIGURE 21.3. An olive oil mill. *(left)* A chute delivers fruits to millstones for crushing. *(right)* A small hydraulic press is used to extract oil from the crushed fruits, which are spread on fiber mats and stacked on a dowel.

used to extract oil by submerging a metal plate in olive paste, to which oil will adhere; the plate is removed and oil drips or "percolates" from the plate into a vessel. Cold-pressing is the most popular method of oil extraction. Finally, the water is separated from the expressed oil by centrifugation, and the final oil may be filtered for clarity.

Storage

Table olives are often stored in brine or acetic/lactic acid solutions prior to processing because pickling vats can process only so many tons of olives at one time. Olives can be stored for several weeks or months in brine or acid solution. Alternatively, green olives can be stored in refrigeration for 4 to 8 weeks, depending on temperature, before they experience chilling injury. Temperatures of 40 to 50°F are used. Once processed, table olives can be stored for about 2 years without loss of quality, similar to other pickled products. Olive oil can be stored many months, particularly if kept from light and heat. It may go rancid, as do other vegetable oils, when exposed to oxygen.

CONTRIBUTION TO DIET

Worldwide, most olives are made into olive oil, with smaller amounts canned in a number of different styles. California production is somewhat anomalous, in that the vast majority is a single product, the black-ripe canned olive. The 2002 utilization for California was as follows:

Product	Percentage of Crop
Canned (black-ripe, green-ripe, and Spanish green styles)	81
Frozen	10
Crushed for oil (culls that cannot be processed as table olives)	6
Dried	4

Olive oil marketing is controlled by an international agreement, overseen by the International Olive Oil Council located in Madrid. Flavor and acidity are the primary determinants of oil quality. Official definitions of olive oil are based upon flavor, acidity, and processing methods used.

- *Virgin oil.* This is minimally processed oil extracted by cold-pressing. It has three subcategories: extra, fine, and ordinary, based largely on acidity and flavor. Extra-virgin olive oil has less than 0.8 percent acidity (by weight from oleic acid), excellent flavor, and often comes from the first pressing of olives. Fine oil is often just termed "virgin" and has acidity of less than 2 percent. Ordinary virgin oil has acidity of up to 3.3 percent. Virgin oils with acidity greater than 3.3 percent are not used for human consumption and are designated "lampante," meaning lamp oil.

- *Refined oil.* This is virgin olive oil refined to remove off-flavors and odors by lye or other treatments, which do not alter the glyceridic structure of the oil. It has acidity of less than 0.3 percent.
- *Blended oil.* This is virgin oil blended with refined oil, with acidity of less than 1 percent. It is labeled "pure" and constitutes the bulk of olive oil sold.
- *Olive-pomace oil or residue oil.* Oil recovered from pressed olive paste by solvents falls into this category. It cannot be submitted to reesterification processes or mixed with other plant oils. Olive-pomace oils are classified as "crude" or "refined," with the latter having lower acidity than the former. A third category, simply "olive-pomace oil," is a blend of refined pomace oil and virgin olive oil with acidity of less than 1 percent.

Per capita consumption of olive oil varies widely by country. In Greece, over 17 liters (4.5 gallons) are consumed per person, the highest per capita rate in the world. Italians and Spaniards consume about 10 liters (2.5 gallons), but most other countries consume less than half a liter (1 pint) per year. Canned olive consumption is far lower but has also increased over time. In the United States, consumption of canned olives was about 1.3 pounds/person in 2003.

Dietary value per 100-gram edible portion:

	Oil (110 g = 7.1 tbsp)	Green Olives (pickled)	Ripe Olives (canned)
Water (%)	0	75.2	80
Calories	884	145	115
Protein (%)	0	1.0	0.8
Fat (%)	100	15.3	10.7
Carbohydrates (%)	0	3.8	6.3
Crude Fiber (%)	0	0.5	3.2
	% U.S. RDA (2,000-calorie diet)		
Vitamin A	0	8	8
Thiamin, B_1	0	1	<1
Riboflavin, B_2	0	<1	0
Niacin	0	1	<1
Vitamin C	0	0	2
Calcium	<1	5	9
Phosphorus	0	<1	<1
Iron	3	3	18
Sodium	<1	65	36
Potassium	<1	1	<1

BIBLIOGRAPHY

Bartolini, G. and R. Petrucelli. 2002. *Classification, origin, diffusion and history of the olive.* Rome: FAO.

Dolamore, A. 1994. *The essential olive oil companion.* New York: Interlink Books.

Ferguson, L., G.S. Sibbett, and G.C. Martin (eds.). 1994. *Olive production manual* (University of California Division of Agriculture and Natural Resources Publication 3353). Davis, CA: University of California, Davis.

Katsoyannos, P. 1992. *Olive pests and their control in the Near East* (FAO Plant Production and Protection Paper 115). Rome: FAO.

Knickerbocker, P. 1997. *Olive oil: From tree to table.* San Francisco: Chronicle Books.

Lavee, S. 1985. *Olea europea*, pp. 423-434. In: A.H. Halevy (ed.), *CRC handbook of flowering*, Volume 3. Boca Raton, FL: CRC Press.

Martinez-Moreno, J.M. 1975. *Manual of olive oil technology.* Rome: FAO.

Pansiot, F.P. 1961. *Improvement in olive cultivation* (FAO Agricultural Studies No. 50). Rome: FAO.

Taylor, J.M. 2000. *The olive in California.* Berkeley, CA: Ten Speed Press.

Tellez Molina, R. (ed.). 1977. *Modern olive production.* Rome: FAO.

Zalom, F.G., R.A. Van Steenwyk, and H.J. Burrack. 2003. *Olive fruit fly* (University of California Statewide IPM Program, Pest Notes Publication 74112). Davis, CA: University of California, Davis.

Chapter 22

Papaya *(Carica papaya)*

TAXONOMY

Papaya belongs to the Caricaceae, a small family of only four genera and 27 to 30 species (some estimate up to 71 species). This family is closely allied with, and formerly part of, the Passifloraceae or passion fruit family. *Carica* contains the majority of the family's species and is indigenous to tropical America. *Carica papaya* L., the papaya of commerce, is called "pawpaw" in some English-speaking countries; however, this is not to be confused with the North American Annonaceous species *Asimina triloba*.

Carica pentagona, the babaco, is similar to papaya but smaller (<10 feet), producing five-angled, seedless fruits reaching 12 inches in length. It is native to higher elevations of the tropics than is papaya and makes a good papaya substitute in regions too cool to grow papaya. *Jacaratia spinosa*, the wild papaya, is found in lowland to premontane moist forests in Costa Rica and adjacent countries; it produces a small, orange, papayalike fruit with white flesh.

Cultivars

Active breeding programs in a number of countries have produced cultivars that match local preferences for fruit size, shape, flesh color, flavor, and other characteristics. 'Hortus Gold' and 'Honey Gold' are gold/yellow-skinned, yellow-fleshed cultivars popular in South Africa; they are twice the size of 'Solo' types grown in Hawaii, but smaller than most papaya grown in tropical America. Fruits from cultivars grown in Central America, such as 'Cartagena', 'Cedro', and 'Santa Cruz Giant', are larger and often cylindrical in shape (Plate 22.1). Larger fruits are also preferred in India, so similar cultivars, such as 'Coorg Honey Dew', are planted there. In Australia, 'Improved Petersen', 'Guinea Gold', and 'Sunnybank' are major cultivars that produce fruits of intermediate size with a yellow flesh color.

The first major cultivar of international importance was the small, pear-shaped 'Solo', which was introduced to Hawaii from Barbados in 1911. The name derives from the relatively small size of the fruit—at 0.5 to 1.5 pounds, it can be eaten by one person, as opposed to the watermelon-sized types grown in Central America and elsewhere. It has been used as a parent in breeding newer cultivars, such as 'Kapoho solo', 'Waimanalo', 'Higgins', and 'Wilder'. The transgenic cultivars 'Sunup' (red flesh) and 'Rainbow' (yellow flesh) have resistance to papaya ringspot virus and were derived from 'Solo' parent lines. 'Solo' types are the most abundant cultivars imported to the United States

ORIGIN, HISTORY OF CULTIVATION

Papaya is native to the tropical region of the Americas, from southern Mexico through the Andes of South America. It was spread to the south by Native Americans, and throughout the

Introduction to Fruit Crops

doi:10.1300/5547_22

Caribbean with Spanish exploration. The Spanish also carried it to Europe and the Pacific Islands. By the middle of the seventeenth century, papaya was distributed pantropically. Papaya was introduced to Hawaii in the 1800s, and Hawaii remains the only state in the United States to produce papaya commercially. A small industry developed in Florida in the first part of the twentieth century but declined rapidly with the introduction of viral diseases that today threaten papaya elsewhere. In fact, the recent decline of the Hawaiian industry was caused primarily by the same pathogen, papaya ringspot virus, that destroyed plants in Florida. However, the disease was overcome by biotechnologists at the University of Hawaii, who inserted a gene into the 'Sunrise' cultivar that conferred resistance to the virus. This made the papaya the first genetically modified fruit crop grown for human consumption. Since 1998, most of the papaya acreage in Hawaii has been changed to genetically modified cultivars.

FOLKLORE, MEDICINAL PROPERTIES, NONFOOD USAGE

Papaya fruit pulp is the basic component of many facial creams, salves, and shampoos. Papaya culls are ground into a puree that can be used for cosmetic purposes; the process also yields seeds for replanting fields.

Papain is one of two proteolytic enzymes found in papaya latex (the other is chymopapain). Papain is extremely useful because it retains proteolytic activity over a wide pH range, unlike other proteases. Thus, it is in more widespread use than bromelain, the proteolytic enzyme found in pineapple juice. Latex is extracted on a commercial scale in East Africa, where the green fruit are "tapped" by making incisions on the fruit surface in the morning, and catching the exuding latex over a period of days. The latex is then dried and ground into powder. The most popular use is as a meat tenderizer. Columbus, on one of his voyages to the Caribbean, noted that natives could consume a large amount of fish and meat without getting indigestion, if unripe papaya was eaten after the meal. Green papaya can be rubbed onto a piece of meat and cooked with it, or the crushed leaves can be wrapped around meat, to achieve the same effect as using a commercial tenderizer. Beef cattle are sometimes injected with papain a half hour before slaughter to tenderize them. Additional uses include beer and juice clarification, wool and silk treatment before dyeing, dehairing hide before tanning, tuna liver oil extraction, face creams and face lift preparations, and in cleansers, such as those used to clean contact lenses. Papain can be used during surgical procedures to dissolve ruptured spinal discs; it is referred to as "nature's scalpel" because it preferentially degrades dead tissue.

Folk medicine uses of fruits, leaves, and the latex may be related to papain content. Fresh latex can be smeared on boils, warts, corns, or freckles to remove them from the skin. It is also used for treating psoriasis and ringworm. Latex is smeared on the uterus to cause miscarriage in India, or sometimes the unripe fruits and/or seeds are eaten to induce miscarriage. Either the latex or plant parts are considered antiseptic, antibacterial, anthelmintic, and amebicidal. Papain is used in remedies for ulcers, diphtheria, swelling, toothache, fever, relief of gas, and sour stomach.

PRODUCTION

World

Total production in 2004, according to FAO statistics, was 6,504,369 MT or 14 billion pounds. Papaya is produced in 54 countries worldwide, on about 900,000 acres. Production has increased 40 percent in the past decade. Yields average 16,000 pounds/acre but vary from 5,000 to over 38,000 pounds/acre. The top ten papaya-producing countries (percent of world production) follow:

1. Brazil (25)
2. Nigeria (15)
3. India (12)
4. Mexico (11)
5. Indonesia (10)
6. Ethiopia (4)
7. Congo (3)
8. Peru (3)
9. Venezuela (3)
10. China (2)

Production fell fivefold in Thailand between 1994 and 1999 due to papaya ringspot virus. Once ranked third in the world, Thailand now is not even among the top ten papaya-producing countries. India is the only country to experience a large increase, about threefold, in the past decade.

United States

Total production in 2004, according to USDA statistics, was 16,136 MT or 35.5 million pounds. Total industry value in 2004 was about $12 million; it fluctuated between $12 million and $19 million prior to 1998. Yield averages about 28,700 pounds/acre, nearly twice the world average. All production is in Hawaii on about 1,400 acres, with another 1,000 acres of nonbearing plants, as the state transitions to new markets and cultivars. Papaya ringspot virus began affecting the industry by 1994, when production had been typically 30,000 MT. Production rebounded somewhat after the introduction of genetically modified cultivars in 1998 but has declined more recently (Figure 22.1). Genetically modified cultivars are resistant to papaya

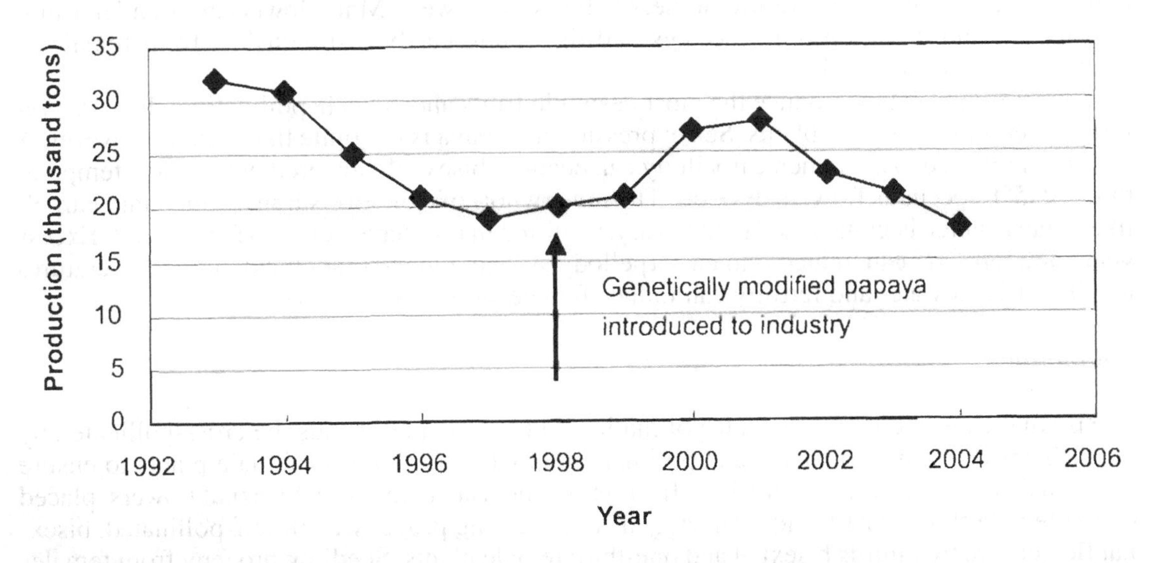

FIGURE 22.1. Production trend for Hawaiian papaya, 1993-2004. The industry declined during the 1990s due to papaya ringspot virus. The arrow shows the point of introduction of genetically modified papaya and the subsequent recovery of the industry, followed by decreases precipitated by other diseases and loss of markets. *Source:* USDA statistics.

ringspot virus but are more susceptible to blackspot fungus (*Phytophthora* spp.). Thus, some growers are reluctant to adopt genetically modified cultivars and continue to grow 'Kapoho', which is virus susceptible. In addition, lucrative export markets, such as Japan, will not accept genetically modified fruits. Thus, prices received for genetically modified cultivars have remained at $0.20 to $0.30 per pound, while countries preferring traditional cultivars have paid up to $0.45 to $0.60 per pound for Hawaiian papaya.

BOTANICAL DESCRIPTION

Plant

The papaya plant is a large, single-stemmed, herbaceous perennial that grows to 30 feet in its native range, but less than 20 feet in cultivation. Leaves are very large (up to 2.5 feet wide), palmately lobed or deeply incised, with entire margins, and petioles of 1 to 3.5 feet in length. Stems resemble a trunk; they are hollow, light green to tan brown, up to 8 inches in diameter, and bear prominent leaf scars. Leaves, stems, and unripe fruits produce copious latex if cut. Plants are fast growing and short-lived, achieving heights of several feet in 1 year when grown from seed (Plate 22.2).

Flowers

Plants are dioecious or hermaphroditic, rarely monoecious, with cultivars producing only female or bisexual (hermaphroditic) flowers used in cultivation. Papayas are sometimes said to be "trioecious," meaning that separate plants bear either male, female, or bisexual flowers. Female and bisexual flowers are waxy, ivory white, and borne on short peduncles in leaf axils along the main stem. Flowers are solitary or borne in small cymes of three individuals. Ovary position is superior. Prior to opening, bisexual flowers are tubular and female flowers are pear shaped. Female flowers lack the bright yellow anthers of bisexual flowers. Male flowers are smaller, trumpet shaped, and borne on long racemes and thus quite easily distinguished from female or bisexual plants (Plate 22.3).

Since bisexual plants produce the most desirable fruits and are self-pollinating, they are preferred over female or male plants. Sex expression in papaya is not quite that simple, however. A condition referred to as stamen carpellody can occur in bisexual-flowered types if low temperatures (<55°F) occur as flowers develop. The inner whorl of stamens is transformed into carpel-like organs, which become fused to the ovary, resulting in misshapen or "cat-faced" fruit. Excess water and fertilizer can induce stamen carpellody as well. On the other hand, high temperatures and insufficient water and fertility can induce female sterility in flowers.

Pollination

Bisexual-flowered plants are self-pollinating, but female plants must be cross-pollinated by either bisexual or male plants. Hand pollination may be practiced for female plants to ensure uniform fruit size, fruit shape, and high fruit set. Anthers are taken from bisexual flowers, placed on the female flower pistil, and then bagged. The seedling progeny from self-pollinated, bisexual flowers are two-thirds bisexual and one-third female plants. Seedling progeny from females pollinated with males are half male and half female plants, whereas seedlings from bisexual plants pollinated by males are one-third each of male, female, and bisexual plants. Since papayas are normally propagated from seed, male plants are removed from plantings to reduce the chance of male seed production.

Fruit

The papaya's fruit is a large, oval to round *berry*, sometimes called "pepolike berry" because it resembles a melon, by having a central seed cavity. Fruits are borne axillary on the main stem, usually singly, but sometimes in small clusters. Fruits weigh from 0.5 pound up to 20 pounds and are green until ripe, when they turn yellow or red-orange. Fruits from female flowers are more round in shape and thin-skinned, whereas fruits from bisexual flowers are the typical pyriform/oval shape with a thicker skin. Flesh is yellow-orange to salmon at maturity, with the edible portion surrounding the large, central seed cavity. Hundreds of small black seeds (3/16 inch), "looking like capers crossed with beluga caviar" (from Schneider, 1986, cited in Chapter 1), are lightly attached to the flesh in the central cavity, each surrounded by a transparent, gelatinous aril (similar to a cantaloupe). Fruit production occurs year-round because flowering is continuous; individual fruits mature in 5 to 9 months, depending on cultivar and temperature. Plants begin bearing in 6 to 12 months.

GENERAL CULTURE

Soils and Climate

Papayas grow on a wide range of well-drained soils with a pH in the range of 5.5 to 7.0. Poor drainage predisposes plants to soilborne diseases. Papayas have a relatively high water requirement and must be provided with irrigation in dry seasons

Papayas perform best in hot, rainy, tropical lowlands. They are rarely cultivated in subtropical climates, due to the impact of cool temperature on fruit growth and maturation. Optimal temperatures are 70 to 90°F, and temperatures as high as 65°F can cause stamen carpellody (see earlier Flowers section). Female plants are less sensitive to stamen carpellody than are bisexual plants, so cooler production areas use dioecious cultivars, such as 'Hortus Gold' or 'Improved Petersen'. Plants are completely intolerant of freezing. High wind also causes damage through fruit loss, leaf damage, or uprooting.

Propagation

Unlike most fruit crops, papayas are seed propagated. Plants are fairly true to type from seed and precocious, so the usual advantages of clonal propagation are diminished in papaya. Since plant sex cannot be determined in seed or seedling stages, several plants or seeds are planted at a single site in the field, roguing all but the bisexual or female plants once they flower. Seeds may be planted directly into planting holes, or seedlings may be transplanted when 6 to 8 inches tall. Direct seeding may result in deeper rooting, more vigorous growth, higher yield, and earlier flowering than is seen with transplanting.

Rootstocks

Papaya has no rootstocks.

Planting Design, Training, Pruning

Plantings are laid out in rows about 6 to 8 feet apart, yielding plant densities of 600 to 1,200 plants/acre. Wider alleys are placed periodically throughout the planting to allow access to larger equipment (Figure 22.2). Plastic mulch is used in some operations to control weeds and retain water. Papayas do not require training and little or no pruning; they form single-stemmed plants naturally. Side shoots, if they occur, are removed, and older leaves are also removed to facilitate harvest.

Pest Problems

Insects

Fruit flies. Several species of tiny flies (*Bactocera, Ceratitis, Anastrepha* spp.) lay eggs just beneath the skin of ripening fruits, potentially leading to wormy fruits. However, fruit damage is usually minimal, and the major problem with fruit flies occurs with exported fruits, which cannot be shipped to countries lacking these pests without postharvest hot-water treatment. Flies do not lay eggs in papaya until about 25 percent of the skin has turned yellow. Papaya fruit fly (*Toxotrypana curvicauda*) is a major hazard in the Caribbean. The larvae are more likely to cause fruit damage here than are other species. Only thick-skinned cultivars are resistant. Control by insecticides is difficult; in some areas, paper bags or newspaper is placed over fruits after flower parts drop.

Papaya webworm. Also called fruit cluster worm, the larvae of this worm (*Homolapalpia dalera*) are found in between adjacent fruits or stems and fruits. They bore into the stems and fruits, opening wounds that allow anthracnose fungus to enter the tissue. Control is by sprays at the first sign of webs, but control is difficult due to the location of the pest within the fruits.

Aphids, whiteflies. At least two species of aphids (*Myzus persicae, Aphis gossypii*) and the papaya whitefly (*Trialeuroides variabilis*) may attack young leaves, shoot tips, and fruits. These insects have piercing, sucking mouth parts and target the sugar-conducting tissues, robbing young leaves, twigs, and fruits of this food source. They also transmit the papaya ringspot virus and other viruses. Whiteflies are generally found underneath leaves, and their secretions promote the growth of sooty mold fungus. Ladybugs and lacewing flies are natural enemies. Sporadic infestations are easily controlled by spot spraying.

FIGURE 22.2. A commercial papaya field on the big island of Hawaii.

Diseases

Papaya ringspot virus. Probably the most severe limitation to papaya production, this virus causes stunting, distortion of leaves, and severe reduction in yield and fruit size (Figure 22.3). It is spread by aphids, which are difficult to control completely. Characteristic circular chlorotic areas appear on leaves and fruits. Fruits borne after symptom development are bitter. There is no treatment for the virus once it enters the plant; infected plants are removed immediately when symptoms appear. Isolated plants are often unaffected because the virus does not survive for long periods inside the aphids, and aphids do not fly very far. Curcurbits (cucumbers, zucchini, squash, etc.) serve as alternate hosts for the virus and thus are eliminated around papaya plantings. 'Cariflora' is a resistant cultivar but lacks the fruit quality of the Hawaiian 'Solo' types. 'Sunup' and 'Rainbow' have been genetically modified to resist the virus and are grown extensively in Hawaii.

FIGURE 22.3. A mature papaya suffering from papaya ringspot virus, a serious limitation to papaya culture.

Anthracnose. The fungus *Colletotrichum gloeosporioides* can rot ripe fruits or attack unripe fruits that have been damaged by insect feeding or other bruising. Generally, it is a problem of only ripe fruits and where insect damage is uncontrolled. Fungicides can be applied at 7- to 10-day intervals for control.

Bunchy top disease. Caused by an unnamed *Mycoplasma* sp., this disease is found in Florida, the Caribbean, and Latin America. Leaves appear yellow, stunted, and stiffened, similar to plants infected with papaya ringspot virus. Leaves will not exude latex in plants infected with bunchy top, but those with papaya ringspot virus will, allowing distinction of the two problems. There is no practical control. The disease is carried by two species of *Empoasca* leafhoppers, but insecticide sprays are futile. Diseased plants are destroyed.

Soilborne diseases. Phytophthora root and stem rot (caused by *Phytophthora parasitica*), pythium root rot (caused by *Pythium* spp.), and damping-off of seedlings (casued by *Pythium, Rhizoctonia, Phytophthora* spp.) are common problems in papaya production because rainfall is usually high in most regions. The phytophthora foot rot organism can also infect the fruits, causing stem end rot. Soil fumigation is sometimes necessary, especially when replanting fields formerly planted with papaya. 'Waimanalo' is resistant to root rot.

HARVEST, POSTHARVEST HANDLING

Maturity

Papayas develop their highest eating quality when the skin is around 80 percent yellow in color. However, for shipping, papayas are harvested when the first hint of yellow coloration appears. Papaya is classified as a climacteric fruit with respect to ripening.

Harvest Method

Fruits are hand harvested carefully to avoid scratching the skin, which would release latex and stain the skin. On taller plants, harvesters use poles or hydraulic lifts to reach fruits. Harvest is a continuous process once plants reach 6 to 12 months of age, and it constitutes about 40 percent of the total labor cost of producing papayas.

Postharvest Handling

To reduce postharvest fruit rot and to kill fruit fly eggs and larvae, papayas are commonly heat-treated postharvest (110-120°F) and then rinsed in cool water. Fungicides also may be used, generally in the wax applied during packing. Radiation treatments, such as "Sure Beam," are used to sterilize fruit fly eggs and larvae in fruits intended for export. Fruits are packed into single-layer boxes (10-15 pounds), often with tissue or foam padding to avoid bruising. Fruits can be cured at 85°F and around 100 percent humidity for better color expression prior to shipping.

Storage

Papaya can be stored at 45 to 55°F for 2 to 4 weeks. Below 50°F, papayas experience chilling injury in prolonged storage. Various protocols have been developed for controlled-atmosphere storage (1-4 percent O_2, 55°F) and hypobaric storage to increase fruit shelf life by several days, but they are not commercially practiced. Papayas are extremely perishable; shelf life at room temperature ranges from 3 to 8 days, depending on storage atmosphere.

CONTRIBUTION TO DIET

Papaya grown in Hawaii is utilized largely for fresh-market sales (96 percent), with small amounts being processed into juices and other packaged foods. Young leaves can be cooked and eaten as a green vegetable. Green or unripe papaya is used as a vegetable or salad garnish as well but must be boiled first to denature the papain in the latex. Papaya seeds are edible, with a taste similar to "nasturtium, watercress, and pepper" (from Schneider, 1986, cited in Chapter 1); they are used as an adulterant of black pepper.

Per capita consumption of papaya is 0.9 pound/year, increasing fourfold since 1980.

Dietary value per 100-gram edible portion:

Water (%)	88
Calories	39
Protein (%)	0.6
Fat (%)	0.1
Carbohydrates (%)	9.8
Crude Fiber (%)	1.8

	% of U.S. RDA (2,000-calorie diet)
Vitamin A	22
Thiamin, B_1	2
Riboflavin, B_2	2
Niacin	2
Vitamin C	103
Calcium	2

Phosphorus	<1
Iron	<1
Sodium	<1
Potassium	7

BIBLIOGRAPHY

Duke, J.A. and J.L. duCellier. 1993. *CRC handbook of alternative cash crops.* Boca Raton, FL: CRC Press.

Lauwers, A. and S. Scharpe. 1997. *Pharmaceutical enzymes: Drugs and the pharmaceutical sciences*, Volume 84. New York: Marcel Dekker.

Malo, S.E. and C.W. Campbell. 1986. *The papaya* (Florida Cooperative Extension Service Fruit Crops Fact Sheet FC-11). Gainesville, FL: University of Florida, Gainesville.

Marler, T.E. 1994. Papaya, pp. 216-224. In: B. Schaffer and P.C. Andersen (eds.), *Handbook of environmental physiology of fruit crops*, Volume 2: *Subtropical and tropical crops.* Boca Raton, FL: CRC Press.

Morton, J.F. 1987. *Fruits of warm climates.* Miami, FL: Julia F. Morton.

Nakasone, H.Y. 1986. Papaya, pp. 277-301. In: S.P. Morselise (ed.), *CRC handbook of fruit set and development.* Boca Raton, FL: CRC Press.

Nakasone, H.Y. and R.E. Paull. 1998. *Tropical fruits.* Wallingford, UK: CAB International.

Olaya, C.I. 1991. *Frutas de America*, pp. 38-51. Barcelona, Spain: Editorial Norma.

Ploetz, R.C., G.A. Zentmyer, W.T. Nishijima, K.G. Rohrbach, and H.D. Ohr (eds.). 1994. *Compendium of tropical fruit diseases.* St. Paul, MN: American Phytopathological Society Press.

Storey, W.B. 1985. *Carica papaya*, pp. 147-165. In: A.H. Halevy (ed.), *CRC handbook of flowering*, Volume 2. Boca Raton, FL: CRC Press.

Sturrock, D. 1959. *Fruits for southern Florida.* Stuart, FL: Southeastern Printing Co.

Chapter 23

Peach *(Prunus persica)*

TAXONOMY

The peach [*Prunus persica* (L.) Batsch] belongs to the Prunoideae subfamily of the Rosaceae, with other species often collectively referred to as "stone fruits." The subgenus *Amygdalus* contains the commercially important peach and almond. In addition, four other species are closely related to peach: *P. davidiana, P. kansuensis, P. ferganensis,* and *P. mira*. The former two species have small fruits, resembling peach, but poor eating quality and are used only as rootstocks. *Prunus ferganensis* is cultivated in the Ferghana Valley region of Tajikistan and Uzbekistan. *Prunus mira* is a shrub form native to southern Tibet, Nepal, and northern India, where it is infrequently cultivated for its fruit.

Ornamental cultivars of peach are used on a small scale as landscape trees. Fully doubled white flowers, deep red flowers, dwarfism, and red-leaved traits have been incorporated in ornamental cultivars.

Cultivars

Thousands of peach cultivars are found worldwide, and far more are grown commercially than is seen with many other tree fruits. One reason for this is the ease with which peaches are bred, unlike cultivars of other tree fruits that are mostly the result of chance selection, not breeding. Precocity and homozygosity result in early bearing of uniform fruits, which are edible from almost all seedlings. Many regions have their own breeding programs to produce cultivars specifically adapted to a particular situation. Cultivars popular 30 years ago within a given region have been or are being replaced by newer ones. Thus, no single cultivar is dominant worldwide, or even nationwide. However, as Scorza and Okie (1990) point out, several cultivars bred in the United States have been adopted by other countries with regularity, and Okie (1998) provides detailed descriptions for many of the U.S. cultivars. Peach cultivars fall into one of three major groups:

1. *Nectarines:* Although labeled and marketed differently from peaches, nectarines are simply fuzzless peaches. A single recessive gene results in the fuzzless condition. Still, *P. persica* var. *nectarina* is sometimes used as a name for this group of cultivars.
2. *Freestone peaches:* These are fresh-market peaches.
3. *Clingstone peaches:* These peaches are used primarily for canning.

The terms *clingstone* and *freestone* refer to the adherence of the mesocarp (flesh) to the endocarp (pit). This degree of adherence per se does not affect canning quality, but firm flesh texture is linked to the clingstone trait, and clingstones retain shape better, have brighter color, and produce clearer juice than do freestones when canned. It is unfortunate that clingstones were not named "firm fleshed" and freestones "melting flesh," since these names would more

doi:10.1300/5547_23

clearly reflect the characteristics of the two groups. What confuses the issue even more is that stone free–ness is also a function of time of maturation for all types of cultivars, with early ripening cultivars tending to be clingstone and later ones freestone. Uniformity of flesh color, texture, and flavor are important determinants of clingstone quality, whereas fruit size and red skin color ("blush") are important for freestone peaches. In general, early ripening cultivars (those which ripen in under 3 months) tend to be of poorer quality than the mid- or late-season cultivars, and they have a greater tendency to be small and to have split pits ("split pit" is a physiological disorder in which the pit splits or shatters before the fruit matures, rendering the fruit unmarketable). Since most peach cultivars have a similar appearance and often cannot be distinguished even by taste, there is little or no consumer preference for, or awareness of, particular cultivars in peach or nectarine. This is not true for many other fruit crops, for which certain cultivars are recognizable and sometimes preferred over others in the marketplace. One recognizable trait that varies in peach and nectarine cultivars is flesh color, being either white or yellow. Production of white-fleshed cultivars has increased over the past decade (Plate 23.1) to revitalize consumer interest in fresh-market peaches and nectarines.

ORIGIN, HISTORY OF CULTIVATION

Peaches were among the first fruit crops domesticated in China about 4,000 years ago. Cultivars grown today derive largely from ecotypes native to southern China, an area with a climate similar to that of the southeastern United States, a major peach-growing region. Peaches were moved to Persia (Iran) along silk trading routes. In fact, the epithet *persica* denotes Persia, which is where Europeans thought peaches originated. Greeks and especially Romans spread the peach throughout Europe and England, starting in 300 to 400 BC. Peaches came to the New World with explorers of the sixteenth and seventeenth centuries, with the Portuguese introducing it to South America and the Spaniards to the northern Florida coast of North America. Native Americans and settlers distributed the peach across North America into southern Canada, and eventually to California, the major production region in the United States today.

Seedlings were cultivated until improved cultivars became available in the 1800s; now, improved cultivars grafted onto a variety of rootstocks are cultivated almost exclusively. Peach germplasm went through a narrow genetic bottleneck when large, firm-fruited cultivars descended from 'Chinese Cling' and 'Shanghai' were introduced in the 1850s. Since that time, several active breeding programs have broadened the genetic base of the peach, adapting it to production regions that range from Ontario to the Guatemalan highlands.

FOLKLORE, MEDICINAL PROPERTIES, NONFOOD USAGE

The symbol of human longevity in Taoist philosophy was the old man (Shou Lu) who appears in illustrations with his finger stuck into the suture of a fuzzy peach, perhaps to symbolize the way to attain a long life. In China, Mother Hsi Wang Mu's peach garden appeared only once every 3,000 years, when the resulting fruits were used to make the gods' elixir of immortality. Similar to other stone fruits, the peach was a symbol of female genitalia in ancient China, a part of the Taoist sexual mysticism. In fact, the Chinese word *tao* means peach, and the word appears in cultivar names of some Chinese selections, such as 'Ta Tao' and 'Tsum Pee Tao'. In Roman mythology, the peach was the fruit of Venus, and St. Albertus Magnus believed that peaches were aphrodisiacs. On a less-sexual note, peaches were associated with immortality and sincerity or truth. A peach with a leaf attached symbolizes the union of the heart and tongue, hence truth.

Peaches were used by ancient Egyptians as offerings to the god of tranquility. Peach is the state flower of Delaware.

As with all members of the genus *Prunus*, peach leaves, flowers, and especially seeds and bark contain cyanogenic glycosides, such as amygdalin and prunasin. These compounds yield cyanide, when the sugar moeity is cleaved, which is of course toxic or lethal in large doses. However, in plant tissues, cyanide is low enough in concentration to be considered therapeutic, particularly for cancer (tumor) treatment, and has been used for this purpose since at least 25 BC. Apricot seeds contain the highest amounts of these cyanogenic compounds, and the cancer drug laetrile is derived from this source (see related discussion in Chapter 4, "Apricot").

Peach bark has been used as an herbal remedy for a wide variety of ailments. It has use in encouraging menstruation in females with delayed menses. It also relieves bladder inflammation and urinary tract problems; functions as a mild laxative; has expectorant activity for the lungs, nose, and throat; and relieves chest pain and spasms. Bark and root extracts contain phloretin, which has antibiotic activity on gram-positive and gram-negative bacteria. Essences of peach and plum are used to flavor cigarettes in some countries.

PRODUCTION

World

Total production in 2004, according to FAO statistics, was 15.561.206 MT or 34 billion pounds. (*Note:* World production data include both peaches and nectarines.) Peaches and nectarines are produced commercially in 71 countries worldwide on about 3.5 million acres. Worldwide average yields are just under 10,000 pounds/acre. Production has increased 44 percent in the past decade, largely due to increases in yield, as acreage has remained constant. Chinese production has increased severalfold since 1980, achieving the first-place spot for the first time in 1993. Prior to that, the United States and Italy led the world in peach and nectarine production. The top ten peach- and nectarine-producing countries (percent of world production) follow:

1. China (42)
2. Italy (13)
3. United States (10)
4. Spain (8)
5. Greece (7)
6. Turkey (3)
7. France (3)
8. Iran (3)
9. Chile (2)
10. Argentina (2)

United States

Total production in 2004, according to USDA statistics, was 1,410,000 MT or 3.1 billion pounds. The total value of the industry is about $548 million, broken down as detailed in the following discussion. Peaches and nectarines are produced commercially in 29 states on about 114,000 acres. Production has remained stagnant for the past ten years, as a decline in acreage has been offset by an increase in yield.

Freestone peaches account for 48 percent of total production. The top freestone peach–producing states (percent of freestone peach crop) follow:

	Crop Value ($ in millions)	Price Received (¢/pound)
1. California (55)	109	14
2. South Carolina (9)	31	28
3. Georgia (7)	33	34
4. New Jersey (4)	23	38
5. Pennsylvania (3)	16	35
6. All other states (22)	<1-15	17-80

Clingstone peaches account for 35 percent of total production, and all are produced in California. The industry value is $141 million; total acreage is 32,000; and average price is $0.13 per pound.

Nectarines account for 18 percent of total production, and almost all are produced in California. The industry value is $86 million; total acreage is 36,500; and average price is $0.17 per pound.

U.S. exports of fresh peaches in 2002 were 120,727 MT or about 9 percent of total production. Another 35,000 MT were exported canned or in fruit salad. Mexico, Canada, and Latin America are the major export destinations. There are no import data for peaches and nectarines, but small quantities are primarily imported from Chile in winter and Mexico in spring.

BOTANICAL DESCRIPTION

Plant

The plant is a vigorous-growing, but small tree with a spreading canopy that usually grows to 6 to 10 feet in cultivation (Figure 23.1). Trees are short-lived, generally living only 15 to 20 years, and even less on sites with a history of peach cultivation. Leaves are linear, with acute tips and finely serrate margins, folded slightly along the midrib, sickle shaped in profile, 2 to 6 inches in length.

Flowers

Peach flowers are light pink to carmine to purplish in color and 1 to 1.5 inches in diameter. The ovary is perigynous and simple (single locule) and surrounded by a hypanthium (Plate 23.2). The color of the inner surface of the hypanthium is indicative of flesh color; whitish green indicates white flesh and gold indicates yellow flesh. Petals can be large and showy or small and curved on margins. Flowers are borne singly on short peduncles (almost sessile) from lateral buds on 1-year-old wood, with

FIGURE 23.1. A peach tree in central Georgia trained to a low, sprawling, open-center form.

usually one to two flower buds per node. Flowers exhibit cleistogamy, pollinating themselves prior to opening.

Pollination

Peach is self-pollinating and so normally grown without a pollinizer in solid blocks. A few old cultivars require pollinizers, for example, 'J.H. Hale'.

Fruit

The peach fruit is a drupe. The bony endocarp (pit) surrounds a single, large, ovate seed; the flesh is the mesocarp, and the skin the exocarp (Plate 23.1). Trees are very precocious, producing some fruits in the second or third year after planting. Peaches require extensive thinning (~80-95 percent of flowers) for proper size development (Figure 23.2). Early thinning increases yield of marketable fruits but is not often practiced because frost is a perennial threat, and thinning before the last frost increases risk of crop loss. There are no chemical thinners for peach that work consistently, but ammonium thiosulfate can be used to desiccate blossoms, and high-pressure water or ropes dragged through the canopy can physically remove blossoms. Usually, thinning is done by hand 30 to 45 days after full bloom, leaving about one fruit per 6 inches of 1-year-old shoot length. Fruits that set the second year are commonly removed to promote tree growth. The first commercial crop is generally harvested in the third year, with maximal yields reached by the fifth and sixth years.

GENERAL CULTURE

Soils and Climate

Deep, well-drained soils are essential for good production and tree longevity.

Peaches are highly susceptible to poor drainage and flooding stress, as are most *Prunus* species. Loamy to moderately sandy soils are best. Sites previously planted with peaches are avoided because they are prone to the "peach tree short life" (PTSL) syndrome, also called peach tree decline, which greatly reduces orchard productivity. Several nematodes attack peach roots, resulting in poor growth and reduced longevity. Ring nematode (*Criconemella xenoplax*)

FIGURE 23.2. Before and after thinning; one fruit is left for every 6 inches of 1-year-old shoot length.

has been implicated as the predisposing agent for PTSL, and these pests move fastest in sandy soils. Irrigation has been beneficial for increasing fruit size, even in humid climates, and is essential on shallow soils or in Mediterranean climates.

Peaches bloom relatively early, before apple, pear, and cherry, and are less cold hardy than many tree fruit species. Frost is a problem in almost all growing areas of the world. Hence, they are considered to be warm-temperate in adaptation and are frequently cultivated in Mediterranean climates. Peach breeding has stretched the limits of adaptation quite broadly. Cultivars have been bred to perform in climates from southern Canada to the tropics, and peaches have a wider range of chilling requirements than any other tree crop. For example, 'Red Ceylon' peach has no chilling requirement and is grown in the tropics; 'La Premier' has a chilling requirement of 1,000 hours. On average, most cultivars have chilling requirements of 600 to 900 hours. Many low-chill cultivars (<500 hours) are available for subtropical areas. Peach flower buds tolerate temperatures as low as –25°F when dormant, and the wood is killed just below this level. Open flowers and young fruitlets are killed by brief exposure to 28°F or below. Peaches do not require cool nights to develop red skin color, as with apples; red color is more a function of cultivar and light exposure. Peaches ripen during the summer months in most climates but develop good quality in regions with cool as well as warm summer temperatures. In Mediterranean climates, peaches suffer much less from disease and insect pressure than when grown in humid climates.

Propagation

Peaches are T- or chip-budded, generally onto seedling rootstocks. In warm climates, pits are planted in fall and seedlings develop enough in spring to be "June-budded," so that the scion reaches marketable size by autumn. In northern climates, pits planted in fall are budded in August when scion wood is entering dormancy, and buds are not forced to grow out until the following spring (Figure 23.3). Unlike many tree fruits, peaches root very well from semi-

FIGURE 23.3. Peach budding. *(left)* Peach pits are planted in fall and reach the appropriate size for budding the following summer. *(center)* Scion wood is collected from current season's growth on disease-free trees, and leaves are trimmed. *(right)* Rootstocks are chip budded near the soil line, and the bud is wrapped to prevent desiccation while the bud takes. The same basic process is used for many tree fruits.

hardwood cuttings, and this provides an inexpensive method of producing own-rooted trees for high-density orchards.

Rootstocks

There are relatively few rootstock options for peach compared to pome fruits, grape, and other temperate tree fruits. The major rootstocks used are briefly described in Table 23.1. There are no compatible dwarfing stocks currently available, although *P. tomentosa* and *P. besseyi* have been used in research plantings (they are short-lived). Semidwarf scion cultivars have been produced by crossing normal and genetic dwarf types, and these may have potential for reducing tree stature in peach orchards. Virtually all rootstocks are grown from seed and are fairly uniform due to the self-pollinating, homozygous nature of peach.

Peach Seedlings

These are the most common source of rootstocks worldwide. Pits of 'Halford', 'Lovell', 'Bailey', and 'Siberian C' are most often used in the United States. The short life–resistant 'Guardian' is rapidly replacing other rootstocks in southeastern states. Pits are readily available from canners, but more uniform seedlings are obtained from pits collected from specially planted mother trees in seed orchards.

Interspecific Hybrids of Peach

'Nemaguard' (*P. persica* × *P. davidiana*) was developed in Georgia for resistance to root-knot nematode (*Meloidogyne* spp.) but is used most often in California, as a different nematode (ring, *Criconomella xenoplax*) and limited longevity have become problems with 'Nemaguard' in southeastern plantings. GF677 ('Amandier') and GF655 are peach-almond crosses from France; they tolerate high pH and are useful on calcareous soils, but they are not popular in the United States.

TABLE 23.1. Common rootstocks used for peach and nectarine.

Rootstock	Characteristics
'Bailey'	Cold-hardy rootstock with good overall performance; best in the northern states
'Guardian'	Vigorous rootstock with resistance to peach tree short life syndrome; slightly more expensive; use in areas where peaches were planted previously
GF677 ('Amandier')	A peach-almond hybrid for adaptation to high-pH soils; highly vigorous; not well adapted for the eastern United States
'Halford'	Good overall rootstock for the northern states
'Lovell'	Moderately invigorating rootstock with fair resistance to peach tree short life
'Nemaguard'	Invigorating rootstock with resistance to root-knot nematode; susceptible to other nematodes and peach tree short life
'Siberian C'	Cold-hardy rootstock used only in the northern tier of states and Canada; short-lived and poor in the southern states

Planting Design, Training, Pruning

Free-standing peach orchards are typically designed using rectangular spacings of 18 × 20 feet (110 trees/acre) or 12 to 15 × 18 feet (161-202 trees/acre) (Figure 23.4). Trellised systems are usually planted at a much higher density, 300 to 500 trees/acre. Pollinizers are not needed, but growers must plant several different cultivars to extend their marketing season because peaches ripen quickly and cannot be stored for more than 2 weeks. Since fruits on a given tree ripen over a period of 7 to 14 days, block size is limited to the area that can be harvested by the available labor in that time period. Cultivars ripening at about 2-week intervals then make up the entire orchard. In the southeastern United States, orchardists commonly grow 20 or more cultivars.

As with apple, there is no shortage of training systems available for peach trees. However, free-standing trees are trained most commonly to the open-center system, with three to five scaffolds radiating from the trunk 18 to 36 inches above the ground. Peaches can be trained to a central-leader system, but this often requires two annual prunings to keep the tops from overgrowing the lower scaffolds (Figure 23.5). In Italy, a temporary central-leader system called *vasetto ritardato* (delayed vase) is employed, in which a leader is left in the tree for the first 3 to 5 years and then re-

FIGURE 23.4. A typical peach orchard in Georgia. Trees are trained to open center and spaced 15 × 20 feet apart. Trees in this photo have been pruned.

FIGURE 23.5. *(above left)* The open-center system is most common for peach. This one died prior to budbreak, but after winter pruning. Note the density of foliage on trees in the background that were pruned the same way. *(above right)* Central-leader trees are somewhat more difficult to train; these are grown on dwarfing rootstock to reduce vigor. *(right)* The Tatura trellis system uses training wires to create a continuous V-shaped canopy down the row.

moved. The leader increases the amount of fruiting in young trees and helps to train the lower scaffolds outward at wide angles. Once the leader is removed, the growth pattern becomes that of an open-center tree.

High-density systems include the Tatura trellis, in-row V, perpendicular V, and palmette, although these are used only to a limited extent in the United States. In Europe and Australia, where trees are longer-lived, the cost of the trellis system can be recouped, and high yields are obtained with these systems. In-row V and palmette systems are variations on the narrow trellis hedgerow theme. The Tatura trellis system, which holds the yield record for peach, uses closely set trees that are allowed to develop only two scaffolds, each growing in opposite directions. The Y-shaped wire trellis serves as a training aid. The perpendicular V system also allows only two scaffolds per tree, and these develop at right angles to the row direction but without a trellis. The Kearney Agricultural Center (KAC) V, developed in California, is a slight variation on the perpendicular V theme that can increase yield per acre by up to 25 percent over the open-center system. Perpendicular V and KAC V systems are being advocated for new orchards because they streamline pruning and other hand labor tasks, and they also have the potential to increase returns on investment as compared to open-center systems.

Peaches can be pruned heavily because they are inherently vigorous, have large fruit size, and hence need fewer fruiting points within the canopy, and bear fruit laterally on 1-year-old wood. Since heavy thinning is essential, severe pruning reduces potential fruit load and thus expedites thinning, but yield is compromised if too much of the thinning is achieved by pruning. Generally, all water sprouts and shoots growing into the center of the tree are removed, and fruiting shoots at the canopy periphery are headed to concentrate bloom and harvest. Some thinning out of fruiting shoots is performed so that the remaining shoots are spaced evenly along scaffolds. Trees are pruned to only 6 to 8 feet in the Southeast but are commonly pruned at 10 feet in California.

Pest Problems

Insects

Plum curculio. Conotrachelus nenuphar adults are gray to dark brown, ⅛ to ¼ inch long snout beetles; grubs are small, yellow-white, and legless; they burrow into the fruit flesh, usually causing the fruits to fall off. Feeding damage by adults alone may cause catfacing (misshapen fruits) or D-shaped brown depressions in the fruit surface. Insecticide sprays are applied beginning at petal fall and then continue at 7- to 10-day intervals throughout spring. This insect is confined to the United States east of the Rocky Mountains and is the major insect pest of peach in the southeastern United States.

Plant bugs (stinkbugs)/catfacing insects. Several species of plant bugs or stinkbugs (*Leptoglossus, Lygus* spp., others) feed on fruits at various times of the year. They are fairly large (½ to ¾ inch), dark brown bugs with shield-shaped bodies and prominent legs. Early feeding results in severe catfacing, or deformed fruits that have extensive indentations. Some fruits may drop if feeding insects are pervasive. Later-season feeding by insects results in shallow, corky lesions that cause fruits to be culled in most cases. Brown rot incidence is enhanced by feeding lesions late in the season. Insecticides are applied at petal fall to shuck split and as needed throughout the season.

Oriental fruit moth. These small moths *(Grapholita molesta)* lay eggs in shoot tips early in the season, and the resulting larvae burrow downward a few inches; wilted and dead expanding leaves at the shoot tips indicate infestation. Later infestation results in fruit loss, as eggs laid on fruits produce ¾-inch, legged, pinkish larvae that burrow through the fruits (Plate 1.7). Although fruit infestation does not occur until fruits are less than half grown, early season sprays are critical for controlling this pest. Insecticides are applied at petal fall and shuck split.

Scale insects. These small (<⅛ inch), round or oval, sedentary insects (*Quadraspidiotus* spp., others) appear as bumps on twigs or branches; they suck the sap from branches and twigs. Dormant oil sprays are applied 1 to 3 weeks before budbreak, combined with insecticide if infestation is heavy.

Leaf rollers. The omnivorous leaf roller *(Platynota stultana)* and fruit tree leaf roller *(Archips argyrospila)* can cause economic damage to peaches in California. These caterpillars use webbing to roll leaves together or onto fruits, where they then feed. Fruit feeding causes scarring or blemishes severe enough to render fruits unmarketable. Bt or other insecticides are applied in early summer for omnivorous leaf roller, but dormant oils, with or without insecticides, depending on the severity of the infestation, are applied for fruit tree leaf rollers.

Borers. Larvae of the peach tree borer and lesser peach tree borer (*Synanthedon* spp.) cause damage to the lower trunk and scaffolds, sometimes killing young trees. The adult moth lays eggs on the tree, and the resulting larvae, which are ½ to 1 inch long with a dark-colored head, bore into the bark to feed. Lower limbs are sprayed thoroughly with insecticide in late summer when the adult moths are laying eggs. Peach twig borers *(Anarsia lineatella)* damage young shoots and fruits. Pheromone mating disruption, or postbloom insecticides if populations are high, can be used for control.

Mites. Of these insects (*Panonychus, Tetranychus* spp., others), the two-spotted mite *(Tetranychus urticae)* and the Pacific spider mite *(T. pacificus)* are the most damaging to peaches, the latter being a threat primarily in California. These mites feed on the lower leaf surfaces (less often the upper) and cause stippling of leaves. Mites have several predators that keep populations in check, but in hot, dry weather, in the absence of predators, miticides must be applied. A dormant oil spray before budbreak reduces initial populations.

Diseases

Brown rot/blossom blight. Monilinia laxa/M. fructicola causes one of the most severe diseases of peach in humid, rainy climates; it causes flowers to rot in some years, and fruits to rot near harvest (see Plate 1.9). The "blossom blight" phase is seldom a problem itself, but it signals potential problems with brown rot of ripening fruits later. A brown, powdery mass of spores in concentric rings is visible around a mushy lesion on the fruits. Fruits infested in a previous year die, shrivel, turn black, and hang on the tree; known as "mummies," these rotted fruits house spores for next year's infection. Removal of mummies in winter or brown-rotted fruits prior to harvest reduces incidence but may not be feasible in large orchards. Fungicide applied as trees bloom controls blossom blight and will reduce brown rot later. During the summer, fungicide sprays may need to be applied at 7- to 10-day intervals up to 1 week before harvest.

Scab. The fungus *Cladosporium carpophilum* causes numerous, small (<⅛ inch), black lesions of the fruit surface nearest the stem end, but it rarely causes fruits to drop or rot outright. The lesions can be distinguished from bacterial spot by the presence of a yellow halo around them. Scab is largely a cosmetic problem but causes culling of fruits or a severe downgrade in quality. Fungicide sprays from the petal fall stage through the next 4 weeks or so (at 7- to 10-day intervals) are used to control scab.

Peach leaf curl. The fungus *Taphrinia deformans* attacks leaves as they emerge from buds in spring; it is a problem in cooler areas and in California, but not in the southeastern United States. Misshapen, reddish to yellowish, thickened, puckered leaves are seen in early spring. Fungicides applied prior to bud swell in late winter kill spores overwintering on twigs and bud scales.

Bacterial spot. This bacterial disease, caused by *Xanthomonas campestris* pv. *pruni,* occurs in most parts of the world, except the Pacific Coast. It is more common in humid, warm climates and in areas with sandy soils. The bacteria can infect leaves, twigs, or fruits, with fruit infections causing black spotting (similar to scab), in milder cases, and sunken, black lesions in more severe cases. Sprays of copper or antibiotic materials are risky and often do not control the disease well. Resistant cultivars exist and are used in areas prone to bacterial spot. Resistant cultivars include 'Biscoe', 'Bounty', 'Candor', 'Clayton', 'Derby', 'Dixired', 'Jerseydawn', 'Newhaven', 'Salem', 'Sentinel', and 'Sweethaven'.

Bacterial canker. This bacterial disease, the result of infection by *Pseudomonas syringae,* causes injury to twigs, branches, and large limbs. Branches/limbs are girdled in severe cases, and they die back quickly. Many suckers are produced from the base of the trunk on severely infected trees because the lower trunk and roots are not killed. It is nearly impossible to control with sprays, as with many bacterial diseases. Of the stone fruits, peaches are among the most tolerant to bacterial canker, and the best defense is a healthy, well-maintained tree. Freezing injury, nematode weakening, other diseases, drought, or waterlogging all predispose trees to infection.

Peach tree short life (PTSL). This is a rootstock- and soil-related syndrome or complex, not a disorder attributable to a specific organism. It is common in the southeastern United States. PTSL is characterized by the sudden death in spring of trees above the soil line that were apparently healthy the previous fall. Profuse suckering usually occurs because roots are not killed. PTSL can be avoided by not replanting trees on old peach cultivation sites, or by using 'Guardian' rootstock, which is tolerant. Predisposing factors, such as low soil pH, hardpans, low nutrient levels, ring nematode buildup, cultivation, fluctuating winter temperatures, fall pruning, and the use of 'Nemaguard' rootstock all accentuate PTSL.

HARVEST, POSTHARVEST HANDLING

Maturity

The best index for peach today is ground (background) color of the skin: red color is a function of cultivar and light exposure and therefore not a good index. Color "chips" developed at Clemson University feature standard colors painted onto cards, and these are used to train harvesters. Ground color of fruit changes from green to straw yellow during ripening, and fruits yellow enough to match color chip 3 (out of 6) are mature enough for harvest. Fruits picked too early (represented by chip 1 or 2) never develop full flavor and texture and may develop a condition called woolly texture or woolliness, characterized by mealy texture and the absence of juice or flavor. Firmness and days from bloom can also be used to estimate maturity, although peaches are notorious for variation in firmness among fruits on the same tree and even between the two cheeks of a single fruit. Peach is classified as a climacteric fruit with respect to ripening.

Harvest Method

Freestone peaches and nectarines are harvested by hand. Usually, trees are picked three to four times at 2- to 3-day intervals, taking only the firm-mature fruits at each picking. Clingstone peaches are mechanically harvested using shake-and-catch systems used for prunes and sour cherries or are hand harvested.

Postharvest Handling

Packing-line operations are standard; fruits are first hydrocooled to reduce temperature; then culled, brushed, and waxed; and then sorted for size. Peaches are defuzzed during the brushing/waxing process (Plate 23.3). Peaches are packed into 25-pound boxes and shipped immediately after harvest, in refrigerated trucks, or after short storage periods, due to poor shelf life.

Storage

Peaches have a short shelf life of about 2 weeks under most conditions. They are not susceptible to chilling injury, so can be stored at 31 to 32°F to maximize shelf life. Several postharvest diseases—anthracnose, gray mold, *Rhizopus*, brown rot, and bacterial spot—occur routinely in storage unless packinghouse fungicide practices are used.

CONTRIBUTION TO DIET

Peaches have few marketing niches other than fresh and canned fruit. Freestone peaches are sold fresh, and clingstones are virtually all canned. About 98 percent of nectarines are marketed fresh. The utilization breakdown is as follows (percent of U.S. peach crop):

Product	Percentage of Crop
Fresh	54
Canned	38
Frozen and dried	8

Per capita consumption is 9.5 pounds/year, considering peach and nectarine together. This is down from 13.2 pounds/year in 1980, as surveys have shown that consumers are weary of peaches and nectarines harvested too early, before flavor develops, to withstand shipping. Poor quality is frequently cited as a reason to buy fruits other than peaches and nectarines in retail markets. A small portion of peaches are harvested "tree ripe," or at least closer to maturity, and marketed in single-layer tray packs.

Dietary value per 100-gram edible portion:

Water (%)	89
Calories	39
Protein (%)	0.9
Fat (%)	0.2
Carbohydrates (%)	9.5
Crude Fiber (%)	1.5

	% of U.S. RDA (2,000-calorie diet)
Vitamin A	6
Thiamin, B_1	2
Riboflavin, B_2	2
Niacin	4
Vitamin C	11
Calcium	<1
Phosphorus	3
Iron	1
Sodium	0
Potassium	5

BIBLIOGRAPHY

Bunyard, E.A. 1938. The history and cultivation of the peach and nectarine. *J. Royal Hort. Sic.* 63:114-121.

Childers, N.F., J.R. Morris, and G.S. Sibbett. 1995. *Modern fruit science,* Tenth edition. Gainesville, FL: Norman F. Childers.

Childers, N.F. and W.B. Sherman (eds.). 1988. *The peach.* Gainesville, FL: Norman F. Childers.

Gur, A. 1985. Rosaceae—Deciduous fruit trees, pp. 355-389. In: A.H. Halevy (ed.), *CRC handbook of flowering,* Volume 1. Boca Raton, FL: CRC Press.

Hedrick, U.P. 1917. *The peaches of New York* (New York Agricultural Experiment Station Report, 1916). Albany, NY: J.B. Lyon Company.

Hesse, C.O. 1975. Peaches, pp. 285-355. In: J. Janick and J.N. Moore (eds.), *Advances in fruit breeding.* West Lafayette, IN: Purdue University Press.

Johnson, R.S. and C.H. Christoso (eds.). 2002. *Proceedings of the 5th international peach symposium* (Acta Horticulturae 592). Belgium: International Society of Horticultural Science.

LaRue, J.H. and R.S. Johnson. 1987. *Peaches, plums and nectarines* (University of California Cooperative Extension Service Publication 3331). Davis, CA: University of California, Davis.

Layne, R.E.C. 1987. Peach rootstocks, pp 185-216. In: R.C. Rom and R.F. Carlson (eds.), *Rootstocks for fruit crops.* New York: John Wiley and Sons.

Myers, S.C. 1989. *Peach production handbook* (Georgia Experiment Station Handbook No. 1). Athens, GA: University of Georgia.

Ogawa, J.M., E.I. Zehr, G.W. Bird, D.F. Ritchie, K. Uriu, and J.K. Uyemoto. 1995. *Compendium of stone fruit diseases.* St. Paul, MN: American Phytopathological Society.

Okie, W.R. 1998. *Handbook of peach and nectarine varieties* (USDA Agricultural Handbook No. 714). Washington, DC: U.S. Department of Agriculture.

Scorza, R. and W.R. Okie. 1990. Peaches *(Prunus)*, pp. 175-232. In: J.N. Moore and J.R. Ballington (eds.), *Genetic resources of temperate fruit and nut crops* (Acta Horticulturae 290). Belgium: International Society for Horticultural Science.

Teskey, B.J.E. and J.S. Shoemaker. 1978. *Tree fruit production.* Third edition. Westport, CT: AVI Publishing.

Watkins, R. 1979. Cherry, plum, peach, apricot, and almond: *Prunus* spp., pp. 242-247. In: N.W. Simmonds (ed.), *Evolution of crop plants.* London: Longman.

Chapter 24

Pear *(Pyrus communis, Pyrus pyrifolia)*

TAXONOMY

Pears are placed in the Rosaceae (rose) family, subfamily Pomoideae, along with apple and quince. The genus *Pyrus* is composed of about 22 species, found in Asia, Europe, and northern Africa. Two major species are commercially cultivated:

1. European pear: *Pyrus communis* L. This species does not occur in nature and possibly derives from *P. caucasia* and *P. nivalis* (snow pear). This is the major pear of commerce.
2. Asian pear: *P. pyrifolia* (Burm. f.) Nak. (syn. *P. serotina* L.). This pear is also called "Japanese" or "Oriental" pear or "Nashi." Grown mostly in the Orient, this fruit has been increasing in popularity in the United States over the past 20 years.

Asian pears are more similar to apples than are European pears; they have hard, crisp, applelike flesh when ripe, unlike the soft, melting flesh of European pears. Also, Asian pears will ripen on trees, as do apples, but European pears are subject to core breakdown if allowed to ripen fully on trees. Asian pears are often touted as being less cold hardy and more resistant to fire blight than European pears (Plate 24.1).

In China, *P. ussuriensis* Maxim., the Ussuri pear, is cultivated for its fruits. Some Asian pear cultivars are complex hybrids between *P. ussuriensis*, *P. pyrifolia*, and *P. bretschneideri*. *Pyrus nivalis*, the snow pear, is used in Europe for cider and perry (alcoholic cider) and may have been hybridized with *P. communis* to produce some cultivars originating in Europe. The popular ornamental pear, *Pyrus calleryana* 'Bradford', is a common feature in southern landscapes. *Pyrus calleryana* is also used as a rootstock for pears in the Southeast because it is well adapted to this area.

Cultivars

Relatively few cultivars of European or Asian pear are grown worldwide. Only about 20 to 25 European and 10 to 20 Asian cultivars represent virtually all of the pears of commerce. Almost all European cultivars were chance seedlings or selections originating in western Europe, mostly in France. All of the Asian cultivars originated in Japan and China (Table 24.1).

'Bartlett' is by far the most widespread European pear in the world. It is also known as 'Williams' and 'Bon Chretien'. 'Bartlett' ripens relatively early, has high productivity and excellent fruit quality, and, unlike most pears, is self-fruitful. Self-fruitfulness simplifies orchard management greatly. 'Bartlett' lends itself to both fresh consumption and canning, whereas few other cultivars are used for dual purposes. There are several sports of 'Bartlett', such as the red-skinned 'Red Bartlett'.

Introduction to Fruit Crops

doi:10.1300/5547_24

TABLE 24.1. Characteristics of major European and Asian pear cultivars.

Cultivar	Production* (%)	Season of harvest	Fruit size	Fruit color	Shelf life (days)
Pyrus communis					
Bartlett [Williams, Bon Cretein]	76	Early	M	Green-yellow	80
Anjou (d'Anjou)	17	Mid	M-L	Green	180
(Buerrè) Bosc	4	Mid	M-L	Russet	100
(Doyenne du) Comice	0.7	Late	L	Green/blush	100
Kieffer	0.4	Late	L	Yellow/blush	120
Clapp Favorite	0.4	Early	L	Yellow/blush	60
Seckel	0.2	Early-Mid	S	Blush	90
(Buerrè) Hardy	0.2	Mid	M	Russet	120
Winter Nelis	0.2	Late	S	Green	200
Pyrus pyrifolia					
Nijisseiki	28	Mid	M	Yellow	130
Kosui	21	Early	M	Brown	60
Chojuro	15	Mid	M	Russet-orange	160
Hosui	14	Mid	L	Russet-orange	60
Shinseiki	5	Early	M	Yellow	180

Source: Modified from Westwood, 1993, and Lombard and Westwood, 1987.

*Production percentages for *P. communis* and *P. pyrifolia* indicate production data in the United States and Japan, respectively.

ORIGIN, HISTORY OF CULTIVATION

As with its relative the apple, the European pear is not found in the wild. Its probable progenitors are native to eastern Europe and Asia Minor near the Mediterranean, but it is not known when they may have hybridized to yield *P. communis*. The European pear has been selected and improved since prehistoric times and was cultivated in Europe in 1000 BC. Pears probably came to the New World with the first settlers on the East Coast and then spread westward with pioneers. When moved to the Pacific Northwest in the 1800s, European pears were able to escape fire blight, a serious bacterial disease that limited pear cultivation in the East. Today, over 90 percent of the pear crop is grown in the Pacific Northwest region and California (Plate 24.2).

Asian pears were domesticated in China about the same time European pears were domesticated in Europe—3,000 years ago. *Pyrus pyrifolia* is native to central and southern China and probably was the first of the Asian pears to be domesticated, as the fruits of the wild trees are edible. Fruits of wild *P. ussuriensis* are astringent, small, and coarse textured; thus, this species was probably hybridized with *P. pyrifolia* prior to domestication. Chinese writings dating from 200 to 1000 BC describe pear propagation and culture. Asian pears moved from China to Japan, Korea, and Taiwan, where they are cultivated commercially today.

FOLKLORE, MEDICINAL PROPERTIES, NONFOOD USAGE

Pear trees are said to be symbolic of comfort and affection in the language of flowers. Pear fruits carry the meaning "give me some hope."

With respect to medicinal properties, pears are similar to apples. Cyanogenic glucosides are found in the seeds, which, as with apple seeds, are poisonous. The bark and roots contain phloretin, an antibiotic-like substance that acts on gram-positive and gram-negative bacteria. Pear juice, similar to apple juice, has been implicated in the cause of chronic, nonspecific diarrhea in infants and children. This stems from the abnormally high levels of fructose and sorbitol relative to glucose, compared to other foods. Pear peel is listed as an ingredient in a Chinese tea for "winter colds without sore throat" by some traditional doctors. The decoction is prepared with ginger, cinnamon, green onion, and black pepper; tangerine peel is added if there is cough, and pear peel is added if there is severe cough.

PRODUCTION

World

Total production in 2004, according to FAO statistics, was 17,909,496 MT or 39.4 billion pounds. Pears are produced commercially in 81 countries on about 4.3 million acres. Production has increased 55 percent in the past decade in response to a 30 percent increase in acreage and a 20 percent increase in yield. Average yields are about 9,100 pounds/acre, ranging from half this amount up to 30,000 pounds/acre in countries practicing intensive cultivation of orchards. Chinese production overtook that of Italy, the former world leader, in 1978. Chinese production has increased over threefold in the past decade and has been responsible for virtually all of the increase in worldwide pear production (Figure 24.1). The top ten pear-producing countries (percent of world production) follow:

1. China (56)
2. Italy (5)
3. United States (5)
4. Spain (4)
5. Argentina (3)
6. Germany (3)
7. Japan (2)
8. South Africa (2)
9. Turkey (2)
10. South Korea (2)

United States

Total production in 2004, according to USDA statistics, was 812,054 MT or 1.8 billion pounds. Pears are produced in nine states on 65,000 acres. Yields average 28,000 pounds/acre, about three times the world average. In 2004, the industry value was $296 million, ranging from $230 million to $310 million over the past decade. Growers received $0.16 per pound in 2002,

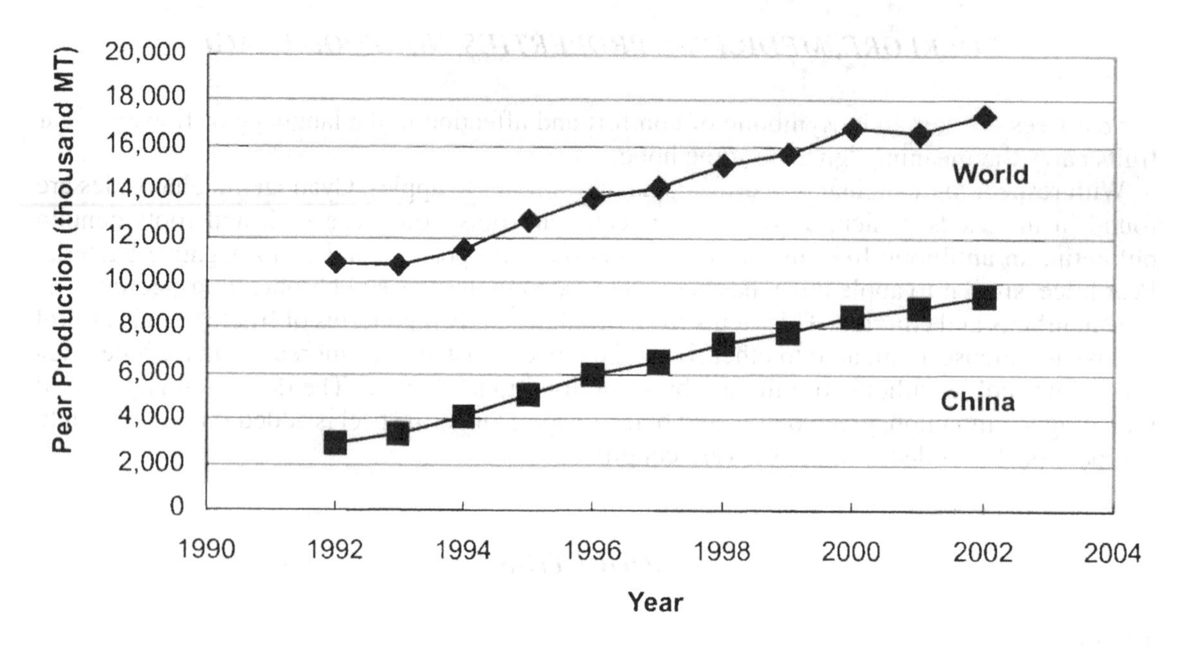

FIGURE 24.1. The 59 percent increase in world pear production over the past decade was precipitated almost entirely by a production increase in China. Production has been relatively stable in other countries. *Source:* FAO statistics.

about average for the range of $0.11 to $0.19 per pound over the past decade. The leading states follow:

1. Washington—43 percent (46 percent 'Bartlett')
2. California—30 percent (82 percent 'Bartlett')
3. Oregon—24 percent (30 percent 'Bartlett')

In 2002, the United States exported 21 percent of its production, 93 percent as fresh fruit, with the remaining being canned, dried, or a component of canned fruit salad. Imports were equivalent to 11 percent of production, 100 percent as fresh fruit.

BOTANICAL DESCRIPTION

Plant

The plant is a medium-sized, upright-growing tree that grows to 30 feet tall but is generally only 8 to 18 feet in cultivation (Plate 24.3). Tree size is heavily dependent on rootstock and training system. Leaves are elliptic/ovate with acute tips, with finely serrate or entire margins, and 2 to 4 inches in length.

Flowers

Flowers are about 1 inch in diameter, with white petals, and similar to apple, except for having much longer pedicels. The ovary is inferior and five-carpellate, as in apple, with the hypanthium fused to the ovary wall. Unlike the cymose inflorescence of apple, the pear inflorescence is corymbose, containing five to seven flowers. Inflorescences are borne from terminal mixed buds on short spurs that occur on 2-year-old and older wood (Plate 24.4).

Pollination

Most cultivars require cross-pollination for commercial fruit set. Some cultivars, such as 'Bartlett', 'Orient', 'Baldwin', 'Kieffer', and 'Spalding', are partially self-fruitful and can be grown as single trees or in solid blocks. Honeybees are the main agent of pollen transfer, although wild bees may contribute somewhat.

Fruit

The fruit is a pyriform (European) or round (Asian) pome (Plate 24.1). As in apple, the fleshy edible portion is derived from hypanthium tissue. There are five central seed cavities, usually bearing two seeds each, as in apple. The flesh contains grit cells (called "brachysclereids"), which are thick-walled, lignified cells that give the European pear its characteristic flesh texture. Asian pears have firm flesh, similar to apple. Heavy skin russeting is typical in many cultivars (e.g., 'Bosc', 'Russet Bartlett'). Russet is the proliferation of corky cells in fruit skin. Russeting is genetically controlled, but some russet may be induced by frost, wet weather in spring, or prolonged periods of dew on the fruit surface. European pears are picked prior to physiological maturity, when still hard and green ("green mature"). If ripened fully on trees, the interior tissues may turn brown and break down; if picked too early, they will not develop full flavor and shrivel in storage sooner. In contrast, Asian pears will ripen normally on the tree. Pears are thinned to one or two fruits per spur, spaced 6 inches apart, usually by hand or less often with the chemicals NAA or NAAm applied postbloom. In some pear cultivars, a high level of natural fruit drop following bloom reduces the need for thinning. Chemical thinning is not as reliable in pear as in apple and often requires touch-up hand thinning. Fruiting begins in 3 to 5 years, depending on rootstock, with dwarf stocks inducing precocity, and seedlings delaying reproductive maturity.

GENERAL CULTURE

Soils and Climate

Pears tolerate heavy, poorly drained soils better than most tree fruits. However, productivity is best on deep, well-drained loams with a pH of 6 to 7.

Pears have similar climatic requirements compared to apples but are much more prone to fire blight and therefore cannot tolerate humid, wet springs. This is the main reason that pears are grown mostly in Mediterranean or arid climates, such as central California, the Hood River Valley of Oregon, and the Yakima Valley of Washington. Some European pear hybrids and Asian pears have good tolerance to fire blight and can be grown in humid areas.

Pears require about 1,000 chill hours to break dormancy, although some cultivars, such as 'Hood', 'Baldwin', 'Orient', and 'Flordahome', and many Asian pears have lower chilling requirements and can be grown as far south as northern Florida. Pears have similar or slightly lower coldhardiness than apples. In general, they can tolerate temperatures of –10 to –20°F without damage. Pear trees bloom 1 to 3 weeks before apple trees and are therefore more prone to frost damage than apple. Open flowers and young fruits are killed by brief exposure to 28°F or less, as with all other tree fruits. Pears mature in as few as 90 days or as many as 200 days. 'Bartlett' and its sports are referred to as "summer pears" because they ripen in July through August in California (115-140 days). "Winter pears" are those harvested in autumn and marketed throughout the winter months, such as 'Anjou', 'Bosc', 'Comice', 'Hardy', 'Winter Nelis', and 'Packham's Triumph'. 'Bartlett' and 'Bosc' reach highest quality when grown in hot climates, and sometimes develop core breakdown and shorter shelf life if the weather preceding harvest is cool.

FIGURE 24.2. A young dwarf pear tree in the field composed of three parts: a dwarfing quince rootstock, an 'Old Home' interstem to function as a compatibility bridge, and a 'Conference' scion.

Propagation

Standard-sized pears are chip- or T-budded onto compatible rootstocks, as with most tree fruits. However, where quince *(Cydonia oblonga)* is used as a dwarfing stock, most pear scions require an interstem for compatibility (of major cultivars, only 'Anjou', 'Comice', and 'Hardy' are compatible with quince, and all Asian cultivars are incompatible). An interstem is simply a short section of the trunk taken from a cultivar that is mutually compatible with the quince below and the scion above. Thus, it forms a "bridge" of sorts between the quince rootstock and the cultivar on top (Figure 24.2). 'Old Home' is normally used as an interstem for 'Bartlett' on quince. This can be accomplished by double budding: a thin slice of stem (only ⅛ inch), similar to a shield bud without the bud, is first inserted into a T-cut on quince stock. The normal shield bud of 'Bartlett' is placed directly over the 'Old Home' stem slice and wrapped. This practice is common in Europe, where pears are often grown on quince stocks. A longer interstem is often used in the United States.

Rootstocks

A number of different rootstocks are available for pear culture, perhaps more than for any other tree fruit, except apple. In any given region, only a few are generally used. The most common rootstocks used worldwide are described in Table 24.2.

Pyrus *Seedlings*

Several species of *Pyrus* are used as rootstocks for pears; this is the most common class of rootstocks worldwide. *Pyrus communis* seedlings ('Winter Nelis' and 'Bartlett') are most common in the United States, Canada, and the Southern Hemisphere. *Pyrus calleryana* is used in the United States, Australia, and China for resistance to fire blight, nematodes, root and crown rot, oak root rot, and crown gall. Seedlings of *P. pyrifolia* are most commonly used for Asian pears.

Clonal Rootstocks

The quince selections Angers quince, E.M. quince A, quince C, and Provence quince are most commonly used in high-density orchards of Europe, due to high yield efficiency and dwarfing. Quince selections are most often propagated by stem cuttings, although root cuttings and suckers are sometimes used. Trees on quince are about 50 to 60 percent the size of trees on

TABLE 24.2. Common rootstocks for pears.

Rootstock	Pear decline	Fire blight	Cold-hardiness	Tree size (% of 'Old Home' clonal)	Yield efficiency	Effect on fruit quality
'Old Home' clonal	resistant	resistant	good	100	low	average
OH × F	resistant	resistant	good	60-100	high	good
Quince	moderately resistant	susceptible	poor	50-60	high	very good
P. communis seedlings	moderately resistant	susceptible	good	90	high	good
P. calleryana seedlings	moderately resistant	resistant	poor	90	high	good
P betulaefolia	resistant	variable	moderate	130	moderate-high	good
P. pyrifolia	susceptible	moderately susceptible	moderate	70	high	average

Source: Modified from Westwood, 1993, and Lombard and Westwood, 1987.

seedling stocks, and they require support due to poor anchorage. Quince is highly susceptible to fire blight, cold tender, and therefore limited to dry climates with mild winters.

Several 'Old Home' × 'Farmingdale' (OH × F) selections were developed in the Pacific Northwest as dwarfing stocks with compatibility with most pear cultivars. They are blight resistant; tolerant of cold damage; resistant to crown rot *(Phytophthora)*, pear decline, and root aphid; but susceptible to nematodes, and they tend to sucker. Trees on OH × F 51 are about as small as trees on quince. Stem cuttings are the most common propagation method.

Planting Design, Training, Pruning

Pear orchard designs are very similar to those in apple orchards. Typically, standard trees are spaced at 25 × 25 feet (70 trees/acre), but hedgerow forms are more common in high-density plantings, with hundreds of trees per acre. Pollinizers are planted in alternate rows or every tenth or fifteenth tree within hedgerows.

The most common training system is central leader for free-standing trees and some form of palmette for trellised orchards. Initial tree training is particularly important with pear because scaffolds tend to grow nearly vertically, causing poor crotch angles and delayed fruiting. Other than this tendency to grow upright, and thereby creating a greater need for limb spreaders, pears are trained in the same way as apples (Figure 24.3).

Pruning is very similar to the approach with apple, since bearing habit and fruit size are about the same. Pear fruits are borne on spurs that may live for several years, producing new fruits in about the same places each year. Therefore, it is important that the spurs receive adequate light to keep them fruitful. Shaded spurs quickly lose their capacity to produce flowers and fruits. Fruiting spurs are found on 2-year-old and older wood, so 1-year-old shoots at the top of the trees can be pruned out each year to avoid shading without sacrificing potential fruiting sites.

FIGURE 24.3. *(left)* Young pear trees being trained to a small central-leader system. Note the upright, narrow-angled shoots, which are typical of pear and make young tree training difficult. *(right)* Limb spreaders are often used as training aids for pear to overcome the tendency to produce narrow-angled scaffolds.

Pruning fire blight–affected shoots requires special care. Blighted twigs are removed as soon as symptoms appear in spring for the best control, although in large orchards this is not feasible. Cuts are made at least 6 inches below the point where symptoms appear, and pruners are dipped intermittently in alcohol or a bleach solution to avoid spreading the bacteria. Diseased prunings are removed from the orchard and burned or buried.

Pest Problems

Insects

Pear psylla. Several species of this pest (*Cacopsylla* spp.) occur worldwide. They are ⅒ inch long, cicadalike insects that feed on leaves, causing tree stunting, reduced fruit size, and early defoliation. Significantly, pyslla is a vector for pear decline. Psyllas secrete honeydew, as do aphids, causing leaves and fruits to be sticky and often blackened by "sooty mold," which grows on honeydew. Insecticide sprays from the early stages of budbreak through the summer may be required because the insect produces three to five generations per year. Asian pears are less susceptible, and cultivars with *P. ussuriensis* or *P. xbretschneideri* parentage are moderately resistant.

Codling moth. Lasperyresia pomonella is one of the worst pests of tree fruits, including pear, worldwide. The moth (¾ inch long, dark color) lays eggs singly on young fruits in midspring, and the larvae (pinkish-brown head, white body, ¾ inch long) burrow through the fruits. Multiple generations per year are possible. Insecticides are applied at petal fall and then biweekly or as needed throughout the summer. Mating disruption with pheromones provides a nonchemical means of control. 'Anjou', 'Bosc', 'Conference', 'Passe Crassane' and 'Comice' are somewhat resistant to infestation late in the season.

Oriental fruit moth. Another common fruit tree pest, these moths *(Grapholita molesta)* lay eggs in shoot tips early in the season, and the resulting larvae burrow downward a few inches. Wilted and dead expanding leaves at the shoot tips indicate infestation. Later infestation results in fruit loss, as eggs laid on fruits produce ½-inch, legged, pinkish larvae that burrow through the fruits. Although fruit infestation does not occur until fruits are less than half grown, early season sprays are critical for controlling this pest. Insecticides are sprayed at petal fall and shortly afterward at regular intervals when the infestation is severe.

Scale insects. Adults (*Quadraspidiotus* spp., others) are small (<⅛ inch), round or oval, sedentary insects that appear as bumps on twigs or branches; they suck the sap from branches and twigs. Scale insects are controlled with dormant oil sprays before budbreak, combined with insecticide if infestation is severe.

Leaf rollers. The omnivorous leaf roller *(Platynota stultana),* fruit tree leaf roller *(Archips argyrospila),* and oblique-banded leaf roller *(Choristoneura rosaceana)* can cause damage to pears in California and lead to economic losses. These caterpillars use webbing to roll leaves together or onto fruits, where they then feed. Fruit feeding causes scarring or blemishes severe enough to render them unmarketable. Bt or other insecticides are applied in early summer for omnivorous leaf roller, but dormant oils, with or without insecticides, are applied for fruit tree and oblique-banded leaf rollers.

Mites. Mites (*Tetranychus, Panonychus* spp.) feed on the lower leaf surface, causing "mite stipple" to the upper leaf surface. If severe, leaf photosynthesis is reduced, and leaves can be shed. Dormant oil sprays, with or without miticide, before budbreak reduce initial populations. In California, mites must be monitored throughout the year, with miticide applications made when specific thresholds are met. Care is taken to avoid killing predators, such as avoiding pyrethroid and organophosphate insecticides. Preventing dusty conditions and water stress are important cultural controls for mites. Asian pears may be more resistant to mites than are European pears. 'Bartlett' and 'Bosc' are somewhat tolerant of rust mite *(Epitrimerus pyri).* Pear leaf blister mite *(Phytoptus pyri)* can damage pear buds, leaves, and fruits but is controlled by sprays for other mites.

Diseases

Fire blight. Caused by *Erwinia amylovera,* fire blight is a severe bacterial disease for virtually all pear cultivars, particularly in warm, wet springs. The bacteria are carried by bees from tree to tree at bloom and can kill all or most of the flowers on a tree if infestation is severe. Highly susceptible cultivars show twig and spur dieback or complete tree death. Flowers, twigs, and leaves often turn black and wilt; shoot tips droop over, giving a distinctive "shepherd's crook" appearance. Resistant cultivars are the best control; of the major cultivars, only 'Kieffer' and 'Seckel' are resistant. Cultivars derived from *P. ×bretschneideri* parentage are resistant, and a few Asian pears are moderately resistant ('Imamura Aki', 'Meigetsu', 'Seuri'). Susceptible cultivars can be managed by spraying a copper-containing fungicide or Bordeaux mixture just prior to bud swell or with streptomycin sulfate at bloom. Sanitation is an extremely important cultural control; infected stems are cut 6 inches below the blackened tissue and removed from the orchard.

Pear scab. Caused by *Venturia pirina,* this disease is very similar to apple scab, occurring most frequently during cool, wet weather at the time of bloom. Spores overwinter on infected leaves beneath trees and are discharged into the air after rains. Spores also may overwinter on twigs. Leaves, fruits, and twigs may be damaged. Lesions on leaves are brown and circular and may occur on either side of the leaf, as in apple. Fruit lesions occur first on the calyx end and then may spread to other areas; lesions are brown or black, coalesce later, and may cause cracking or misshapen fruits. Fungicide sprays from the "green tip" stage (when a tiny bit of green tissue is visible at bud tips after they swell) until 2 weeks after petal fall control scab. Covering, removal, or destruction of fallen leaves may somewhat reduce the incidence of the disease. 'Bartlett', 'Conference', and 'Dr. Jules Guyot' are reported to be resistant, as is *P. ussuriensis.*

Blossom blast or pseudomonas canker. This disease is caused by the bacterium *Pseudomonas syringae* pv. *syringae,* which causes several diseases of fruit crops throughout the world. It enters the flower tissue at the base of trichomes or through wounds and kills flowers and entire spurs, if cool, wet conditions occur at bloom. Flowers or young fruits are blackened, and the disease looks similar to fire blight. Another form of infection does not kill flowers but can cause black lesions on the calyx end of fruits and poor fruit set. Copper-containing sprays applied when trees are dormant can be used to reduce populations, but this produces variable results. Frost generally accentuates damage. Good site selection to avoid frost, dew, or long periods of wetting at bloom can reduce incidence of the disease. Few cultivars show resistance. 'Hardy', 'Comice', and 'Forelle' are more tolerant than other European types, but all Asian cultivars are susceptible.

Powdery mildew. The fungus *Podosphaera leucotricha*, which causes powdery mildew, is a problem in drier areas of the world, such as the Pacific Northwest and Western Europe. Leaves, flowers, and fruits are affected and may fall off if the disease is pervasive. Its name comes from the grayish-white, feltlike patches of fungus that occur on the lower leaf surface and all over flowers and young fruits. Shoot growth of young trees is reduced, and fruits are cosmetically affected in cases of flower infection. Fungicides applied as flower buds become distinct, but before they open, are effective in controlling mildew. Sprays may need to be repeated until shoot growth ceases on highly susceptible cultivars in summer. 'Winter Nelis' and 'Comice' are moderately resistant European types.

Pear decline. This disease is caused by an unnamed *Mycoplasma* organism. This bacteria-like organism is vectored by pear psylla and grafting. It invades the phloem tissue of a tolerant scion and slowly travels to the graft union, where it causes bark necrosis and eventual tree death. Young trees are killed quickly, but mature trees may decline in vigor and die over several years. The disease is easily controlled by resistant rootstocks: *P. betulaefolia;* quince; most *P. communis*, OH, and OH × F selections. Trees grafted onto *P. pyrifolia*, *P. ussuriensis*, and *P. calleryana* are susceptible. Budding with certified disease-free bud wood is important. Antibiotics or insecticide sprays to control psylla are futile.

HARVEST, POSTHARVEST HANDLING

Maturity

European pears are harvested when "firm mature," stored immediately, and then allowed to ripen for several days prior to fresh consumption. When fully ripe, European pears have the typical "melting" flesh texture and full development of flavor. Asian pears, however, are similar to apples; they are harvested closer to physiological ripeness and placed in cold storage. Flesh firmness is the single most reliable indicator of pear maturity. Firmness in the range of 10 to 15 pounds, as measured by a pressure tester, is desirable for most cultivars. This is often combined with days from full bloom or degree-days for greater accuracy. In lieu of a pressure tester, one may observe when fruit lenticels turn from white to brown, and when skin color begins to lighten. Pears are classified as climacteric fruit with respect to ripening.

Harvest Method

Pears for are picked by hand several times over a 10- to 20-day period.

Postharvest Handling

Standard packing-line procedures are used for pear: hydrocooling, washing, culling, waxing, sorting, and packing. Quality grade is based on size and appearance of skin; greater prices are obtained for larger fruits and those with minimal surface blemishes.

Storage

Pears are stored just below freezing for 2 to 7 months and then ripened for a few days at 70 to 75°F, prior to consumption or canning. Unlike apples, pears are not subject to chilling injury. USDA data show that cold-storage holdings of 'Bartlett' are depleted by February, although other cultivars (later ripening) generally last until May or June of the year following harvest.

Storage life depends on cultivar and, to some extent, on growing conditions and harvest date. Pears grown in cool climates generally have shorter storage lives, and those picked too immature shrivel or scald. Overmature fruits experience core breakdown. Storage life can be extended by controlled-atmosphere (CA) storage (3 percent O_2 and 2-5 percent CO_2), but this is not economically justified in most cases, except with some late-season pears that do not ripen normally in prolonged cold storage. There is less of an advantage to CA storage with fruits that can be stored just below freezing. Temperatures of 31°F reduce respiration sufficiently without the need for altered atmospheres.

Although pears are not susceptible to chilling injury, as are apples, they are susceptible to freezing injury (<30°F), soft and superficial scalds, water core, cork spot, and CO_2 injury ("brown heart") during storage. Several postharvest diseases—black mold *(Alternaria)*, gray mold *(Botrytis)*, blue mold *(Penicillium)*, brown rot *(Monolinia)*, and others common to fruit crops—occur routinely in storage unless packinghouse fungicide practices are used.

CONTRIBUTION TO DIET

In the United States, 55 to 60 percent of pear production is marketed as fresh fruits, and the bulk of the processed fruits are canned. The proportion of fresh-fruit production has been gradually increasing since the mid-1980s. A small number of pear fruits are processed into fruit cocktail mixes or dried. In Europe, the alcoholic cider perry is made from *P. nivalis*, the snow pear.

Per capita consumption of pears in the United States was 5.6 pounds/year in 2004; 55 percent of consumption is fresh fruit, and the remainder is canned. Overall consumption of pears is down 23 percent since 1980, as a 46 percent decline in canned consumption has partially offset a 19 percent increase in fresh consumption.

Dietary value per 100-gram edible portion:

	European Pear	Asian Pear
Water (%)	84	88
Calories	58	42
Protein (%)	0.4	0.5
Fat (%)	0.1	0.2
Carbohydrates (%)	15.5	10.6
Crude Fiber (%)	3.1	3.6
	% of U.S. RDA (2,000-calorie diet)	
Vitamin A	<1	0
Thiamin, B_1	1	<1
Riboflavin, B_2	2	<1
Niacin	1	1
Vitamin C	7	6
Calcium	1	<1
Phosphorus	2	2
Iron	1	0
Sodium	<1	0
Potassium	3	4

BIBLIOGRAPHY

Bell, R.L. 1990. Pears *(Pyrus)*, pp. 655-698. In: J.N. Moore and J.R. Ballington (eds.), *Genetic resources of temperate fruit and nut crops* (Acta Horticulturae 290). Belgium: International Society for Horticultural Science.

Childers, N.F., J.R. Morris, and G. S. Sibbett. 1995. *Modern fruit science,* Tenth edition. Gainesville, FL: Norman F. Childers.

Forshey, C.G., D.C. Elfving, and R.L. Stebbins. 1992. *Training and pruning apple and pear trees.* Alexandria, VA: American Society of Horticultural Science.

Gur, A. 1985. Rosaceae—Deciduous fruit trees, pp. 355-389. In: A.H. Halevy (ed.), *CRC handbook of flowering,* Volume I. Boca Raton, FL: CRC Press.

Hedrick, U.P., G.H. Howe, O.M. Taylor, E.H. Francis, and H.B. Tukey. 1921. *The pears of New York* (New York Department of Agriculture 29th Annual Report, Vol. 2). Albany, NY: J. B. Lyon Company.

Jackson, J.E. 2003. *Biology of apples and pears.* Cambridge, UK: Cambridge University Press.

Jones, A.L. and H.S. Aldwinckle. 1990. *Compendium of apple and pear diseases.* St. Paul, MN: American Phytopathological Society.

Layne, R.E.C. and H.A. Quamme. 1975. Pears, pp. 38-70. In: J. Janck and J.N. Moore (eds.), *Advances in fruit breeding.* West Lafayette, IN: Purdue University Press.

Lombard, P., J. Hull, and M.N. Westwood. 1980. Pear cultivars of North America. *Fruit Var. J.* 34:74-83.

Lombard, P.B. and M.N. Westwood. 1987. Pear rootstocks, pp. 145-183. In: R.C. Rom and R.F. Carlson (eds.), *Rootstocks for fruit crops.* New York: John Wiley and Sons.

Teskey, B.J.E. and J.S. Shoemaker. 1978. *Tree fruit production.* Third edition. Westport, CT: AVI Publishing.

Van der Zwet, T. and N.F. Childers (eds.). 1982. *The pear.* Gainesville, FL: Norman F. Childers.

Westwood, M.N. 1993. *Temperate zone pomology.* Third edition. Portland, OR: Timber Press.

White, A.G., D. Cranwell, B. Drewitt, C. Hale, N. Lallu, K. Marsh, and J. Walker. 1990. *Nashi—Asian pear in New Zealand.* Wellington, NZ: DSIR Publ.

Chapter 25

Pecan *(Carya illinoensis)*

TAXONOMY

The pecan is a member of the Juglandaceae family, along with walnuts, but is more closely related to hickories than walnuts. The current name of the pecan is *Carya illinoensis* (Wangenh.) K. Koch, although in the past 100 years it has been referred to as *Juglans pecan, Juglans illinoensis, Hicoria pecan*, and *Carya pecan*. The name "pecan" comes from the Native American word *pacane,* meaning "nut so hard as to require a stone to crack."

There are about 20 *Carya* species, a few of which produce edible nuts. Relative to pecan, they generally have low yields, thick, hard shells, small kernels, and astringent flavor. There are more named cultivars of shagbark hickory *(Carya ovata)* than for any other *Carya* species, except pecan. It is also among the most cold hardy of hickories. Nuts are similar to those of pecan, with thicker shells and more rounded in shape. Cultivars include 'Bridgewater', 'Davis', 'Porter', 'Weschcke', 'Wilcox', 'Wurth', and 'Yoder'. Shellbark hickory *(C. lacinosa)* is similar to shagbark hickory, but its nuts are larger with thicker shells. Cultivars include 'Bradley', 'Dewey Moore', 'Fayette', 'Henry', 'Keystone', 'Lindauer', 'Nieman', and 'Stephens'.

Shagbark and shellbark hickories are good pecan replacements in the northern Midwest, mid-Atlantic, and New England states. "Hicans" are hybrids between pecan and shagbark hickory or another *Carya* species; there are about 12 named cultivars, for example, 'Burton', 'Hartmann', 'Hershey', and 'James'. Plants produce low-quality, poorly filled nuts and generally have low yields relative to pecan.

Cultivars

Over 1,000 pecan cultivars are documented in the literature, but only four make up over 50 percent of the orchard cultivation in the United States. 'Stuart' is the leading cultivar in the southeastern United States, accounting for 22 to 26 percent of production. It does poorly in arid areas and is somewhat tolerant of fungal scab. It is protogynous (see the Botanical Description section) and lacks precocity (requires 7 to 8 years to bear nuts). 'Stuart' became the leading cultivar by having good nut size (52 nuts/pound), high yield (1,000-1,500 pounds/acre), good coldhardiness, and, initially, tolerance of scab. Since its introduction, its scab resistance has broken down somewhat. The nuts have good shelling characteristics.

'Desirable' is moderately scab resistant and accounts for 9 to 11 percent of U.S. production (Plate 25.1). It is popular in the eastern United States. Being protandrous, it makes a good pollinizer for protogynous cultivars. It is more precocious than 'Stuart' and has larger nuts (47 nuts/pound). 'Western', also called 'Western Schley', is the most popular cultivar in the western United States. It is scab susceptible, so it is not grown in eastern states. It makes up 12 to 15 percent of U.S. production. 'Western' is protandrous but exhibits good overlap between male and female flower function; thus, a high degree of self-pollination is possible (see Pollination in the

Introduction to Fruit Crops

doi:10.1300/5547_25

Botanical Description section). It is more precocious than 'Stuart' (begins bearing at 6 years) but has a smaller nut (64 nuts/pound).

'Wichita' is second to 'Western' in importance from Texas westward, accounting for 7 to 11 percent of production. As with 'Western', it too is scab susceptible. It is protogynous and a good pollinizer for 'Western'. Fairly precocious and high yielding, 'Wichita' produces many nuts per cluster and is more prone to alternate bearing as a result. The nuts are moderate in size (57 nuts/pound), with plump kernels of high quality.

There is a group of northern pecan cultivars that mature in shorter periods than do standard cultivars, and these can be grown in the Midwest and Northeast. They have been characterized as low yielding and variable in quality. Examples include 'Colby', 'Fritz', 'Giles', 'Greenriver', 'Kanza', 'Major', 'Peruque', 'Posey', 'Starking', and 'Witte'. 'Major' is touted to be the best of the northern cultivars, with good production and disease resistance. 'Kanza' may replace 'Major' in the future due to its larger nut size.

ORIGIN, HISTORY OF CULTIVATION

Pecan is the most important *native* North American orchard species, particularly in terms of acreage (note that few fruit crops are native to North America, and despite the large area of cultivation, pecan is not considered an important fruit crop worldwide). Pecan trees are native to the Mississippi River watershed, from Illinois in the north to eastern Texas, and eastward to Alabama in the Gulf states. Pecans were used by Native Americans at least 8,000 years ago in Texas. Crows may have contributed to the selection and distribution process before humans, as they select thin-shelled nuts and can carry them several miles. The first budded trees were produced in Louisiana in the mid-1800s, and orchards were established throughout the native range, eastward to Georgia and westward in Texas. The Georgia industry began in the early 1900s, as 5-acre tracts of retirement property. These small orchards were consolidated over time to yield fewer, but much larger, farms with appropriate economies of scale. The first commercial shipment of pecans originated in Georgia in 1917, when orchards were already producing high-quality nuts graded according to size and color. The high level of cultivation the state currently enjoys began in 1970. Georgia has since been the leading producer of "improved pecans," although collection of pecans from wild trees in Texas and Oklahoma exceeds Georgia's production in some years.

FOLKLORE, MEDICINAL PROPERTIES, NONFOOD USAGE

Pecans are recent domesticates and have relatively few alternative uses. As an astringent, pecans are folk remedies for blood ailments, dyspepsia, fever, flu, hepatitis, leucorrhea, malaria, and stomachache. Pecan oil is used in drugs, essential oils, and cosmetics. Leaves and fruits contain juglone, which is allelopathic and fungitoxic. The presence of another fungitoxic substance, linalool, has also been reported. Pecan pollen is a common allergen. Pecan shells stored for long periods may contain aflatoxin, a dangerous poison.

Pecan and other hickory wood is ranked third among hardwoods used for furniture in the United States, behind black walnut and black cherry. Hickory wood is used for furniture, tool handles, skis, gymnastic bars, flooring material (for gymnasiums, roller skating rinks), piano construction, interior trim, dowels, pallets, and ladder rungs. It is famous for its use in smoking meats and cheeses and is sometimes combined with charcoal to impart a distinctive flavor to grilled foods.

PRODUCTION

World

Reliable data are not available, as pecan is primarily a North American crop. Pecans are lumped together with at least seven other species of nuts, including macadamia, in FAO statistics. Most non-U.S. production occurs in northern Mexico, and the USDA Foreign Agricultural Service reported 34,000 and 64,000 MT in 2003 and 2004, suggesting that Mexico is now producing up to about half that of the United States. Brazil, Australia, South Africa, and Israel produce small quantities (generally cultivated on 10,000 or fewer acres).

United States

Total production in 2004, according to USDA statistics, was 82,273 MT or 181 million pounds. Pecans show heavy alternate bearing, and 2004 was an "off" year; more typically, 95,000 to 160,000 MT are produced (Figure 25.1). Due to record prices, the industry value was $301 million, despite low production, and has ranged between $134 million and $330 million over the past decade.

Pecans are separated into two categories based on intensity of cultivation: *improved* and *native and seedling*. Improved pecans are produced from superior cultivars in orchard settings, while native and seedling production occurs largely from wild-harvested native stands of trees in Oklahoma and Texas. Native and seedling cultivation ranges from no management inputs, except harvest, to cultivation of seedling trees in orchards with inputs of fertilizers and pesticides. Yield, quality, and prices received are much greater for improved than for native and seedling pecans.

Total pecan acreage has been estimated at over 1.4 million. This is far greater than areas under grape and citrus cultivation, two of the most extensively planted fruit crops in the United States.

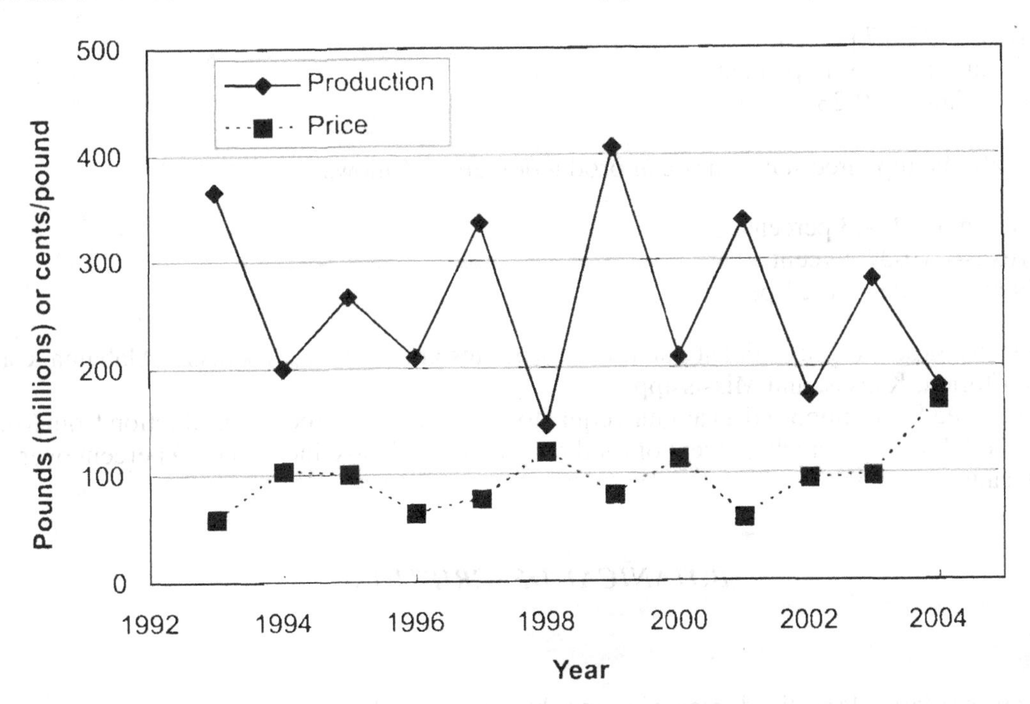

FIGURE 25.1. Alternate bearing in pecans shows up in nationwide production data. Note that price generally increases in "off" years when production is low, and vice versa. *Source:* USDA statistics.

About 60 percent of this is native and seedling, and 40 percent improved, but the industry is moving toward greater production from improved cultivars.

Yield data vary widely with cultivar, location, and cultivation method, but yields are among the lowest of nut crops grown in the United States. Native and seedling pecans produce only a few hundred pounds per acre, or amounts not justifying harvest in "off" years. In well-managed orchards, yields average over 1,000 pounds/acre, with "on" year amounts of 2,000 to 2,500 pounds/acre.

Improved

Total production in 2004 was 60,191 MT or 132 million pounds. Improved cultivars are valued at $247 million, with growers receiving $1.86 per pound. In 7 of the past 10 years, price per pound has averaged over $1.00 for improved cultivars. This is a great improvement over the $0.60 to $0.80 per pound received throughout the 1980s. Improved pecans are produced in 13 of the 14 states that grow pecans; the leading states follow:

1. Georgia—27-37 percent
2. New Mexico—25-28 percent
3. Texas—15-20 percent

Native and Seedling

Total production in 2004 was 22,081 MT or 49 million pounds, with a corresponding industry value of $55 million. Prices received were $1.12 per pound in 2004, significantly higher than the normal prices of $0.49 to $0.75 per pound. Native and seedling pecans are produced in 11 of 14 pecan-producing states, with leading states as follows:

1. Texas—17-47 percent
2. Oklahoma—15-45 percent
3. Louisiana—10-28 percent

Overall, the top three states in pecan production are as follows:

1. Georgia—25-43 percent
2. Texas—19-27 percent
3. New Mexico—16-21 percent

Other states producing significant quantities of pecans are Arizona, Arkansas, Alabama, California, Florida, Kansas, and Mississippi.

The United States imported an amount equal to 44 percent of domestic production from Mexico in 2002. Exports were 40 percent of total production and have increased 89 percent over the past decade.

BOTANICAL DESCRIPTION

Plant

Pecans are large, long-lived trees with upright, vase-shaped crowns. Pecans are the largest orchard trees in the United States, reaching 60 to 80 feet tall, with densities as low as 10 to 15 trees/acre (Figure 25.2). Leaves are odd-pinnately compound, with generally 7 to 17 leaflets. Leaflets are lanceolate to obovate, with the proximal half of the blade generally smaller than the distal half (asymmetrical); margins are serrate. Current season's wood has solid pith in *Carya*, whereas chambered pith occurs in *Juglans*.

FIGURE 25.2. Mature pecan trees in an orchard in Georgia. Note the open, spreading canopy shape.

Flowers

Pecans are monoecious; male inflorescences are pendulous spikes, commonly referred to as "catkins" (Plate 25.2). Catkins appear to be borne laterally in groups of two to three on 1-year-old wood but are actually produced on extremely short, aborted shoots of current season's growth. Catkins are 2 to 6 inches in length and are shorter in protandrous than in protogynous cultivars (see Pollination). The female inflorescences are also spikes, borne terminally on current season's growth. Both male and female flowers are green in color, with individual flowers about ⅛ inch in size. Flowers are incomplete, and both sexes lack petals and sepals. The males are basically groups of anthers subtended by bracts, and the females are merely an ovary with a large, feathery stigma at the distal end, with almost no length of style in between. The ovary wall is fused to bract or involucre tissue in females, and the outer ovary wall plus the involucre become the fleshy shuck of the fruits.

Pollination

In pecans, the period of time over which pollen is shed commonly does not coincide with the period of female flower receptivity to pollen. Thus, pecans often require another cultivar for pollination because the timing of the male and female functions is different on any given tree. This condition is referred to as "dichogamy" and is common in trees in the Juglandaceae family. The two possible forms are *protandry* (means "first male"), for which pollen shed begins before the female flower becomes receptive, and *protogyny* (means "first female"), for which female flower receptivity precedes pollen shedding. A good pair of cultivars for cross-pollination will contain a protandrous and a protogynous cultivar. In most cases, there is some overlap in the functional periods, allowing for some degree of self-pollination. There is a tendency for protandrous cultivars to have greater overlap or *incomplete dichogamy* than is seen with protogynous cultivars, resulting in more self-pollination in protandrous cultivars. Some cultivars have no overlap of pollen shed and female receptivity whatsoever (i.e., they exhibit *complete dichogamy*), and these are sometimes termed *functionally dioecious* because the tree is either functionally male or functionally female at any given time, never both. (This term confuses the fact that all pecans are truly monoecious, not dioecous.) Adding to the complexity of the issue, environment can influence the degree of dichogamy expressed by a given cultivar and, less commonly, the type of dichogamy. For example, 'Barton' consistently begins pollen shed 3 days before the stigma receptivity period in Las Cruces, New Mexico, but sheds pollen 10 days before in Brownwood, Texas. In Georgia, 'Barton' becomes protogynous, and pollen shed begins 5 days *after* stigma receptivity. This underscores the importance of choice of pollinizer(s) in pecan culture.

Pecans, walnuts, and many other nut crops are wind pollinated, and the pollen may travel for miles on a strong wind. Also, each catkin, or male inflorescence, contains over 1 million pollen grains, and it takes only one to pollinate a female flower and make a nut.

Fruit

Nuts are borne in terminal clusters of two to ten individuals on current season's growth (Plate 25.3). The green, fleshy shuck is composed of bract and ovarian tissue (exo- and mesocarp) and surrounds the nut until maturity. The shuck dehisces along four regular sutures, unlike in walnuts, which lack sutures. Nuts are ovoid, 1 to 2 inches long, and contain a large embryo composed mostly of two cotyledons or kernels. Nut size ranges from 27 to 100 nuts/pound; "good" size is considered 50 nuts/pound. Cotyledons are separated by a thin, papery central plate extending from the inner layer of the endocarp. Fruiting occurs 4 to 6 years after transplanting, although maximal yields may not be achieved for 20 or more years.

Alternate bearing is common in pecan (see Figure 25.1). This is when the tree bears a large crop one year, followed by a small crop the next. This is a natural tendency in pecan and many other nut trees. Commercial growers sometimes shake developing nuts from the trees in a high-crop year to reduce the alternate-bearing habit. However, fluctuations in disease incidence, insect populations, and climatic conditions from year to year tend naturally to synchronize trees in an orchard or an entire region into the "on" and "off" year cycle. Some theorize that a large crop in "on" years depletes carbohydrate reserves in the trees, resulting in poor flowering (and therefore yield) the following year. In the "off" year, carbohydrates can be replenished, allowing for high yield the next year. Thus, as young trees begin bearing, all it takes is one good or bad year to begin the alternate-bearing cycle, which then perpetuates itself. Others theorize that growth regulators in the developing nuts actually inhibit flower initiation, so in high-crop years, few flowers can be set for the following year's crop (and vice versa). The growth regulator theory has been shown to be true for apple, but neither theory has been proved or disproved in pecan.

Alternate bearing is common for most nut tree species and may be a natural protection against seed predation. High crops year after year would allow buildup of high populations of insects, birds, or mammals that eat the nuts. Producing a low crop of nuts every so often would decrease the food available for the predators and, consequently, put a check on their population growth. So, a low-production year for the trees results in a lower population of predators the next year, when nut production is high. This behavior is similar to predator-prey relationships commonly seen in nature. Pecan growers are thus attempting to reduce an innate tendency that has been programmed into the trees over millennia.

GENERAL CULTURE

Soils and Climate

Pecan trees prefer light- to medium-textured soils at a pH of 5.5 to 6.0, but they can be grown on soils with a higher clay content and higher pH. Soil depth should be several feet, with the water table below the root zone during the growing season. Pecans are native to floodplains and river bottoms and have an inherently high water requirement. Irrigation has been shown to increase yields even in the humid southeastern United States, and it is essential for western orchards.

Pecans perform equally well in arid and humid climates, if appropriate cultivars are selected and trees are provided with adequate soil water and disease management. Disease pressure is worse in humid climates, as for most horticultural crops. The commercial "pecan belt" is the tier of southern states stretching from Georgia to Arizona. Pecans form natural forests along riverbanks of the Mississippi Valley in the south-central United States, so a warm-temperate climate is optimal for growth.

Coldhardiness is adequate for tree growth as far north as Wisconsin and New England, although young trees may be killed outright by temperatures of –5 to 0°F. Frost is rarely a problem, due to late budbreak, although emerging shoots can be killed by relatively mild subfreezing temperatures (around 28°F). Heat requirement is perhaps more important than coldhardiness in limiting pecan production to the southern United States. Nuts do not "fill" properly in northern areas, due to inadequate temperatures for photosynthesis during the critical kernel growth period in autumn. Most pecan cultivars require growing seasons of 180 to 220 days. Northern pecan cultivars are those which mature in short periods. Chilling requirement is relatively short, around 500 hours, but sometimes pecans break bud normally without any chilling (as in parts of Mexico and Israel). Pecans are among the last tree crops to resume growth in spring, apparently due to a high postrest heat requirement.

High humidity and rainfall during the early stages of development promote pecan scab, a serious disease that can cause extensive crop loss if not controlled.

Propagation

Grafting is the most popular method of propagation. Rootstocks are often whip-and-tongue grafted in February through March using 1-year-old scions. The graft is made below the ground line in southern areas, although trees propagated this way experience more winter injury to trunks than those budded or grafted higher. The "banana graft" is simple and becoming increasingly popular for pecans (Figure 25.3).

The banana graft requires an active rootstock and dormant scion wood. Seedlings are often budded or grafted in situ after 2 to 3 years of growth in the orchard, as this is less costly and risky than transplanting relatively large trees from nurseries.

Patch budding is the most common budding technique. This is usually done in late summer using buds from current season's wood; buds are forced the following spring.

Rootstocks

Pecan seedlings are the most common rootstocks. To date, no clonal or size-controlling stocks have been introduced to the industry, largely due to difficulty in vegetative propagation of the species. Certain cultivars are said to produce superior rootstocks, such as 'Riverside' and

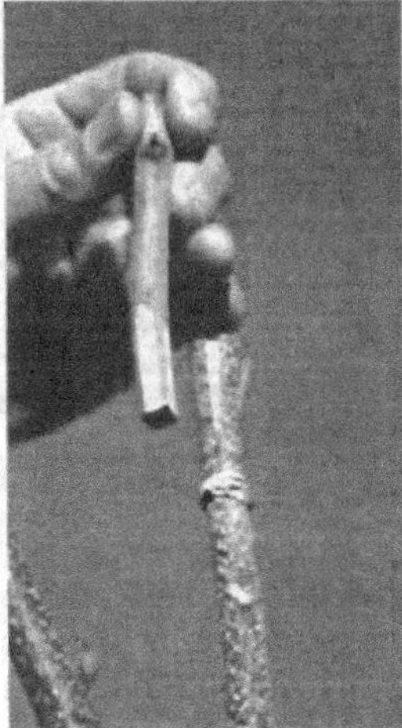

FIGURE 25.3. Banana grafting of pecan in the orchard. Seedling trees are planted in situ and allowed to grow for 1 or more years prior to grafting. *(left)* The stock is prepared by making four vertical cuts through the bark, peeling it like a banana, and removing the plug of wood remaining. *(right)* Scion wood is prepared with four complementary cuts and inserted into the stock.

'Apache' for western trees, and 'Stuart', 'Curtis', or 'Success' in eastern regions. In the orchard, pecan trees are highly variable, compared to other tree crops, because the seedling rootstocks are highly heterozygous. In its native range along river bottoms, bitter pecan *(Carya aquatica)* trees are sometimes grafted with pecan. The trees are said to be slightly dwarfed compared to native trees arising from seed.

Planting Design, Training, Pruning

Trees are often planted at densities of 36 to 100 trees/acre, gradually removing alternate rows when crowding occurs, yielding final densities as low as 14 trees/acre. Normal spacings of 56 × 56 feet, giving 14 trees/acre, are accomplished by planting in patterns of 40 × 40 feet and then removing alternate diagonal rows later. The use of temporary trees greatly increases the profitability of pecan growing because early returns (per acre) from young orchards are increased greatly (Figure 25.4).

Pollinizers are planted every ninth and eleventh row, so that they are not removed when crowding occurs. As a general rule, 10 percent of the trees should be pollinizers. In isolated orchards, two pollinizers are needed—one early and one late. In orchards near other orchards or wild trees, pollinizers are less important, as adequate pollen from several cultivars is available to the main cultivar.

During the first 3 years, trees are trained to a central leader, with scaffolds at 12- to 18-inch intervals, the lowest one beginning at 4 to 6 feet. After 3 years, very little pruning is necessary because trees form a natural, vase-shaped canopy through repeated forking of branches. Only dead, diseased, or broken limbs are removed on a regular basis. Since pecans bear nuts terminally on current season's growth, any pruning of limbs reduces yield.

Pest Problems

Insects

Pecan nut casebearer. Acrobasis nuxvorella is primarily an early season pest in the western United States; it rarely requires sprays in the Southeast. The larvae of this moth bore into developing nuts in spring, feeding on the contents of several nuts prior to pupation. Adults emerge in

FIGURE 25.4 *(left)* An orchard in Georgia where closely spaced trees have become crowded and some are being removed. *(right)* A 12-year-old orchard in Texas planted at high density produces over 2,000 pounds per acre, but it is rapidly approaching the stage where tree removal will be necessary.

June to July and lay eggs for a second generation, which is usually less damaging because larvae require fewer of the larger nuts for development at this stage. Third and fourth generations may occur but are even less damaging. The key to control is to time sprays to kill the first-generation larvae in spring (April to May). In most years, no action is required unless greater than about 5 percent of the nut clusters show damage in late spring.

Aphids. Black pecan aphids (*Melanocallis caryaefoliae*) and yellow pecan aphids (*Monelliopsis pecanis*) are sucking insects that reduce tree vigor by extraction of photosynthate. Severe infestations can cause defoliation in summer and reduce yield and nut filling. Perhaps more deleterious than feeding per se is the development of sooty mold (caused by *Polychaeton* spp.), which reduces photosynthesis at the critical nut-filling stage in late summer. Current evidence suggests that spraying for aphids actually causes greater aphid infestations, perhaps due to elimination of natural predators. Therefore, many growers do not spray insecticides in summer unless absolutely necessary for control of other pecan insects. 'Cape Fear' may be immune, and 'Owens', 'Wichita', 'Kiowa', 'Desirable', 'Shoshoni', 'Cherokee', and 'Chickasaw' show less than 10 percent damage by black pecan aphid.

Hickory shuckworm. Primarily a late-season pest, the larvae of *Laspeyresia caryana* can infest nuts as early as June, when they tunnel through the entire fruits, causing drop. More commonly, they tunnel into the shuck at shell hardening, causing poor nut development, since the shuck is important in nut filling. Populations are higher in older, established orchards/growing regions, but scant in newer western plantings. Sprays are timed from pheromone trap levels from the beginning of shell hardening. Nuts are cut crosswise during the summer to determine when shell hardening has begun. Black-light traps can reduce adult populations somewhat if hung in trees. Removal of old shucks and debris from beneath trees is helpful in reducing overwintering insects. No cultivars are immune, but 'Cape Fear', 'GraBohls', 'Chicasaw', and 'Cherokee' showed less than 40 percent fruit infestation in one study, with other cultivars showing 40 to 80 percent infestation.

Stinkbugs/plant bugs. The insects *Leptoglossus phyllopus* (leaf-footed), *Nezara viridula* (southern green), and *Euschistus servus* (brown) puncture the cotyledons at the "water stage," causing abortion of the nuts. Later feeding causes kernel spot—small brown spots on the surface of kernels, which have an off-flavor. It is particularly a problem on 'Schley' and 'Wichita', but 'Gloria Grande' showed no injury in one study. The bugs are large, brown or gray, and shield shaped; they are easily seen. These insects are generally not a problem requiring action. In orchards with a history of problems, sprays are applied when insects are visible in and around nut clusters.

Pecan weevil. This pest (*Curculio caryae*) is important in all pecan areas, exclusive of the Gulf Coast. However, the infestation may be sporadic because insects do not travel far from the trees from which they emerge. Both adults and larvae cause damage; adults drill holes in shucks for oviposition in the early stage of growth, and if the cotyledons are punctured, nuts drop. The larvae feed directly on maturing nuts, which do not drop but become "stick tights" or "pops" (empty shells) at harvest. Indirect or cultural control is achieved through early harvest, as insects become active only in late summer. Some cultivars, such as 'Johnson', mature before weevils emerge from the soil, escaping damage completely. Removal of infested nuts as they fall can reduce populations. Otherwise, spraying insecticide on the lower part of the trees and on the ground when nuts have fallen, or beginning in August, can provide some control.

Diseases

Pecan scab. Cladosporium caryigenum causes pecan scab, by far the most serious disease of pecans. Lack of control can seriously reduce yields in every year, particularly in wet years in the southeastern United States. The fungus attacks expanding tissues only, beginning at budbreak.

Infections appear as olive-brown/black spots on the undersides of leaves; trees defoliate prematurely if infections are severe. Spores from leaf and twig infections provide inoculum for shuck infection, beginning in June, which causes the nuts to drop or shrivel on the peduncles. There are no truly resistant cultivars, since there are several races of scab, but 'Desirable', 'Farley', 'Gloria Grande', 'Elliott', and 'Stuart' are resistant to many races. Removal of old shucks and leaf stems by shaking the limbs prior to budbreak may reduce the inoculum available for reinfection in spring. In the Southeast, sprays are absolutely required for control. Starting at budbreak, sprays are applied at 2- to 3-week intervals until late summer or are targeted to periods of prolonged dew or rainfall. The disease is not a problem in the arid western states.

Downy spot. This disease, caused by *Mycosphaerella caryigena*, is considered the second most destructive to pecan and occurs throughout the pecan belt. The disease occurs in arid areas (and dry years) with equal or greater incidence than that seen in humid areas (wet years). Yellow or whitish lesions occur on the lower leaf surface in late spring or early summer. As with other fungal diseases, premature defoliation can severely affect yield. Several cultivars exhibit immunity to downy spot: 'Burkett', 'Caspiana', 'Cline', 'Dependable', 'Frotscher', 'Mahan', 'Odom', 'Pabst', 'Sabine', 'Success', and 'Williamson'. Downy spot is controlled as a result of early sprays for scab, around the prepollination period.

Vein spot. This disease is caused by a fungus *(Gnomonia nerviseda)* and is unusual in that it attacks vascular tissues, not the interveinal areas most commonly affected by fungi. In May to June, the pathogen creates tiny lesions on leaf veins that later coalesce into brown or black spots, similar to those seen with pecan scab, but glossier. Premature defoliation may occur in severe cases. All major cultivars are susceptible. Sprays applied around pollination time or shortly after (April to June) will control vein spot.

Other diseases. Stem end blight is apparently a fungal disease, but it remains unclear whether *Botryosphaeria ribis, Glomerella cingulata,* or *Phytophthora cactorum* (or all) is involved. Lesions are generally black sunken spots at the proximal end of the fruits and peduncles. Shuck dieback is a self-descriptive disease of unknown etiology that causes nut drop 3 to 4 weeks prior to shuck split. Powdery mildew (caused by *Micosphaera alni*) and zonate leaf spot (caused by *Cristulariella pyramidalis*) can be problematic on the leaves of trees grown in the southeastern United States. Virtually all fungal diseases are controlled by fungicide sprays for scab, and they generally require individual attention in arid areas where scab is not a problem.

HARVEST, POSTHARVEST HANDLING

Maturity

Pecans can be harvested when the shucks loosen from the shell or when shucks split.

Harvest Method

Pecans are harvested with trunk or limb shakers, depending on tree age. Nuts are collected on tarps or mats in less sophisticated orchards; nuts are wind-rowed and swept in larger, more sophisticated plantings (Figure 25.5).

Postharvest Handling, Storage

Nuts are separated from shucks, leaves, and other debris and then dried in forced-air heaters to a moisture content of about 10 to 15 percent. Nuts can be stored for about 4 months at room temperature before becoming rancid but last 1 to 2 years when stored at 34°F.

FIGURE 25.5. Pecan-harvesting equipment. *(above left)* Trunks or individual limbs are shaken, depending on tree size. *(above right)* A windrow machine consolidates nuts scattered on the ground into rows, *(right)* over which a sweeper is run to collect nuts. Leaves and other debris are separated from the nuts by blowers

CONTRIBUTION TO DIET

Most of the pecan crop is sold shelled and used in baking and confectioneries, particularly around the Thanksgiving and Christmas holidays. Per capita consumption of pecans was 0.38 pound/year in 2004, the lowest in over 20 years. More typically, Americans consume 0.4 to 0.5 pound/year.

Dietary value per 100-gram edible portion:

Water (%)	4
Calories	691
Protein (%)	9.2
Fat (%)	72
Carbohydrates (%)	13.9
Crude Fiber (%)	9.6

	% of U.S. RDA (2,000-calorie diet)
Vitamin A	1
Thiamin, B_1	44
Riboflavin, B_2	8
Niacin	6
Vitamin C	2
Calcium	7
Phosphorus	40
Iron	14
Sodium	0
Potassium	12

BIBLIOGRAPHY

Brison, F.R. 1974. *Pecan culture.* College Station, TX: Texas Pecan Growers Association.

Crocker, T.F. 1982. *Commercial pecan production in Georgia* (University of Georgia Cooperative Extension Service Bulletin 609). Athens, GA: University of Georgia.

Duke, J.A. 2001. *Handbook of nuts.* Boca Raton, FL: CRC Press.

Gast, K.L.B. 2003. Marketing, pp. 363-370. In: D.W. Fulbright (ed.), *Nut tree culture in North America,* Volume 1. East Lansing, MI: Northern Nut Growers Association.

Hanna, J.D. 1987. Pecan rootstocks, pp. 401-410. In: R.C. Rom and R.F. Carlson (eds.), *Rootstocks for fruit crops.* New York: John Wiley and Sons.

Madden, G. 1979. Pecan, pp. 13-34. In: R.A. Jaynes (ed.), *Nut tree culture in North America.* Hamden, CT: Northern Nut Growers Association.

Madden, G., F.R. Brison, and J.C. McDaniel. 1969. Pecans, pp. 163-189. In: R.A. Jaynes (ed.), *Handbook of North American nut trees.* Knoxville, TN: North American Nut Growers Association.

Reid, W. and K.L. Hunt. 2003. Pecan production in the Midwest, pp. 107-116. In: D.W. Fulbright (ed.), *Nut tree culture in North America.* Hamden, CT: Northern Nut Growers Association.

Rosengarten, F. 1984. *The book of edible nuts.* New York: Walker and Co.

Sparks, D.S. 1992. *Pecan cultivars—The orchard's foundation.* Watkinsville, GA: Pecan Production Innovations.

Teviotdale, B.L., T.J. Michailides, and J.W. Pscheidt. 2002. *Compendium of nut crop diseases in temperate zones.* St. Paul, MN: American Phytopathological Society.

Thompson, T.E. and L.J. Grauke. 1990. Pecans and hickories *(Carya),* pp. 837-904. In: J.N. Moore and J.R. Ballington (eds.), *Genetic resources of temperate fruit and nut crops* (Acta Horticulturae 290). Belgium: International Society for Horticultural Science.

Thompson, T.E. and G. Madden. 2003. Pecans, pp. 79-106. In: D.W. Fulbright (ed.), *Nut tree culture in North America.* Hamden, CT: Northern Nut Growers Association.

Thompson, T.E. and F. Young. 1985. *Pecan cultivars, past and present.* College Station, TX: Texas Pecan Growers Association.

Woodroof, J.G. 1967. *Tree nuts: Production, processing, products,* Volume 2. Westport, CT: AVI Publishing.

Chapter 26

Pineapple *(Ananas comosus)*

TAXONOMY

The pineapple, *Ananas comosus* Merr., is a member of the Bromeliaceae, a large, diverse family of about 2,000 species. Bromeliads, with the exception of one species *(Pitcairnia feliciana)*, are native to only the New World, largely in the tropics. Most of the 50 or so genera are composed of epiphytic species, the exception being *Ananas* (pineapple) and a few others, such as *Bromelia* and *Pitcairnia*. Family members are further distinguished by being herbaceous and rosette forming, with stellate hairs on appendages and colored floral bracts. The family contains hundreds of taxa used as ornamentals in greenhouses or subtropical areas: *Billbergia, Vresia, Nidularium, Aechmea, Guzmania, Pitcairnia*, and *Tillandsia. Tillandsia usneoides* is Spanish moss native to the Gulf states. The related species *Ananas ananassoides* and *A. bracteaus* have been used to a limited extent in pineapple breeding. Formerly, the pineapple was named *Ananas sativus, Bromelia ananas*, or *Bromelia comosus*. As with the banana and palms, it is one of the few important fruiting monocots.

Cultivars

Cultivated pineapple went through an astoundingly narrow genetic bottleneck during domestication. The predominant cultivar in production is 'Smooth Cayenne', which was selected by Venezuelan Indians for its attractive, flavorful, seedless fruits and, more important, the lack of spines on leaves. These characteristics, along with large fruit size and high yield, ensured that it became the most desirable form of the plant for worldwide dissemination. Furthermore, the ease of vegetative propagation and lack of seed production contributed to preserving its genetic constitution. It is highly susceptible to mealybug wilt disease and nematodes and has been supplanted by superior mutations in many areas: 'Hilo' in Hawaii, 'Sarawak' in Malaysia, and 'St. Michael' in the Azores are all 'Smooth Cayenne' sports. Its fruits are used for both canning and fresh-market sales.

Four other groups of cultivars are recognized: 'Queen', 'Spanish', 'Abacaxi', and 'Maipure'. 'Queen' and its derivatives are grown in South Africa and Australia for fresh-market fruits. Its small fruits have a higher sugar-acid ratio and ship better than those of 'Smooth Cayenne'. 'Spanish' contains 'Red Spanish', probably the second most important cultivar of pineapple, and the major fresh-market cultivar of the Caribbean. 'Red Spanish' is more disease resistant, and its fruits ship better than those of 'Smooth Cayenne', and they are smaller, rounder, redder, and have fewer eyes. The 'Abacaxi' and 'Maipure' groups are of local interest only in tropical America. Only 'Maipure' has spineless leaves, as does 'Smooth Cayenne'.

Introduction to Fruit Crops

doi:10.1300/5547_26

ORIGIN, HISTORY OF CULTIVATION

The pineapple is native to dry forest or thorn scrub vegetation regions of South America, although its exact origin is disputed. Older sources placed the center of diversity in southern Brazil and Paraguay, but more recent study suggests it may be northern Brazil, Colombia, and Venezuela. In part, the confusion stems from distribution of cultivated types by native peoples throughout tropical America and the Caribbean prior to the arrival of Columbus. Carib Indians probably distributed it to Guadeloupe, where it was collected by Columbus in 1493. The pineapple was then taken to Europe and distributed to the Pacific Islands, India, and Africa by Spaniards and the Portuguese explorers of the sixteenth and seventeenth centuries.

The first commercial plantation was established on Oahu in 1885, and Hawaii produced most of the world's pineapple until the 1960s, when urbanization and scarcity of labor forced production elsewhere, particularly to the Philippines. The Hawaiian industry has continued its slow decline over the past decade and now produces only 2 percent of the world's pineapple. Florida produced pineapple for a short period at the turn of the twentieth century but freezes in 1917 devastated the industry. Pineapple was first canned in 1888 in Malaysia, and canned fruits were first exported from Singapore in about 1900. Today, Southeast Asia still dominates world production, but large amounts are also produced in Latin America and Africa.

FOLKLORE, MEDICINAL PROPERTIES, NONFOOD USAGE

Pineapples are international symbols of welcome and are symbolic of saying, "You are perfect," when presented to someone. Gateposts and door frames are often topped with sculptures of pineapple as a welcome sign. The name "pineapple" derives from the Spanish word for the fruit *piña,* which means pine. Spanish explorers thought the fruit looked like a pinecone.

Bromelain is a proteolytic enzyme obtained from pineapple juice that has several uses. Primarily, it is used to tenderize meat, similar to papain, the proteolytic enzyme derived from green papaya. It also chill-proofs beer, stabilizes latex paint, and tans leather. Being a protease, it can cause dermatitis on exposed skin. Several folk remedies of pineapple may stem from its bromelain content: as an anti-inflammatory, for muscle relaxation, and in the treatment of warts, abscesses, bruises, and ulcers. Pineapple workers are said to end up having no fingerprints due to the proteolytic action of bromelain on the skin. Juice from unripe fruits is said to be purgative and also anthelmintic, perhaps due to the known nematicidal effects of bromelain.

Pineapple leaves contain strong fibers that are used to make coarse textiles, thread, casting nets, and even underwear(!). 'Perolera', a cultivar in the 'Maipure' group, is said to be best for this purpose because it has long, wide, fibrous leaves.

PRODUCTION

World

Total production in 2004, according to FAO statistics, was 15,287,413 MT or 34 billion pounds. Worldwide, 82 countries produce pineapple in economic quantities on about 2.1 million acres. Production has increased 22 percent over the past decade, due to an increase in acreage, as yields have been stable for at least 10 years. Average yields are 16,000 pounds/acre, with a huge range from only 5,000 to over 64,000 pounds/acre (Costa Rica). Production has dropped 28 percent in the leading country, Thailand, over the past decade. However, substantial increases in all major countries, except Brazil, have allowed production to increase steadily in recent years. The top ten pineapple-producing countries (percent of world production) follow:

1. Thailand (11)
2. Philippines (11)
3. China (10)
4. Brazil (10)
5. India (9)
6. Nigeria (6)
7. Costa Rica (5)
8. Mexico (5)
9. Indonesia (5)
10. Kenya (4)

United States

Total production in 2004, according to USDA statistics, was 195,450 MT or 430 million pounds. The industry value is $80 million. All production is in Hawaii, with some on all islands, but Kauai produces the most. Current acreage stands at 13,000, with yields of about 33,000 pounds/acre, double the world average. Prices received by growers have increased slightly to the current $0.19 per pound, with prices of $0.32 per pound for fresh fruits and $0.07 per pound for processed. Hawaiian production has been declining steadily over the past decade (Figure 26.1) and is down over 50 percent since 1980.

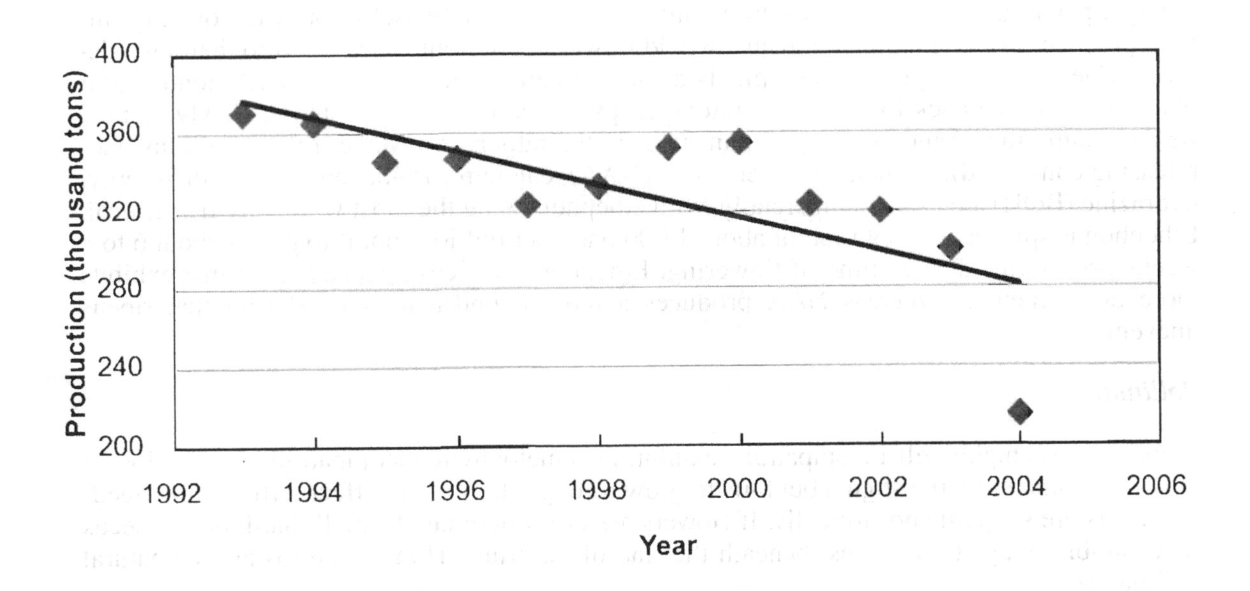

FIGURE 26.1. The trend in Hawaiian pineapple production over the past decade shows a decline of about 8,600 tons per year, with a good deal of annual variation. In 1980, the state produced about 600,000 tons, almost triple the 2004 level. *Source:* USDA statistics.

BOTANICAL DESCRIPTION

Plant

Pineapples are rosette-forming, herbaceous monocots, 2 to 4 feet tall and 3 to 4 feet wide (Plate 26.1). Stems are short (12 inches) and inconspicuous in the center of the rosette of long (20-72 inches), linear leaves. The leaves have spines at the tips and margins, except in 'Smooth Cayenne' types, which lack marginal spines. Leaves are spirally arranged on stems and have axillary buds at their base that can produce lateral shoots called "suckers" or "ratoons." Lateral shoots borne in axils at the base of the inflorescence are called "slips." These shoots are used as planting stock for propagating the next crop. An alternative method is for plants to be trimmed back after harvest, leaving one sucker to grow in place of the original plant, yielding a "ratoon crop," or a second crop from the same field. Since the root system is intact in this case, suckers develop into fruiting plants faster than when fields are completely renovated.

Flowers

Individual flowers are small (½ to 1 inch), purple-red, subtended by a single yellow, green, or red bract, borne laterally on the rachis of a spike of 100 to 200 individuals (Plate 26.2). Flowers open from the base to the apex, with one or more flowers opening each day over a period of about a month. The inflorescence is subtended by a whorl of small red bracts, which signal the onset of flowering. The apex of inflorescence is vegetative, becoming the "crown" on the fruit, which can be used for propagation.

Bromeliads are unusual plants in that flowering can be induced by chemicals: in nature, the gaseous hormone ethylene initiates flowering. This was discovered accidentally in the Azores in 1874, when smoke from burning fields was observed to elicit uniform flowering in adjacent, younger plantings in about 6 weeks. Floral induction is extremely useful for synchronizing cultural practices because individual plants would flower at different times if left to their own devices. Chemicals are applied when plants achieve a certain size, about 6 to 12 months after planting or cutting back to suckers in ratoon crops (around the 30-leaf stage). Newly planted fields require up to 6 months longer than fields in the ratoon crop stage. Ethephon (ethylene-releasing compound), naphthalene acetic acid (NAA), calcium carbide, and beta-hydroxyethyl hydrazine (BOH) are used commercially, with ethephon being the most widely used chemical. Ethephon is sprayed as a solution of about 1,000 parts per million (ppm) to plants about 6 to 8 weeks prior to the desired time of flowering. Ethephon produces a squarer fruit that exhibits more even ripening, whereas NAA produces a more cylindrical, pointed fruit that ripens unevenly.

Pollination

Pineapple is highly self-incompatible, exhibiting gametophytic incompatibility, whereby the pollen germinates on the stigma but fails to grow through the style to effect fertilization. Seedless fruits are set parthenocarpically. If flowers are cross-pollinated, small, hard, brown seeds (undesirable) may be found just beneath the rind of the fruits. Hummingbirds are the natural pollinators.

Fruit

The fruit is a large, cylindrical multiple of berries, formed from the fusion of adjacent flowers on the spike and the development of ovaries in the fleshy, edible portion (Figure 26.2). Another name for this type of fruit is syncarp. The core is the fleshy rachis of the spike, often fibrous and

FIGURE 26.2. *(left)* An immature fruit a few weeks after flowering, with a developing vegetative crown. Note also the developing lateral shoot at the base of the fruit, called a slip, often used as propagation material. *(above)* A firm-mature pineapple ready for harvest and long-distance shipping from Latin America to Europe.

unpalatable. The fruit is covered with a waxy, leathery rind, made up of hexagonal "eyes," arranged spirally, that denote the position of individual flowers. One fruit per plant is produced, and the shoot it was borne on dies back or is cut off. Fruits require about 6 months from floral induction to harvest. Total production time is 15 to 18 months from transplanting, or about 12 months for a ratoon crop.

GENERAL CULTURE

Soils and Climate

Well-drained sandy loams, with a pH of 4.5 to 6.5 are best. Rooting depth is only about 2 feet, so soil depth is not important. Drainage is essential because pineapples are intolerant of waterlogging and are grown in climates with high rainfall. Also, fumigant permeates sandy soils better. Highly organic soils are also suitable. Fumigation is practiced routinely, as nematodes are serious problems in most growing areas. Pineapples are naturally drought tolerant by virtue of exhibiting Crassulacean acid metabolism (CAM) photosynthesis, opening their stomata at night for carbon dioxide (CO_2) uptake, and closing them during the day. They can be grown in seasonal wet/dry areas of the tropics that do not support less-water-efficient crops. However, supplemental irrigation is often provided to maximize yield in dry seasons. Black plastic mulch is used in intensive plantings for weed control and conservation of soil moisture.

Pineapples are restricted to hot, tropical lowlands with temperatures above 65°F. Optimal growth temperatures are between 75 and 90°F. Low temperature slows growth, delays harvest, and causes fruits to have a lower sugar-acid ratio. Freezing kills pineapple outright, and temperatures below 50°F for long periods of time can cause minor leaf injury. Precipitation varies widely among growing regions, with as little as 24 inches and well over 100 inches in the rainy tropics. Rainfall is often unevenly distributed in the wet/dry tropics, so periods of drought may occur.

Propagation

Plants are vegetatively propagated from crowns or axillary shoots arising from either the base of the fruit (slips) or the base of the plant (suckers or ratoons) (Figures 26.2, 26.3). Unrooted shoots are placed into the soil and root in situ. The size of the planting material affects production; larger suckers give the highest yield and reduce time from planting to harvest. Plant materials for establishing new fields are obtained from old fields after harvest, prior to clearing or burning.

FIGURE 26.3. A newly propagated field of pineapple showing plant spacing and the condition of the propagation material. Note the black plastic mulch for weed control.

Rootstocks

Pineapple has no rootstocks.

Planting Design, Training, Pruning

Pineapple is often grown on large, vertically integrated plantations that maintain fields in all stages of development, from soil preparation to harvest (Plate 26.3). The whole operation is generally highly mechanized. Plants are usually grown on two-row beds, spaced 12 inches apart in rows, and with 1 to 2 feet between rows on a bed. Alleys between beds are 2 to 4 feet wide. Plant densities of 18,000 to 24,000 per acre are obtained this way (Plate 26.3, right). Suckers are planted by hand or by machines that set up to 50,000 plants/day. Black plastic mulch is used in larger operations. Soil fumigation is also routinely practiced prior to planting, due to problems with nematodes. Machinery similar to that used for plasticulture strawberry and vegetable production is used to mechanize all aspects of field preparation and planting.

Plantings are renovated entirely after harvest, or one to three additional crops are obtained by cutting back plants to suckers. Disease problems and smaller fruit size occur in subsequent crop cycles, so most commercial operations renovate after one crop or obtain only one ratoon crop.

Pest Problems

Insects/Nematodes

Nematodes. Pineapple is extremely sensitive to nematode feeding on roots. Lesion *(Pratylenchus brachyurus)*, reniform *(Rotylenchulus reniformis)*, and root-knot *(Meloidogyne javanica)* nematodes are the main pests, but several other species may occur in various production regions. In all cases, feeding on root tips causes darkened lesions, necrosis, root stunting, and/or galling, which in turn stunts or kills the plants. Fumigation of soil prior to planting is routinely practiced, but some nematode losses are inevitable in plantations where pineapple is replanted in the same soil. In addition to fumigants, nematicides may be applied through drip irrigation in cases where a second or subsequent crop is desired. Dipping planting materials in nematicides can reduce carryover from previous fields. Crop rotation with nonhost species, such as pangola grass *(Digitaria decumbens)*, can greatly reduce nematode populations but ties up valuable land for years.

Scale and mealybugs. Pineapple scale (*Diaspis bromeliae*) is widespread in the industry but often controlled by natural enemies. Small infestations can be troublesome, not because of direct plant damage, but due to strict phytosanitary requirements for exportation of fresh fruits. Two species of mealybugs (*Dysmicoccus brevipes* and *D. neobrevipes*) also feed on pineapple. Similar to scale insects, they suck plant sap and debilitate plants. Mealybugs are a larger threat, as the vectors of mealybug wilt, a serious disease of pineapple of uncertain etiology (see Diseases).

Others. Several other minor pests may require treatment in pineapple. Mites can be serious pests in dry areas or during dry seasons. The pineapple red mite (*Dolichotetranychus floridanus*) feeds on vegetative tissues, and the pineapple fruit mite (*Steneotarsonemus ananas*) feeds on floral bracts, fruits, and other plant parts. Two lepidopterous pests, *Batrachedra methesoni* and *Thecla basiliodes*, can damage the fruits through larval feeding; they are restricted to Latin America and the Caribbean region. Pineapple is no longer considered a fruit fly host, although at one time fresh fruits had to be fumigated prior to shipment.

Diseases

Mealybug wilt. This severe disease is present almost everywhere in the pineapple industry, except in parts of Thailand. Infected plants first develop reddish leaves and then become chlorotic or necrotic and are severely stunted. The pineapple wilt virus has been associated with the disease but also occurs in healthy plants. Mealybugs (*Dysmicoccus* spp.) vector the disease and are suspected of producing a toxin that may interact with the virus to produce symptoms. Mealybugs are tended by ants, which move them to new hosts, protect them from parasites and predators, and, in return, feed on their honeydew. Thus, control measures often target the ants with insecticides or baits to prevent disease spread.

Soilborne diseases. Top rot (syn. heart rot) and root rot (*Phytophthora cinnamomi* and *P. nicotianae* var. *parasitica*), butt rot (*Thielaviopsis paradoxa*), and wilts caused by *Fusarium* spp. are serious problems worldwide. Most occur as a result of excess rainfall or poor soil drainage, and all organisms destroy the stem base and/or roots, eventually killing plants. The practices of soil fumigation and growing plants on raised beds reduce or eliminate these problems. Fungicide dips for planting stock are used successfully to avoid introduction of the pathogen to new fields.

Black rot. The same organism that attacks the plant base is also an important postharvest disease of fresh fruits. The fungus *Chalara paradoxa* (syn. *Thielaviopsis paradoxa*) enters through wounds caused by bruising and rough handling during harvest. The fruits become soft, watery, and darkened in the area near the wound. Control is achieved by prevention of bruising and immediate postharvest fungicide treatments of fruits.

HARVEST, POSTHARVEST HANDLING

Maturity

Color change of the fruit exterior from green to yellow is the most common method of determining maturity. Crops harvested during the warm season are picked when the basal eyes show a pale green color. During the cool season, fruits mature more slowly and thus are not picked until eyes at the base begin to yellow. Cool-season fruits tend to be lower in sugar and higher in acid than warm-season fruits. Hawaiian standards call for fruits to have a minimum of 12 percent solids (sugar content), flat eyes, and a large, well-formed crown. For canning, fruits are allowed to reach a more advanced stage prior to harvest, about one-half to three-quarters yellow.

Ethephon, applied when fruits are ripening, synchronizes maturity and eliminates the need for multiple pickings. Pineapple is classified as a nonclimacteric fruit with respect to ripening.

Harvest Method

For fresh markets, fruits are hand harvested. Fruits are cut or broken off stalks and carried on the back or placed on a conveyor to load them onto trucks or into bins. Mechanical harvest is used for canning in some plantations; two conveyors, one above the other, harvest the fruits by breaking them off and carrying them to de-crowners by the lower conveyor.

Postharvest Handling

Fresh fruits are washed and waxed prior to boxing. Fungicides for postharvest disease prevention are in the wax. A short section of flower stalk is left to protect the base of the fruits during shipment. For canning, fruits are de-crowned, cored, and peeled and then sliced and canned. Slices damaged during processing are sold as chunks. About 60 percent of the fruit is recovered as slices or chunks, and the remaining flesh on the core and skin is crushed for juice. The residue left over, called "pineapple bran," is used as livestock feed. Frozen pineapple develops off-flavors; this is why most processed fruits are canned.

Storage

Pineapples can be stored for up to 4 weeks at temperatures of 45°F or warmer. Chilling injury is common at temperatures of less than 45°F and shows up only after removal from refrigeration as flesh browning and crown injury. Chilling injury is exacerbated by drought stress, heavy rainfall at harvest, nonoptimal fruit maturity at harvest, shade, fertilizer, and growth regulators. The recommended storage and shipping temperature range is 47 to 50°F.

CONTRIBUTION TO DIET

About 35 to 45 percent of the Hawaiian crop is sold fresh, and the majority is processed into canned slices, pieces, juices, and fruit cocktail. In addition to the fruits, the tender shoots and terminal buds (inflorescences) are eaten in salads, similar to heart o' palm. Per capita consumption was at 12.9 pounds/year in 2004; it peaked at 15 pounds/year in the early 1990s and has been dropping to the 1980s' levels recently. Consumption is evenly divided among juice, canned, and fresh products, and fresh consumption has increased threefold since 1980 at the expense of canned consumption (down 25 percent); pineapple juice consumption has been stable since 1980.

Dietary value per 100-gram edible portion:

Water (%)	86
Calories	48
Protein (%)	0.5
Fat (%)	0.1
Carbohydrates (%)	12.6
Crude Fiber (%)	1.4

	% of U.S. RDA (2,000-calorie diet)
Vitamin A	1
Thiamin, B_1	5
Riboflavin, B_2	2
Niacin	2
Vitamin C	60
Calcium	1
Phosphorus	1
Iron	2
Sodium	0
Potassium	3

BIBLIOGRAPHY

Bartholomew, D.P. and E.P. Malezieux. 1994. Pineapple, pp. 243-291. In: B. Schaffer and P.C. Andersen (eds.). *Handbook of environmental physiology of fruit crops*, Volume 2: *Subtropical and tropical crops*. Boca Raton, FL: CRC Press.

Bartholomew, D.P., R.E. Paull, and K.G. Rohrbach (eds.). 2003. *The pineapple: Botany, production, and uses*. Wallingford, UK: CABI.

Boucher, D.H. 1983. Pineapple (Pina), pp. 101-103. In: D.H. Janzen (ed.). *Costa Rican natural history*. Chicago, IL: University of Chicago Press.

Collins, J.L. 1960. *The pineapple: Botany, cultivation, and utilization*. London: Leonard Hill Books.

Duke, J.A. and J.L. duCellier. 1993. *CRC handbook of alternative cash crops*. Boca Raton, FL: CRC Press.

Leal, F. and G. Coppens D'Eeckenbrugge. 1996. Pineapple, pp. 515-557. In: J. Janick and J.N. Moore (eds.), *Fruit breeding*, Volume 1: *Tree and tropical fruits*. New York: John Wiley and Sons.

Malo, S.E. and C.W. Campbell. 1984. *The pineapple* (Florida Cooperative Extension Seriers Fruit Crops Fact Sheet FC-7). Gainesville, FL: University of florida, Gainesville.

Morton, J.F. 1987. *Fruits of warm climates*. Miami, FL: Julia F. Morton.

Nakasone, H.Y. and R.E. Paull. 1998. *Tropical fruits*. Wallingford, UK: CAB International.

Olaya, C.I. 1991. *Frutas de America*, pp. 62-77. Barcelona, Spain: Editorial Norma.

Ploetz, R.C., G.A. Zentmyer, W.T. Nishijima, K.G. Rohrbach, and H.D. Ohr (eds.). 1994. *Compendium of tropical fruit diseases*. St. Paul, MN: American Phytopathological Society Press.

Chapter 27

Pistachio *(Pistacia vera)*

TAXONOMY

The cultivated pistachio, *Pistacia vera* L., is a member of the Anacardiaceae or cashew family. Other important members of this family include cashew, mango, mombins (*Spondias* spp.), poison ivy, poison oak, and sumac. The genus *Pistacia* contains only 11 species, of which *P. vera* is by far the most economically important.

Cultivars

Pistachio trees are dioecious, meaning that there are separate male and female trees. Both are required for nut production, but only the females produce nuts. The standard male cultivar is 'Peters', named after A. B. Peters, its discoverer. It is the closest to a universal pollinizer of all male cultivars, and the primary pollinizer for 'Kerman', the main female cultivar. 'Chico' is an excellent pollinizer for females that are earlier blooming than 'Kerman', such as 'Red Aleppo'.

The pistachio industry is reminiscent of sour cherry in that commercial production in the United States is based on a single cultivar. 'Kerman' accounts for about 99 percent of all pistachio production in California, due to its relatively large size and crisp kernel texture. It is named after the major pistachio region of Iran and was introduced in the 1950s after testing for many years in California. It produces high yields of large nuts but has a strong alternate-bearing habit, produces many blanks (empty nuts), and produces nuts with unsplit shells (undesirable from a market standpoint).

'Joley' is another female cultivar, released from the University of California program in 1980. It is slightly smaller than 'Kerman' but comes into bearing sooner and lacks the problems of poor shell splitting and high blank production. It has been planted in New Mexico. Other female cultivars include 'Red Aleppo', 'Bronte', 'Trabonella', and 'Rashti', all of which produce smaller nuts than 'Kerman'.

ORIGIN, HISTORY OF CULTIVATION

The pistachio is native to Asia Minor, from the islands of the Mediterranean in the west to India in the east. It probably developed in interior desert areas, since it requires long, hot summers for fruit maturation, is drought and salt tolerant, and yet has a high winter chilling requirement.

The pistachio has been considered a delicacy since the beginning of recorded history and has been cultivated for centuries throughout its native range. It was introduced to California in 1854, but commercial plantings did not develop until 1970. Full production from those orchards was realized only recently. Major barriers to development of an earlier California industry were lack of good cultivars, lack of cultural information for this peculiar-growing tree, and dominance of the world market by Iran. Many California orchardists were investing in almonds as a tax shelter

Introduction to Fruit Crops

doi:10.1300/5547_27

until legislation in the 1960s put an end to that. Tax incentives encouraged some of those folks to switch to pistachios, and they did quite well when the Ayatollah Khomeini took Americans hostage in the U.S. embassy in Iran in 1979. Iran was the world's leading pistachio producer at that time, and the lack of exportation from this country following Khomeini's revolution sent prices up for California nuts. California production skyrocketed, as thousands of acres of new orchards were planted and research provided insight on better tree husbandry. By the 1990s, increases in U.S. production had all but eliminated the need for pistachio imports. Today, pistachios are grown primarily in California, although there is some production in Arizona, New Mexico, and western Texas.

FOLKLORE, MEDICINAL PROPERTIES, NONFOOD USAGE

Pistachios have been reported as a remedy for: sclerosis of the liver; abdominal ailments; abscesses, bruises, and sores; chest ailments; circulation problems; and other problems. Powdered pistachio root in oil is used for children's cough in Algeria. Leaves were used to enhance fertility in Lebanon, and Arabs consider the nuts an aphrodisiac. A gum from *Pistacia mutica,* the Turk terebinth, is used to make chewing gum in Iran. It is said that gums and mastics prevent periodontal disease and relieve toothache. The gum is also used as a blood-clotting agent in Europe and the Middle East. Nut husks are used in India for dyeing or tanning. The husks are composted and used as fertilizer on many farms with processing plants. Pistachio wood is used for carving, cabinetwork, and firewood.

PRODUCTION

World

Total production in 2004, according to FAO statistics, was 466,634 MT or 1.0 billion pounds. Pistachios are produced commercially in 18 countries on 1.1 million acres. Production has increased by 36 percent since 1994, due to a 30 percent increase in acreage and a 5 percent increase in yield. Since 1980, production has increased sixfold, with every major country experiencing large increases. Worldwide average yields are 930 pounds/acre, ranging widely from a few hundred pounds to over 3,700 pounds/acre in the United States. The top ten pistachio-producing countries (percent of world production) follow:

1. Iran (41)
2. United States (34)
3. Syria (9)
4. China (7)
5. Turkey (6)
6. Greece (2)
7. Afghanistan (<1)
8. Italy (<1)
9. Uzbekistan (<1)
10. Tunisia (<1)

United States

Total production in 2004, according to USDA statistics, was 158,182 MT or 348 million pounds. Virtually all pistachio production is in California (statistics are for California only). A small amount is also produced in Arizona, New Mexico, and western Texas. Pistachio production has increased in every measurable way, including price paid to growers, over the past decade. The industry value is $438 million, up from $163 million in 1993. Bearing acreage is 93,000, up from 57,000. California has averaged 13,000 to 24,000 nonbearing acres over the past decade, but nonbearing acreage for 2004 was not reported. Yield is the highest in the world at 3,740 pounds/acre, also a record high. Yield is generally 1,700 to 3,300 pounds/acre, as pistachios exhibit strong alternate bearing (Figure 27.1). Price paid to growers was $1.26 per pound, increasing from prices of $0.92 to $1.16 per pound received over the past decade.

The United States has imported an amount of pistachios equal to less than 1 percent of domestic production for over a decade. In 2003, the United States exported about 15 percent of its production, valued at $91 million, representing a doubling over the past decade.

BOTANICAL DESCRIPTION

Plant

Pistachios are small- to medium-sized dioecious trees, obtaining heights of around 20 feet in their native range, but generally smaller in cultivation (Plate 27.1). Unusually strong apical dominance in vegetative and reproductive buds but weak apical control make the tree leggy and difficult to train. Leaves are pinnately compound, generally with five leaflets, each broadly oval, with entire margins and obtuse tips.

Flowers

Male and female flowers are borne on separate plants (Plate 27.2). In both sexes, flower buds are conspicuously larger than vegetative buds. Inflorescences of several hundred tiny, brownish-

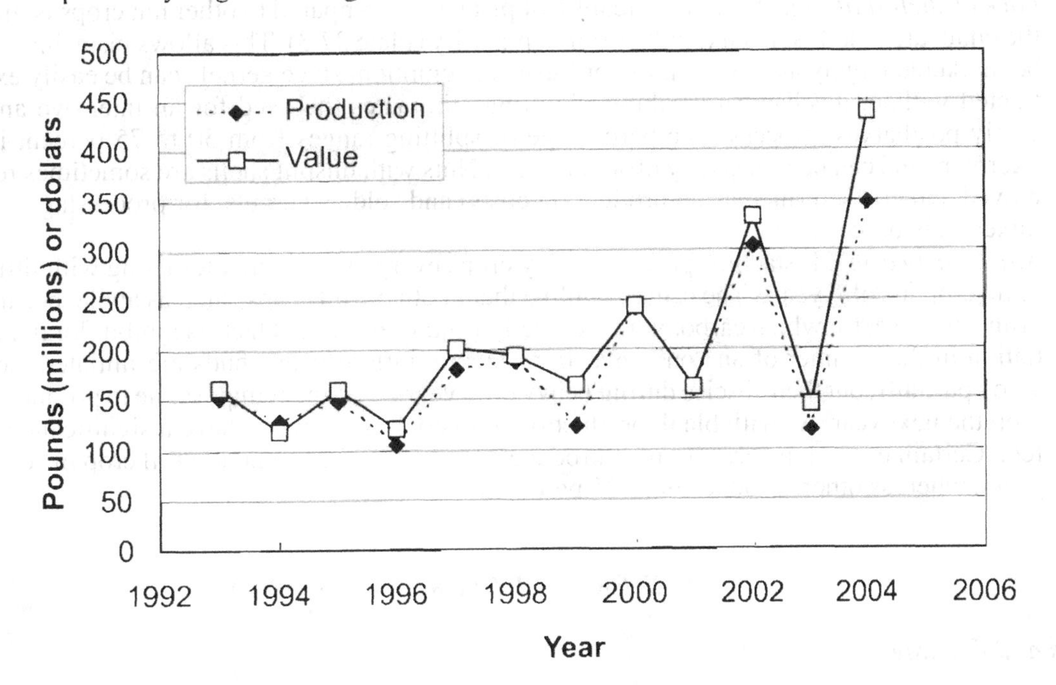

FIGURE 27.1. The value and production of pistachios in California, showing a tendency for alternate bearing superimposed on an overall increasing trend. *Source:* USDA statistics.

green flowers are borne laterally on 1-year-old wood. Flowers lack petals and have up to five sepals. Males have five stamens, and females lack stamens and have a single tricarpellate superior ovary. The inflorescence is a panicle in both cases, with about 13 primary branches, each bearing 1 terminal and 5 to 19 lateral flowers. Most fruit set occurs from terminal flowers. Fruit set averages about 10 percent.

Pollination

Pistachios are wind pollinated. Although bees may be attracted to male flowers for pollen, they are not attracted to females because they lack pollen, nectar, and petals. Pollinizers (males) are often planted in the center of a 3 × 3 square of females, yielding an 8:1 ratio, or sometimes ratios up to 24:1 are used if additional male trees are planted around the borders of the orchard.

Fruit

Botanically, the fruit is a (semidry) drupe, although marketed as a nut. The hull is thin and fleshy, pale tan in color, with a red blush at maturity (Plate 27.3). Fruiting occurs 4 to 5 years after transplanting, but the first economically significant crops are obtained in 7 to 8 years. Yield in the "on" year is expected to be 50 to 100 pounds/tree, for a per acre yield of 1.5 to 2.0 tons. There are three disorders related to fruiting in pistachio:

1. *Blank production.* "Blanks" are fruits without kernels, primarily caused by embryo abortion and subsequent fruit development. Blank percentage averages around 20 percent in 'Kerman'. The amount of blank production is apparently controlled by rootstock, since individual trees produce consistent percentages of blanks from year to year. Selection of superior rootstocks and development of a clonal rootstock propagation technique may be future means of reducing the problem.
2. *Lack of shell splitting.* One unique feature of pistachios compared to other nut crops is that the endocarp (shell) splits naturally prior to maturity (Plate 27.3). This allows pistachios to be marketed largely as in-shell nuts for fresh consumption, since kernels can be easily extracted without mechanical cracking. Splitting eliminates the need for an intensive and costly postharvest process. The percentage of splitting ranges from 50 to 75 percent in 'Kerman' and can be reduced by drought stress. Nuts with unsplit shells are sometimes removed and cracked (increasing production costs) and sold as kernels, for processing into mixed nuts or ice cream.
3. *Alternate bearing.* Pistachios produce heavy crops every other year, alternating with little or no crop in "off" years. The cause is unlike that in other fruit crops, such as apple, pecan, prune, or olive, for which carbohydrate depletion and/or hormonal factors inhibit floral initiation in the summer of an "on" year. In pistachio, inflorescence buds are initiated, develop partially, but then abscise during heavy crop years, making it impossible to produce a crop the next year. As with blank production, rootstock is known to have a significant effect. Certain trees in the orchard may produce as much as 75 percent of a full crop in "off" years, whereas others produce only 25 percent.

GENERAL CULTURE

Soils and Climate

Pistachio production is best on deep, well-drained loams with high lime content and pH. Pistachios are more tolerant of alkaline (Na^+) and saline soil than are most tree crops. In Italy,

Greece, and Iran, much of the production is nonirrigated throughout the rainless summer months, and thus yields are low. In California, irrigation greatly increases production, and many growers provide up to 3 acre-feet of water throughout the summer. Soils with a history of verticillium wilt (such as poorly drained soil, old cotton land) are avoided or fumigated prior to planting.

Pistachios are truly Mediterranean in adaptation; they thrive in the hot, dry, desertlike conditions, such as those found in the Central Valley region of California. Trees are highly disease prone, and rain or even high humidity during the spring and summer promotes severely debilitating diseases. Rain at harvest greatly increases the chances of aflatoxin contamination of the harvested nuts. Aflatoxin, a potent carcinogen, is toxic to mammals and forms in fungus-infected nuts. Under cool conditions at higher altitudes, kernel development is poor. Strong, desiccating winds and/or rainfall during bloom can reduce fruit set by interfering with pollination.

Chilling requirement is relatively long, around 1,000 hours. Thus, arid climates with rather cool, prolonged winters are needed for pistachios. Coldhardiness ranges around 5 to 15°F in midwinter in California but has been reported to reach 0°F in Iran. Among rootstocks, *P. terebinthus* is the most hardy, followed by *P. atlantica*, and then hybrids UCB#1 and 'Pioneer Gold II' (*P. atlantica* × *P. integerrima*), with *P. integerrima* (called 'Pioneer Gold') being the least hardy.

Propagation

Pistachios are most commonly T- or chip-budded onto 1-year-old seedling rootstocks, either in September on first-year seedlings or in April on second-year seedlings. Bud wood is taken from juvenile trees to ensure buds are vegetative, or it can be taken from severely pruned, vigorous older trees. Since transplanting often damages or kills rootstocks, they are grown in containers to lessen transplant shock, and budding is deferred until after successful establishment of the stock in the orchard (Figure 27.2).

Rootstocks

Seedlings of *P. atlantica* and *P. terebinthus* were traditionally the most popular stocks, with the former more readily available and easier to bud. *Pistacia terebinthus* is more cold hardy and

FIGURE 27.2. *(left)* The trunk of a 2-year-old pistachio tree that has been budded in the field. Note the relatively high bud union compared to other fruit crops. The trunk of the scion has been painted white to avoid sunburn injury. *(right)* Trees are staked to promote height growth and headed at 3 to 4 feet to promote scaffold development for open-center training.

thus used by growers in colder areas. Both are susceptible to verticillium wilt, and many orchards originally planted with these stocks have died. *Pistacia integerrima* and hybrids of this species with *P. atlantica* are vigorous stocks with resistance to verticillium wilt. Now the preferred rootstocks in California, these are called 'Pioneer Gold' and 'Pioneer Gold II', respectively, after Pioneer Nursery in California. However, coldhardiness is poor. In Sicily and Turkey, wild seedlings of *P. atlantica* or *P. vera* are simply field-budded to desired cultivars already in place.

Currently, there is no method of clonal propagation for stocks, which results in trees being nonuniform in the orchard, due to great genetic variability among seedlings. Large differences in yield, shell splitting, blank nut production, and alternate bearing among trees in an orchard are apparently rootstock related.

Planting Design, Training, Pruning

Most commercial orchards in California are planted to square or triangular arrangements, with a variety of spacings, ranging from 11 to 30 feet between trees. Many growers use filler trees at spacings of 11 to 22 × 22 to 24 feet, removing every other row at 12 to 15 years. Pollinizer ratios vary from 8:1 to 12:1 (Figure 27.3).

Training during the first 5 years is important for establishment of a full canopy of fruiting wood (Figure 27.2). Due to strong apical dominance, heading cuts must be used to stimulate branching; otherwise, long, weak branches develop and the tree gradually becomes poorly shaped and low yielding. Also, older pistachios produce very few lateral vegetative buds, and thus it becomes very difficult to induce branching by pruning older trees.

Ungrafted rootstocks are planted in orchards and budded in late summer. Buds are forced to grow immediately, producing a trunk in the same year. Trees are often staked to increase growth rate and also headed to induce branching at no lower than 3 feet above the soil surface, to facilitate trunk shaker attachment. Scaffolds selected from the resulting branches are headed at 30 inches to stimulate further branching and canopy development. One of the scaffolds can be rather erect, and thus a "modified leader," to allow height growth to continue, but trees eventually develop the open-center pattern, as in peaches or apricots.

Pest Problems

Insects

Stinkbugs or plant bugs. Large, shield-shaped, brown or gray insects pierce the hull and shell of very young fruits, causing staining, distortion, and possibly drop of nuts. The damage extends

FIGURE 27.3. *(left)* A widely spaced pistachio orchard in Sicily composed of 80-year-old trees. *(right)* A California orchard planted at twice the density.

to the mesocarp (hull) and endocarp (shell), contrary to the skin-only damage implied by the name. Damage occurs anytime from just after bloom to 1 month preharvest. The symptoms of brown to black spotting/staining of the shell, sometimes accompanied by kernel necrosis, originally were thought to be caused by a physiological disorder. Now it is known to be caused by leaf-footed plant bugs (*Leptoglossus lypealis* and *L. occidentalis*) and several species of stinkbugs and other plant bugs (e.g., *Thyanta pallidovirens, Lygus hesperus*). Early season feeding usually results in fruit drop, but late-season feeding causes the staining and underdevelopment of fruits, which remain on the tree. Insecticide sprays after bloom are often required, but sprays can cause outbreaks of scale, perhaps due to the elimination of the scale's natural insect predators. In Iran and Syria, nuts with stained shells and kernels are sorted postharvest and dyed red to disguise the lesions.

Navel orangeworm. The larvae of this moth (*Amyelosis transitella*) appear late in the season. They enter the nuts through cracks in the hulls and then bore into and eat through the kernels. Several larvae may invade the same nut. The pest overwinters in damaged nuts ("mummies"), either on trees or on the ground. Control is obtained by removal of mummified fruits and early harvesting. Otherwise, insecticides are applied when skin cracking begins, since this is the stage when egg laying is most likely.

Citrus flat mite. Citrus flat mites (*Brevipalpus lewisi*) are reddish brown or pink. They overwinter under bark and bud scales and then feed on fruits in early spring, causing a blotchy discoloration. High populations may render entire fruit clusters unmarketable. Sulfur compounds applied as the mites emerge to begin feeding in spring can control the pest. A dormant oil spray, possibly combined with an insecticide in late winter, may kill some of the overwintering adults and reduce populations.

Diseases

Verticillium wilt. Caused by *Verticillium dahliae*, this is the most serious fungal disease of pistachio in California. The fungus has a broad host range, is widespread, and can persist in soil for many years. Roots are infected and the fungus eventually plugs the water-conducting tissues, causing wilting and tree death. Trees are especially susceptible when grown on sites where cotton, tomato, pepper, or eggplant (*Verticillium* hosts) were previously planted. Soil fumigation or fungicide drenches are expensive but can eliminate the fungus. The disease is primarily controlled by the use of *P. integerrima* rootstocks, which are resistant.

Botryosphaeria fruit panicle and shoot blight. The organism *Botryosphaeria dothidea* may attack flowers, young shoots, buds, leaves, and fruits in years with rainfall during bloom. It is worse when high temperatures accompany rains, or if overhead irrigation keeps trees moist when it is warm. Control is achieved by avoiding overhead irrigation and through sanitation of organs affected the previous year, where the fungus overwinters. Fungicides are relatively ineffective.

Botrytis shoot blight. Caused by *Botrytis cineria*, another broad-host fungus, this disease affects many fruit crops, usually attacking the flowers and fruits. In pistachio, the fungus enters through flower clusters, grows down into the stems, and then girdles them. In contrast to *Botryosphaeria*, *Botrytis* is favored by rain and cool temperatures at bloom. Fungicide sprays at bloom are used to control the disease in years of late-spring rains.

Phytophthora root rot. Several species of the fungus *Phytophthora* attack the roots and trunks of pistachio plants in poorly drained soils. Avoiding heavy, poor-draining soils is the primary means of control. Proper tree care to ensure vigor and good overall health will increase the natural resistance of the trees.

HARVEST, POSTHARVEST HANDLING

Maturity

The skin changes from translucent to opaque, and the hull becomes loosened from the shell when nuts are mature. Highest quality is obtained when harvest occurs within 7 to 10 days of this stage. Later harvest results in a greater number of stained shells and more severe navel orangeworm infestation.

FIGURE 27.4. A pistachio-processing facility, consisting of hull-removing equipment and forced-air dryers *(at left).*

Harvest Method

Young trees (<10 years) are generally too fragile to be harvested by trunk shakers, so these are harvested by hand knocking nuts onto canvas sheets. Mature trees are harvested by conventional shake harvest equipment used for almonds.

Postharvest Handling, Storage

Harvested fruits must be hulled and dried within 24 hours to avoid shell staining and aflatoxin contamination. Fruits are fed through machines that have two parallel, rubberized belts rotating at different speeds. Hulled nuts are floated and washed, and hulls are often composted and used for fertilizer. Nuts are dried in large forced-air driers to a moisture content of around 5 percent and then stored in large bins until roasting (Figure 27.4). Roasting and salting are done on stainless steel racks. Nuts first pass under spray nozzles of salt solution and then are roasted for 20 minutes at 250°F. Nuts are cooled in forced-air ventilators and allowed to attain room temperature for 24 hours prior to bagging. As with most nut crops, pistachios can be stored for months in a dried state.

CONTRIBUTION TO DIET

Pistachios are marketed primarily as in-shell nuts, due to the ease with which nuts can be opened without cracking. This accounts for 75 to 80 percent of pistachio utilization. The remainder are marketed as shelled nuts, which are also sold fresh or processed into candies, baked goods, and ice cream. The nut hulls are made into marmalade in Iran.

Per capita consumption of pistachios was 0.29 pound/year in 2004, increasing sixfold since 1980.

Dietary value per 100-gram edible portion:

Water (%)	4
Calories	557
Protein (%)	20.6
Fat (%)	44.4
Carbohydrates (%)	28
Crude Fiber (%)	10.3

	% of U.S. RDA (2,000-calorie diet)
Vitamin A	11
Thiamin, B_1	58
Riboflavin, B_2	9
Niacin	6
Vitamin C	8
Calcium	11
Phosphorus	70
Iron	23
Sodium	0
Potassium	29

BIBLIOGRAPHY

California Pistachio Commission. 2003. *Pistachios* (California Foundation for Agriculture in the Classroom Fact Sheet). Davis, CA: University of California, Davis.

Crane, J.C. 1984. Pistachio production problems. *Fruit Var. J.* 38:74-85.

Crane, J.C. and J. Maranto. 1988. *Pistachio production* (University of California Division of Agriculture and Natural Resources Publication 2279). Davis, CA: University of California, Davis.

Duke, J.A. 2001. *Handbook of nuts*. Boca Raton, FL: CRC Press.

Joley, L.E. 1969. Pistachio, pp. 348-364. In: R.A. Jaynes (ed.), *Handbook of North American nut trees*. Knoxville, TN: North American Nut Growers Association.

Kaska, N. (ed.). 1995. *First international symposium on pistachio nut* (Acta Horiculturae 419). Belgium: International Society for Horticultural Science.

Rosengarten, F. 1984. *The book of edible nuts*. New York: Walker and Co.

Sheibani, A. 1995. Pistachio production in Iran, pp. 165-174. In: N. Kaska (ed.), *First international symposium on pistachio nut* (Acta Horticulturae 419). Belgium: International Society for Horticultural Science.

Socias, R. (ed.). 2001. *Proceedings of the 3rd international symposium on pistachios and almonds* (Acta Horticulturae 591). Belgium: International Society for Horticultural Science.

Spiegel-Roy, *P. pistacia*, pp. 88-93. In: A.H. Halevy (ed.), *CRC handbook of flowering*, Volume 4. Boca Raton, FL: CRC Press.

Teviotdale, B.L., T.J. Michailides, and J.W. Pscheidt. 2002. *Compendium of nut crop diseases in temperate zones*. St. Paul, MN: American Phytopathological Society.

Woodroof, J.G. 1967. *Tree nuts: Production, processing, products*, Volume 2. Westport, CT: AVI Publishing.

Chapter 28

Plum *(Prunus domestica, Prunus salicina)*

TAXONOMY

Plums are placed within the Prunoideae subfamily of the Rosaceae, which contains all of the stone fruits, such as peach, cherry, and apricot. The subgenus *Prunophora* contains plums and apricots. Recent hybrids between plums and apricots are said to produce finer fruits than either parent. A "plumcot" is 50 percent plum, 50 percent apricot; an "aprium" is 75 percent apricot, 25 percent plum; and the most popular hybrid, the "pluot," is 75 percent plum, 25 percent apricot. Some pluots are marketed in stores as "dinosaur eggs," due to their odd, mottled appearance (Plate 28.1).

Ornamental forms of plum are used as landscape trees. Traits for doubled flowers (many petals), red leaves, and a range of flowering colors and times have been selected, adding to the natural beauty of plums. Most of these species do not form fruits, or if they do, they are of poor quality. *Prunus cerasifera,* the myrobalan plum, and its hybrids are the most often used; cultivars include 'Blireiana', 'Cistena', 'Krauter Vesuvius', 'Frankthrees', 'Newport', and 'Thundercloud'.

Plums are the most taxonomically diverse of the stone fruits, and they are adapted to a broad range of climatic and soil factors. The two sections within the *Prunophora*—the *Euprunus* and the *Prunocerasus*—contain about 42 species of plums. Although many species produce edible fruits, the plums of commerce are derived from the Euprunus section.

- Section *Euprunus* ("true plums"):
  1. European plums—*Prunus domestica* L. Worldwide, this is one of the main species grown. Trees are taller, more upright, longer lived, more cold hardy, and later blooming than Japanese plums. Fruits are generally oval, smaller, and more variable in color than with Japanese plums. In the United States, *P. domestica* is used for prunes, fruit cocktail, or other products and rarely eaten fresh.
  2. Japanese plums—*P. salicina* Lindl. and hybrids. These are the most common fresh-market plums in the United States. They are larger, rounder (or heart shaped), and firmer than European plums and are primarily grown for fresh consumption. Hybrids of Japanese plum with several species of Native American plums (section *Prunocerasus*) are grown in the eastern United States because the hybrids tolerate conditions there better than the pure *P. salicina* cultivars.
  3. Damson, bullace, 'St. Julien', and mirabelle plums—*P. insititia* L. These are the small, wild plums native to Europe, cultivated there prior to the introduction of *P. domestica.* The 'St. Julien' types are used as dwarf rootstocks for plums. Fruits are small and oval (~1 inch), purple and clingstone for damsons, and yellow and freestone for mirabelles, with heavy bloom. They are used primarily for jams, jellies, and preserves.
- Section *Prunocerasus* ("plum-cherries"): Native American plums (e.g., *P. americana, P. angustifolia, P. hortulana, P. munsoniana, P. maritima*). These are wild, small trees or

Introduction to Fruit Crops

doi:10.1300/5547_28

shrubs found mostly in the eastern United States that produce small, round, edible fruits. They are used mostly for jams, jellies, and preserves by local people. They have been important in plum breeding, particularly for increasing disease resistance of *P. salicina*.

European Plum Cultivars

These are placed into four groups, based mostly on fruit color and/or size and use (processed or fresh). Most were introduced from Europe (Plate 28.2).

1. Reine Claude or greengage—These are round, green, or golden plums used for canning and fresh-market sales; cultivars include 'Reine Claude', 'Imperial Gage', and 'Hand'.
2. Yellow egg—These are large, yellow, oval plums primarily used for canning; cultivars include 'Yellow Egg' and 'Golden Drop'.
3. Lombard—These large, oval, red or pink plums are used for fresh-market sales in Western Europe; cultivars include 'Victoria', 'Lombard', and 'Pond'.
4. Prunes—The fruits of these oval, dark blue or purple, freestone cultivars are dried postharvest; cultivars include 'French' (syn. Agen, Petite d'Agen, Prune d'Agen), 'Stanley', 'Italian' (syn. 'Fellenberg'), 'Blufre', and 'President'.

In the United States, prune cultivars dominate the production of *P. domestica*. 'Stanley', introduced from New York in 1926, has become one of the most popular cultivars worldwide. It is self-fruitful, unlike many other plums, has high yield, and exhibits a strongly spur bearing habit. 'Pozegaca' (called 'German Prune' in the United States) is popular in Europe for its small, but excellent quality sweet fruits. There are several sports or selections of the most popular cultivars.

Japanese Plum Cultivars

Many Japanese cultivars are the result of breeding, in contrast to European plums, for which chance selection has played a larger role. Luther Burbank, the famous plant breeder, produced some of these cultivars. 'Santa Rosa', 'Burbank', 'Shiro', 'Beauty', 'Gold', 'Methley', 'Red Beaut', and 'Ozark Premier' are grown in several countries. In addition, 'Friar' and 'Simka' are popular in the United States.

ORIGIN, HISTORY OF CULTIVATION

P. domestica

This species is native to western Asia, in the Caucasus Mountains adjacent to the Caspian Sea. Genetic analysis indicates that it is a relatively recent hybrid of wild parents. The spontaneous chromosome doubling of a hybrid between *P. cerasifera* (diploid) and *P. spinosa* (tetraploid) may have produced this hexaploid species. This may have occurred around 2,000 years ago, evidenced by the lack of *P. domestica* seeds in the excavations of Pompeii, where the seeds of most other Old World fruits were found. This plum was brought to North America by Spanish missionaries (West Coast) and English colonists (East Coast). Today, most production is in western states, with the vast majority in California's Central Valley, where climate disfavors disease and fruit cracking due to rainfall.

P. salicina

Contrary to the name, this species originated in China, where it was cultivated for thousands of years. It was brought to Japan 200 to 400 years ago, where it then spread around the world, being falsely called "Japanese plum." In the United States, it is grown primarily in California, and it is the major fresh-market plum seen in grocery stores.

FOLKLORE, MEDICINAL PROPERTIES, NONFOOD USAGE

Wild plum trees are symbolic of independence. Plum is the national flower of Taiwan, and its flowers are often depicted in Asian art.

As with all stone fruits, plum leaves, flowers, and especially seeds and bark contain toxic compounds that generate cyanide, which is of course toxic or lethal in large doses. However, in plant tissues, cyanide is low enough in concentration to be considered therapeutic, particularly for cancer (tumor) treatment, and has been used for this purpose since at least 25 BC. Prunes and prune juice are commonly used as natural laxatives. Phloretin is an antibiotic-like compound found in bark and root extracts; in concentrated form, phloretin can kill certain bacteria.

PRODUCTION

World

Total production in 2004, according to FAO statistics, was 9,836,859 MT or 21.6 billion pounds. Plums are produced commercially in 81 countries on 6.4 million acres. Production has increased 46 percent over the past decade, due to a 49 percent increase in acreage. Yields average only 3,400 pounds/acre worldwide, the lowest of any stone fruit, but can reach 12,000 to 15,000 pounds/acre in countries with intensive orchards. Chinese production has doubled in the past decade and alone is responsible for the vast majority of the increase in worldwide production. Chinese production surpassed that of the United States, the former world leader, in the mid-1980s. The top ten plum-producing countries (percent of world production) follow:

1. China (45)
2. United States (7)
3. Serbia (6)
4. Romania (6)
5. Germany (5)
6. France (2)
7. Chile (2)
8. Turkey (2)
9. Spain (2)
10. Italy (2)

United States

Total production in 2004, according to USDA statistics, was 327,900 MT or 656 million pounds. The 2004 year was unusually low; production is generally double that reported here due to record-low prune production (Figure 28.1). The industry value was $153 million in 2004,

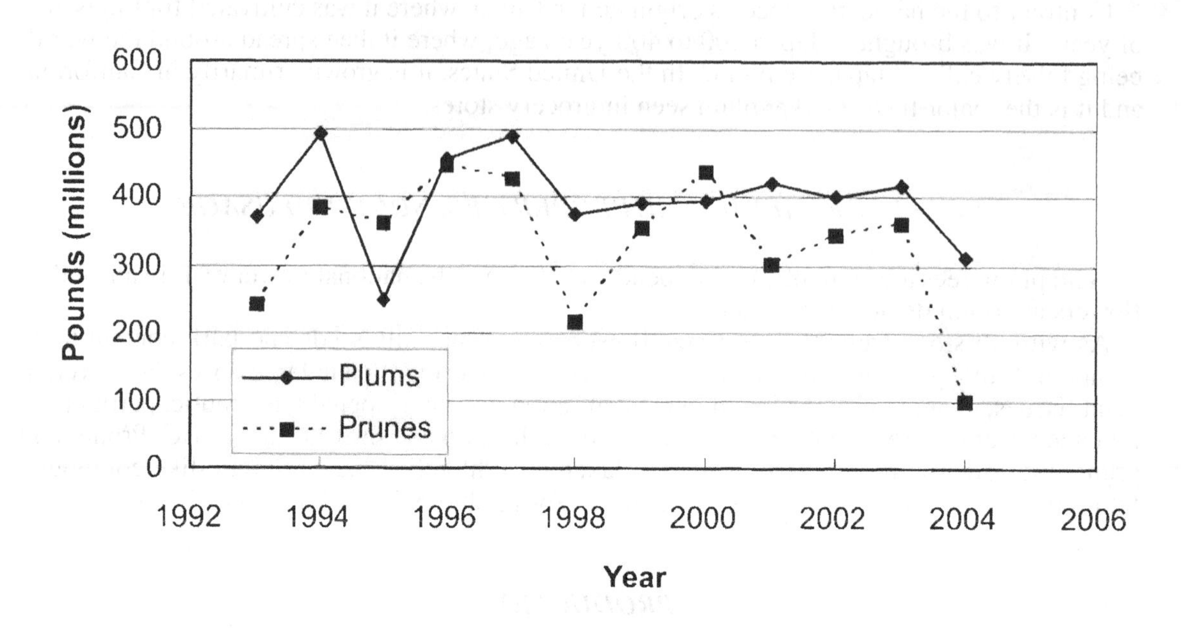

FIGURE 28.1. Long-term average prune and Japanese plum production in California has been stable over time, but both crops exhibit high year-to-year variation. *Source:* USDA statistics.

well below the range of $200 million to $320 million over the past decade. Most commercial plums, exclusive of roadside stand sales, are produced in only five states:

1. California—95 percent
2. Oregon—2 percent
3. Washington—1-1.5 percent
4. Michigan—<1 percent
5. Idaho—<1 percent

California produces 95 percent of all plums in the United States. The two main categories are as follows:

1. Japanese plums: Total production in 2004 was 141,820 MT or 312 million pounds. The value and acreage are $74 million and 36,000 acres, respectively. Price per pound fluctuates inversely with production (high price in low production years). Growers received $0.26 per pound in 2004, but this has ranged from $0.15 to $0.47 per pound over the past decade. (A small amount of European plums sold fresh may be included in these figures.)
2. Prunes, *dried basis:* Total production in 2004 was 67,180 MT or 148 million pounds, valued at $79 million ($78 million to $210 million over the past decade). Current acreage is 70,000 and has been declining by 2,000 acres/year over the past decade. Price per pound has steadily declined from $0.55 to $0.39 per pound since 1993 but returned to $0.75 per pound in 2004 due to unusually low production. One pound of dried prunes is derived from 2.7 pounds of fresh fruits, so on a fresh basis, there were 400 million pounds of fruits produced, for which growers normally receive $0.15 per pound.

Outside of California, production is mostly European plums, with just over half marketed fresh, and the remainder dried for prunes, canned, or frozen. The non-California industry value is $4 million to $8 million.

The United States exported 36 percent of production in 2002, with roughly equal tonnages of fresh and dried (prunes), and less than 1 percent in fruit salad. Major destinations include Japan, Germany, Italy, Belgium, Hong Kong, United Kingdom, Canada, and the Netherlands. Imports were equivalent to 5 percent of production, 90 percent as fresh fruits, largely from the Southern Hemisphere in winter months.

BOTANICAL DESCRIPTION

Plant

Plums grow on small- to medium-sized trees, similar to but more erect growing than peach. European plum trees are larger and more erect than those for Japanese plums. Leaves are ovate or elliptic with acute or obtuse tips, short petioles, and crenulate margins. Japanese plum trees have rougher bark, more persistent spurs, and more numerous flowers than do European plum trees. They are also more precocious, disease resistant, and vigorous than are the European plum trees (Plate 28.3; Figure 28.2).

Flowers

Flowers are similar in morphology to those of peach, but they are white, smaller, and with longer pedicels. Flowers are borne mostly in umbel-like clusters of two to three individuals on short spurs, and solitary or two to three in axils of 1-year-old wood. Japanese plum trees produce thousands of flowers and appear snow covered when in full bloom. European plum trees do not produce as many flowers and are slower to come into bearing than are Japanese plum trees. European types bloom much later than Japanese types and are therefore less frost prone.

FIGURE 28.2. Japanese plums require the heaviest thinning of any stone fruit: *(left)* before thinning and *(right)* after thinning.

Pollination

Honeybees are the major pollinator. For Japanese plums, pollinizers are necessary for commercial production of most cultivars. 'Bruce', 'AU Producer', 'Beauty', 'Santa Rosa' (and its sports), 'Simka', 'Casselman', and 'Methley' do not require cross-pollination. In *P. domestica*, about half of the major cultivars require pollinizers, but most of the major prunes produced in the United States do not. 'Stanley', 'French', 'Italian', and 'Sugar' prune types are self-fruitful.

Fruit

The plum fruit is a drupe that is oval shaped in European types, and round to conical in Japanese types (Plates 28.1, 28.2). Native American types and Japanese hybrids often have smaller, round fruits. Bloom (epicuticular wax) is usually present on glabrous surface (thus, the fruit surface is termed "glaucous"). The pit is generally smooth, not furrowed, as in peach, and encloses a single seed. Trees may flower on laterals in their second year, but substantial bearing does not begin until 4 years. Fruits are borne on short spurs that arise from 2-year-old or older wood or axillary on 1-year-old wood. European plums are less precocious than Japanese types, with little spur and fruit production until years 3 to 4 in European types, but flowering and a light crop in year 2 are not uncommon for Japanese plums. Plums require 2.5 to 6 months for fruit development, with most Japanese plums ripening over relatively short periods (~3 months), and some prune and canning cultivars ripening in autumn.

Thinning is necessary for proper size development for Japanese plums, which can produce about 100,000 flowers per tree, of which only 1 percent should set for a properly sized crop. Fruits are thinned by leaving one to two per spur, or about one fruit per 5 inches of shoot. On laterals, fruits are spaced 4 to 6 inches apart. Thinning is done by hand at pit hardening, or with poles or shakers. Thinning is not always necessary for European plums, particularly prunes, since they are not as floriferous and fruit set is generally lighter (Figure 28.2).

GENERAL CULTURE

Soils and Climate

Deep, well-drained soils with a pH of 5.5 to 6.5 give best results. However, plums are the most tolerant of all stone fruits with respect to heavy soils and waterlogging. Also, plum rootstocks may tolerate drought better than rootstocks of peach. Some rootstocks are available for high-pH conditions.

Plums are adapted to a wide range of climatic conditions; at least some cultivars can be grown in almost every state in the United States. European plums have a more northern adaptation, and Japanese plums do better in southern areas of the temperate zone or in Mediterranean climates. Commercially, Japanese plums and prunes are grown where rainfall during the growing season is minimal, and humidity is low, to prevent diseases; this is why most production is in California. Growers in the eastern United States have a more difficult task with respect to disease management due to higher rainfall and humidity.

Coldhardiness is excellent for European plums, similar to apple and pear, but Japanese plums are less cold hardy (similar to peach). European plums can be grown farther north than Japanese types. Plum flower buds tolerate –15 to –30°F when dormant, and the wood is killed just below this level. Open flowers and young fruitlets of both species are killed by brief exposure to 28°F or below in spring. Japanese plums bloom early, generally with or before the earliest peaches (late February to early March in the South, a month later in the North), and are therefore prone to frost damage. European plums bloom later, after peaches and with sweet cherries. Plums have chilling requirements ranging from 550 to 800 hours for Japanese plums, and greater than 1,000 for European plums. Therefore, "low chill" cultivars must be grown in areas such as the Gulf Coast, Florida, and southern Texas, where less than 600 hours of chilling occurs each winter. The Japanese cultivars 'Robusto' and 'Segundo' are low chill, but there are no low-chill European cultivars.

Rainfall and high humidity during the growing season can reduce production by accentuating diseases, such as brown rot, and causing fruit cracking, as occurs in cherry and other thin-skinned fruits. Prunes often crack, even in the dry conditions of California, and are not grown commercially where rainfall occurs at harvest (Figure 28.3). In arid areas, plums are irrigated during the growing season with about 2 acre-feet of water dispersed throughout the season.

FIGURE 28.3. Cracking in prune occurs from too much rain or irrigation near harvest.

Propagation

Plums are T- or chip-budded onto rootstocks, as are other stone fruits.

Rootstocks

Since plum scions are genetically diverse, many different species/selections are used as rootstocks (Table 28.1). Next to apple, plum rootstocks are the most well characterized and researched, and more clonal material is available for them than for any of the other stone fruits. However, in the United States, 'Myrobalan 29C' (*P. cerasifera*) and 'Marianna 2624', a hybrid between a myrobalan and a native American plum, are used most frequently because they are widely compatible with most cultivars. 'Myrobalan 29C' produces large trees with slightly delayed ripening and is not particularly resistant to diseases or other root-related problems. 'Marianna 2624' produces a somewhat smaller tree with slightly earlier ripening and is resistant to a number of problems confronting other stocks. Peach seedling rootstocks 'Nemaguard', 'Lovell', and 'Rancho Resistant' are used for plum in some areas to gain nematode and bacterial canker resistance, and to reduce size moderately, compared to myrobalan, and slightly, compared to marianna.

Although myrobalan rootstocks are often grown from seed, several selections of myrobalan are vegetatively propagated: 'Myrobalan B', 'Myrobalan GF31', 'Myrobalan 29C', 'Myrobalan 5-Q', and 'Myrobalan 2-7'. Marianna is also propagated by seeds or cuttings, and several selections of this are clonally propagated: 'Marianna 2624, 'Marianna 4001', and 'Marianna GF8-1'.

In Europe, several *P. domestica* selections are used as rootstocks, including 'Brompton', 'Black Damas', and 'Ackermanns'. Other *Prunus* species such as *P. spinosa* (sloe), *P. insititia* (the 'St. Julien' series), and *P. ussuriensis* are used in Eastern Europe and Russia.

TABLE 28.1. Some commonly used plum rootstocks and their characteristics.

Rootstock	Tree size	Yield	Compatibility	Resistances	Susceptibilities
'Ackermann's Marunke' *(P. domestica)*	semidwarf	low	European	—	Cold, suckering
'Black Damas' *(P. domestica)*	large	high	European	Bacterial canker	Suckering
'Brompton' *(P. domestica)*	very large	medium	European	Cold	Bacterial canker, Sharka virus carrier
Common Mussel *(P. domestica)*	semidwarf-large	high	European	—	Bacterial canker, drought, suckering
'St. Julien A' *(P. insititia)*	semidwarf	medium	European	Cold	Suckering
Pixy *(P. insititia)*	dwarf	high	European	Bacterial canker	Drought
Myrobalan seedling *(P. cerasifera)*	large	high	European and Japanese	Drought, root and crown rot	Bacterial canker, cold, oak root rot, root-knot nematode, prune brown-line disease, peachtree borer
'Myrobalan 29C' *(P. cerasifera)*	very large	high	European and Japanese	Root and crown rot, root-knot nematode	Poor anchorage when young, suckering, peachtree borer
Marianna seedling *(P. cerasifera × P. munsoniana)*	semidwarf-large	high	European and Japanese	Rosette virus	Bacterial canker, nematodes
'Marianna 2624' *(P. cerasifera × P. munsoniana)*	semidwarf-large	high	European and Japanese	Waterlogging, oak root rot, prune brown-line disease, root-knot nematode, Stanley decline	Bacterial canker (can outgrow), peachtree borer, some suckering
Peach *(P. persica)*	large	high	European and Japanese	Root-knot nematode, bacterial canker	Waterlogging, oak root rot, root and crown rot, prune brown-line disease, sharka virus

Source: Adapted from Okie, 1987.

Planting Design, Training, Pruning

Japanese plums, similar to peaches, are typically small, spreading trees. Therefore, they are planted at relatively close in-row spacings (10-20 feet), leaving about 18 to 20 feet between rows, depending on equipment size. Pollinizers are planted in alternate-row arrangements or distributed about every third tree in every third row. The larger European plum requires wider spacings than does the Japanese plum, in many cases, and those grown for prunes do not require pollinizers.

Plums are generally trained to the open-center system, but they are usually more upright than peach due to their natural growth habit and less need for light exposure for color development. In formative years, upright types may need to be headed during active growth to encourage branching. In California, trees intended to be harvested mechanically are headed no lower than 3 feet (Figure 28.4).

Pruning during formative years is light; interior branches and waterspouts are thinned, and growing scaffolds are headed to induce branching. At maturity, vigorous upright shoots are removed, since fruiting occurs increasingly on spurs on older wood as trees age. Trees are topped annually to maintain a height of 10 to 15 feet; mechanically harvested trees can be taller than Japanese types that are hand harvested. Some renewal of fruiting wood is necessary because spurs live about 5 to 8 years.

Pest Problems

Insects

Plum curculio. Conotrachelus nenuphar adults are gray to dark brown, ⅛- to ¼-inch-long snout beetles; grubs are small, yellow-white, and legless, and they burrow into the fruit flesh, usually causing the fruit to fall off. Feeding damage by adults alone may cause catfacing (misshapen fruits) or D-shaped brown depressions in the fruit surface. Insecticide sprays are applied starting at petal fall and then at 7- to 10-day intervals through the spring. The insect is confined to the United States east of the Rocky Mountains.

Codling moth. Lasperyresia pomonella is a widespread insect problem of fruit crops, though it is not as severe a problem with plums as with pome fruits. The moth (¾ inch long, dark color) lays eggs singly on young fruits in midspring, and the larvae (pinkish-brown head, white body, ¾ inch long) tunnel through the fruits to the pit. Tunnel entrances exude frass. Sophisticated growth models are used to predict adult emergence, and insecticides are timed to moth flights. Larvae spend most of their time deep inside of fruits, so control measures are not targeted at them.

Aphids. Mealy plum aphid (*Hyalopterus pruni*) and leaf curl plum aphid (*Brachycaudus helichrysi*) attack plums in California, causing cupping and malformation of leaves, and excret-

FIGURE 28.4. *(left)* A Japanese plum hybrid in full bloom, trained to an open-center system. *(right)* A prune orchard in California also trained to an open-center system, with higher crotches to allow shaker attachment.

ing honeydew that may cause fruit cracking. Control is achieved by dormant sprays targeted at overwintering eggs, although natural predators may control low populations.

Plant bugs (stinkbugs)/catfacing insects. Several species of plant bugs or stinkbugs (*Leptoglossus, Lygus* spp., others) feed on fruits at various times of the year. They are fairly large (½ to ¾ inch), dark brown bugs with shield-shaped bodies and prominent legs. Early feeding results in severe catfacing, or deformed fruits that have extensive indentations. Some fruits may drop if feeding is severe. Later-season feeding results in shallow, corky lesions that are less of a concern. Brown rot incidence is enhanced by feeding lesions late in the season. Insecticides are applied at petal fall to shuck split and as needed throughout the season.

Oriental fruit moth. These small moths (*Grapholita molesta*) lay eggs in shoot tips early in the season, and larvae burrow downward a few inches; wilted and dead expanding leaves at the shoot tips indicate infestation. Later infestation results in fruit loss, as eggs laid on fruits produce ¾-inch, legged, pinkish larvae that burrow through the fruits. Although fruit infestation does not occur until fruits are less than half grown, early season sprays are critical for controlling this pest. Insecticides are applied at petal fall and shuck split.

Omnivorous leaf roller. This caterpillar (*Platynota stultana*) has a whitish translucent body and a brown head. It rolls leaves and webs them together or onto fruits. Fruit feeding causes scarring or blemishes severe enough to render fruits unmarketable or to cause the drop of young fruits. Bt or other insecticides are applied in early summer when caterpillars are still small.

Borers. Larvae of the peach tree borer (*Synanthedon exitosa*) cause damage to the lower trunk and scaffolds, sometimes killing young trees. The adult moth lays eggs on the tree, and the larvae bore into the bark to feed. The larva is ½ to 1 inch long with a dark-colored head. Lower limbs are sprayed thoroughly with insecticide in late summer when the adult moths are laying eggs. Peach twig borers (*Anarsia lineatella*) damage young shoots and cause fruit damage. Pheromone mating disruption can be used for control, or postbloom insecticides can be applied if populations are high.

Western flower thrips. These tiny (<1/10 inch) insects (*Frankliniella occidentalis*) infest flowers and feed on ovaries and young fruits, causing mostly cosmetic damage to fruits (scarring, spotting, blotching). However, damaged fruits are rendered unmarketable. Damage is worse if weather is cool at bloom. Insecticides are applied at petal fall.

Mites. Two-spotted and European red mites (*Panonychus, Tetranychus* spp., others) can be particularly damaging to plums—more so than for peach due to glabrous fruit surfaces and thinner leaves. A dormant oil spray before budbreak reduces initial populations. In hot, dry weather, miticides are applied when counts exceed about ten mites per leaf after checking 100 leaves.

Diseases

Brown rot/blossom blight. The same fungi (*Monolinia laxa* and *M. fructicola*) cause flowers to rot in some years and fruits to rot near harvest. Blossom blight is seldom a problem itself, but it signals potential problems with brown rot of ripening fruits later. A brown, powdery mass of spores in concentric rings is visible around a mushy lesion on the fruits. European plums, in general, are more resistant to brown rot than are Japanese plums. Fruits infested in the previous year die, shrivel, turn black, and hang on the tree—called "mummies"—and house spores for next year's infection. They should be removed from trees in winter. Also, brown-rotted fruits should be removed from trees as soon as the disease manifests. Fungicide applied as trees bloom will control blossom blight and reduce brown rot later. In rainy climates fungicide sprays may be needed at 7- to 10-day intervals up to 1 week before harvest.

Root and crown rot. At least ten species of *Phytophthora* can infect plums and other stone fruits; no region of the world is free of these fungi. Infected trees are weakened and eventually die, with younger trees dying in the same season as the onset of infection, but older ones living a few more seasons. The stunting, chlorosis, defoliation, and dieback symptoms mimic those of other soilborne maladies. *Phytophthora* fungi need wet soils to proliferate. Thus, areas with well-drained soil are not affected. Trees can be planted on berms or beds to improve drainage. Rootstocks differ in resistance and are the main form of disease control in many situations. Japanese plum, marianna, and myrobalan are more resistant than other stone fruits. European plum is more susceptible than Japanese plum.

Bacterial spot. Another general stone fruit problem, bacterial spot (caused by *Xanthomonas campestris* pv. *pruni*) is particularly serious on plums. It occurs in most parts of the world, except the Pacific Coast, and is more common in humid, warm climates with sandy soils. The bacterium attacks leaves, fruits, and twigs. Leaves exhibit the characteristic "shot-holed" appearance, since infection occurs before leaf expansion is complete, and lesions fail to grow. Leaves often turn yellow and drop, decreasing tree vigor and yields. Fruit lesions are irregularly shaped dark pits, about 1/16 to 1/4 inch in diameter, that resemble open sores, as the center of the lesions cracks open and sometimes produces gum. Twig cankers are the source of inoculum from year to year, and they can kill entire branches over a 2- to 3-year period. Cankers are deep, with flaky margins. Sprays of copper or antibiotic materials are risky and often do not control the disease well. European plums are less susceptible than Japanese plums. Resistant Japanese cultivars exist and should be used in areas prone to bacterial spot: 'AU Amber', 'AU Crimson', 'Bruce', 'Earlisweet', 'Robusta', 'Queensland', and 'Explorer'.

Bacterial canker. As described for peach and other stone fruits, *Pseudomonas syringae* is a major problem worldwide. The bacteria kill buds and create large, perennial cankers that girdle limbs or kill entire trees. This is nearly impossible to control with sprays, as with many bacterial diseases. Japanese plum hybrids are fairly resistant. Trees budded high on 'Myrobalan B' (resistant) are used to avoid tree losses due to canker. Freezing injury, nematode weakening, other diseases, drought, or waterlogging all predispose trees to infection, so healthy trees are a good defense.

Black knot. Caused by the fungus *Apiosporina morbosa,* black knot affects only woody tissues, generally twigs, causing black, brittle, elongated swellings. Affected twigs and branches may die, and entire trees may be killed over a period of years if severe. The disease is found throughout North America but is worst in the East, where humidity and rainfall are high. The disease is difficult to control once established; knots are pruned out, but this is expensive. Removing alternate hosts, such as wild plums, from around the orchard reduces disease pressure. Fungicide application for other diseases during shoot elongation can control black knot. Resistance is found in 'President' European plum and 'Shiro' and 'Santa Rosa' Japanese plums.

Plum pox (syn. sharka) virus. Symptoms include severe fruit malformation and sporadic occurrence of ring-shaped chlorotic spots on leaves. Fruits may have irregular spots or rings and be dry and flavorless. The virus spreads by infected budwood and aphids. The disease is a serious problem in Europe and was introduced to North America in the mid-1990s. Quarantine and eradication measures have effectively reduced its spread outside of a few isolated areas, and the stone fruit industry in the United States has not been affected to date. Resistant cultivars include 'Frontier' and 'Underwood' (Japanese) and 'Stanley' and 'Early Rivers' (European).

HARVEST, POSTHARVEST HANDLING

Maturity

A variety of indices are used for plum maturity, depending on use, species/cultivar, and location. Japanese plums and European plums for fresh-market sales are harvested based on skin color and firmness, although sugar content and sugar-acid ratio have been used. Sugars range from 14 to 17 percent for shipping mature plums, while firmness is 10 to 20 pounds, depending on cultivar. European plums intended for drying are harvested at a more mature stage than are fresh plums. Flesh color, firmness, and sugar content are the most reliable indicators. Days from maturity are used as a guideline for when to start sampling. Flesh color turns from green-yellow to amber during ripening, and solids reach 25 to 35 percent. Firmness of 1 to 2 pounds is optimal. Plums are classified as climacteric fruits with respect to ripening.

Harvest Method

Plums for fresh consumption must be hand harvested, and they require two to four pickings over a 7- to 10-day period for optimal maturity, as with peach. Prunes for canning or drying are harvested by shake-and-catch methods, as with sour cherries.

Postharvest Handling

Fresh plums are handled similarly to peaches postharvest, but with more care, as they are more susceptible to bruising. They are picked into buckets with padded bottoms, gently poured into shallow bulk bins, and marketed in small cardboard boxes or single-layer flats to prevent crushing. Size grades vary with location but range from about 1$\frac{11}{16}$ to 2$\frac{1}{2}$ inches. Hydrocooling is essential to extend postharvest life. Prunes used to be dried in the sun, similar to raisins, but now are dried in forced-air tunnels for a more uniform product.

Storage

Plums have similar storage characteristics and problems as peaches, cherries, and apricots. They can be stored about 2 to 4 weeks at 32°F and 90 percent relative humidity. Neither species is susceptible to chilling injury in normal storage conditions, although flesh translucence and browning may occur in some cultivars if storage is prolonged. Brown rot, rhizopus rot, and blue and gray molds are the most common storage problems. Once prunes are dried, they are relatively resistant to postharvest diseases and can last for months.

CONTRIBUTION TO DIET

Most Japanese plums are marketed as fresh fruits. European plums have a much wider variety of uses. In California, almost all European plums are dried for prunes. In other plum-producing states, utilization is reported as follows:

Product	Percentage of Crop
Fresh	30-50
Canned	20-25
Dried (prunes)	17-25
Frozen	1-8

Plums are used for jellies, jams, and preserves; brandy and cognac; pies, cakes, and tarts; and in confectionery. Per capita consumption of plums and prunes is 2.5 pounds/year, down 39 percent since 1980. Prune consumption is 0.9 pound/year, and it seems to have peaked at 1.5 to 2.0 pounds/year in the late 1980s and has trended downward since. Fresh consumption (1.1 pounds/year) is down 27 percent, while canned and juice consumption is down fourfold in the same period.

Dietary value per 100-gram edible portion:

	Fresh Plum	Prune
Water (%)	87	31
Calories	46	240
Protein (%)	0.7	2.2
Fat (%)	0.3	0.4
Carbohydrates (%)	11.4	63.9
Crude Fiber (%)	1.4	7.1
	% of U.S. RDA (2,000-calorie diet)	
Vitamin A	7	16
Thiamin, B_1	2	3
Riboflavin, B_2	2	11
Niacin	2	9
Vitamin C	16	1
Calcium	<1	4
Phosphorus	2	10
Iron	1	5
Sodium	0	<1
Potassium	4	21

BIBLIOGRAPHY

Crane, M.B. 1949. *The origin of the garden plum* (Fruit Year Book No. 3). London: Royal Horticultural Society.

Gur, A. 1985. Rosaceae—Deciduous fruit trees, pp. 355-389. In: A.H. Halevy (ed.), *CRC handbook of flowering*, Volume 1. Boca Raton, FL: CRC Press.

Hartman, W. (ed.). 1994. *Fifth international symposium on plum and prune genetics, breeding, and pomology* (Acta Horticulturae 359). Belgium: International Society for Horticultural Science.

Hedrick, U.P. 1911. *The plums of New York.* (New York Agricultural Experiment Station 18th Annual Report 3, Part 2). Albany, NY: J. B. Lyon Company.

LaRue, J.H. and R.S. Johnson. 1987. *Peaches, plums and nectarines* (University of California Cooperative Extension Service Publication 3331). Davis, CA: University of California, Davis.

Ogawa, J.M., E.I. Zehr, G.W. Bird, D.F. Ritchie, K. Uriu, and J.K. Uyemoto. 1995. *Compendium of stone fruit diseases.* St. Paul, MN: American Phytopathological Society.

Okie, W.R. 1987. Plum rootstocks, pp. 321-360. In: R.C. Rom and R.F. Carlson (eds.), *Rootstocks for fruit crops.* New York: John Wiley and Sons.

Ramming, D.W. and V. Cociu. 1990. Plums (*Prunus*), pp. 233-288. In: J.N. Moore and J.R. Ballington (eds.), *Genetic resources of temperate fruit and nut crops* (Acta Horticulturae 290). Belgium: International Society for Horticultural Science.

Ramos, D.E. (ed.). 1981. *Prune orchard management* (University of California Division of Agriculture and Natural Resources Special Publication 3260). Davis, CA: University of California, Davis.

Teskey, B.J.E. and J.S. Shoemaker. 1978. *Tree fruit production,* Third edition. Westport, CT: AVI Publishing.

Watkins, R. 1979. Cherry, plum, peach, apricot, and almond; *Prunus* spp., pp. 242-247. In: N.W. Simmonds (ed.), *Evolution of crop plants.* London: Longman.

Weinberger, J.H. 1975. Plum, pp. 336-347. In: J. Janick and J.N. Moore (eds.), *Advances in fruit breeding.* West Lafayette, IN: Purdue University Press.

Wight, W.F. 1915. *Native American species of* Prunus (USDA Bulletin 179). Washington, DC: U.S. Government Printing Office.

Chapter 29

Strawberry *(Fragaria ×ananassa)*

TAXONOMY

The cultivated strawberry, *Fragaria ×ananassa* Duch., is a member of the Rosaceae, subfamily Rosoideae, along with blackberries and raspberries. The Rosoideae subfamily also includes roses (*Rosa* spp.); the mock or Indian strawberry *(Duchesnea indica)*, a naturalized weedy species of the eastern United States; and the cinquefoils (*Potentilla* spp.). There are about 34 other species of *Fragaria* found in Asia, North and South America, and Europe, of which two are cultivated commercially for their fruits: *Fragaria moschata*, the musky or hautboy strawberry, and *F. vesca*, the wood or alpine strawberry. These species were cultivated for centuries, but there is very little production of them today, due to the success of *Fragaria ×ananassa*.

The name "strawberry" may have derived from the practice of using straw mulch for cultivation many years ago, or it may have come from the Anglo-Saxon word *strew*, meaning "to spread," as strawberry plants spread by runners. *Strewbery*, or a similar word, was changed to *strawberry* in English.

Cultivars

Cultivars have been produced by plant breeders to fit particular environmental or marketing niches, and, generally, no single cultivar is grown worldwide or even nationwide. Two main types of cultivars are recognized, based on environmental control of flowering:

1. *Short day or June bearing.* These are short-day plants, requiring photoperiods under 14 hours and/or temperatures under 60°F for floral initiation. As the name suggests, they generally bear fruits in the spring only, unless grown in areas with cool summer temperatures, such as the coastal areas of central California.
2. *Everbearing.* These cultivars fruit throughout the growing season provided temperatures are not too high. The continuous-flowering trait was obtained by breeding strawberries with *F. virginiana glauca*, *F. vesca*, and other alpine-type strawberries. There are two types of everbearers:
   a. *Long day.* Lengthening days promote more or less continuous flowering throughout the summer, provided temperatures are not too high.
   b. *Day neutral.* Photoperiod has no effect on flowering. These cultivars also will flower several times per year, but they would do so in short as well as long days.

Temperature may interact or override the photoperiodic effect in all types of strawberries. Basically, cool temperatures promote flowering and warm temperatures inhibit flowering. Temperature sensitivity is greatest in short-day cultivars, and least in the day-neutral cultivars.

The short-day types dominate the industry in most areas of the world. However, day-neutral types are increasing in importance in the United States, which leads the world in strawberry pro-

Introduction to Fruit Crops

doi:10.1300/5547_29

duction. Major California cultivars in 2002 were 'Camarosa' (40 percent of acreage), 'Diamante' (25 percent), 'Aromas' (2.5 percent), and 'Selva' (2 percent). About one-third of California's acreage is made up of minor cultivars and those produced by private breeders, which are unavailable to the general public. Florida, the second-place state, also grows 'Camarosa' and 'Selva', and 'Oso Grande', 'Sweet Charlie', and 'Rosa Linda'. Larger companies employ plant breeders who continually produce new cultivars with large, firm fruits and long shelf life.

ORIGIN, HISTORY OF CULTIVATION

Prior to the relatively recent development of *F.* ×*ananassa*, wood strawberries *(F. vesca)* and musky strawberries *(F. moschata)* were cultivated in Europe and Russia for centuries. Wood strawberries are small and have both red- and white-colored fruits. Musky strawberries are light red to purple and are said to have a strong vinous flavor, similar to that of muscat grapes. These species were largely supplanted by cultivation of *F.* ×*ananassa* over the past 250 years.

In many fruit crops, the precise origin and time of domestication of the species are not well known. Modern cultivated strawberries represent an anomaly in that we know almost precisely when and where the species arose. In 1714, a French spy named Frezier returned from duty in Peru and Chile with five plants of *Fragaria chiloensis*, a large-fruited species native to coastal areas of South America. These plants were female and unfruitful unless cross-pollinated by other species with perfect flowers, such as *F. virginiana*, the Virginian strawberry. Interplanting of *F. virginiana* (male) with *F. chiloensis* (female) in the Brittany region led to production of hybrid seedlings that came to be known as pineapple or pine strawberries, progenitors of the modern cultivated strawberries. French botanist Antoine Duchesne published a book in 1766 detailing the origin of the pine strawberry, and he became the authority for the botanical name—*Fragaria* ×*ananassa* Duch. Around that time, pine strawberry culture spread to other European nations, particularly England, where the first cultivars were produced during the early 1800s. Selections of these were imported to America in the 1800s and provided the germplasm for more cultivars. The Rocky Mountain strawberry, *F. ovalis*, has been used to a limited extent in strawberry breeding, in addition to the Chilean and Virginian strawberries. Breeding programs in California and the southeastern United States are still actively producing new cultivars today.

Native American strawberries were enjoyed by early settlers in the eastern United States, and in the early 1800s, *F.* ×*ananassa* cultivars were brought to America from Europe. Some of these plants found their way to the Pajaro Valley in northern California, where a superior plant was selected from an old strawberry patch. The cultivar, called 'Sweet Briar', became the basis of the California industry, sited near Watsonville, south of San Francisco, in the early 1900s. This area is still the main strawberry region in California, with some production in other areas of coastal California.

FOLKLORE, MEDICINAL PROPERTIES, NONFOOD USAGE

Wild strawberries were collected for many years prior to domestication of *Fragaria* ×*ananassa*. Roots and leaves were made into lotions and gargles in England and used for fastening loose teeth. Fruit juice was used for mouth ulcers. Using a different species, Indians in Western Washington made a tea from leaves and used it against diarrhea. Fruits of the parent species *(F. virginiana* and *F. chiloensis)* were considered folk remedies for diarrhea, gonorrhea, gout, stomachache, and kidney stones. They were used as astringents, diuretics, and mild laxatives. Strawberries are symbolic of "perfect excellence."

Medicinally, strawberries have been shown to kill certain viruses (polio and herpes) in vitro, and they may block the formation of nitosamines, which can cause cancer. Strawberries contain relatively high quantities of ellagic acid, which has a wide range of biological activity only recently discovered:

- Ellagic acid inhibits cancer or mutation induced by such chemicals as benzopyrene or aflatoxin in rats, mice, and human tissue. This may be through inhibition of metabolic activation of carcinogens, by binding to active forms of carcinogens to form harmless complexes, by occupation of sites in DNA that may react with carcinogens, or by stimulating beneficial enzymes. Several folk remedies for cancer were based on strawberry. One study showed that cancer among the elderly was reduced threefold by eating strawberries.
- Ellagic acid may inhibit adsorption of HIV onto cells and HIV enzyme activity on DNA.
- Ellagic acid may bind to such metals as magnesium and calcium, chelating them and forming insoluble aggregates.
- Blood clotting is promoted, and hemorrhage inhibited, by ellagic acid, perhaps through aggregating and metal-chelating activity.
- Ellagic acid and similar compounds function as antioxidants and free-radical scavengers.
- Ellagic acid has shown allelopathic and growth regulator activity in plants. It may inhibit plant growth and seed germination and growth of nitrogen-fixing bacteria. Ellagitannins may inhibit heart-rotting fungi in white oak. Ellagic acid itself may reduce rooting in cuttings and may inhibit degradation of indoleacetic acid (IAA) in plants.
- Ellagic acid can inhibit the growth of certain insect larvae and deters feeding of several insect species.

In addition to strawberry, ellagic acid is also found elsewhere in the Rosaceae, including the genera *Rubus* (blackberry and raspberry), *Malus* (apple), *Amelanchier* (Juneberry), and *Rosa*, and in the Fagaceae and Saxifragaceae families.

PRODUCTION

World

Total production in 2004, according to FAO statistics, was 3,113,840 MT or 6.9 billion pounds. Strawberries are produced in 73 countries worldwide on 529,000 acres. Production has increased 18 percent since 1994 in response to increases in yield, since acreage has been stable for at least a decade. Strawberries are among the highest yielding of all fruit crops. Average yields worldwide are just under 13,000 pounds/acre but approach 40,000 pounds/acre in the United States, the most productive country. Many tree crops with much greater leaf area per unit of land cannot produce fruit yields as high as strawberry. The top ten strawberry-producing countries (percent of world production) follow:

1. United States (27)
2. Spain (9)
3. South Korea (7)
4. Japan (7)
5. Mexico (5)
6. Italy (5)
7. Russia (5)
8. Turkey (5)
9. Poland (4)
10. Germany (3)

United States

Total production in 2004, according to USDA statistics, was 1,006,227 MT or 2.2 billion pounds. Production has increased 34 percent since 1994, while total acreage has increased only 9 percent. Yield was 42,900 pounds/acre in 2004, increasing by about 30 percent in the past decade (Figure 29.1). The industry value is one of the highest among fruits grown in the United States, $1.47 billion, steadily increasing from $836 million in 1994. Prices paid to growers are also good compared to those for other fruit crops. Fresh berries received $0.66 per pound in 2004, higher than the $0.47 to $0.65 per pound received over the past decade. Processed berries received $0.28 per pound in 2004, within the average range of $0.20 to $0.33 per pound. Note that almond production value is about $1 billion, for plants grown on over 500,000 acres; thus, strawberry has a greater crop value than almond, although grown on less than one-tenth the acreage.

The leading strawberry-producing states are California and Florida, with about eight other states producing commercial quantities of strawberries. Many states have small pick-your-own operations and are not counted here.

1. California accounts for 85 to 90 percent of production, with 65 to 75 percent for fresh markets, and the remainder processed. Acreage is 33,200, and the industry value is about $1.2 billion.
2. Florida accounts for 7 percent of total production, with 100 percent marketed fresh during the winter months. Acreage is 7,100, and the industry value is $178 million. Prices received by Florida growers were $1.09 per pound versus $0.62 per pound for growers in California, largely due to the premium paid for winter berries from Florida.
3. Other states (MI, NY, NC, OH, OR, PA, VA, WA, WI) account for 4 percent of production, split fairly evenly, with 53 percent fresh and 47 percent processed. Acreage is 13,000, value about $74 million.

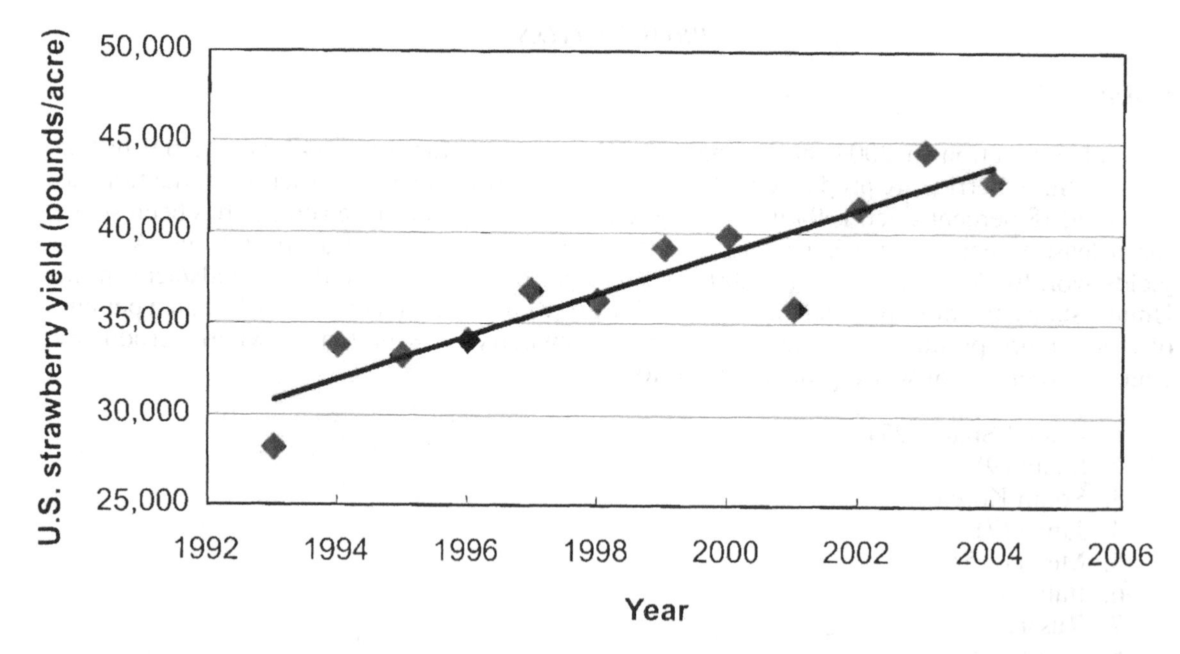

FIGURE 29.1. The trend of strawberry yield in the United States over the past decade. Through breeding of better cultivars and technical improvements in culture, the United States has been able to increase yields by an astounding 1,200 pounds/acre each year. *Source:* USDA statistics.

In 2003, the United States imported an amount of strawberries equal to 4 percent of domestic production, valued at $60 million. Exports were 10 percent of production, valued at $196 million.

FIGURE 29.2. A strawberry plant producing a runner with two daughter plants *(at left)*.

BOTANICAL DESCRIPTION

Plant

Strawberries are perennial, stoloniferous herbs, meaning that they spread via stolons or "runners" (Figure 29.2). The leaves have three leaflets and arise from the "crown" (a reduced stem in the center of the plant). Leaflets are ovate or broadly oval, obtuse, and dentate or coarsely serrate. The runners produce "daughter" plants at every other node, particularly in the summer, and these root where they touch the ground and become independent plants.

Flowers

Flowers are white, about 1 inch across, with 25 to 30 yellow stamens and 50 to 500 pistils on a raised, yellow, conical receptacle. Borne on a dichasial cyme, the centermost terminal flower opens first and is largest, producing the largest fruit. Subordinate flowers are smaller, have fewer pistils, and produce smaller fruits. Flowering occurs over several weeks, and plants may have ripe fruits, developing fruits, and flowers all at once (Plate 29.1).

Pollination

Most cultivars are self-fruitful and therefore do not need cross-pollination for fruit set. However, bee activity is beneficial in transferring pollen to stigmas in an individual flower. A few hundred pollination events must take place to produce a well-formed berry.

Fruit

The strawberry is an accessory fruit, since the edible portion is nonovarian in origin (it is largely swollen receptacle tissue). The true fruits that contain the seeds of the strawberry are achenes, which are similar to tiny sunflower seeds. The achenes are the numerous, tiny, ellipsoid specks that cover the fruit surface (Plate 29.2). They are essential to fruit development because they produce growth regulators that enhance growth of the underlying fleshy tissue. Areas on the fruit surface devoid of functional achenes do not grow, causing irregularly shaped fruits (see Plate 1.2). Ultimate fruit size and shape are therefore heavily dependent on achene set and, hence, pollination.

Fruiting begins in the spring after fall planting and continues for 3 or more years, although declining in size and quality after the first year. Fruits mature rapidly; ripening occurs in 20 to 50 days after pollination.

GENERAL CULTURE

Soils and Climate

Sandy to loamy soils with good drainage and pH of 5 to 7 are best for strawberries. Depth is not important in most cultural systems because plants are often grown for only 1 to 4 years and develop only shallow root systems. Strawberries are most often grown on 6 to 10 inches of soil in raised beds to improve drainage. Strawberries are among the most sensitive of crop plants to salinity.

Soil fumigation is regularly practiced for strawberries because several soilborne diseases and nematodes are problems. For this reason, sandy, light soils are best, since fumigation works best with large pore size. Methyl bromide has been used routinely for fumigation, although it was phased out of production in the United States in 2005. The strawberry industry has received critical-use exemptions from the phaseout from the Environmental Protection Agency, extending the practice through 2006, at least. Methyl bromide has a high ozone depletion potential, in addition to being highly toxic to virtually all organisms. Alternative soil treatments, such as solarization (heating soil with sunlight), and other fumigants are being tested as replacements for methyl bromide (Figure 29.3).

Strawberries have a cool-temperate climate preference. Strawberries have relatively low optimal temperatures for growth and fruiting, compared to other fruit crops; temperatures of 50 to 80°F are best. This is why strawberries do so well in central Florida in winter, and in coastal California from spring to fall. Although the everbearing cultivars will continue to flower at warmer temperatures, fruit quality decreases as temperature increases during summer. Short days and cool temperatures favor reproductive growth, whereas long days and warm temperatures favor leaf and runner production.

Although strawberries are adapted to cool climates, they are sensitive to winter freezing. Crowns are killed at 10 to 20°F. Soil provides some protection, but mulch or snow must be provided in areas with harsh winters. Frost is a serious problem in strawberry production for two reasons: (1) plants are low to the ground, where air is coldest on calm clear nights, and (2) early blooming habit. Crop losses may occur even when the official temperature does not go below 32°F; official temperatures are measured 5 feet above the ground and can be several degrees higher than the temperature of exposed strawberry flowers at ground level. Protection is usually provided by overhead irrigation; the freezing of water on plants keeps them between 28 to 32°F,

FIGURE 29.3. Strawberries are often grown on raised beds to improve root zone drainage. *(left)* Specialized equipment is used to form the beds, lay irrigation lines, and add plastic mulch in one pass. *(right)* The fumigant methyl bromide is used to kill weed seed, nematodes, and fungi that would otherwise severely reduce productivity. Methyl bromide was phased out of production in January 2005, but growers received a critical-use exemption allowing them to use it through 2006.

preventing damage. Plastic row covers, pulled over the plant beds prior to freeze nights, are also used. Strawberries have low chilling requirements to break dormancy, less than 500 hours at or below 45°F in many cases. The low chilling requirement combined with low optimal growth temperature means that strawberries tend to resume growth and bloom earlier than most other crops.

Irrigation is essential in all climates because rooting depth is shallow and plants are most often grown on plastic mulch. Drip irrigation beneath the plastic delivers water, and often fertilizer, directly to plant roots. Overhead irrigation is also used for evaporative cooling at transplanting and for frost protection in spring.

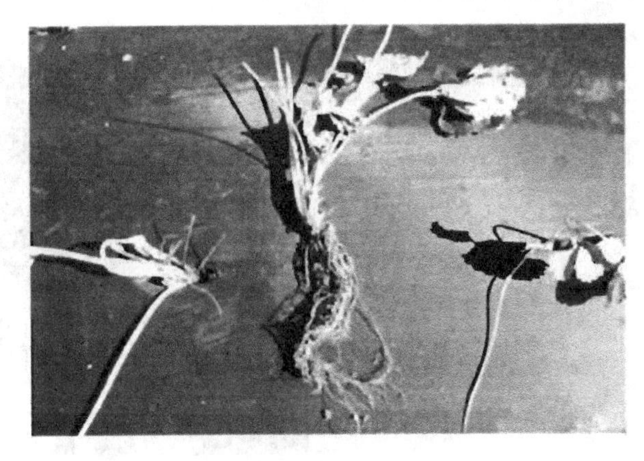

FIGURE 29.4. Strawberry daughter plants are taken from rooted runners and sold as propagation material. This one is bare rooted, but the use of "plugs," or plants rooted in small volumes of soil, is increasing.

Propagation

Strawberry plants sold to growers most often come from nurseries that specialize in strawberry production. Propagation is by division of daughter plants (Figure 29.4). However, due to virus, disease, and nematode problems, special guidelines are used in nurseries to ensure healthy daughter plants. Mother plants are often produced through tissue culture to cleanse them of viruses. After virus indexing, these plants are set in fumigated sites with no prehistory of soilborne problems.

Rootstocks

Strawberry has no rootstocks.

Planting Design, Training, Pruning

Planting systems fall into two main categories: matted-row and annual-hill systems. Annual-hill systems are typical of commercial production; they are fruited once and renovated and usually intensively managed under "plasticulture." Matted-row systems are more typical of home gardens and processing operations; they are cropped for up to 4 years and then renovated (Figure 29.5). Whether in the matted-row or annual-hill system, strawberries are typically planted on raised beds with 6 to 10 inches of soil to improve drainage around roots and provide discrete row middles for cultural practices and picking.

Matted-row plantings are typically established in early spring, deblossomed, and encouraged to grow vegetatively for the first season. Plants are spaced 1 to 2 feet apart initially, but they produce daughter plants on runners and fill in the entire bed by the end of the summer. Obviously, plastic mulch cannot be used if daughter plants are to root, so weed control becomes an issue. Aisles between beds are tilled to keep daughter plants and weeds out of the beds. Plants are then fruited for two to three seasons following establishment; they are renovated when weed and disease problems become too intense, and/or when fruit quality and yield become poor. Leaves can be mowed off plants after harvest to eliminate diseased foliage and facilitate weed control and mulching. Cheaper establishment costs are traded off for higher maintenance and labor costs later.

FIGURE 29.5. The two methods of strawberry culture: *(above left)* annual hill and *(above right, and right)* matted row. Plastic mulch in the annual-hill system prevents weeds, keeps berries clean, and warms the soil in late winter, which forces plants into early growth. *(above left)* A double-row bed system in January after fall establishment of plants. *(above right)* A matted-row planting of 'Totem' strawberries used for processing in Oregon. The grower obtains 3 to 4 years of production and then renovates the site. *(right)* A matted-row bed in a backyard garden.

Annual-hill plantings are established in the fall, harvested in late winter and spring, and renovated during the summer. "Plasticulture" is the most common form of annual-hill culture. It is the most-intensive, highest-yielding system used in most commercial growing regions. Custom machines form beds, apply fertilizer underneath, lay drip irrigation tubing, and cover with plastic mulch in one pass through the field (Figure 29.3). Freshly dug, nondormant plants are set through holes in the plastic in September and October. Spacings are typically 12 to 14 inches between plants in a row, with two rows per bed. Beds are typically 2.5 feet wide, on 4.5- to 5-foot centers, leaving a small area between rows for traffic. Plants must be irrigated intermittently with overhead sprinklers throughout the first week of establishment to reduce transplant shock. New leaves are produced in the fall, before plants go dormant during winter. Root growth will occur during the winter, since soil is kept warm by the black plastic. Plants bloom in late winter to early spring, depending on location, and are harvested over several weeks until summer heat precludes production. Plantings can be held over for another year, but maintenance costs increase, fruit size is reduced, and beds are more difficult to harvest. In Florida, plantings are harvested from December through April, depending on time of establishment and cultivar. In California, harvest begins in March, and some production continues throughout summer to fall.

Pest Problems

Insects

Lygus (plant) bugs. These bugs (*Lygus* spp.) cause catfacing in a manner similar to that seen in peaches; they feed early on growing achenes, killing them and thus stopping fruit enlargement in spotty regions of the fruits. The malformed fruits are often culled. Losses can be as high as 100 percent. The economic threshold for spraying is one to two nymphs per plant, but lygus bugs are difficult to control. There is little information on cultivar susceptibility, except that later cultivars are typically more severely affected than early season ones. Since adults overwinter in debris, annual-hill systems experience less damage than perennial matted rows.

Strawberry bud weevil or clipper. Anthonomous signatus is a dark red-brown weevil or snout beetle, about 1/10 inch long. It lays an egg in a developing flower and then girdles the flower stem, so that the flower appears broken, wilted, and dangling. The larva feeds in the wilted flower bud. The insect overwinters in the dead leaves and mulch beneath the plants. It also attacks brambles and blueberries, damaging them in much the same way. Yield losses can be 50 to 100 percent, and the economic threshold is only one female per 40 feet of row or one clipped bud per 18 inches of row. Control is through insecticides and the use of annual cropping systems, which experience less damage because overwintering is reduced. If the insect was a problem in previous years, insecticides are applied just prior to bloom. In matted-row culture, sprays are applied after bloom ceases, and dead leaves and mulch are removed after harvest.

Root weevils. At least 20 species of root weevils attack strawberry; most are of the genera *Otiorhynchus, Nemocestes,* and *Dyslobus*. Notching of leaves by adults is not injurious, but feeding of larvae on roots and crowns can kill plants and sometimes force the premature termination of perennial plantings. Carbofuran insecticide is effective in controlling the weevil but can be used only after harvest. Annual cropping systems experience fewer problems with this pest.

Strawberry leaf roller. The larvae of this small red-brown moth *(Ancylis comptana fragariae)* feed on leaves and roll them into tubes. The larvae are 1/2 inch long, green caterpillars. Insecticides can be applied in April through May, when large populations occur.

Mites. Spider mites, two-spotted mites (*Tetranychus* spp.), and cyclamen mites *(Steneotarsonemus pallidus)* can be serious pests in dry climates or in hot, dry weather. Mites are tiny (<1/10 inch) spiders that congregate on the underside of leaves, feeding on the lower leaf surface; this causes a characteristic "mite stipple" or bronzed appearance to the upper leaf surface. If the infestation is severe, their feeding will reduce leaf photosynthesis and ultimately yield. Cyclamen mites feed on young, expanding leaves near the crown and can be very debilitating. It is also difficult to control; it reproduces quickly and is difficult to reach with sprays. Control is accomplished with miticides applied when visible injury occurs or scouting reveals high damaging populations. A few cultivars (e.g., 'Cardinal', 'Heidi', 'Lassen') show some mite resistance.

Diseases

Leaf spot. One of the most common diseases of strawberry worldwide, leaf spot, caused by *Mycosphaerella fragariae*, was formerly the most serious disease, until resistant cultivars were produced. The lesions can occur on all plant parts, including fruits, but are most obvious and debilitating on leaves. Lesions are small (1/8-inch diameter) spots with white centers and purple-black margins; they may coalesce or be irregularly shaped, depending on cultivar susceptibility. Lesions on fruits appear as shallow black spots. Control is through the use of resistant cultivars (e.g., 'Apollo', 'Daybreak', 'Lassen', 'Quinault', 'Southland', 'Tribute', 'Tristar') and fungicide programs. Angular leaf spot (caused by *Xanthomonas fragariae*) is a bacterial disease that also produces leaf spotting.

Red stele. Phytophthora fragariae causes this serious disease in areas with high rainfall or heavy, poorly drained soils. Roots are rotted from apices toward the crown; the stele becomes red in the area above the rot. Lateral roots are destroyed, giving the roots a "rat-tail" appearance. Severely affected plants are stunted and "hug" the ground. The disease usually occurs as irregular patches throughout the fields. Since the zoospores must have free water to move, control is through provision of good drainage and avoidance of soil compaction. Raised-bed culture helps reduce the disease, and fumigation limits the seriousness of this and other soilborne diseases. Resistant cultivars are abundant, including 'Aberdeen', 'Allstar', 'Darrow', 'Earliglow', 'Olympus', 'Quinault', 'Redchief', 'Surecrop', 'Totem', and 'Tristar'.

Verticillium wilt. Verticillium albo-altrum is responsible for this wilt, the same problem that occurs on fruit trees, crop plants, and, especially, members of the Solanaceae in arid, irrigated regions. The organism attacks roots and crowns, eventually killing plants. Fumigation and resistant cultivars (several, e.g., 'Surecrop', 'Vermillion', 'Wiltguard') have essentially made this disease a minor problem, but the phaseout of methyl bromide for fumigation may allow it to become a problem once again.

Anthracnose. Four to five species of fungi (*Colletotrichum, Gleosporium* spp.) can cause what is perhaps the most serious disease of strawberry in the southeastern United States, especially in Florida. Fungicides alone are relatively ineffective in controlling the disease, so care must be taken to avoid it altogether in warm, humid regions. It is a particular problem in southern nurseries. Four manifestations are recognized: (1) typical anthracnose lesions on stolons and petioles, (2) crown rot, (3) fruit rot, and (4) black leaf spot. Black leaf spot is usually the first indication that the pathogen is present, followed by anthracnose of stolons and petioles. The most severe problems, however, are crown and fruit rots. Fruit rot is difficult to control, sometimes requiring removal of all ripening fruits and repeated spray applications; entire fields sometimes have to be abandoned. Crown rot kills entire plants, usually by a sudden wilt and collapse, and therefore reduces productivity per acre. Control is achieved by not introducing the fungus to the site, by buying plants from northern nurseries where the fungus does not exist, or by the use of resistant cultivars, although few are available (e.g., 'Dover', 'Heidi', 'Narcissa', 'Solana'). The fungus overwinters on dead plants, which should be tilled in or removed and burned.

Fruit rots. Many fungi can rot ripening or harvested fruits—gray mold *(Botrytis cineria)* is one of the most common. Rhizopus rot (caused by *Rhizopus* spp.) causes fruits to leak fluid, distinguishing it from gray mold, which causes shriveled, but dry berries. Leather rot *(Phytophthora cactorum)* and mucor fruit rot (*Mucor* spp.) also cause crop losses. Anthracnose *(Colletotrichum fragariae)* causes black spotting of fruits and, as mentioned earlier, can turn into an uncontrollable problem. Frequent harvests to remove ripe and overripe fruits quickly helps control these diseases, although fungicide sprays are generally unavoidable with damp weather. Botrytis resistance is found in 'Allstar', 'Columbia', 'Earliglow', 'Holiday', 'Lester', 'Northwest', 'Shuksan', 'Tioga', 'Totem', and 'Tyee'.

Viruses. Strawberries are susceptible to a number of virus diseases, which often cause no outward symptoms but still reduce vigor and yield. The best defense is to purchase virus-free planting stock. Aphid-borne viruses are still a threat, even when using virus-free plants, but annual-hill culture has reduced the significance of these viruses.

HARVEST, POSTHARVEST HANDLING

Maturity

Strawberries are ripe when red color covers the fruits. Berries for shipping markets are picked when fruits have a white tip. At this stage, they are firm enough to ship and have good flavor. For processing, fruits are allowed to become fully ripe, so sugars can accumulate to their fullest. Strawberries are classified as nonclimacteric fruits with respect to ripening.

Harvest Method, Postharvest Handling

Strawberries are harvested by hand, whether for fresh-market sales or processing. For fresh markets, caps (calyx) are left intact; for processing, fruits are plugged (caps removed). Strawberries are picked directly into shipping containers to avoid handling damage (Plate 29.3).

Storage

Strawberries are extremely perishable, having a maximum storage life of only 5 to 7 days at 32°F and 95 percent relative humidity. They are either processed or shipped immediately. Strawberries are not susceptible to chilling injury.

CONTRIBUTION TO DIET

About 80 percent of the strawberries grown in the United States are sold fresh, and the remainder are processed (mostly frozen). Strawberries are higher in vitamin C than many citrus fruits. Per capita consumption of strawberries in 2004 was 7 pounds/year, 5.5 pounds as fresh fruit and 1.5 pounds as frozen fruit. Consumption has increased 118 percent since 1980.

Dietary value per 100-gram edible portion:

Water (%)	91
Calories	32
Protein (%)	0.7
Fat (%)	0.3
Carbohydrates (%)	7.7
Crude Fiber (%)	2

	% of U.S. RDA (2,000-calorie diet)
Vitamin A	<1
Thiamin, B_1	2
Riboflavin, B_2	1
Niacin	2
Vitamin C	98
Calcium	2
Phosphorus	3
Iron	2
Sodium	0
Potassium	4

BIBLIOGRAPHY

Bowling, B.L. 2000. *The berry grower's companion*. Portland, OR: Timber Press.

Childers, N.F. (ed.). 1980. *The strawberry: Cultivars to marketing*. Gainesville, FL: Norman F. Childers.

Dale, A. and J.J. Luby. 1991. *The strawberry into the 21st century*. Portland, OR: Timber Press.

Darrow, G.M. 1966. *The strawberry: History, breeding, and physiology*. New York: Holt, Rinehart and Winston.

Duke, J.A. and J.L. duCellier. 1993. *CRC handbook of alternative cash crops*. Boca Raton, FL: CRC Press.

Galletta, G.J. and R.S. Bringhurst. 1990. Strawberry management, pp. 83-156. In: G.J. Galleta and D.G. Himelrick (eds.), *Small fruit crop management*. Englewood Cliffs, NJ: Prentice-Hall, Inc.

Guttridge, C.G. 1985. *Fragaria xananassa*, pp. 16-33. In: A.H. Halevy (ed.), *CRC handbook of flowering*, Volume 3. Boca Raton, FL: CRC Press.

Hancock, J.F. 1999. *Strawberries*. Wallingford, UK: CAB International.

Hancock, H.F., J.L. Maas, C.H. Shanks, P.J. Breen, and J.J. Luby. 1990. Strawberries *(Fragaria)*. pp. 489-546. In: J.N. Moore and J.R. Ballington (eds.), *Genetic resources of temperate fruit and nut crops* (Acta Horticulturae 290). Belgium: International Society for Horticultural Science.
Maas, J.L. (ed.). 1984. *Compendium of strawberry diseases.* St Paul, MN: American Phytopathological Society.
Shoemaker, J.S. 1978. *Small fruit culture,* Fifth edition. Westport, CT: AVI Publishing.
Turner, D. and K. Muir. 1985. *The handbook of soft fruit growing.* London: Croom Helm.
Welch, N.C. 1989. *Strawberry production in California* (University of California Division of Agriculture and Natural Resources Leaflet 2959). Davis, CA: University of California, Davis.

Chapter 30

Walnut (*Juglans* spp.)

TAXONOMY

Walnuts are members of the relatively small Juglandaceae family, containing about 60 species, 21 of which are placed in the genus *Juglans*. Nuts from all species are edible, although none is as large and easily cracked as the Persian or English walnut, *Juglans regia* L. The name "walnut" derives from the corruption of "Gaul nut," coined by the French. The genus is divided into four sections:

1. Section *Juglans: Juglans regia* L. is the only species. Persian walnut is the major walnut of commerce, and Carpathian walnuts are just cold-hardy strains of the same species. *Juglans regia* is native to the Carpathian Mountains, eastward to Korea, but was brought to Europe through Persia in early recorded history. The name "English walnut" came from transport of walnuts from Asia Minor to England aboard English boats. Carpathian walnuts originated in Ukraine, Russia, Czechoslovakia, and Germany. Carpathian walnuts can be grown further north than Persian walnuts, or in climates with more variable winters, such as in the eastern United States. Persian walnuts are large, thin shelled, and four celled, with a smooth, irregularly dehiscent husk that separates easily from the nut.
2. Section *Rhysocaryon:* This section includes the black walnuts. Of the 16 species in the genus, *J. nigra* L. , the eastern black walnut, is the most important (Figure 30.1). It is prized for its high-quality wood more than for its nuts; the wood is among the best-quality hardwoods used for furniture. Most of the nut crop is harvested from wild trees located in the south-central and southeastern United States, although some improved cultivars are grown. Kernels are of excellent quality but make up as little as 12 percent of the weight of the nuts, for wild nuts, and only 30 to 40 percent for the improved cultivars. Nuts are difficult to crack due to thick shells and adherent hulls; shells are used in plastics, glue extenders, sandblast cleaners, and metal polishers.

   Black walnuts have indehiscent, thick, adherent husks, four-celled nuts, and thick shells that are ridged. Other important black walnut species are *J. californica* (Southern California black walnut), *J. hindsii* (Northern California black walnut), *J. major* (Arizona black walnut), *J. microcarpa* (Texas black walnut), but, of these, only *J. hindsii* and its hybrids are of significance (as rootstocks for Persian walnuts).
3. Section *Trachycaryon: Juglans cineria* L., the butternut, is the only species in this section (Figure 30.1). Native to the eastern United States, this species has the most northern distribution of North American *Juglans* species. Butternuts are used in New England to make maple-butternut candy, which is sold at roadside stands on a very limited scale. Nuts have very thick shells with eight ridges, indehiscent husks with four ribs, and two-celled nuts with small kernels that tend to shatter when cracked. At least 25 cultivars have been selected, mostly for ease of cracking.

Introduction to Fruit Crops

doi:10.1300/5547_30

4. Section *Cardiocaryon: Juglans ailantifolia* var. *cordiformis*, the heartnut, is a native of Japan, and the only variant in this group producing easily cracked nuts. Other species include *J. ailantifolia* (Japanese walnut), *J. cathayensis* (Chinese walnut), and *J. mandshurica* (Manchurian walnut). In this section, nuts are two celled with four to eight ridges, have indehiscent husks, and are borne on long clusters of several nuts.

Persian Walnut Cultivars

A successful breeding program has greatly changed the way Persian walnuts are grown. Older cultivars, such as 'Hartley' and 'Franquette', bear fruits only on terminals of branches; therefore, any pruning reduces yield, so trees become very large. Newer cultivars, produced by breeders at the University of California, Davis, have 'Payne' parentage and bear a large proportion of fruits from lateral buds; thus, they lend themselves to hedgerow planting and mechanical pruning (Plate 30.1). 'Payne' is still a major cultivar and has the advantage of being self-pollinating. 'Hartley' is an old, terminal-bearing, but productive cultivar, and as of the mid-1990s, it was still the major cultivar in terms of acreage. It produces large nuts with light kernels and attractive shells, and it leafs out late enough in spring to avoid walnut blight. Other 'Payne'-type cultivars popular in California are 'Chandler', 'Serr', 'Vina', 'Ashley', 'Tehama', 'Pedro', 'Sunland', and 'Howard'. 'Chandler' is increasing most rapidly in terms of popularity. In 1995, there were over 15,000 nonbearing acres of this cultivar, more acres than were in production at that time.

Carpathian Walnut Cultivars

These are cold-hardy cultivars of *J. regia* that can be grown outside of California, in the Pacific Northwest or the mid-Atlantic and southeastern United States. 'Cascade' has large nuts with up to 56 percent kernel, and it is rated number one by the North American Nut Grower's Association. Other Carpathian walnuts include 'Colby', 'Fickes', 'Gratiot', 'Hansen', 'Lake', 'Mesa', and 'Somers'. 'Manregian' is a Carpathian walnut that is used as a rootstock.

Eastern Black Walnut Cultivars

Several selections have been made from chance seedlings of eastern black walnut, largely for nut-cracking characteristics, since shells are thick and kernels are difficult to extract. 'Beck',

FIGURE 30.1. Developing fruits of the eastern black walnut *(left)* and the butternut *(right)*, close relatives of the Persian walnut.

'Bowser', 'Emma Kay', 'Ohio', and 'Sparrow' are easily cracked. Those showing resistance to anthracnose, a debilitating fungal disease of black walnut, include 'Clermont', 'Ohio', and 'Sparrow'. 'Deming's Purple' is an ornamental cultivar with purple leaves and purple hues to the flowers, shucks, and even kernels.

ORIGIN, HISTORY OF CULTIVATION

Juglans regia has its origins in Eastern Europe, Asia Minor, and points eastward to the Himalayan Mountains. However, there are native *Juglans* species in North, Central, and South America as well as Europe and Asia. Although native to Europe, it probably was not utilized there until improved forms were imported from Persia. Romans spread cultivation throughout southern Europe. The species came to the New World with English settlers, and to California via missionaries. Today, walnut production is almost entirely located in the San Joaquin and Sacramento Valleys of California, where over 5,000 growers and 52 processors (marketers) make up a highly organized and productive industry.

Black walnuts have been appreciated by Native Americans for millennia, but only recently have they been cultivated. Initial interest was in the wood, which is the finest native hardwood available in the United States. Most wood is used for veneer and in furniture making. About 20,000 tons of black walnuts are harvested and sold each year from the Missouri area, where they grow wild. By-products of the nuts are more important than the kernels themselves, particularly since kernels are very difficult to extract whole. The shells are used for a variety of cleaning and polishing operations. The green hulls or shucks are used to make tinctures helpful in treating intestinal parasites and fungal conditions.

FOLKLORE, MEDICINAL PROPERTIES, NONFOOD USAGE

Walnuts have a number of medicinal and nonfood uses, as well as some toxic properties. Juglone is excreted by the roots of black walnut and other walnuts and is toxic to many other plants (i.e., it is allelopathic). Even dead roots can release juglone for years after the tree is gone. Black walnuts should be kept away from most other plants, particularly vegetable gardens. The most susceptible plants include asparagus, cabbage, eggplant, pepper, tomato, potato, apple, pear, blueberries, blackberries, azaleas and rhododendrons, some pines, silver maple, ornamental cherries and crabapples, crocus, some chrysanthemums, columbine, lilies, and petunia. Juglone from black walnut fruits and bark acts against dermatomycosis (skin fungi), being first used for this purpose by Greeks and Romans. *Juglans insularis* is used in Cuba as an herb decoction in bath water to treat various skin diseases of children. Walnut bark is used as a dentifice in Pakistan. Black walnut tincture, an extract made with grain alcohol, is derived from fresh green hulls of the black walnut tree. It is said to kill adult and developmental stages of at least 100 parasites. It is touted as a great antiseptic that is high in iodine, and excellent for the treatment of any kind of fungal condition. It also is a good vermifuge for pinworm, ringworm, and other parasites, and it even removes warts and treats psoriasis. The tincture is generally used in conjunction with wormwood and cloves as part of a complete parasite-eradication program. In 2000, four fluid ounces cost $39 from an Internet herb shop!

Ellagic acid is found in walnut leaves and fruits; it is being studied for use as a cancer therapy drug, in addition to having many other biological effects (see Chapter 29, "Strawberry," for more on ellagic acid).

Shells are ground and used as antiskid agents for tires, blasting grit, activated carbon, and sometimes an adulterant of spices. The husk yields a valuable oil and a yellow dye when pressed; the oil is used in soaps, paints, and dyes. The oil from walnut kernels is high in unsaturated fats and can be used in cooking. The wood is heavy and fine-grained and is used mostly for furniture and gun stocks.

PRODUCTION

World

Total production in 2004, according to FAO statistics, was 1,491,152 MT or 3.3 billion pounds. Walnuts are produced commercially in 48 countries on 1.6 million acres. Production has increased 42 percent since 1994, due to a 22 percent increase in acreage and a 16 percent increase in yield. Yields average about 2,100 pounds/acre. Chinese production has increased over threefold since 1980, and China overtook the United States as the world leader in walnut production in 1994. The top ten walnut-producing countries (percent of world production) follow:

1. China (28)
2. United States (20)
3. Iran (11)
4. Turkey (8)
5. Ukraine (5)
6. Romania (2)
7. India (2)
8. France (2)
9. Egypt (2)
10. Serbia and Montenegro (2)

United States

Total production in 2003, according to USDA statistics, was 296,000 MT or 652 million pounds of in-shell nuts. The industry value is $375 million, fluctuating between $275 million and $380 million over the past decade. There were 213,000 acres of walnuts in California's Central Valley in 2003, which is up 13 percent since 1994. Yield averages 3,060 pounds/acre, but some orchards can produce up to 6,000 pounds/acre. The price paid to growers is $0.58 per pound, on the low end of the range of $0.50 to $0.79 per pound received over the past decade.

California is the only state to produce Persian walnuts. Sun Diamond, a California cooperative, controls 50 percent of production. A federal marketing order established the Walnut Marketing Board in 1949, and state legislation created the California Walnut Commission in 1987; these two organizations regulate and promote the industry. Oregon had 12,000 acres in 1960, but high winds in 1962 blew down half of the trees, and a severe freeze in 1972 killed half of the remaining trees. Oregon produced only 500 tons of walnuts worth $355,000 in 1977, the last year records were kept.

The United States imports trivial amounts of walnuts, less than 0.1 percent of production in 2003. Exports are equivalent to 18 and 10 percent of total production for in-shell and shelled nuts, respectively, and have increased almost tenfold since the early 1980s. Exports are valued at over $200 million. Principal destination countries are Japan, Germany, Spain, Italy, and Israel.

BOTANICAL DESCRIPTION

Plant

The walnut is a medium to large tree that grows to 100 feet in nature but only 20 to 50 feet in cultivation, with a spreading crown. Leaves are compound and composed of seven to nine leaflets, which have prominent, herring-bone venation. Leaflets are ovate, with pointed tips and smooth margins. Hickories, including pecan, look very similar to walnuts. Current season's twigs have solid pith in pecans, whereas the pith is chambered or has numerous, tiny partitions running horizontally across the central pith in walnuts. Also, the shuck or husk of Persian walnut does not split into four *regular* sections at maturity, as it does in pecan and hickories; it splits, but *irregularly*.

Black walnuts are the largest of the North American walnuts, potentially reaching 150 feet in height and 6 feet in diameter. They grow very slowly, compared to many other tree species, and are not often over 100 feet in cultivation. Leaves are compound and composed of 13 to 23 leaflets, which have prominent, herring-bone venation. Leaflets are ovate, with pointed tips and finely toothed margins (see Figure 30.1, left).

Flower

All *Juglans* species are monoecious, bearing male and female reproductive organs on separate flowers on the same tree. Although *J. regia* is self-fertile, it is dichogamous, either protandrous or protogynous, depending on cultivar. Catkins (male inflorescences) are borne laterally on 1-year-old wood, and pistillate flowers are borne at the tips of terminal or lateral shoots on current season's wood, in spikes of typically two to three flowers. Catkins are longer than those of pecan by an inch or so, and they are borne singly, rather than in small clusters (Plate 30.2). Each catkin may produce 2 million pollen grains. Pistillate flowers lack visible sepals and petals and are pubescent, small, and green. The females are merely an ovary with large, feathery stigma. The ovary wall is fused to bract or involucre tissue in females, and the outer ovary wall plus the involucre become the fleshy shuck of the fruit.

Pollination

Walnuts are similar to pecans in that the time of pollen shedding does not always overlap well with the time of female flower receptivity to pollen. Hence, although most walnuts are self-fertile, they sometimes require another cultivar for pollination, since the timing of the functions of male and female flowers is different. This condition is referred to as "dichogamy." Two possible forms of dichogamy are "protandry," meaning "first male"), in which the pollen is shed before the female flower becomes receptive, and "protogyny," meaning "first female," in which the female flower becomes receptive to pollen before pollen is shed. Most walnuts are protandrous. Some cultivars, such as 'Payne', are only slightly dichogamous and can be planted in solid blocks because they are self-pollinating.

In the early 1990s, California researchers discovered that too much cross-pollination resulted in a condition known as "pistillate flower abscission" (PFA). This is the drop of female flowers just as they begin enlarging in spring. Experiments showed that shaking catkins from pollinizers prior to pollen shed, or removing pollinizers entirely, increased yield by 26 and 86 percent, respectively. Cultivars that suffer from PFA ('Serr', 'Chandler', 'Chico', 'Ashley', 'Vina') are now recommended to be planted without pollinizers or with reduced pollinizer frequency. In existing orchards, trunk shakers can be used to reduce pollen load in spring.

As with most catkin-bearing species, walnuts are wind pollinated. Pollen will carry about 250 to 300 feet in mature orchards, so the minimum pollinizer arrangement would be about every tenth row (cross-wind). However, higher ratios may be used if pollinizers also bear high-quality nuts.

Fruit

The walnut fruit is a nut. Nuts are borne singly or in clusters of two to three on shoot tips. A green, fleshy shuck surrounds the nut, which splits irregularly at maturity. In Persian walnut, the shuck is easily separated from the nut shell, in contrast to black walnuts, in which the shuck is adherent. The shell is rough, wrinkled or furrowed, and thin. Nuts are ovoid to round, ½ to 2 inches in diameter, containing two kernels separated by a thin, papery central plate, extending from the inner layer of the shell (Plate 30.3). There are fewer nuts per cluster, but more nut clusters per tree, compared to pecan.

Walnuts and most tree nuts are relatively slow to bear. Often a 4- to 6-year juvenile period must pass, and the first significant yields occur at 8 to 10 years. Maximal yields may not be achieved for 10 or more years. Alternate bearing occurs in walnuts but is generally less severe than in pecan or pistachio (see Chapter 25, "Pecan," for complete description). Lateral-bearing Persian walnuts bear earlier than the older, terminal-bearing cultivars. Male flowers generally appear later in life than do female flowers, so fruiting may be delayed by lack of pollen in early years, not lack of female flowers. Nut size ranges from 32 to 45 nuts per pound in Persian walnut, and 22 to 35 nuts per pound in black walnut. Walnuts are mature at 4.5 to 5 months after flowering and are harvested in September and October.

GENERAL CULTURE

Soils and Climate

The best soils are deep, well-drained silt loams with pH 6 to 8, as found in the Central Valley of California. Walnuts and related species are generally deep rooted (9-12 feet), and strongly taprooted, if no restrictive layers are present in soils.

Irrigation is necessary in the arid climates where walnuts are grown. Up to 4 acre-feet of water is applied per year in California. Walnuts are extremely intolerant of soil flooding, with growth reductions noted in seedlings in as little as 24 hours. A condition termed "apoplexy," a sudden wilting and shedding of foliage, may occur in summer following flood irrigation, especially on coarse-textured soils. This is generally thought to be a flooding stress that follows the stresses of drought and high soil (and air) temperatures. Walnuts are also extremely intolerant of salinity, experiencing yield reductions when salts reach about 1,500 parts per million (ppm). They are sensitive to boron, chloride, and carbonates often contained in saline waters.

As the dominance of the California industry suggests, Persian walnuts are best adapted to Mediterranean climates, with dry, hot summers and mild winters. Coldhardiness is a major limiting factor for Persian walnut. In Iran, the native range of the Persian walnut, trees have been reported to survive temperatures of –5 to –40°F in midwinter. However, cold acclimation must occur slowly in North America because trees can be severely injured or killed by much higher temperatures in late autumn. In California, trees are considered cold hardy to 12 to 15°F. Carpathian cultivars are far more cold hardy and quicker to acclimate than are Persian cultivars. Chilling requirement is highly variable in walnut, from 400 to 1,600 hours. With the need for late-leafing characteristics to avoid walnut blight, cultivars with higher chill requirements have been produced. Rainfall in spring at budbreak, or in midsummer, greatly increases the severity of walnut blight. Newer cultivars have been selected to leaf out after the last rains of spring occur in California. High temperatures (100-110°F) and/or high sunlight can cause sunburn, darkening, and shriveling of kernels, especially at the top of trees. This is another reason that production is north of the areas where summer heat of this magnitude is common. However, temperatures too cool in summer can result in inadequate kernel development or "nut fill."

Propagation

Common methods are (1) whip grafting and (2) ring or patch budding. This is commonly done in spring on 1-year-old seedling rootstocks in nurseries, but some growers prefer to plant ungrafted stocks and then graft in the orchard, after the rootstock becomes established. All rootstocks are produced from open-pollinated seed.

Rootstocks

Persian walnut seedlings *(J. regia)* are the most popular rootstock worldwide, and in areas where blackline disease is a problem. 'Manregian' is the selection most tolerant of blackline disease (see Diseases, under Pest Problems). Trees usually lack vigor and yield efficiency in California when propagated on this rootstock, so it is not used there. It is also more sensitive to salinity and flooding and less tolerant of root and crown rots than is the Northern California black walnut.

Northern California black walnut *(J. hindsii)* is the most common stock for Persian walnut in California but has not been used elsewhere, perhaps due to its susceptibility to blackline disease. It has coarse, furrowed, dark bark, and thus the graft union is very distinct when Persian walnut is the scion (Figure 30.2). Popular attributes include low cost and good availability, good vigor and yield, and some tolerance to root and crown rots. However, it is intolerant of heavy, high-pH soils, poor soil aeration, most types of *Phytophthora* root rot, and blackline disease.

FIGURE 30.2. The graft union is clearly visible when Persian walnut is grafted onto California black walnut *(J. hindsii)*, one of the most common rootstocks for walnut in California.

'Paradox' *(J. hindsii × J. regia)* is a hybrid of Persian and Northern California black walnut that is generally superior to its parents in several traits. Therefore, 'Paradox' is the most preferred rootstock in California, but high variation among seedlings and susceptibility to blackline disease limit its use somewhat.

Eastern black walnut *(J. nigra)* seedlings are the most common rootstocks for improved cultivars of eastern black walnut. To date, no superior rootstocks have been introduced. It is compatible with Persian walnut cultivars, but they have low yield efficiency and lack vigor under California conditions. In the eastern United States, *J. nigra* can be used for Carpathian walnuts, where it produces a smaller tree with greater tolerance of nematodes and root and crown rots, as compared to those mentioned previously.

Planting Design, Training, Pruning

FIGURE 30.3. A Persian walnut orchard in California. These trees have been pruned back once to alleviate crowding.

Traditional orchards are usually planted at densities of 50 to 70 trees/acre initially and then thinned to 30 to 50 trees/acre (30 × 30 feet or 40 × 40 feet), as crowding occurs over time (Figure 30.3). Pollinizer cultivars are planted in solid rows (cross-wind) at selected intervals, usually at a 9:1 ratio. Hedgerow orchards of lateral-bearing cultivars can be planted at higher densities (100 trees/acre) and do not require thinning of trees, since periodic hedging and topping maintains tree size. Hedgerows are planted at 10 to 15 feet between plants in each row and 20 to 25 feet between rows. Mechanical hedging is performed every 2 to 4 years, and trees are topped, as needed, when they reach the same height as the row spacing (20-25 feet).

Traditionally spaced trees are trained to a modified central-leader system, otherwise referred to as "delayed open center," allowing four to five scaffolds to develop on the central leader before its removal. Limbs below 6 feet are removed to facilitate equipment movement and harvest (trunk shaking when young).

Pest Problems

Insects

Codling moth. Laspeyresia pomonella is a pest of many tree crops, including walnuts. The adult moth overwinters in a cocoon under the bark or in debris. It first lays eggs on or near fruits in early spring, and there can be two to three generations per year. Larvae bore into and eat the contents of developing nuts, causing nut drop in early infestations, but nuts attacked later often do not drop. Natural enemies are insufficient for control in most cases. Sprays are timed from trapping adult insects to determine when egg laying is occurring. The first spray is usually timed when the nuts are ⅜ to ½ inch in diameter. Subsequent generations are hard to predict and control. A mating disruption pheromone can be used to prevent mating and egg laying.

Aphids. The walnut aphid *(Chromaphis juglandicola)* and the dusky-veined aphids *(Callipterus juglandis)* are sucking insects that reduce tree vigor through the extraction of photosynthate. Severe infestations can cause defoliation in summer and reduce yield and nut filling. Perhaps more deleterious than feeding per se is the development of sooty mold, a black fungal coating on leaves, which reduces photosynthesis at the critical nut-filling stage in late summer. Spraying for aphids may actually cause greater aphid infestations due to elimination of predators, such as ladybugs, wasps, and lacewings. Therefore, insecticides are not applied in summer unless absolutely necessary for control of other insects.

Navel orangeworm. The larvae of this moth *(Amyelois transitella)* appear late in the season. They enter the nuts through cracks in the shuck and then bore into and eat through the kernels.

Several larvae may invade the same nut. This feature, and the presence of webbing, distinguish it from codling moth damage. The pest overwinters in damaged nuts ("mummies") that remain in trees or on the ground, or in any intact fruits on the ground. Insecticide applications are not fully effective for control. Some natural enemies (wasps) exist but cannot control the pest completely. Early harvest is the best way to avoid damage. Mummified nuts are shaken from trees in winter, and all debris beneath the trees is removed to eliminate the overwintering insects.

Walnut husk fly. The larvae (maggots) of this insect (*Rhagoletis* spp.) feed on the husks of walnuts and cause shell and sometimes kernel staining, reducing quality grade. Husk tissue breakdown and subsequent decay organisms cause the staining. Black walnut is the preferred host of the fly, but most cultivars of Persian walnut are also susceptible. Control is difficult because emergence of adults in midsummer is rather unpredictable. Insecticides give inconsistent control, but when mixed with protein baits to attract adult flies, they have been successful in some cases. Since it is a mid- to late-season pest, early maturing cultivars, such as 'Ashley', avoid damage in most years.

Diseases

Walnut blight. This bacterial disease (caused by *Xanthomonas campestris* pv. *juglandis*) is estimated to cause greater economic losses than any other pest problem of walnut in California. It is caused by a bacteria that survives on dormant buds, catkins, and twig cankers. It causes irregular black lesions on catkins, fruits, twigs, and leaves. Rain or sprinkler irrigation spreads the disease. Early leafing cultivars are most severely affected, and the disease tends to be more severe in Northern California. Sprays of copper or Bordeaux mixture (lime and copper sulfate) can be applied, starting at bloom of female flowers, and continued as rainy, wet weather persists. On large trees, this is not feasible. Late-leafing cultivars, such as 'Chandler' and 'Franquette', avoid the rainy period in California, and thus the disease. 'Serr', 'Howard', and 'Hartley' are less affected than 'Payne' and 'Ashley'.

Blackline. This is caused by a hypersensitive reaction of the rootstock to invasion by the cherry leaf roll virus from the top of the tree. The virus is pollen, graft, and seed transmitted, making it exceptionally difficult to control. Persian walnuts are tolerant, so they remain productive while virus particles make their way to the graft union over a period of years. At the graft union, the rootstock resists infection by a hypersensitive reaction, whereby a thin layer of cells die, girdling the tree. Suckering may occur prior to tree death. Trees may remain productive for many years after infection. The disease is rootstock specific, with trees on all rootstocks, except *Juglans regia* 'Manregian', susceptible. Infested trees are removed to reduce the chance of virus spread.

Root and crown rots. Crown rot (caused by up to eight *Phytophthora* species) has become a serious problem in California due to mechanization, use of flood irrigation, and multiorchard management, all of which spread these soilborne diseases. The fungus causes cankers at the crown, eventually debilitating or killing trees. Oak root rot (caused by *Armillaria mallea*) is similar but can persist in the soil for decades, and it is difficult to eliminate once introduced. 'Paradox' rootstock is somewhat tolerant to crown rot, although, strangely, neither of its parents are tolerant. Oak root/crown rot is not a problem for trees grafted to *Juglans hindsii*, which is resistant. Good soil drainage will reduce disease incidence.

Crown gall. Crown gall (caused by *Agrobacterium tumefaciens*) is a bacterial disease that causes large tumors to form at the tree crown, debilitating the tree. The galls are soft and spongy. It is primarily a problem on young trees, as older trees can fight off the bacterium. It is less serious than phytophthora root rot in California, primarily due to improved nursery practices. Since the bacterium enters through wounds, careful handling of stocks, good sanitation, and dipping roots into solutions of antagonistic, nonpathogenic bacteria have limited this disease.

Anthracnose. This disease, caused by a fungus *(Gnomonia leptostyla)*, attacks new, expanding green tissues and causes premature defoliation, low yield, and poor nut fill. It is much worse on eastern black walnut than on Persian walnut. Selection of resistant cultivars is the best means of control. 'Hartley' may be severely affected, whereas other Persian walnuts are not as badly affected. 'Clermont', 'Sparrow', and 'Ohio' eastern black walnut cultivars are resistant to anthracnose.

HARVEST, POSTHARVEST HANDLING

Maturity

Persian walnuts are mature 1 to 4 weeks prior to hull dehiscence, when packing tissue (inner layer of shuck) becomes brown. However, they are harvested at the beginning of shuck split, when the seedcoat is a light tan color (market preference). Ethephon has been used to advance maturity and shuck split.

Harvest Method

For Persian walnut, trunk or limb shakers are used, depending on tree size. A windrow machine places the nuts into narrow rows to be picked up by a sweeper. Nuts are collected in large bins and taken to the processing plant. The process is analogous to that used in pecan (see Chapter 25, "Pecan," for photos). Eastern black walnuts are generally picked up from the ground after falling from trees. They may be mature slightly before this time but would be difficult to remove from tall trees.

Postharvest Handling

Freshly harvested Persian walnuts are removed from hulls and dried in forced-air dryers at 100 to 109°F, until 8 percent moisture content is achieved. In-shell nuts are bleached and sold fresh or shelled and marketed as kernels. Eastern black walnuts are exceedingly difficult to remove from their shucks and crack from the thick shells; often, only pieces of the kernels are recovered.

Storage

Dried nuts can be stored for about 4 months at room temperature before becoming rancid, but they last 1 to 2 years when stored frozen.

CONTRIBUTION TO DIET

Walnuts are marketed primarily as shelled kernels (75 percent), with the remainder as in-shell nuts. Shelled kernels are generally processed into baked goods, candies, cereals, and other snack foods. Per capita consumption of walnuts is 0.55 pound/year, the highest rate since 1980, but there is no clear trend in walnut consumption over the past 25 years.

Dietary value per 100-gram edible portion:

	Persian Walnut	Black Walnut
Water (%)	4	4
Calories	654	618
Protein (%)	15.2	24.1
Fat (%)	65.2	59
Carbohydrates (%)	13.7	9.9
Crude Fiber (%)	6.7	6.8
	% of U.S. RDA (2,000-calorie diet)	
Vitamin A	<1	1
Thiamin, B_1	23	4
Riboflavin, B_2	9	8
Niacin	6	2
Vitamin C	2	3
Calcium	10	6
Phosphorus	49	73
Iron	16	17
Sodium	<1	<1
Potassium	13	15

BIBLIOGRAPHY

Duke, J.A. 2001. *Handbook of nuts.* Boca Raton, FL: CRC Press.

Forde, H.I. 1979. Persian walnuts in the western United States, pp. 84-97. In: R.A. Jaynes (ed.). *Nut tree culture in North America.* Hamden, CT: Northern Nut Growers Association.

Funk, D.T. 1979. Black walnuts for nuts and timber, pp. 35-50. In: R.A. Jaynes (ed.), *Nut tree culture in North America.* Hamden, CT: Northern Nut Growers Association.

Grimo, E. 1979. Carpathian (Persian) walnuts, pp. 74-83. In: R.A. Jaynes (ed.), *Nut tree culture in North America.* Hamden, CT: Northern Nut Growers Association.

McGranahan, G.H. and P.B. Catlin. 1987. *Juglans* rootstocks, pp. 411-450. In: R.C. Rom and R.F. Carlson (eds.), *Rootstocks for fruit crops.* New York: John Wiley and Sons.

McGranahan, G.H. and C. Leslie. 1990. Walnuts (*Juglans*), pp. 905-952. In: J.N. Moore and J.R. Ballington (eds.), *Genetic resources of temperate fruit and nut crops* (Acta Horticulturae 290). Belgium: International Society for Horticultural Science.

O'Rourke, F.L.S. 1969. The carpathian (Persian) walnut, pp. 232-239. In: R.A. Jaynes (ed.), *Handbook of North American nut trees.* Knoxville, TN: North American Nut Growers Association.

Ramos, D., G. McGranahan, and L. Hendricks. 1984. Walnuts. *Fruit Var. J.* 38:112-120.

Rosengarten, F. 1984. *The book of edible nuts.* New York: Walker and Co.

Serr, E.F. 1969. Persian walnuts in the western states, pp. 240-263. In: R.A. Jaynes (ed.), *Handbook of North American nut trees.* Knoxville, TN: North American Nut Growers Association.

Teviotdale, B.L., T.J. Michailides, and J.W. Pscheidt. 2002. *Compendium of nut crop diseases in temperate zones.* St. Paul, MN: American Phytopathological Society.

Woodroof, J.G. 1967. *Tree nuts: Production, processing, products.* Volume 2. Westport, CT: AVI Publishing.

Zarger, T.G. 1969. Black walnuts—As nut trees, pp. 203-211. In: R.A. Jaynes (ed.), *Handbook of North American nut trees.* Knoxville, TN: North American Nut Growers Association.

Appendix A

Common and Scientific Names of Fruit Crops

Common Name	Scientific Name	Family
Abiu	*Pouteria caimito* Radlk.	Sapotaceae
Abyssinian banana	*Ensete ventricosum* (Welw.) Cheeseman	Musaceae
Acerola	*Malpighia glabra* L.	Malpighiaceae
Achiote	*Bixa orellana* L.	Bixaceae
African breadfruit	*Treculia africana* Decne.	Moraceae
African mangosteen	*Garcinia livingstonei* T. Anderson	Clusiaceae
African oil palm	*Elaeis guineensis* Jacq.	Arecaceae
African pear	*Dacryodes edulis* (G. Don) H.J. Lam.	Burseraceae
African wild mango	*Irvingia gabonensis* (Aubrey-Lecomte ex O'Rorke) Baill.	Irvingiaceae
Aguaje palm	*Mauritia flexuosa* L. f.	Arecaceae
Akebia	*Akebia quinata* (Thunb.) Decne.	Lardizabalaceae
Akee	*Blighia sapida* K. Konig	Sapindaceae
Alemow	*Citrus macrophylla* Wester	Rutaceae
Allegheny chinkapin	*Castanea pumila* Mill.	Fagaceae
Allspice	*Pimenta dioica* Merr.	Myrtaceae
Almond	*Prunus dulcis* (Mill.) D.A. Webb	Rosaceae
Alpine strawberry	*Fragaria vesca* L.	Rosaceae
Amazon tree grape	*Pouroma cercopiaefolia* Mart.	Moraceae
Ambarella	*Spondias dulcis* Parkins or *S. cytherea* Sonn.	Anacardiaceae
American bunch grape	*Vitis labrusca* L.	Vitaceae
American chestnut	*Castanea dentata* Borkh.	Fagaceae
American gooseberry	*Ribes hirtellum* Michx.	Saxifragaceae
American hazel	*Corylus americana* Marsh.	Betulaceae
American oil palm	*Elaeis oleifera* (Kunth.) Cortes	Arecaceae
American persimmon	*Diospyros virginiana* L.	Ebenaceae
American plum	*Prunus americana* Marsh.	Rosaceae
Amra (Ambra)	*Spondias pinnata* Kurz. or *S. mangifera* Willd.	Anacardiaceae
Andean blackberry	*Rubus glaucus* Benth.	Rosaceae
Andean walnut	*Juglans neotropica* Diels	Juglandaceae
Annatto	*Bixa orellana* L.	Bixaceae
Ansu apricot	*Prunus armeniaca* var. *ansu* Maxim.	Rosaceae
Apple	*Malus domestica* Borkh.	Rosaceae
Apricot	*Prunus armeniaca* L.	Rosaceae
Aprium	*Prunus armeniaca* L. × *P. salicina* Lindl.	Rosaceae
Arabica coffee	*Coffea arabica* L.	Rubiaceae

doi:10.1300/5547_31

Common Name	Scientific Name	Family
Araca-boi (Araza)	*Eugenia stipitata* McVaugh	Myrtaceae
Arctic berry	*Rubus arcticus* L. or *R. stellatus* Sm.	Rosaceae
Asian pear	*Pyrus serotina* L. or *P. pyrifolia* (Burm. f.) Nak.	Rosaceae
Atemoya	*Annona cherimola* × *A. squamosa*	Annonaceae
Avocado	*Persea americana* Mill.	Lauraceae
Babaco	*Carica pentagona* Hielb.	Caricaceae
Bachang (or gray) mango	*Mangifera foetida* Lour.	Anacardiaceae
Bacuri (Bacur)	*Platonia insignis* Mart.	Clusiaceae
Bael fruit	*Aegle marmelos* Correa	Rutaceae
Banana	*Musa acuminata* Colla or *Musa* spp.	Musaceae
Banana passionfruit	*Passiflora mollissima* Bailey	Passifloraceae
Barbados cherry	*Malpighia glabra* L.	Malpighiaceae
Barbados gooseberry	*Pereskia aculeata* Mill.	Cactaceae
Batoko plum	*Flacourtia inermis* Roxb.	Flacourtiaceae
Beach plum	*Prunus maritima* Marsh.	Rosaceae
Beach strawberry	*Fragaria chiloensis* Duch.	Rosaceae
Bearberry	*Arctostaphylos uva-ursi* L. Spreng.	Ericaceae
Bearss lime	*Citrus latifolia* Tan.	Rutaceae
Bell-apple	*Passiflora laurifolia* L.	Passifloraceae
Ber	*Zizyphus jujuba* Mill.	Rhamnaceae
Bergamot	*Citrus aurantium* var. *bergamia* Wight & Arn.	Rutaceae
Betel nut (palm)	*Areca catechu* L.	Arecaceae
Bignay	*Antidesma bunius* Spreng.	Euphorbiaceae
Big-tree plum	*Prunus mexicana* S. Wats.	Rosaceae
Bilberry	*Vaccinium myrtillus* L.	Ericaceae
Bilimbi	*Averrhoa bilimbi* L.	Oxalidaceae
Binjai	*Mangifera caesia* Jack	Anacardiaceae
Bird cherry	*Prunus padus* L.	Rosaceae
Biriba	*Rollinia mucosa* Baill.	Annonaceae
Bitter almond	*Prunus dulcis* var. *amara* (Mill.) D.A. Webb	Rosaceae
Bitter pecan	*Carya aquatica* Nutt.	Juglandaceae
Blackberry	*Rubus* spp. Focke	Rosaceae
Black chokeberry	*Aronia melanocarpa* (Michx.) Elliott	Rosaceae
Black currant	*Ribes nigrum* L.	Saxifragaceae
Black huckleberry	*Gaylussacia baccata* (Wang.) K. Koch	Ericaceae
Black mulberry	*Morus nigra* L.	Moraceae
Black persimmon	*Diospyros texana* Scheele	Ebenaceae
Black raspberry	*Rubus occidentalis* L.	Rosaceae
Black sapote	*Diospyros digyna* Jacq.	Ebenaceae
Black walnut	*Juglans nigra* L., other *Juglans* spp.	Juglandaceae
Blood orange	*Citrus sinensis* (L.) Osb.	Rutaceae
Blueberry	*Vaccinium corymbosum* L., *V. ashei* Reade, *V. angustifolium* Ait.	Ericaceae
Blue honeysuckle	*Lonicera caerulea* L. var. *edulis* Turcz. ex Herder	Caprifoliaceae
Blue(-crown) passionflower	*Passiflora caerulea* L.	Passifloraceae
Box huckleberry	*Gaylussacia brachycera* (Michx.) A. Gray	Ericaceae
Boysenberry	Loganberry × unknown blackberry	Rosaceae
Brazilian cherry	*Eugenia uniflora* L.	Myrtaceae
Brazilian guava	*Psidium guineense* Sw.	Myrtaceae

Common Name	Scientific Name	Family
Brazil nut	*Bertholletia excelsa* Hump. & Bonpl.	Lecythidaceae
Breadfruit (breadnut)	*Artocarpus altilis* Fosb.	Moraceae
Buffalo currant	*Ribes odoratum* H.L. Wendl.	Saxifragaceae
Bullace plum	*Prunus insititia* L.	Rosaceae
Bullock's heart	*Annona reticulata* L.	Annonaceae
Bush mango	*Irvingia gabonensis* (Aubrey-Lecomte ex O'Rorke) Baill.	Irvingiaceae
Butternut	*Juglans cineria* L.	Juglandaceae
Cacao	*Theobroma cacao* L.	Sterculiaceae
Calamondin (Calamansi)	× *Citrofortunella mitis* J. Ingram & H.E. Moore	Rutaceae
California blackberry	*Rubus ursinus* Cham. & Schlect	Rosaceae
Camu-camu	*Myrciaria dubia* McVaugh	Myrtaceae
Canada blueberry	*Vaccinium myrtilloides* Michx.	Ericaceae
Canada plum	*Prunus nigra* Ait.	Rosaceae
Canistel	*Pouteria campechiana* Baehni	Rutaceae
Cape gooseberry	*Physalis peruviana* L.	Solanaceae
Capuacu	*Theobroma grandiflorum* K. Schum.	Sterculiaceae
Capulin cherry	*Prunus serotina* Ehrh. ssp. *capuli* (Cav.) McVaugh	Rosaceae
Carambola	*Averrhoa carambola* L.	Oxalidaceae
Cardamon	*Elettaria cardamomum* L.	Zingiberaceae
Carissa	*Carissa macrocarpa* A. DC.	Apocynaceae
Carpathian walnut	*Juglans regia* L.	Juglandaceae
Cas	*Psidium friedrichsthalianum* Ndz.	Myrtaceae
Cashew (nut and apple)	*Anacardium occidentale* L.	Anacardiaceae
Cattley guava	*Psidium cattleianum* Sabine	Myrtaceae
Cayenne cherry	*Eugenia uniflora* L.	Myrtaceae
Ceriman	*Monstera deliciosa* Liebm.	Araceae
Champedak	*Artocarpus integer* Merr.	Moraceae
Che	*Maclura tricuspidata* Carriere	Moraceae
Cherimoya	*Annona cherimola* Mill.	Annonaceae
Cherry-of-the-Rio-Grande	*Eugenia involucrata* DC.	Myrtaceae
Chickasaw plum	*Prunus angustifolia* Marsh.	Rosaceae
Chico sapote (Chicle)	*Manilkara zapota* (L.) P. Royen	Sapotaceae
Chilean myrtle	*Luma apiculata* (DC.) Burret	Myrtaceae
Chilean strawberry	*Fragaria chiloensis* Duch.	Rosaceae
Chinese cherry	*Prunus pseudocerasus* Lindl.	Rosaceae
Chinese chestnut	*Castanea mollissima* Bl.	Fagaceae
Chinese mulberry	*Maclura tricuspidata* Carr.	Moraceae
Chinese walnut	*Juglans cathayensis* Dode	Juglandaceae
Chocolate pudding fruit	*Diospyros digyna* Jacq.	Ebenaceae
Chocolate tree	*Theobroma cacao* L.	Sterculiaceae
Citrange	*Poncirus trifoliata* (L.) Raf.× *Citrus sinensis* (L.) Osb.	Rutaceae
Citron	*Citrus medica* L.	Rutaceae
Citrumelo	*Poncirus trifoliata* (L.) Raf. × *Citrus paradisi* Macfad.	Rutaceae
Cloudberry	*Rubus chamaemorus* L.	Rosaceae
Cocoa	*Theobroma cacao* L.	Sterculiaceae
Cocona	*Solanum sessifolium* Dunal.	Solanaceae
Coconut	*Cocos nucifera* L.	Arecaceae
Coco plum	*Chrysobalanus icaco* L.	Chrysobalanaceae

Common Name	Scientific Name	Family
Coffee	*Coffea arabica* L. or *C. canephora* Pierre ex Froehner	Rubiaceae
Cola nut	*Cola nitida* (Vent.) Schott & Endl.	Sterculiaceae
Conch-apple	*Passiflora maliformis* L.	Passifloraceae
Concord grape	*Vitis labrusca* L.	Vitaceae
Congo coffee	*Coffea congensis* A. Froehner	Rubiaceae
Cornelian cherry	*Cornus mas* L.	Cornaceae
Costa Rican guava	*Psidium friedrichsthalianum* Ndz.	Myrtaceae
Cowberry	*Vaccinium vitis idaea* L.	Ericaceae
Crabapple	*Malus* spp. Mill.	Rosaceae
Cranberry	*Vaccinium macrocarpon* Ait.	Ericaceae
Cupuassu	*Theobroma grandiflorum* K. Schum.	Sterculiaceae
Curuba de indio	*Passiflora mixta* L.	Passifloraceae
Custard apple	*Annona reticulata* L.	Annonaceae
Cut-leaf blackberry	*Rubus lacinatus* Willd.	Rosaceae
Daidai	*Citrus aurantium* L.	Rutaceae
Damson (plum)	*Prunus insititia* L.	Rosaceae
Dangleberry	*Gaylussacia frondosa* Torr. & A. Gray	Ericaceae
Date	*Phoenix dactylifera* L.	Arecaceae
Dewberry (American)	*Rubus flagellaris* Willd.	Rosaceae
Dewberry (European)	*Rubus caesius* L.	Rosaceae
Dewberry (Southern)	*Rubus trivialis* Michx. or *R. hispidus* L. or *R. canadensis* L.	Rosaceae
Downy cherry	*Prunus tomentosa* Thunb.	Rosaceae
Downy currant	*Ribes spicatum* E. Robson	Saxifragaceae
Dragon fruit	*Hylocereus undatus* Britt. & Rose	Cactaceae
Duke cherry	*Prunus avium* L. × *P. cerasus* L. or *Prunus* ×*gondouinii* Rehd.	Rosaceae
Durian	*Durio zibethinus* L.	Bombacaceae
Dwarf huckleberry	*Gaylussacia dumosa* Torr. & A. Gray	Ericaceae
Eastern black walnut	*Juglans nigra* L.	Juglandaceae
Egg fruit	*Pouteria campechiana* Baehni	Rutaceae
Elderberry (American)	*Sambucus canadensis* L.	Caprifoliaceae
Emblic	*Phyllanthus emblica* L.	Euphorbiaceae
English walnut	*Juglans regia* L.	Juglandaceae
Ensete	*Ensete ventricosum* (Welw.) Cheeseman	Musaceae
European chestnut	*Castanea sativa* Mill.	Fagaceae
European grape	*Vitis vinifera* L.	Vitaceae
European mountain ash	*Sorbus aucuparia* L.	Rosaceae
European pear	*Pyrus communis* L.	Rosaceae
European plum	*Prunus domestica* L.	Rosaceae
European strawberry	*Fragaria vesca* L. ssp. *vesca*	Rosaceae
Evergreen blackberry	*Rubus lacinatus* Willd.	Rosaceae
Feijoa	*Feijoa sellowiana* Berg.	Myrtaceae
Fig	*Ficus carica* L.	Moraceae
Filbert	*Corylus avellana* L.	Betulaceae
Flatwoods plum	*Prunus umbellata* Ell.	Rosaceae
Flowering quince (Chinese)	*Chaenomeles speciosa* (Sweet) Nak.	Rosaceae
Flowering quince (Japanese)	*Chaenomeles japonica* (Thunb.) Lindl. & Spach.	Rosaceae
Fox grape	*Vitis labrusca* L.	Vitaceae
Frost grape	*Vitis vulpina* L.	Vitaceae

Common Name	Scientific Name	Family
Genipap	*Genipa americana* L.	Rubiaceae
Giant Colombian blackberry	*Rubus nubigenus* Kunth.	Rosaceae
Giant granadilla	*Passiflora quadrangularis* L.	Passifloraceae
Ginkgo	*Ginkgo biloba* L.	Ginkgoaceae
Golden berry	*Physalis peruviana* L.	Solanaceae
Golden currant	*Ribes aureum* Pursh.	Saxifragaceae
Gooseberry (American)	*Ribes hirtellum* Michx.	Saxifragaceae
Gooseberry (European)	*Ribes uva-crispa* L. or *Grossularia reclinata* (L.) Mill.	Saxifragaceae
Goumi	*Eleagnus multiflora* Thunb.	Elaeagnaceae
Governor's plum	*Flacourtia indica* Merr.	Flacourtiaceae
Grape	*Vitis vinifera* L., *V. labrusca* L., *V. rotundifolia* Michx.	Vitaceae
Grapefruit	*Citrus paradisi* Macfad.	Rutaceae
Greengage plum	*Prunus domestica* L.	Rosaceae
Ground cherry	*Physalis pruinosa* L.	Solanaceae
Ground cherry	*Prunus fruticosa* Pall.	Rosaceae
Grumichama	*Eugenia brasiliensis* Lam. or *Eugenia dombeyi* Skeels	Myrtaceae
Guanabana	*Annona muricata* L.	Annonaceae
Guava	*Psidium guajava* L.	Myrtaceae
Hardy kiwi	*Actinidia arguta* Planch. ex Miq.	Actinidiaceae
Hautboy strawberry	*Fragaria moschata* Duch.	Rosaceae
Hazelnut	*Corylus avellana* L.	Betulaceae
Heartnut	*Juglans ailantifolia* Carr. var. *cordiformis*	Juglandaceae
Hican	Pecan × *Carya* spp. Nutt.	Juglandaceae
Highbush blackberry	*Rubus allegheniensis* Porter	Rosaceae
Highbush blueberry	*Vaccinium corymbosum* L.	Ericaceae
Highbush cranberry	*Viburnum trilobum* Marsh.	Caprifoliaceae
Himalayan blackberry	*Rubus procerus* PJ. Mull. or *R. rugosus* Sm.	Rosaceae
Hog plum	*Spondias* spp. L.	Anacardiaceae
Horned melon	*Cucumis metuliferus* E. Mey. ex Naudin	Cucurbitaceae
Hortulan plum	*Prunus hortulana* L.H. Bailey	Rosaceae
Huckleberry	*Gaylussacia* spp.	Ericaceae
Husk tomato	*Physalis philadelphica* Lam.	Solanaceae
Icaco	*Chrysobalanus icaco* L.	Chrysobalanaceae
Ichang papeda	*Citrus ichangensis* Swing.	Rutaceae
Ilama	*Annona diversifolia* Safford	Annonaceae
Imbe	*Garcinia livingstonei* T. Anderson	Clusiaceae
Imbu	*Spondias tuberosa* Arruda	Anacardiaceae
Indian fig	*Opuntia ficus-indica* (L.) Mill.	Cactaceae
Indian jujube	*Zizyphus mauritiana* Lam.	Rhamnaceae
Inga	*Inga edulis* C. Mart.	Fabaceae
Jaboticaba	*Myrciaria cauliflora* Berg.	Myrtaceae
Jackfruit	*Artocarpus heterophyllus* Lam.	Moraceae
Jamaica cherry	*Muntingia calabura* L.	Elaeocarpaceae
Jambolan	*Syzygium cumini* Skeels	Myrtaceae
Japanese apricot	*Prunus mume* Sieb. & Zucc.	Rosaceae
Japanese chestnut	*Castanea crenata* Sieb. & Zucc.	Fagaceae
Japanese pear	*Pyrus serotina* L. or *P. pyrifolia* (Burm. f.) Nak.	Rosaceae
Japanese plum	*Prunus salicina* Lindl.	Rosaceae

Common Name	Scientific Name	Family
Japanese walnut	*Juglans ailantifolia* Carr.	Juglandaceae
Java apple	*Syzygium samarangense* Merr. & Perry	Myrtaceae
Jelly palm	*Butia capitata* (Mart.) Becc.	Arecaceae
Jostaberry	*Ribes* ×*nidigrolaria* Rud. Bauer & A. Bauer	Saxifragaceae
Jujube (Chinese)	*Zizyphus jujuba* Mill.	Rhamnaceae
July grape	*Vitis rupestris* Scheele	Vitaceae
Juneberry	*Amelanchier alnifolia* Nutt.	Rosaceae
Karanda	*Carissa congesta* Wight	Apocynaceae
Kei apple	*Dovyalis caffra* (Hook f. & Harv.) Warb.	Flacourtiaceae
Key lime	*Citrus aurantifolia* Swing.	Rutaceae
King's acre berry	*Rubus rusticanus* Mercier × *R. idaeus* L.	Rosaceae
Kinnikinnick	*Arctostaphylos uva-ursi* L. Spreng.	Ericaceae
Kiwano (melon)	*Cucumis metuliferus* E. Mey. ex Naudin	Cucurbitaceae
Kiwi (kiwifruit)	*Actinidia deliciosa* C.F. Liang & A.R. Ferguson	Actinidiaceae
Kola nut	*Cola nitida* (Vent.) Schott & Endl.	Sterculiaceae
Kolomikta kiwi	*Actinidia kolomikta* Maxim.	Actinidiaceae
Kumquat	*Fortunella* spp. Swing.	Rutaceae
Kuwini	*Mangifera odorata* Griff.	Anacardiaceae
Langsat (Lanzone)	*Lansium domesticum* Corr.	Meliaceae
Leafy-flowered blackberry	*Rubus frondosus* Bigel	Rosaceae
Lemon	*Citrus limon* Burm. f.	Rutaceae
Liberian coffee	*Coffea liberica* Bull.	Rubiaceae
Lignonberry	*Vaccinium vitis idaea* L.	Ericaceae
Lime	*Citrus aurantifolia* Swing.	Rutaceae
Limequat	*Citrus aurantifolia* Swing. × *Fortunella* spp. Swing.	Rutaceae
Loganberry	*Rubus vitifolius* Cham. & Schlect × *R. idaeus* L.	Rosaceae
Longan	*Dimocarpus longana* Lour.	Sapindaceae
Loquat	*Eriobotrya japonica* Lindl.	Rosaceae
Lowbush blueberry	*Vaccinium angustifolium* Ait., *V. myrtilloides* Michx.	Ericaceae
Lucuma (Lucumo)	*Pouteria lucuma* Ruiz & Pavon	Sapotaceae
Lychee (Litchi)	*Litchi chinensis* Sonn.	Sapindaceae
Mabolo	*Diospyros blancoi* A. DC.	Ebenaceae
Macadamia (rough shell)	*Macadamia tetraphylla* L. Johnson	Proteaceae
Macadamia (smooth shell)	*Macadamia integrifolia* Maiden & Betche	Proteaceae
Mace	*Myristica fragrans* Houtt.	Myristicaceae
Madruno	*Garcinia madruno* (Kunth.) Hammel	Clusiaceae
Mahaleb cherry	*Prunus mahaleb* L.	Rosaceae
Mahdi (-berry)	*Rubus rusticanus* Mercier × *R. idaeus* L.	Rosaceae
Malay apple	*Syzigium malaccense* Merr. & Perry	Myrtaceae
Mamey apple	*Mammea americana* L.	Clusiaceae
Mamey sapote	*Pouteria sapota* (Jacq.) H.E. Moore & Stearn	Sapotaceae
Mamoncillo	*Meliococcus bijigatus* Jacq.	Sapindaceae
Manchurian apricot	*Prunus mandshurica* (Maxim.) Koehne	Rosaceae
Manchurian walnut	*Juglans mandshurica* Maxim.	Juglandaceae
Mandarin	*Citrus reticulata* Blanco	Rutaceae
Mangaba	*Hancornia speciosa* Gomez	Apocynaceae
Mango	*Mangifera indica* L.	Anacardiaceae

Common Name	Scientific Name	Family
Mangosteen	*Garcinia mangostana* L.	Clusiaceae
Marula	*Sclereocarya birrea* (A. Rich.) Hochst. ssp. *caffra*	Anacardiaceae
Mayhaw	*Crataegus aestivalis* (Walter) Torr. & A. Gray, *C. rufula* Sarg.	Rosaceae
Maypop	*Passiflora incarnata* L.	Passifloraceae
Mediterranean mandarin	*Citrus deliciosa* Tan.	Rutaceae
Medlar	*Mespilus germanica* L.	Rosaceae
Mexican husk tomato	*Physalis ixocarpa* Brot.	Solanaceae
Mexican plum	*Prunus mexicana* S. Wats.	Rosaceae
Mirabelle	*Prunus insititia* L.	Rosaceae
Miracle fruit (miraculous berry)	*Synsepalum dulcifum* Daniell & S. Bell	Sapotaceae
Mission prickly pear	*Opuntia megacantha* Salm-Dyck	Cactaceae
Mocambo	*Theobroma bicolor* Bonpl.	Sterculiaceae
Mossberry	*Vaccinium oxycoccus* L.	Ericaceae
Mountain apple	*Syzigium malaccense* Merr. & Perry	Myrtaceae
Mountain cranberry	*Vaccinium vitis idaea* L.	Ericaceae
Mountain grape	*Vitis rupestris* Scheele	Vitaceae
Mountain papaya	*Carica candamarcensis* Hook. f. or *C. pubescens* Koch	Caricaceae
Mountain soursop	*Annona montana* Macfad.	Annonaceae
Mulberry	*Morus* spp. L.	Moraceae
Muscadine grape	*Vitis rotundifolia* Michx.	Vitaceae
Musky strawberry	*Fragaria moschata* Duch.	Rosaceae
Myrobalan plum	*Prunus cerasifera* Ehrh.	Rosaceae
Myrtle (berry)	*Myrtus communis* L.	Myrtaceae
Mysore raspberry	*Rubus niveus* Thunb.	Rosaceae
Nance	*Byrsonima crassifolia* (L.) Kunth.	Malpighiaceae
Nanking cherry	*Prunus tomentosa* Thunb.	Rosaceae
Naranjilla	*Solanum quitoense* Lam.	Solanaceae
Naseberry	*Manilkara zapota* (L.) P. Royen	Sapotaceae
Nashi	*Pyrus serotina* L. or *P. pyrifolia* (Burm. f.) Nak.	Rosaceae
Natal plum	*Carissa macrocarpa* A. DC.	Apocynaceae
Navel orange	*Citrus sinensis* (L.) Osb.	Rutaceae
Nectarberry	*Rubus* spp. L.	Rosaceae
Nectarine	*Prunus persica* (L.) Batsch.	Rosaceae
Nessberry	*Rubus rubristeus* Rydb. × *R. idaeus* L.	Rosaceae
Noni	*Morinda citrifolia* L.	Rubiaceae
Nutmeg	*Myristica fragrans* Houtt.	Myristicaceae
Oil palm	*Elaeis guineensis* Jacq.	Arecaceae
Olive	*Olea europaea* L.	Oleaceae
Ollalieberry	*Rubus* spp. L.	Rosaceae
Orange	*Citrus sinensis* (L.) Osb.	Rutaceae
Orangelo	*Citrus sinensis* (L.) Osb. × *C. paradisi* Macfad.	Rutaceae
Oriental pear	*Pyrus serotina* L. or *P. pyrifolia* (Burm. f.) Nak.	Rosaceae
Otaheite apple	*Spondias dulcis* Parkins or *S. cytherea* Sonn.	Anacardiaceae
Otaheite gooseberry	*Phyllanthus acidus* Skeels	Euphorbiaceae
Otaheite orange	*Citrus limonia* Osb.	Rutaceae

Common name	Scientific name	Family
Pacific plum	*Prunus subcordata* Benth.	Rosaceae
Palestine apple	*Solanum incanum* L.	Solanaceae
Palestine sweet lime	*Citrus limetta* Risso	Rutaceae
Papaya	*Carica papaya* L.	Caricaceae
Para nut	*Bertholletia excelsa* Hump. & Bonpl.	Lecythidaceae
Passionfruit	*Passiflora edulis* Sims	Passifloraceae
Pawpaw	*Asimina triloba* Dunal.	Annonaceae
Peach	*Prunus persica* (L.) Batsch.	Rosaceae
Peach palm	*Bactris gaesipes* Kunth.	Arecaceae
Pear	*Pyrus communis* L.	Rosaceae
Pecan	*Carya illinoensis* (Wang.) K. Koch	Juglandaceae
Pejibaye	*Bactris gaesipes* Kunth.	Arecaceae
Pepper	*Piper nigrum* L.	Piperaceae
Perry pear	*Pyrus nivalis* Jacq.	Rosaceae
Persian walnut	*Juglans regia* L.	Juglandaceae
Persimmon	*Diospyros kaki* L.	Ebenaceae
Phalsa	*Grewia subinaequalis* DC.	Tiliaceae
Phenomenal berry	*Rubus ursinus* Cham. & Schlect × *R. idaeus* L.	Rosaceae
Pigeon grape	*Vitis aestivalis* Michx.	Vitaceae
Pimento (pimenta)	*Pimenta dioica* Merr.	Myrtaceae
Pindo palm	*Butia capitata* (Mart.) Becc.	Arecaceae
Pineapple	*Ananas comosus* Merr.	Bromeliaceae
Pineapple guava	*Feijoa sellowiana* Berg.	Myrtaceae
Pistachio	*Pistacia vera* L.	Anacardiaceae
Pitanga	*Eugenia uniflora* L.	Myrtaceae
Pithaya	*Hylocereus undatus* Britt. & Rose	Cactaceae
Pitomba	*Eugenia luschnanthiana* Klotzsch ex O. Berg.	Myrtaceae
Plantain	*Musa ×paradisiaca* L.	Musaceae
Plum	*Prunus domestica* L. or *P. salicina* Lindl.	Rosaceae
Plumcot	*Prunus armeniaca* L. × *P. salicina* Lindl.	Rosaceae
Pluot	*Prunus armeniaca* L. × *P. salicina* Lindl.	Rosaceae
Pomegranate	*Punica granatum* L.	Punicaceae
Pond apple	*Annona glabra* L.	Annonaceae
Prickly pear	*Opuntia ficus-indica* (L.) Mill.	Cactaceae
Prune	*Prunus domestica* L.	Rosaceae
Pulasan	*Nephelium mutabile* Blume	Sapindaceae
Pummelo (pommelo)	*Citrus maxima* Merr. or *C. grandis* (L.) Osb.	Rutaceae
Purple apricot	*Prunus ×dasycarpa* Ehrh.	Rosaceae
Purple chokeberry	*Aronia ×prunifolia* (Marsh.) Rehd.	Rosaceae
Purple mombin	*Spondias purpurea* L.	Anacardiaceae
Purple passionfruit	*Passiflora edulis* Sims	Passifloraceae
Quince	*Cydonia oblonga* Mill.	Rosaceae
Rabbiteye blueberry	*Vaccinium ashei* Reade	Ericaceae
Rambutan	*Nephelium lappaceum* L.	Sapindaceae
Ramontchi	*Flacourtia indica* Merr.	Flacourtiaceae
Rangpur lime	*Citrus limonia* Osb.	Rutaceae
Red chokeberry	*Aronia arbutifolia* (L.) Pers.	Rosaceae

Common Name	Scientific Name	Family
Red currant	*Ribes sativum* Syme or *R. rubrum* L.	Saxifragaceae
Red mombin	*Spondias purpurea* L.	Anacardiaceae
Red raspberry (American)	*Rubus strigosus* L. or *Rubus idaeus* L. var. *strigosus* (Michx.) Maxim.	Rosaceae
Red raspberry (European)	*Rubus idaeus* L.	Rosaceae
Riverbank grape	*Vitis riparia* Michx.	Vitaceae
Rock grape	*Vitis rupestris* Scheele	Vitaceae
Robusta coffee	*Coffea canephora* Pierre ex Froehner	Rubiaceae
Rose apple	*Syzygium jambos* Alston	Myrtaceae
Rough lemon	*Citrus jambhiri* Lush.	Rutaceae
Rough-shell macadamia	*Macadamia tetraphylla* L. Johnson	Proteaceae
Rumberry	*Myrciaria floribunda* Berg.	Myrtaceae
Salal	*Gaultheria shallon* Pursh.	Ericaceae
Salmonberry	*Rubus spectabilis* Pursh.	Rosaceae
Sand blackberry	*Rubus cuneifolius* Pursh.	Rosaceae
Sand cherry	*Prunus pumila* L.	Rosaceae
Santol	*Sandorium koetjape* Merr.	Meliaceae
Sapodilla	*Manilkara zapota* (L.) P. Royen	Sapotaceae
Sapote	*Pouteria sapota* (Jacq.) H.E. Moore & Stern	Sapotaceae
Sarvis (-tree)	*Amelanchier alnifolia* Nutt.	Rosaceae
Saskatoon berry	*Amelanchier alnifolia* Nutt.	Rosaceae
Satsuma	*Citrus unshiu* Marc.	Rutaceae
Scarlet strawberry	*Fragaria virginiana* Duch.	Rosaceae
Seaberry	*Hippophae rhamnoides* L.	Elaeagnaceae
Sea buckthorn	*Hippophae rhamnoides* L.	Elaeagnaceae
Sea grape	*Coccoloba uvifera* L.	Polygonaceae
Serviceberry	*Amelanchier alnifolia* Nutt.	Rosaceae
Seville orange	*Citrus aurantium* L.	Rutaceae
Shadblow	*Amelanchier alnifolia* Nutt.	Rosaceae
Shaddock	*Citrus maxima* Merr. or *C. grandis* (L.) Osb.	Rutaceae
Shagbark hickory	*Carya ovata* (Mill.) K. Koch	Juglandaceae
Shellbark hickory	*Carya lacinosa* (F. Michx.) Loudon	Juglandaceae
Sierra plum	*Prunus subcordata* Benth.	Rosaceae
Sloe	*Prunus spinosa* L.	Rosaceae
Small cranberry	*Vaccinium oxycoccus* L.	Ericaceae
Smooth-shell macadamia	*Macadamia integrifolia* Maiden & Betche	Proteaceae
Snot apple	*Rollinia mucosa* Baill.	Annonaceae
Snow pear	*Pyrus nivalis* Jacq.	Rosaceae
Sodom apple	*Solanum incanum* L.	Solanaceae
Soncoya	*Annona purpurea* Moc. & Sesse	Annonaceae
Sour cherry	*Prunus cerasus* L.	Rosaceae
Sour orange	*Citrus aurantium* L.	Rutaceae
Soursop	*Annona muricata* L.	Annonaceae
Sourtop blueberry	*Vaccinium myrtilloides* Michx.	Ericaceae
Southern highbush blueberry	*Vaccinium corymbosum* L. × *Vaccinium* spp.	Ericaceae
Spanish lime	*Meliococcus bijigatus* Jacq.	Sapindaceae
Spanish plum	*Spondias purpurea* L.	Anacardiaceae
Star anise tree	*Illicium verum* Hook. f.	Illiciaceae

Common Name	Scientific Name	Family
Star apple	*Chrysophyllum cainito* L.	Sapotaceae
Star fruit	*Averrhoa carambola* L.	Oxalidaceae
Strawberry	*Fragaria* ×*ananassa* Duch.	Rosaceae
Strawberry pear	*Hylocereus undatus* Britt. & Rose	Cactaceae
Strawberry tomato	*Physalis pruinosa* L.	Solanaceae
St. Julien plum	*Prunus insititia* L.	Rosaceae
Sugar apple	*Annona squamosa* L.	Annonaceae
Summer grape	*Vitis aestivalis* Michx.	Vitaceae
Sunberry	*Rubus ursinus* Cham. & Schlect × *R. idaeus* L.	Rosaceae
Surinam cherry	*Eugenia uniflora* L.	Myrtaceae
Sweet calabash	*Passiflora maliformis* L.	Passifloraceae
Sweet cherry	*Prunus avium* L.	Rosaceae
Sweet granadilla	*Passiflora ligularis* Juss.	Passifloraceae
Sweet lemon	*Citrus limetta* Risso	Rutaceae
Sweet lime	*Citrus limettioides* Tan.	Rutaceae
Sweet orange	*Citrus sinensis* (L.) Osb.	Rutaceae
Sweetsop	*Annona squamosa* L.	Annonaceae
Tachibana orange	*Citrus tachibana* Tan.	Rutaceae
Tahiti lime	*Citrus aurantifolia* Swing.	Rutaceae
Tall blackberry	*Rubus argutus* Link	Rosaceae
Tamarillo	*Cyphomandra betacea* Senot.	Solanaceae
Tamarind	*Tamarindus indica* L.	Fabaceae
Tangelo	*Citrus reticulata* Blanco × *C. paradisi* Macfad.	Rutaceae
Tangerine	*Citrus reticulata* Blanco	Rutaceae
Tangor	*Citrus reticulata* Blanco × *C. sinensis* (L.) Osb.	Rutaceae
Tayberry	*Rubus vitifolius* Cham. & Schlect × *R. idaeus* L.	Rosaceae
Thimbleberry	*Rubus parviflorus* Nutt.	Rosaceae
Tree tomato	*Cyphomandra betacea* Senot.	Solanaceae
Trifoliate orange	*Poncirus trifoliata* (L.) Raf.	Rutaceae
Tummelberry	*Rubus vitifolius* Cham. & Schlect × *R. idaeus* L.	Rosaceae
Tuna cardona	*Opuntia streptacantha* Lem.	Cactaceae
Turkish hazel tree	*Corylus colurna* L.	Betulaceae
Ungurahui palm	*Oenocarpus bataua* Mart. var. *bataua*	Arecaceae
Ussuri pear	*Pyrus ussuriensis* Maxim.	Rosaceae
Veitchberry	*Rubus rusticanus* Mercier × *R. idaeus* L.	Rosaceae
Velvet apple	*Diospyros blancoi* A. DC.	Ebenaceae
Virginian strawberry	*Fragaria virginiana* Duch.	Rosaceae
Volkamer lemon	*Citrus limonia* Osb.	Rutaceae
Walnut	*Juglans regia* L., other *Juglans* spp.	Juglandaceae
Wampee	*Clausena lansium* Skeels	Rutaceae
Water apple	*Syzygium aqueum* Alst.	Myrtaceae
Water lemon	*Passiflora laurifolia* L.	Passifloraceae
Wax jambu	*Syzygium samarangense* Merr. & Perry	Myrtaceae
Western sand cherry	*Prunus besseyi* Bailey	Rosaceae
West Indian lime	*Citrus aurantifolia* Swing.	Rutaceae
White currant	*Ribes sativum* Syme or *R. rubrum* L.	Saxifragaceae
White mulberry	*Morus alba* L.	Moraceae
White sapote	*Casimiroa edulis* Llave & Lex.	Rutaceae
Whortleberry	*Vaccinium myrtillus* L.	Ericaceae

Common Name	Scientific Name	Family
Wild goose plum	*Prunus munsoniana* W. Wright & Hedrick	Rosaceae
Wild papaya	*Jacaratia spinosa* (Aubl.) A. DC.	Caricaceae
Wild service tree	*Sorbus torminalis* (L.) Crantz	Rosaceae
Wild soursop	*Annona montana* Macfad.	Annonaceae
Willowleaf mandarin	*Citrus deliciosa* Tan.	Rutaceae
Wineberry	*Rubus phoenicolasius* Maxim.	Rosaceae
Wine grape	*Vitis vinifera* L.	Vitaceae
Winter grape	*Vitis aestivalis* Michx.	Vitaceae
Wolfberry (-vine)	*Lycium chinense* Mill.	Solanaceae
Wood apple	*Feronia limonia* Swing.	Rutaceae
Wood (-land) strawberry	*Fragaria vesca* L. ssp. *americana*	Rosaceae
Yellow mombin	*Spondias mombin* L.	Anacardiaceae
Yellow passionfruit	*Passiflora edulis* f. *flavicarpa* Deg.	Passifloraceae
Yellow sapote	*Pouteria campechiana* Baehni	Sapotaceae
Youngberry	*Rubus* spp. L.	Rosaceae
Yuzu	*Citrus junos* Sieb. ex Tan.	Rutaceae

Appendix B

Useful Conversion Factors

Imperial	Metric
1 acre	0.40 hectare
1 acre-foot	1,233,346 liters
1 bushel (50 pounds, 32 quarts)	22.7 kilograms, 30.4 liters
1 bushel per acre	56.1 kilograms per hectare
1 cubic inch	16.39 milliliters
1 cubic foot	28.32 liters
1 foot	0.305 meter
1 gallon	3.78 liters
1 inch	2.54 centimeters
1 mile	1.61 kilometers
1 ounce (weight)	28.3 grams
1 ounce (volume)	29.6 milliliters
1 pound	0.45 kilogram
1 pound per acre	1.12 kilograms per hectare
1 quart	0.95 liter
1 square foot	929 square centimeters
1 square inch	6.45 square centimeters
1 square yard	0.84 square meter
1 square mile	2.59 square kilometers, 259 hectares
1 ton (short ton)	0.91 metric ton
1 ton per acre	2.24 metric tons per hectare
1 yard	0.91 meter

Introduction to Fruit Crops

doi:10.1300/5547_32

Appendix C

List of Illustrations

PLATES

Introduction to Fruit Crops

doi:10.1300/5547_33

FIGURES

EXHIBIT AND TABLES

Glossary

abscission: Separation of an organ from a plant.

accessory: A term given mostly to aggregate fruits in which the conspicuous and often edible portion is nonovarian in origin. Example: strawberry, where the receptacle is the fleshy, edible portion and the true fruit an achene.

achene: A dry, indehiscent, one- or two-seeded fruit, with generally thin pericarp loosely attached to the seed(s). Fruit is generally small and comes from a unicarpellate ovary. Example: sunflower, where the thin, black pericarp (shell) surrounds the single seed.

acre: Unit of land area equal to 43,560 square feet.

adventitious roots, buds: Organs that arise from tissues unrelated to the normal hypocotyl- and radicle-derived meristems; i.e., roots or buds arising from the vascular cambium, cortex, phloem parenchyma, or wound callus.

aggregate: A fruit derived from two or more ovaries contained within a single flower: may contain nonovarian tissue. Example: blackberry and raspberry, where each fruit is an aggregate of drupelets.

air layering: A vegetative propagation technique that involves rooting of a stem cutting while still attached to the parent plant. An area of bark is often removed from the stem where roots are desired, treated with rooting hormone, and covered with moist potting media to encourage rooting. Synonym: marcottage.

albedo: The mesocarp of a citrus fruit; the white, spongy tissue lying between the colored peel and the juice sacs.

alternate bearing: High production of fruit one year, followed by low production the next. Common in nut trees. Synonyms: biennial bearing, irregular bearing.

androecium: The stamens of a flower, collectively.

anthelmintic: A medicinal property acting against parasitic worms.

anther: The swollen, apical, pollen-bearing section of the stamen.

anthesis: Time of flower expansion when pollination takes place.

anthocyanin: A class of water-soluble pigments responsible for the red, purple, and blue coloration of flowers and fruits.

apomixis, apomictic seed: Reproduction of a plant through a seed, wherein the embryo has arisen clonally from nucellar or integument tissue and is genetically identical to the parent plant. Synonym: agamospermy.

asexual: Without sex, as in vegetative reproduction or propagation.

Introduction to Fruit Crops

doi:10.1300/5547_34

auxin: A class of plant growth regulators used in fruit crops; primary uses are to root cuttings, chemically thin fruits, and prevent preharvest drop of fruits.

axil: The angle formed at the point of insertion of a leaf to a stem.

axillary bud: A bud found in a leaf axil.

bacteria: Single-celled prokaryotic organisms, some of which are fruit pathogens.

bearing habit: The position of the flower buds, with respect to the type and age of wood. Example: spur-bearing trees produce flowers (thus fruits) on short, long-lived, lateral branches called spurs.

bench graftage: Grafting technique whereby a union is made at a propagation bench (as opposed to in the field) with special tools or saws, often prior to root formation on the rootstock. Common with grape.

berry: A fleshy, indehiscent, one- to many-seeded fruit with more or less homogeneous texture throughout, derived from a single, superior ovary. Example: grape. Term is often misused. An epigynous berry is the same but derived from an *inferior* ovary (example: blueberry).

bisexual: Having both sexes present. Synonyms: hermaphroditic, perfect.

bloom: Epicuticular wax found on the surface of a plant organ, particularly fruits. Also, used as synonym for flower or anthesis. *See also* FLOWER; ANTHESIS.

Bordeaux mixture: A mixture of copper sulfate and hydrated lime used as a fungicide and bactericide.

bract: A modified leaf structure that subtends a flower or inflorescence.

brambles: A group of fruit crops in the genus *Rubus*. Includes blackberries, raspberries, and hybrids thereof.

Brix (°Brix): A measure of sugar content in fruit juices; 1 °Brix is the same as about 1 percent sugar (weight per unit volume).

Brix-acid ratio: The ratio of sugar-to-acid content of fruit juice when both are expressed as a percent. Synonym: sugar-acid ratio.

Bt: Abbreviation for *Bacillus thuringiensis,* a bacterium used as a biological control agent for the larvae of insects in the order Lepidoptera (moths, butterflies).

bud: An undeveloped vegetative shoot, flower, or inflorescence; borne laterally or terminally on stems.

budding: Means of vegetative propagation whereby the scion is reduced to a single bud, usually axillary, which is inserted into a rootstock. Main types used for fruit trees are T-budding and chip budding.

bur (also burr): A spiny appendage; the involucre of a chestnut fruit.

burrknot: Concentration of preformed root initials on a stem or the trunk; can cause partial girdling and stunting of the tree.

bushel: A unit of measurement of fruit yield, equal to about 50 pounds. Generally, wooden baskets or boxes are used, with a volume of about $\frac{1}{28}$ cubic meter.

callus: Undifferentiated, swollen tissue, often found at the site of a wound.

calyptra: A hood or lid; specifically, the corolla of a grape flower.

calyx: The sepals, collectively.

CAM plant: A plant that utilizes a photosynthetic pathway known as Crassulacean acid metabolism, or CAM for short. Example: pineapple.

cane: An elongated, flexible stem, coming from the ground, as in *Rubus* and blueberry, or from older wood, as in grape.

canopy: The foliage- and fruit-bearing portion of a crop plant.

capsule: A dry, dehiscent, one- to many-seeded fruit from a compound ovary (compound = multiple locules within the ovary). Subcategories are organized by dehiscence, for example, circumscissile, loculicidal, septicidal. Often irregularly shaped. Example: poppy.

carpel: The megasporophyll, or structure enclosing the ovules (seeds). If a simple ovary, then the carpel and ovary are the same structure. If a compound ovary, then it is composed of two or more carpels.

caryopsis: Small, indehiscent, one-seeded fruit, often dry and mealy at maturity, with the pericarp fused to the seedcoat. Synonym: grain. Example: small grains, such as wheat, maize, and barley.

CA storage: *See* CONTROLLED-ATMOSPHERE STORAGE.

catfacing: Injury to fruit, generally from insects, resulting in severe distortion of fruit shape. Occurs when an area of the surface of a young fruit is injured, killing the tissue, but surrounding healthy tissue continues to grow and develop.

catkin: A slender, flexible, pendulous spike inflorescence.

cauliflory (cauliflorous): A term applied to plants that bear flowers and fruits on main stems or trunks, such as cacao, jackfruit, and jaboticaba.

central leader: A tree-training system in which a main central bole (the "leader") extends from the trunk to the top of the tree. At intervals along the central leader, tiers of fruiting scaffolds are trained, with the lowest tier extending the farthest, and the upper tier extending only a few feet, giving an overall shape similar to that of a Christmas tree.

chasmogamous: Flowers that must open before pollination; opposite of cleistogamous. Often seen in cross-pollinating species.

chilling injury: Injury from prolonged exposure to low, nonfreezing temperatures. Commonly affects tropical fruits. Results in discoloration, pitting, and flesh breakdown in susceptible species.

chilling requirement: The amount of time of exposure to cool, nonfreezing temperatures during winter that is necessary to allow normal budbreak and development the following spring. Measured in hours at or below 45°F, from approximately leaf drop in fall throughout the winter.

chimera: A partial mutation of a plant or plant organ in which mutated and normal tissues coexist in distinct patterns: periclinal, mericlinal, sectorial. In periclinal chimeras, a thin layer of mutated tissue completely surrounds nonmutated interior tissues; mericlinal chimeras are similar but the mutated tissue layer only partially surrounds the inner, normal tissue. In sectorial chimeras, the mutation exists from the epidermis to the center of the organ, but only in a portion of the organ.

chlorosis: Yellowing of the foliage.

circumscissile capsule: A dry, dehiscent, one- to many-seeded fruit from a single, compound ovary, opening at the equator, with the top separating like a lid. Synonym: pyxis.

cleistogamous (cleistogamy): Self-pollinating nature of closed flowers; opposite of chasmogamous.

climacteric: A period of fruit maturation characterized by a sudden increase in respiration and ethylene production; often accompanied by fruit softening and color change.

climacteric fruit: A type of fruit that exhibits a sudden increase in respiration and ethylene production during ripening. Examples: apple, banana.

clingstone: Adherence of the flesh (mesocarp) to the pit (endocarp) in a drupe or stone fruit. Also used to denote a group of peach cultivars used for canning that carry this trait.

clonal: Said of a plant derived from some form of vegetative propagation. Example: a clonal rootstock is one produced by layerage, division, or cuttage, but not by seed.

clone: A cultivar or selection of a plant produced by vegetative propagation.

coldhardiness (cold hardy): Minimum temperature tolerance of a plant, usually given in °C or °F; exposure to lower temperatures causes death or severe injury.

compatibility (congeniality): Capable of coexisting; said of a rootstock and scion that unite and form a healthy, long-lived tree.

complete flower: A flower having a calyx (sepals), corolla (petals), androecium (anthers), and gynoecium (pistil).

compound: An organ composed of multiple parts, as with a compound leaf, which has a blade composed of leaflets, or a compound ovary that is divided into two or more carpels.

compound spadix: An inflorescence composed of multiple fleshy spikes; the inflorescence type of many palms.

controlled-atmosphere storage: Subjecting fruit to low oxygen and high carbon dioxide during cold storage to extend postharvest life. Abbreviated as "CA storage."

cordon: A permanent, horizontally trained limb; most commonly used in grapes to denote the major scaffolds.

corm: A shortened, vertically oriented, solid underground stem.

corolla: The petals, collectively.

corymb: An indeterminate inflorescence where flowers are born on a plane or slight arc. The pedicels of lowermost flowers are elongated and those of uppermost flowers shortened, so that all flowers are displayed at about the same height (example: pear). Similar in shape to an umbel but with pedicels arising from different points on the main axis.

cross-compatible: Pollen of one plant is capable of fertilizing ovules of another, genetically distinct plant.

cross-incompatible: Said of genetically distinct plants that cannot fertilize one another.

cross-pollination: Pollen transfer between the anther and stigma of two genetically distinct plants.

cross-sterile: Inability of a plant to produce fruits with viable seed when cross-pollinated by another genetically distinct plant; does not preclude parthenocarpic fruit set and development.

cross-unfruitful: A genetically distinct plant is used as a pollinizer for another plant, and the latter fails to produce a commercial crop.

crown: A compressed stem, such as that of a strawberry plant, or the vegetative apex of a pineapple fruit.

culling (cull): Removal of defective or unmarketable fruits postharvest.

cultivar: A cultivated variety or subspecies of a plant. Often used interchangeably with "variety" in the field.

cutting: Organ isolated from a plant prior to root (stem cutting) or shoot (root cutting) development.

cyme: A determinate inflorescence in which the central flower opens first. Usually rounded, as in apple, or flat.

day-neutral: Lacking a photoperiodic response. Day-neutral strawberry cultivars produce flowers and fruits regardless of daylength.

deciduous: Plant that loses its foliage annually.

dehiscence: Natural splitting or opening of organs (dry fruits, anthers, hulls), causing the contents to be released.

delayed dormant spray: A pesticide application applied to trees when buds have begun to swell, but new tissues are not yet fully exposed.

delayed incompatibility: A situation in which a rootstock and scion appear to be compatible and grow vigorously for several years, after which time the tree declines and may eventually break cleanly at the graft union.

determinate: Term applied to inflorescences having the topmost or central flower appearing and maturing first in the blooming sequence.

dichasial cyme: A cymose inflorescence where the lateral axes fork and contain multiple flowers (example: strawberry). Synonym: dichasium.

dichogamy: Prevention of self-pollination by temporal separation of pollen shed and stigma receptivity. Includes protandry, protogyny, and synchronous protogyny. *See also* PROTANDRY; PROTOGYNY; SYNCHRONOUS PROTOGYNY.

dicot: A plant in the class Dicotyledoneae, characterized by having two seed leaves or cotyledons.

dioecious: When staminate and pistillate flowers are borne on separate plants, as with, for example, date, pistachio, and kiwifruit.

diploid: Having two sets of chromosomes.

division: A form of propagation whereby a complete plant develops while still attached to the parent plant.

dormancy: A temporary suspension of visible growth in organs containing meristems; occurs each winter in temperate fruit trees.

dormant oil: An important spray material for most fruit trees. Emulsifiable oil is mixed with water and applied to trees before buds swell, killing many overwintering pests.

dormant pruning: Pruning during the dormant season.

double budding: Placing a thin sheath of wood between the rootstock and scion bud to serve as a compatibility bridge. Results in a tree with an extremely short interstem.

double fertilization: In angiosperms, the union of one generative nucleus from a pollen grain with the egg nucleus, yielding a $2n$ zygote; *and* the union of the other generative nucleus from a pollen grain with the two polar nuclei, yielding $3n$ endosperm.

drupe: A fleshy, indehiscent, usually one-seeded fruit with a hard, woody endocarp surrounding the seed. The "stone" (endocarp), often confused for the seed, is called a pit or pyrene. A drupelet = small drupe. Example: peach.

dwarfing rootstock: A rootstock that reduces the potential size and/or vigor of the scion relative to a seedling rootstock.

effective pollination period: The window of time that pollination can occur and effect fertilization. Can be calculated as the longevity of the embryo sac minus the time required for pollen germination on the stigma and pollen tube growth.

egg nucleus: One of eight nuclei contained within the embryo sac. It fuses with a generative nucleus from a pollen grain and develops into the zygote and ultimately the embryo within a seed.

embryo sac: The eight-nucleate cell in the ovule, containing the egg and polar nuclei, that develops into the embryo and endosperm.

endocarp: The innermost tissue layer of the ovary; often becoming specialized, similar to the pit of a peach.

endosperm: The nutritive tissue surrounding the embryo in a seed; created by fusion of one generative nucleus with two polar nuclei during double fertilization. Often absorbed by the cotyledons during seed maturation in dicots.

endosperm incompatibility: Prevention of successful double fertilization caused by the failure of one generative nucleus to unite with the polar nuclei and produce endosperm.

enology: The study of winemaking.

epigynous: Inferior ovary position. The point of attachment of the sepals, petals, and stamens is *above* the ovary.

epigynous berry: A berrylike fruit derived from an inferior ovary.

ethephon: An ethylene-releasing chemical, used to stimulate fruit ripening and abscission in many crops, or to stimulate flowering in bromeliads.

even pinnate: Even number of leaflets in a pinnately compound leaf.

everbearing strawberry: A cultivar of strawberry with either a long-day or day-neutral flowering response, resulting in flower and fruit production throughout the spring and summer months.

exocarp: Outermost tissue layer of the ovary; often becoming all or part of the fruit skin or peel.

feathered: Said of a nursery tree having branches. *Contrast* WHIP.

fecundity: The capacity of a plant for production of great quantities of viable seed.

fertilization: The union of a male gamete contained in the pollen grain with the female gamete (egg nucleus) in the ovule. Synonym: syngamy.

filament: The stalk of a stamen; organ that holds the anther at its tip.

flavedo: The exocarp of a citrus fruit. The thin, colored part of the rind, containing the oil glands.

floral initiation: The first discernable change from a vegetative apex to a reproductive apex; biochemical in nature; not visible under microscope.

floricane: A flowering branch in brambles; a 1-year-old cane on which shoots producing flowers and fruits arise. Floricanes represent the second year of a biennial cane's life cycle; they die naturally after fruit production.

flower: The reproductive organ in angiosperms, bearing pistils and/or stamens.

follicle: A dry, dehiscent, simple, one- to many-seeded fruit dehiscing via only one suture. Example: milkweed.

freestone: Separation of the flesh (mesocarp) from the pit (endocarp) in a drupe or stone fruit. Also used to denote a group of peach cultivars used for fresh-market sales that carry this trait.

frost pocket: A low-lying, frost-prone area of the landscape.

fruit: A matured ovary plus any associated parts (such as receptacle, calyx, peduncle, or corolla tissue).

fruit crop: A perennial, edible crop for which the economic product is the true botanical fruit or is derived therefrom.

fruitlet: An immature or diminutive fruit.

fruit set: Persistence and development of an ovary after flowering.

frutescent: Describes a woody plant that is shrubby in habit.

fumigant: A volatile pesticide used to treat soil or storage facilities for pests (example: methyl bromide).

functionally dioecious: Term applied to a monoecious, dichogamous plant when there is a complete temporal separation of pollen shed and stigma receptivity.

fungi (singular: fungus): A highly diverse group of heterotrophic organisms that digest food externally. The fungi are the most common class of fruit pathogens, causing fruit rot, leaf spot, root rot, and other diseases.

fungicide: A chemical used to kill fungi.

generative nucleus (nuclei, plural): The genetic material of the pollen-bearing plant that goes on to unite with either the egg nucleus or the polar nuclei in the process of double fertilization; each pollen grain contains two generative nuclei.

genome: A complete set of chromosomes of a given plant.

germplasm: The genetic variability of a population of organisms. Physically, any plant part containing genetic information that can be used by nurserymen, breeders, or genetic engineers to improve or alter a plant species.

gibberellic acid (GA): A class of growth hormone (gibberellins) that stimulates cell elongation and growth in general. Uses in fruit crops include bloom delay, reduction of flowering, and parthenocarpic fruit set.

girdling: Removal of a complete strip of bark from a stem or branch to impede phloem transport. Sometimes used to improve fruit set or size.

glabrous: Smooth, hairless surface.

glaucous: A smooth surface covered with epicuticular wax.

grafting: Means of vegetative propagation for which the scion is one or more buds attached to a section of stem; methods of uniting stock and scion are more numerous and detailed than those for budding.

graft union: The point at which rootstock and scion are united.

grain: The fruit type of many members of the grass family. *See* CARYOPSIS.

grove: A chance or natural planting of fruit crops; generally not designed or uniformly spaced. Term often incorrectly used for "orchard." *Contrast* ORCHARD.

growing degree-hours: An unit of measure of the amount of hours above 40°F (4.5°C) that a plant receives during dormancy after its chilling requirement is satisfied.

ground color: The background color of fruit skin, generally green or yellow. Contrast with the blush color or the skin near maturity, which is generally red.

gummosis: Exudation of gums and resins, generally from the wood of a tree; a nonspecific malady, caused by fungi, bacteria, insects, mechanical damage, and so forth, frequently seen in stone fruits.

gynoecium: The female parts of the flower; the pistils, collectively.

half-inferior ovary: *See* PERIGYNOUS: OVARY.

haploid: Having one set of chromosomes, as in the pollen and egg cells of diploid plants.

heading back: Pruning a limb at a point somewhere along its length. Some portion of the limb remains.

heading cut: The type of pruning cut made when heading back a limb. *See* HEADING BACK.

hectare: Unit of land area equal to 10,000 square meters, or about 2.5 acres.

herbicide: A chemical used to kill plants.

hermaphroditic: Having anthers and pistils in the same flower. Synonyms: perfect, bisexual.

hesperidium: A fleshy, indehiscent fruit with a leathery or hard rind; flesh is divided into several segments by thin septa. Example: citrus.

heterodichogamy: Having flowers of different sexes that are functional at different times on the same plant.

heterogamy: Having flowers of differing sex on the same plant (monoecious).

high-density orchard: An orchard planting design with several hundred to thousands of trees per acre. Trees are spaced much closer than in conventional orchards.

hull: The outer covering of a fruit or a seed, derived from floral, ovarian, or bract tissue; often used as a synonym for "shuck," in nut crops, or "husk," in grain crops.

husk: *See* HULL; SHUCK.

hybrid: The offspring of a cross between two genetically distinct individuals, often two different cultivars.

hydrocooling: Initial postharvest handling practice in which cold water is used to rapidly reduce fruit temperature.

hypanthium: The floral cup; fusion of the bases of the sepals, petals, and stamens into a continuous, cup-shaped structure surrounding the ovary. In common usage, the hypanthium of stone fruits is termed the "shuck." *See* SHUCK; SHUCK SPLIT.

hyphae: The filamentous, branched bodies of fungal organisms, collectively.

hypogynous: Superior ovary position. The point of attachment of the sepals, petals, and stamens is *below* the ovary.

IAA: Indoleacetic acid. *See* AUXIN.

imperfect flower: Lacking one of the two sex organs, the androecium or gynoecium.

incompatibility: With reference to pollination, inability to successfully fertilize an ovule and produce seed. With reference to grafting, a scion-rootstock combination incapable of coexisting. *See also* DELAYED INCOMPATIBILITY; ENDOSPERM INCOMPATIBILITY; NUCLEAR INCOMPATIBILITY; POLLEN TUBE FAILURE.

incomplete flower: Lacking one or more of the following: calyx (sepals), corolla (petals), androecium, gynoecium.

indehiscent: Not splitting open at maturity.

indeterminate: Descriptive term applied to inflorescences having the topmost or central flower appearing and maturing last in the blooming sequence.

inferior ovary: Epigynous ovary position. *See* EPIGYNOUS.

inflorescence: A cluster of flowers.

infructescence: A ripened inflorescence.

inoculum (inocula, plural): The material by which plant diseases are spread to new host plants. Examples: fungal spores, bacterial cells.

insecticide: A chemical used to kill insects.

integrated pest management (IPM): A pest management method that utilizes all techniques of pest control (e.g., cultural, biological, chemical) in an integrated fashion to keep pest populations below an economic threshold level.

integuments: The outermost part of the embryo sac, which develops into the seedcoat.

internode: The space between two nodes on a stem.

interspecific: Between two species, such as a hybrid.

interstock (interstem): A stem piece inserted between a rootstock and a scion.

involucre: Bract(s) subtending a flower or inflorescence; can be leafy (hazelnut), fleshy (walnut), or woody (cup of an oak acorn).

juice vesicle: The saclike structure, composing the bulk of the endocarp of a hesperidium (citrus fruit), that contains the juice.

junebearing strawberry: A cultivar of strawberry within a group that has a short-day flowering response, producing fruits during the spring months exclusively, and stolons during the summer months.

june drop: Shedding of fruits in May or June due to competition for resources; reduced by thinning.

juvenility: Stage of life cycle at which flowering/fruiting is not yet possible.

lateral: An elongated shoot coming from a main shoot; also used to denote a fruit bearing habit whereby fruits arise from axillary buds on elongated shoots.

layer: A plant produced by a form of layerage.

layering: A vegetative propagation technique whereby a new, complete plant is produced from an organ while still attached to the mother plant. Example: mound layering, in which shoots are rooted by mounding soil around the base of an existing plant.

leader: The main axis or central bole of a tree.

legume: A dry, dehiscent, simple, one- to many-seeded fruit, with dehiscence via two sutures, ventral and dorsal. Example: bean. Synonym: pod.

liana: A woody, climbing vine.

locule: The seed cavity; the central and often open portion of a carpel.

loculicidal capsule: A dry, dehiscent, one- to many-seeded fruit from a single compound ovary, opening through the locules.

loment: A dry, dehiscent, simple, many-seeded fruit that is constricted between the seeds; dehiscence is via two sutures, as with a legume.

long-day plant: One that initiates flowers when daylength exceeds some critical value.

low-chill cultivar: A cultivar that has an unusually low chilling requirement relative to the average for the crop. Allows cultivation in more southern locales.

mechanical control: Any number of methods whereby agricultural pests are physically killed or removed from the crop. Example: mowing, tillage, and burning of weeds; hand removal of insects or their eggs.

Mediterranean climate: An excellent climate for fruit production that is characterized by warm, rainless summers with low humidity and high light intensity, and cool winters with rainy, often foggy weather and infrequent, mild freezes. Typical in countries that border the Mediterranean sea, and in other temperate areas, such as California, South Africa, and Chile.

mericlinal chimera: *See* CHIMERA.

mesocarp: The middle tissue layer of a fruit or an ovary.

metaxenia: Effects of the pollen source on the fruit tissues exclusive of the seeds.

metric ton (MT): A unit of measure equal to 2,200 pounds or 1,000 kilograms.

mites: A class of arthropods related to spiders (class Arachnida, order Acarina). Generally small, with two or four pairs of legs; most injure by superficial feeding on leaves, fruits, and stems.

mixed bud: A bud possessing both flowers and leaves, as opposed to a simple bud that contains *either* flowers *or* leaves.

monocot: A plant in the class Monocotyledoneae, characterized by having only one seed leaf or cotyledon.

monoecious: A plant having separate male and female flowers.

monopodial: A growth habit in which a plant produces a single main stem or trunk (example: palm tree).

mound layering: A vegetative propagation technique where media are mounded at the base of a plant to induce rooting of shoots; commonly used to produce clonal rootstocks in crops such as apple. Synonym: stool layering.

mulch: A layer of material, generally organic or plastic, placed on the ground to control weeds.

multiple fruit: Fruit produced by the fusion or adherence of two or more ovaries arising from different flowers; a fused, matured inflorescence. Examples: pineapple, mulberry. Sometimes termed a "syncarp" (synonym: syncarpium) if it contains nonovarian tissues, as it often does.

mummified fruit: Diseased fruit that dries, shrivels, and remains attached to the plant, serving as a source of inoculum for future infection. Mummified fruits are referred to as "mummies" in common usage.

mycoplasma: A small bacterium that lacks cell walls (class Mollicutes); the genus *Xylella* contains pathogens that affect some fruit crops.

NAA: Naphthalene acetic acid; a synthetic auxin used to thin fruitlets in the spring or, ironically, to prevent fruit abscission just before harvest.

NAAm: Naphthalene acetamide; a chemical analogue of NAA with similar function.

necrosis (necrotic): Death of organs or tissues.

nectary: Organ that secretes nectar, usually at the base of the ovary in flowers.

nectary disc: A raised, prominent nectary, as in the flowers of *Citrus* species.

nematocide: A chemical that kills nematodes.

nematode: A microscopic, nonsegmented roundworm; a root pest of fruit crops.

node: The point of origin of buds or leaves on the stem.

nonclimacteric fruit: A type of fruit that does not exhibit a sudden increase in respiration and ethylene production during ripening, but shows gradual changes in metabolism, sugar and acid content, firmness and color change during maturation. Examples: citrus, strawberry.

nucellar embryony: Adventitious embryos arise from the nucellus, yielding embryos genetically identical to the mother plant.

nucellus: Maternal ($2n$) tissue surrounding the embryo sac in an ovule.

nuclear incompatibility: Lack of seed and possibly fruit set due to lack of fusion of the egg and generative nuclei during double fertilization.

nut: A dry, indehiscent, one-seeded (usually) fruit with a hard exterior. Examples: pecan, walnut.

odd pinnate: Odd number of leaflets in a pinnately compound leaf.

open center: A tree-training system in which three to five main structural limbs (scaffolds) radiate in all directions from a stout trunk 1 to 3 feet tall. No structural limbs grow in the center of the canopy. Often used with trees having weak apical control that tend to form rounded canopies naturally (examples: peach, plum, apricot, almond).

orchard: A regularly spaced, geometric planting of fruit crops. *Contrast* GROVE.

organic acids: Natural acidic compounds found in fruit pulp and juice; most common are citric acid and malic acid, but dozens of forms exist. They make fruit tangy and/or sour.

ossiculus: Synonyms: pit, pyrene. *See* PYRENE.

ovary: The swollen base of the pistil, containing the ovules; matures into the fruit.

overwinter: To persist through the winter, generally in an inactive or dormant state.

ovule: An immature seed; the embryo sac surrounded by the integuments and nucellus.

own-rooted scion: Fruit tree propagated vegetatively by layering or cuttings. The tree's root system is produced adventitiously from mature scion wood.

palmate: Radiating out from a common point, as with leaflets in palmately compound leaves.

panicle: A multiply branched, indeterminate inflorescence with two or more orders of laterals (example: mango).

parthenocarpy (parthenocarpic): Fruit development in the absence of fertilization and seed production. In vegetative parthenocarpy, pollination need not occur for fruit set, whereas in stimulative parthenocarpy, pollination stimulates fruit set, yet fertilization does not occur.

pathogen: An agent that causes plant disease.

pedicel: The stalk of a flower.

peduncle: The stalk supporting the entire inflorescence or a single fruit.

peel: The outer covering of a fruit, often composed of exocarp tissue.

penetrometer: Device for measuring the firmness of fruit flesh.

pepo: A fleshy fruit from a compound, inferior ovary, with a thick, tough rind. Distinguished from a hesperidium by these characteristics: parietal placentation instead of axile and origin in an inferior ovary. Example: watermelon (most Curcurbitaceae).

percent kernel: The percent, by weight, of a nut that is the edible portion or kernel. Expressed as: [kernel weight/(kernel + shell weight)] × 100. Synonym: shelling percentage.

perennial: A plant capable of living more than two years.

perfect flower: A flower having both male and female parts. Synonym: hermaphroditic flower.

perianth: The calyx and corolla together.

pericarp: The wall of the fruit derived from ovary tissue. Divided into three histogenic layers: exocarp—outermost; mesocarp—middle; endocarp—innermost.

periclinal chimera: *See* CHIMERA.

perigynous: Half-inferior ovary position. The point of attachment of the sepals, petals, and stamens surrounds the ovary.

pest: Any organism that reduces yield and/or crop quality.

pesticide: A chemical used to kill weeds, insects, fungal diseases, nematodes, or other pests.

petal: A member of the corolla; a floral appendage, often showy.

petal fall: Stage of floral development after anthesis when petals abscise. Often used as a visual cue for spray timing.

petiole: The stalk that attaches a leaf to a stem.

pheromone: A chemical produced by insects that is used for mating and communication.

phloem: The photosynthate-conducting tissue of plants, located just beneath the bark and outside of the wood of a stem.

photosynthate: End products of photosynthesis; sugars mostly.

phyllotaxis (also phyllotaxy): The arrangement of leaves on a stem. Denoted as a fraction; the numerator is the number of revolutions around the stem, and the denominator is the number of nodes between two leaves with the same vertical orientation.

phytotoxic: A chemical that is poisonous to crop plants.

pilose: Covered with soft, long hairs.

pinnate: Having subdivisions arranged oppositely along a main axis in pairs; as in pinnately compound leaves.

pistil: The female reproductive organ of the flower, composed of the stigma, style, and ovary.

pistillate: Said of a flower or plant containing the gynoecium or pistil(s).

pit: The woody endocarp or stone in a drupe. Also "pyrene" or "ossiculus."

placenta (placentation): Portion of the ovary to which the ovules are attached. Arranged in several ways: axile, basal, free central, or parietal.

pod: *See* LEGUME.

polar nuclei: Two of the eight nuclei contained in a mature embryo sac that form the endosperm of the seed after uniting with one generative nucleus from the pollen.

pollen: The male gametophyte of an angiosperm or gymnosperm; tiny structure carrying haploid, generative nuclei.

pollen tube: An elongated, narrow, tubular structure arising from a germinated pollen grain. It grows through the style and carries the generative nuclei toward the egg sac of the ovule.

pollen tube failure: Inability of viable pollen to grow at all or rapidly enough in the style for the generative nuclei to reach the embryo sac and successfully fertilize the egg.

pollination: Transfer of pollen from the anther to the stigma.

pollinator: An agent of pollen transfer, generally honeybees, or the wind in wind-pollinated species.

pollinizer (also pollenizer): With reference to cross-pollinated species, a cultivar that functions as a source of compatible pollen.

polyembryony: Two or more embryos arising from a single seed.

polygamous: Having unisexual and bisexual flowers on the same plant.

polygamodioecious: Primarily dioecious, but having some bisexual flowers or flowers of the opposite sex on the same plant.

polygamomonoecious: Primarily monoecious, but having some bisexual flowers.

pome: A fleshy, indehiscent fruit from an inferior compound ovary, generally having a cartilaginous endocarp; the fleshy receptacle or hypanthium completely enclosing and fused to the pericarp. Example: apple.

pome fruits: A group of crops having a pome as a fruit type; members of the Pomoideae, subfamily of the Rosaceae, which includes, among others, apple, pear, and quince.

pomologist: One who studies fruit culture.

pomology: The study of fruit culture.

poricidal capsule: A dry, dehiscent, one- to many-seeded fruit from a single compound ovary, opening through pores or flaps.

precocious (precocity): Advanced in development; said of a species with a short period of juvenility, as with a plant that flowers and bears fruits at a young age.

primocane: The current season's shoot that comes from the ground in brambles; vegetative in most cases, except primocane-fruiting cultivars.

propagule: A cell, tissue, or organ that gives rise to a new plant. Examples: seed, cutting, rhizome, layer.

protandry (protandrous): Pollen is shed before the stigma is receptive.

protogyny (protogynous): The stigma is receptive before pollen is shed. Synonym: metandry.

pruning: Removing limbs or stems from a plant.

pseudofruit: Literally, a "false fruit." A fruitlike structure of nonovarian origin that may function in seed dispersal (example: cashew apple). Synonym: pseudocarp.

pseudostem: Literally, a "false stem." The main axis of a banana or plantain plant, composed of sheathing petiole bases that form a supportive, cylindrical structure resembling a tree trunk.

pubescent: Possessing fine hairs on the surface.

pulp: The edible or juice-containing portion of the fruit.

pulvinus (pulvini, plural): A swollen or enlarged point at the base of a petiole that functions in controlling leaf movement and/or orientation.

pyrene: The hard pit of a stone fruit; bony endocarp. Synonym: ossiculus.

pyriform: Pear-shaped.

pyxis: A circumscissile capsule.

quiescence: A dormant condition brought on by unfavorable environmental conditions, not internal factors.

raceme: An indeterminate inflorescence with only one order of branching. Pedicels of flowers are about the same length.

rachis: The main or primary axis of a compound leaf or inflorescence.

ratoon crop: A crop derived after harvesting the original crop and cutting plants back to side shoots.

receptacle: The base of the flower; point of attachment of other flower parts.

reflexed: Curved backward or downward.

refractometer: Device for measuring the soluble solids (sugar) content of fruit juice.

rest: *See* DORMANCY.

rhizome: A horizontally oriented underground stem; often used as a propagule.

ringing: *See* GIRDLING.

rootstock: The root system, the bottom part of a grafted tree. Synonyms: stock, understock.

runner: *See* STOLON.

russet: Tan or brownish, coarse-textured area on fruit skin. May be genetic, as in 'Bosc' pear, or may be induced by frost, dew, or injury to the fruit skin.

samara: A dry, indehiscent, winged fruit.

scaffold: Large, permanent, lateral limb of a fruit tree that produces fruiting wood.

schizocarp: A dry, dehiscent fruit from a compound ovary; fruit splits into one-seeded segments at maturity, but carpels do not dehisce to release seeds.

scion: The aboveground portion of a plant propagated by graftage.

scion rooting: Rooting of the scion portion of the tree as a result of burying the graft union.

scoring: Making shallow knife cuts around the circumference of a stem or branch to impede phloem transport. Similar to girdling, but without removal of a strip of bark.

sectorial chimera: *See* CHIMERA.

seed: A mature ovule. Contains the embryo, endosperm (only remnants in most dicots), and the seedcoat(s).

seedling rootstock: A rootstock propagated by seed.

selection: A variety or form of a species exhibiting desirable traits. Selections taken from the wild or breeding populations may be used directly as a new cultivar or as a parent in a breeding program.

self-compatible: Capable of successful fertilization and seed production when pollinated with its own pollen.

self-fruitful: Capable of producing a commercial crop of fruits when self-pollinated.

self-incompatible: Incapable of successful fertilization and seed production when pollinated with its own pollen, despite having viable pollen and egg nuclei. Some seeds (and thus fruits) will be set in most self-incompatible species when self-pollinated.

self-pollination: Pollen transfer from the anther to the stigma within the same flower or plant genotype.

self-sterile: Lacking either pollen or egg nuclei that are viable; cannot produce any viable seed when self-pollinated. *Contrast* SELF-INCOMPATIBLE.

self-unfruitful: Incapable of setting commercial crops of fruits when self-pollinated.

sepal: One unit of the calyx. A unit of the outermost whorl of appendages in a flower.

septicidal capsule: A dry, dehiscent, one- to many-seeded fruit from a single compound ovary, opening along the septa.

septum (septa, plural): The partition separating the locules of a compound ovary.

sessile: A leaf, flower, or fruit attached directly to the plant, with no stalk or stem.

shelling percentage: The percentage of a nut's weight contributed by the kernel. Synonym: percent kernel.

short-day plant: One that initiates flowers when daylength falls below some critical value.

shrub: Several-stemmed woody plant; renewal growth from the base or crown.

shuck: The outer covering of a fruit or seed; often fleshy in nut crops, such as pecan and walnut. Shucks are composed of involucre, ovarian, and/or floral tissues. In common usage, a synonym for "hull" or "husk."

shuck split: The stage of development when the shuck dehisces from the developing fruit. In nut crops, shuck split occurs at maturity; in stone fruits, common usage signifies the stage at which the hypanthium is split open by the enlarging ovary a few weeks after bloom.

silicle: A dry, dehiscent fruit from a two-carpellate ovary, *less than twice as long as wide,* with the carpels separated by a thin, translucent septum (replum). Common in the Brassicaceae.

silique: A dry, dehiscent fruit from a two-carpellate ovary, *more than twice as long as wide,* with the carpels separated by a thin, translucent septum (replum). Common in the Brassicaceae.

simple: Undivided, as in a leaf blade or ovary, or containing only one type of structure, such as a simple bud containing only flowers or vegetative shoots.

slip: A lateral shoot borne on the rachis just below a pineapple fruit; used for propagation.

small fruits: A group of taxonomically diverse fruit crops, generally shrubs, herbs, or vines, that produce small, soft fruits. Includes grapes, blueberries, brambles, strawberries, currants, gooseberries, and several others.

solitary: Occurring by itself, not in a cluster (as with flowers).

spadix: A fleshy spike with numerous, tiny flowers.

spathe: A bract subtending or ensheathing an inflorescence, usually a spadix or compound spadix.

spike: An indeterminate, elongated, unbranched inflorescence, with flowers that lack pedicels (example: pineapple).

spindlebush: A type of central-leader training system with numerous, shortened scaffolds arising along the leader.

sport: A mutant strain of a cultivar. *See* STRAIN.

spur: A short, slow-growing, lateral branch generally bearing flower buds.

stamen: The male reproductive organ of a flower; composed of the filament (stalk) and the anther (apical portion).

staminate: Said of a flower or plant that produces only male reproductive structures.

staminode: A nonfunctional (sterile) stamen, often modified in form. Synonym: staminodium.

stenospermocarpy: Seedless fruit development as a result of fertilization followed by seed abortion.

stigma: The terminal portion of the style, often swollen, flattened, domed, or feathery. Functions in catching pollen and allowing for its germination. Synonym: stigmatic surface.

stion: A plant propagated by graftage; st̲ock + sci̲o̲n̲. Shorthand designation: scion/stock. Example: 'Granny Smith'/M.9 designates a stion composed of the scion cultivar 'Granny Smith' on M.9 rootstock; 'Granny Smith'/M.9/MM.111 designates a stion composed of the scion cultivar 'Granny Smith' with an M.9 interstock and MM.111 rootstock.

stipules: A pair of generally inconspicuous appendages at the base of the petiole.

stock: Abbreviation for rootstock.

stock plant: A plant used as a source of propagation material.

stolon: A horizontal, creeping stem that roots at the node or tip, producing a new plant. Synonym: runner.

stone fruits: A group of fruit crops belonging to the genus *Prunus* and having the fruit type of drupe. Examples: peach, plum, cherry, and apricot.

stool: A mounded planting site.

stool bed: A raised nursery row or plot used for mound layering.

strain: A sub-subspecies of a plant; a form. A variant of a cultivar that is nearly identical, not deserving full cultivar status.

stratification (chilling): Exposure of seed to cool (40-50°F) temperatures in the presence of moisture for 30 to 180 days to break seed dormancy and induce uniform germination and seedling development.

strig: Term applied to clusters of fruits in currants and gooseberries.

style: The part of the pistil between the stigma and the ovary; often slender, elongated.

sucker (suckering): Shoot arising from the plant base or rootstock adventitiously, either from roots, stems, or the trunk; undesirable usually, except when occurring on stock plants used for propagation.

suffrutescent: A shrub that is slightly woody at the base, but mostly composed of herbaceous stems.

sugar-acid ratio: The percentage of sugar divided by the percentage of acid in fruit juice. Synonym: Brix-acid ratio.

summer pruning: Pruning during the growing season.

superior ovary: Hypogynous ovary position. *See* HYPOGYNOUS; OVARY.

suture: The cleft or line of dehiscence on a fruit or other plant organ.

syconium: A fleshy, multiple fruit composed mostly of an inverted, hollow receptacle containing many individual flowers. Access to flowers is provided by a small hole or ostiole (synonym: eye) at the fruit tip. Example: fig, for which the true fruits are drupelets (a multiple of drupelets or a syncarp).

sympodial: A growth habit characterized by frequent branching to form multiple trunks or stems and a spreading canopy.

syncarp: A multiple fruit composed of many fruitlets plus a fleshy inflorescence axis. Example: mulberry. Synonym: syncarpium.

synchronous protogyny: When all open flowers on the same plant or cultivar have functional female, but not male, parts and later become functionally male. A perfect flower condition that favors outcrossing (example: avocado).

take: Successful result of budding or grafting.

taxonomy: The study of the classification and naming of organisms.

temperate zone: Geographic region between 23.5° and 50° latitude having distinct seasons.

tendril: A slender, elongated, twining organ used for climbing in vines.

tepal: A unit of the perianth in species having indistinguishable sepals and petals.

testa: The seedcoat.

tetraploid: Having four sets of chromosomes.

thinning: Partial removal of flowers or fruitlets to increase the ultimate size of the remaining fruits. Accomplished mechanically, by chemicals, or by hand.

thinning cut: The type of pruning cut made when a limb is removed at its point of origin. *See* THINNING OUT.

thinning out: Removal of a limb or stem at its base; no portion of the stem remains.

threshold (economic threshold): The level of pest infestation that causes economic losses and often prompts growers to take action to protect the crop.

thyrse: An inflorescence resembling a panicle in which the lateral axes are cymose (determinate) and the main axis is indeterminate; a cymose panicle.

tip layering: A vegetative propagation technique in which the tip of an aerial shoot roots after contact and/or partial burial in the soil surrounding the mother plant.

topworking: Replacement of the original scion with a different one by grafting onto an established tree.

training: Obtaining a particular form for a tree or shrub by pruning, tying, bending, or staking limbs in various orientations.

training system: A particular tree or shrub form.

tree: Large woody plant, usually with a main stem or trunk. Renewal growth generally from the top of the canopy, not the trunk or ground.

trioecious: Bearing male, female, and perfect flowers on separate plants.

triploid: Having three sets of chromosomes; triploid plants are often sterile and produce seedless fruits.

umbel: An indeterminate inflorescence in which flowers are borne on a plane or slight arc, resembling an umbrella. The pedicels of all flowers originate at the same place on the main axis.

understock: *See* ROOTSTOCK.

union (also graft union): The point where the graft was made in a two-part tree.

unisexual: Having only one sex, as in an imperfect flower.

urceolate: Urn-shaped.

utricle: A small, dry, one-seeded, indehiscent fruit with a thin, bladderlike wall. Example: spinach.

variety: A subspecies or cultivar of a plant.

vegetative bud: A bud producing leaves and stems initially, not flowers.

vegetative propagation: Producing a plant asexually, through cuttings, layers, grafting or budding, or tissue culture, *not* by sexual seed. However, apomictic seed can be considered a form of vegetative propagation. Results in plants genetically identical to the parent.

vine: A plant that is not self-supporting, climbing by means of twining stems or tendrils.

vineyard: A regular, geometric planting of grapevines.

viroid: A virus that lacks a protein coat.

virus: A member of the smallest class of plant pathogens. Viruses generally consist of a nucleic acid core and a protein coat.

viticulture: The study of grape cultivation.

vivipary: Precocious germination of seed prior to fruit abscission or harvest (undesirable).

water sprout: A vigorous, upright, generally vegetative shoot arising from a scaffold or main tree trunk.

weed: A plant out of place; competes with the crop for growth resources.

whip: An unbranched nursery tree; scion is a single shoot.

windrow: A long row of plant material and nuts piled between rows of nut trees in an orchard after shake harvesting; facilitates nut collection and debris separation by mechanical harvesters.

xenia: Effects of the pollen source on the tissues within the seed.

xylem: The water-conducting tissue of plants. The wood of trees, shrubs, and vines.

yield: A measure of crop output, generally in weight per unit of land area, such as pounds per acre or kilograms per hectare.

yield efficiency: A measure of fruit production efficiency, using unit of fruit weight per unit of plant size, such as kilograms of fruit per square centimeter of trunk cross-sectional area, or kilograms per cubic meter of canopy volume.

zygote: The diploid cell formed by fusion of the egg with one generative nucleus; the progenitor of the embryo within the seed.

Index

Page numbers followed by the letter "t" indicate tables; those followed by the letter "f" indicate figures; and those followed by the letter "e" indicate exhibits. The abbreviation "pl." indicates a color plate located in the gallery following Chapter 1.

Introduction to Fruit Crops

doi:10.1300/5547_35